T0337871

Mechanobiology

Mechanobiology

Exploitation for Medical Benefit

Edited by Simon C. F. Rawlinson

Queen Mary University of London, Institute of Dentistry, England

Published by John Wiley & Sons, Inc., Hoboken, New Jersey

Published simultaneously in Canada

For general information on our other products and services or for technical support, please contact our Customer Care Department within the United States at (800) 762-2974, outside the United States at (317) 572-3993 or fax (317) 572-4002.

Wiley also publishes its books in a variety of electronic formats. Some content that appears in print may not be available in electronic formats. For more information about Wiley products, visit our web site at www.wiley.com.

Library of Congress Cataloging-in-Publication data applied for

ISBN: 9781118966143

Printed and bound in Malaysia by Vivar Printing Sdn Bhd

Cover credits: Background: Mimi Haddon/Gettyimages

10 9 8 7 6 5 4 3 2 1

Contents

List of Contributors

Kartik Balachandran
Department of Biomedical Engineering
University of Arkansas, Fayetteville
AR, USA

Christoph Ballestrem
Wellcome Trust Centre for Cell-Matrix
Research, Faculty of Life Sciences
University of Manchester
Manchester, UK

Charlie Campion
Institute of Bioengineering and School of
Engineering and Materials Science
Queen Mary University of London
London, UK

Naomi C. Chesler
Department of Biomedical Engineering
University of Wisconsin–Madison
Madison, WI, USA

Yanan Du
Department of Biomedical Engineering
School of Medicine,
Tsinghua University
Beijing, China

Kian F. Eichholz
Trinity Centre for Bioengineering
Trinity Biomedical Sciences Institute
and Department of Mechanical and
Manufacturing Engineering
School of Engineering
Trinity College Dublin, Dublin
Ireland
Department of Mechanical, Aeronautical
and Biomedical Engineering
Centre for Applied Biomedical
Engineering Research
Materials and Surface Science Institute
University of Limerick, Limerick
Ireland

Alicia J. El Haj
Institute for Science and Technology in
Medicine, University of Keele
Staffordshire, UK

Winston Elliott
Department of Mechanical Engineering
University of Colorado at Boulder
Boulder, CO, USA

Nicholas D. Evans
Centre for Human Development, Stem
Cells and Regeneration, Institute for Life
Sciences, University of Southampton
Faculty of Medicine, Institute for
Developmental Sciences
University of Southampton
Southampton, UK
Bioengineering Sciences Group
Faculty of Engineering and the
Environment, University of
Southampton, Southampton, UK

Gabriel L. Galea
Developmental Biology of Birth Defects
UCL Great Ormond Street Institute of
Child Health,
London, UK

Ulysse Gaspard
Department of Gynecology
Liège University Hospital
Liège, Belgium

Julien E. Gautrot
Institute of Bioengineering and School of Engineering and Materials Science
Queen Mary University of London
London, UK

Hamish T. J. Gilbert
Wellcome Trust Centre for Cell-Matrix Research, Faculty of Biology
Medicine and Health
University of Manchester
Manchester, UK

K. Jane Grande-Allen
Department of Bioengineering
Rice University, Houston, TX, USA

James R. Henstock
Institute for Ageing and Chronic Disease
University of Liverpool, Liverpool
UK

Karin A. Hing
Institute of Bioengineering and
School of Engineering and
Materials Science
Queen Mary University of London
London, UK

David A. Hoey
Trinity Centre for Bioengineering
Trinity Biomedical Sciences Institute
and Department of Mechanical and
Manufacturing Engineering
School of Engineering, Trinity College
Dublin, Dublin, Ireland
Department of Mechanical, Aeronautical
and Biomedical Engineering
Centre for Applied Biomedical
Engineering Research, Materials and
Surface Science Institute
University of Limerick, Limerick
Ireland

Chao-Kai Hsu
Department of Dermatology, National
Cheng Kung University Hospital
College of Medicine, National Cheng
Kung University, Tainan, Taiwan
Institute of Clinical Medicine, College
of Medicine, National Cheng Kung
University, Tainan, Taiwan
International Research Center of
Wound Repair and Regeneration
National Cheng Kung University
Tainan, Taiwan

Chenyu Huang
Department of Plastic, Reconstructive,
and Aesthetic Surgery, Beijing Tsinghua
Changgung Hospital, Beijing, China
Medical Center, Tsinghua University
Beijing, China

Philippe Humbert
University of Franche-Comté, Besançon
France
Department of Dermatology, University
Hospital Saint-Jacques, Besançon, France
Inserm Research Unit U645, IFR133
Besançon, France

Hanna Isaksson
Department of Biomedical Engineering
Lund University, Lund, Sweden

Amir Keshmiri
Engineering and Materials Research
Centre, Manchester Metropolitan
University, Manchester, UK
School of Mechanical, Aerospace and
Civil Engineering, the University of
Manchester, Manchester, UK

Hanifeh Khayyeri
Department of Biomedical Engineering
Lund University, Lund, Sweden

Franziska Lausecker
Wellcome Trust Centre for Cell-Matrix
Research, Faculty of Life Sciences
University of Manchester

Manchester, UK
Institute of Human Development
Faculty of Human Sciences
University of Manchester
Manchester, UK

Rachel Lennon
Wellcome Trust Centre for Cell-Matrix Research, Faculty of Life Sciences
University of Manchester
Manchester, UK
Institute of Human Development
Faculty of Human Sciences
University of Manchester
Manchester, UK
Department of Paediatric Nephrology
Central Manchester University Hospitals NHS Foundation Trust (CMFT)
Manchester Academic Health Science Centre (MAHSC), Manchester, UK

Stefania Marcotti
Department of Materials Science and Engineering, INSIGNEO Institute for In Silico Medicine, Sheffield, UK

Lee B. Meakin
School of Veterinary Sciences
University of Bristol, Bristol, UK

Keiji Naruse
Department of Cardiovascular Physiology, Graduate School of Medicine Dentistry and Pharmaceutical Sciences
Okayama University, Okayama, Japan

Andromeda M. Nauli
Department of Pharmaceutical Sciences
Marshall B. Ketchum University, Fullerton
CA, USA

Surya M. Nauli
Department of Biomedical & Pharmaceutical Sciences
Chapman University, Irvine
CA, USA

Rei Ogawa
Department of Plastic, Reconstructive, and Aesthetic Surgery, Nippon Medical School, Tokyo, Japan

Hulin Piao
Department of Cardiovascular Physiology, Graduate School of Medicine Dentistry and Pharmaceutical Sciences
Okayama University, Okayama
Japan
Department of Cardiovascular Surgery
The Second Affiliated Hospital of Jilin University, Changchun, China

Gérald E. Piérard
Laboratory of Skin Bioengineering and Imaging (LABIC), Department of Clinical Sciences, University of Liège
Liège, Belgium
University of Franche-Comté, Besançon
France

Sébastien L. Piérard
Telecommunications and Imaging Laboratory INTELSIG, Montefiore Institute, University of Liège, Liège
Belgium

Claudine Piérard-Franchimont
Department of Dermatopathology
Unilab Lg, Liège University Hospital
Liège, Belgium
Department of Dermatology
Regional Hospital of Huy, Huy
Belgium

Andrew A. Pitsillides
Comparative Biomedical Sciences
The Royal Veterinary College
London, UK

Andrea S. Pollard
Comparative Biomedical Sciences
The Royal Veterinary College
London, UK

Patrick J. Prendergast
Trinity Centre for Bioengineering
School of Engineering, Trinity College
Dublin, Dublin, Ireland

Daniel Puperi
Department of Bioengineering
Rice University, Houston, TX, USA

Prashanth Ravishankar
Department of Biomedical Engineering
University of Arkansas, Fayetteville
AR, USA

Simon C. F. Rawlinson
Centre for Oral Growth and
Development, Institute of Dentistry
Barts and The London School of
Medicine and Dentistry
London, UK

Caretta J. Reese
Department of Biomedical &
Pharmaceutical Sciences
Chapman University, Irvine
CA, USA

Gwendolen C. Reilly
Department of Materials Science and
Engineering, INSIGNEO Institute for
In Silico Medicine, Sheffield, UK

Hitomi Sano
Department of Plastic, Reconstructive
and Aesthetic Surgery, Nippon Medical
School, Tokyo, Japan

Rinzhin T. Sherpa
Department of Biomedical &
Pharmaceutical Sciences
Chapman University, Irvine
CA, USA

Joe Swift
Wellcome Trust Centre for Cell-Matrix
Research, Faculty of Biology
Medicine and Health
University of Manchester
Manchester, UK

Ken Takahashi
Department of Cardiovascular
Physiology, Graduate School of Medicine
Dentistry and Pharmaceutical Sciences
Okayama University, Okayama
Japan

Wei Tan
Department of Mechanical Engineering
University of Colorado at Boulder
Boulder, CO, USA

Lian Tian
Department of Medicine, Queen's
University, Kingston, ON, Canada

Camelia G. Tusan
Centre for Human Development, Stem
Cells and Regeneration, Institute for Life
Sciences, University of Southampton
Faculty of Medicine, Institute for
Developmental Sciences
University of Southampton
Southampton, UK
Bioengineering Sciences Group
Faculty of Engineering and the
Environment, University of
Southampton, Southampton, UK

Zhijie Wang
Department of Biomedical Engineering
University of Wisconsin–Madison
Madison, WI, USA

Preface

Mechanobiology is the study of how tissues and cells interact with, and respond to, the physical environment, either through direct contact with a substrate via cell attachments or through cell-surface perturbation by a varying extracellular situation/climate.

The vast majority of cells are subjected to a fluctuating physical environment – and this is not restricted to the animal kingdom. In response to increased loading conditions (bending), the branches of trees compensate with new wood formation. Interestingly, though, conifers and hardwoods respond to this increased bending differently: conifers tend to produce "tension wood" on the upper part of the bough, whereas hardwoods produce "compression wood" on the lower surface – two distinct solutions to one problem.

This volume attempts to briefly introduce the topic of mechanobiology in humans to a broad audience, with the intention of making the phenomenon more widely recognized and demonstrating its relevance to medicine. It covers three broad topics: (i) recognition of the mechanical environment by extracellular matrix (ECM) and primary cilium, (ii) selected tissue types, and (iii) physical, computational/substrate models and the use of such findings in practice.

Obviously, the list of chapters for each topic is not exhaustive – there are too many examples, and this volume therefore can only be an introduction. The tissue types discussed are some of the more immediately recognizable as being subjected to mechanical forces, though a few are less obvious.

One important question is, given that most biology is subjected to the mechanical environment, how can we best reproduce that in experimental conditions? Would the effect of a compound be influenced if the tissue/cells were subjected to their normal physiological environment at the time of application? Such questions need to be at least acknowledged, if not accommodated within experimental design.

I hope the volume generates interest in, and appreciation of, this emerging field with those considering a career in science or medicine.

Finally, I would like to thank all the contributing authors to this manuscript. They have all devoted their time to writing their chapters and have focused on presenting their ideas clearly and logically to the target audience.

Simon C. F. Rawlinson

1

Extracellular Matrix Structure and Stem Cell Mechanosensing

Nicholas D. Evans and Camelia G. Tusan

Centre for Human Development, Stem Cells and Regeneration, Institute for Life Sciences, University of Southampton, Faculty of Medicine, Institute for Developmental Sciences, University of Southampton, Southampton, UK
Bioengineering Sciences Group, Faculty of Engineering and the Environment, University of Southampton, Southampton, UK

1.1 Mechanobiology

An ability to sense the external environment is a fundamental property of life. All organisms must be able to interpret their surroundings and respond in a way that helps them survive – for example, by feeding, moving, and reproducing. The ability to sense also allows organisms to communicate with one another. Communication and cooperation were the primary driving forces that led to the evolution of complex multicellular organisms from simpler unicellular organisms. Evidence of this remains in many of the signaling pathways found in mammals that promote cell arrangements during development, which evolved from primordial chemical signals that unicellular organisms used to communicate with one another (King et al. 2003). Cells in our mammalian bodies are experts at communicating with one another using chemicals, and our physiology is completely dependent on this, from the precisely orchestrated cascades of growth factors during development to the hormones necessary for homeostasis and the immune mechanisms fundamental to repelling microbes.

Cells can also interact with one another by direct contact. Cells express characteristic surface proteins of various types, most prominently the cadherins, which allow them to determine whether they have a close neighbor.

Organisms are not just aggregates of cells – cells also make materials that provide structural support and knit groups of cells together. This material is called "extracellular matrix" (ECM). Again, the ECM is rich in chemical information for cells, provided in the three-dimensional information encoded in the myriad proteins that may be deposited there. In this way, cells can communicate with one another not only in space, but also over relatively long periods of time, with insoluble ECM having a much longer half-life that secreted soluble cues (Damon et al. 1989).

But this is not the whole story. The environment is not solely open to sensing by chemical means. Consider what we think of as our own senses: sight, sound, smell, taste, and touch. Smell and taste are perhaps the most analogous to the cellular sensing mechanisms just

Mechanobiology: Exploitation for Medical Benefit, First Edition. Edited by Simon C. F. Rawlinson.

described, while sight is a somewhat more specialized form of sensing, based on the ability of certain cellular molecules to become altered by the absorption of electromagnetic radiation. Sound and touch are also fundamental sensations, the former a specialized type of the latter, based on our ability to detect the mechanical force of the interaction of matter with our bodies. This property is generally referred to as "mechanosensitivity," the study of which is known as "mechanobiology." But despite the importance of these senses, for many years they remained relatively under-researched in the field of biological sciences, and were limited to some fascinating, specialist examples. One such example is the hair cells of the inner ear, which transduce movement into neural signals that can be interpreted by the central nervous system (CNS) (Lumpkin et al. 2010). These cells not only detect vibrations in materials of particular wavelengths that we understand as sound, but are also able to act as accelerometers – detecting acceleration due to physical movement or the continuous acceleration resulting from the earth's gravity. In addition, a similar system is thought to be present in the skeleton. Astronauts who experience long periods of reduced acceleration in the microgravity of the earth's orbit suffer from a reduced bone mass on return to earth (Sibonga et al. 2007). A prevailing hypothesis (yet to be universally accepted) is that osteocytes within the bone matrix, like the hair cells of the inner ear, are able to detect and respond to acceleration (Klein-Nulend et al. 1995). Evidence for this comes from the observation that bones remodel in response to mechanical stress, tending to increase in density (and strength) in regions where the applied stress is the greatest, an effect unambiguously demonstrated in the forearms of professional tennis players (Figure 1.1), where bone thickness is greater in the dominant arm (Ducher et al. 2005).

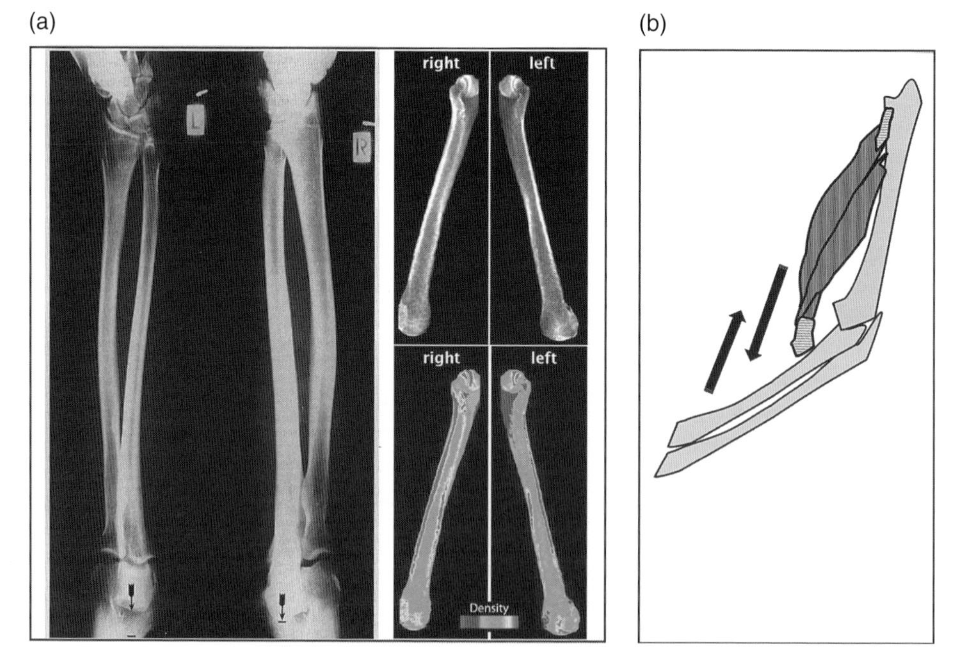

Figure 1.1 Bone growth and development are affected by mechanical stress. (a) The response of tissues to mechanical stimulation can clearly be seen in the arms of a professional tennis player. The bone thickness and density are greater in the dominant right arm. (b) On hitting the ball with the racket, the skeletal muscle pulls against the bones, causing them to rebuild and become denser. *Source:* x-ray images reproduced from Krahl et al. (1994) and Taylor et al. (2000).

Aside from these specific examples of mechanosensing, it is increasingly evident that all cells retain intrinsic mechanisms for sensing the mechanical properties of the environment around them. And this property has fundamental repercussions in almost all aspects of physiology and disease. In the context of human health and well-being, one aspect of mechanobiology that continues to receive special attention is its effect on stem cells.

1.2 Stem Cells

Stem cells are cells that can divide to make more copies of themselves, or which can differentiate into two or more specialized cell types. The concept of the stem cell emerged from ideas about both evolutionary and developmental biology in the late 19th century, generally with the notion that cell lineages, either throughout evolution or in the development of an organism, followed a family tree-like pattern of descent, with the putative stem cell at the top (Maehle 2011). This concept was brought into sharp focus in the mid-20th century with the work of a succession of experimental biologists who characterized "haematopoetic stem cells." These cells were shown to have enormous plasticity and replicative power, and to completely reconstitute the immune systems of animals lacking a working one (the immune systems of these animals had been destroyed with radiation), supporting the early ideas of proponents of the stem cell hypothesis, such as Pappeheim (Figure 1.2a) (Ramalho-Santos and Willenbring 2007). Today, the concept of the stem cell has spread throughout organismal biology, with stem cells identified in most if not all organs and tissues of the mammalian body. Some are amenable to extraction and culture in *in vitro* or *ex vivo* conditions and can be studied relatively easily, but some must be studied *in situ*. In the latter case, stem cells are known to occupy specific locations where they retain their stem-like properties. There, they have the correct provision of extracellular signals necessary to keep them in a state primed to divide and produce more functional descendants in normal homeostasis or in case of disease or injury. Such regions are called stem cell "niches," and the characteristics of such niches are vital to understanding how stem cells are regulated in normal and disease processes (Figure 1.2b).

Of particular interest is the pluripotent stem cell – so called because it has the ability to generate all of the cell types found in the adult organism. These cells, like cancer cells, divide indefinitely, making them a highly attractive source for cell replacement therapy, for example in diseases where the loss of a particular cell or tissue causes the severe effect of the disease. Originally, pluripotent stem cells were synonymous with embryonic stem cells (ESCs), but now it is known that cells with such properties can be artificially engineered from many adult somatic cell types – these are called "induced pluripotent stem cells" (iPSCs) (Takahashi and Yamanaka 2006). ESCs, which exist only transiently in development, can be extracted from the early blastocyst of the developing embryo and kept in an undifferentiated, developmentally frozen state by growing them in a precisely defined medium containing a cocktail of chemicals (Evans and Kaufman 1981; Thomson 1998). Similar conditions are required for iPSCs. On exposure to the right chemicals, at the correct concentrations, and at the appropriate time, such cells can be directed to differentiate to various lineages (e.g., pancreatic β cells, dopaminergic neurons, and hepatocytes). Controlling this is, of course, key to the utility of iPSCs in medicine – producing an adequate number of functional cells is necessary if they are to fulfill their intended medical use.

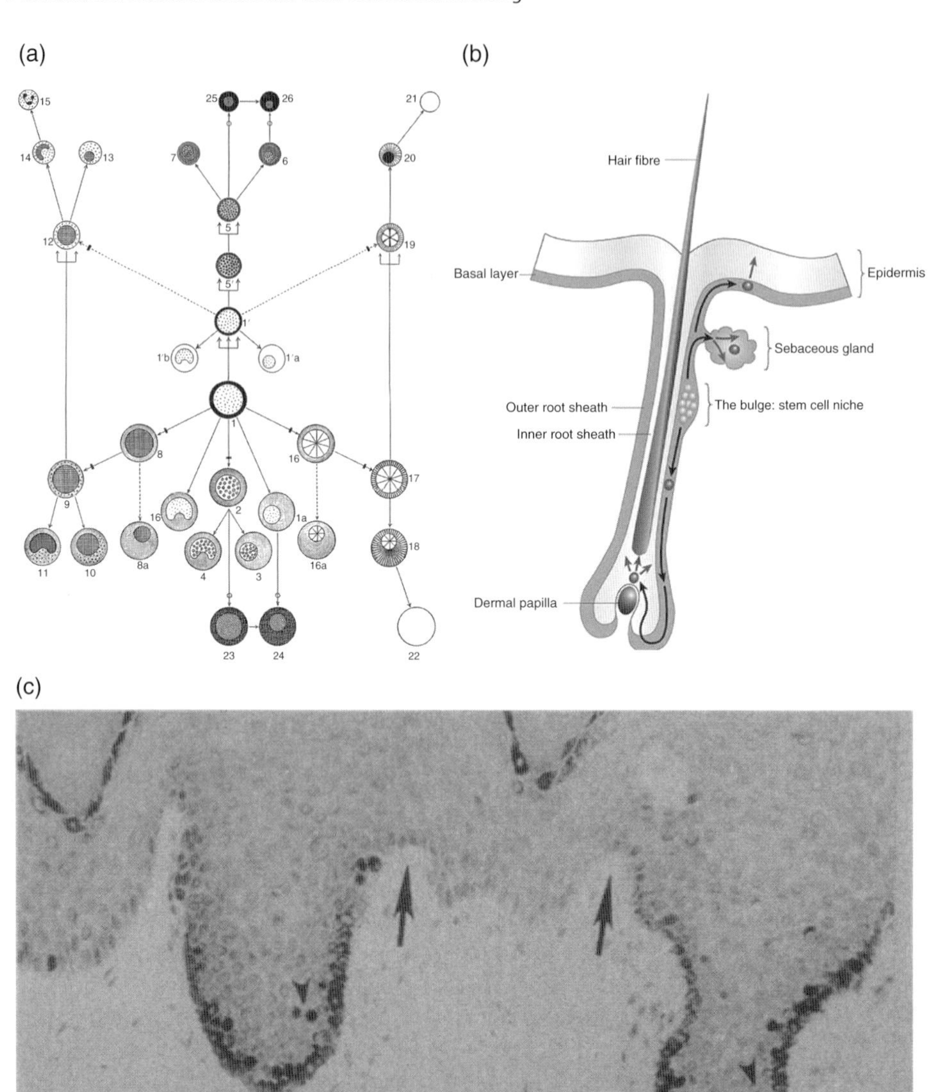

Figure 1.2 Stem cells and their niches. (a) Artur Pappenheim's hypothesis of hematopoiesis from 1905. The center cell, designated a "stem cell," represents the common progenitor of the entire blood system. (b) Stem cells exist in "niches" throughout the body, one of the best characterized being the bulge of the hair follicle. They become active during the anagen phase of the hair follicle cycle, replenishing many of the cell types that contribute to the follicle. Mechanical microenvironments such as topography may provide specific extracellular signals vital for keeping the stem cell in normal homeostasis. (c) Skin stem cells have been postulated to inhabit the rete ridge regions of the basal layer of the epidermis, formed by the epithelial morphology. *Source:* (a) reproduced from Ramalho-Santos and Willenbring (2007); (b) reproduced from Reya and Clevers (2005). Reproduced with permission of Nature Publishing Group; (c) reproduced from Lavker and Sun (1982).

As implied earlier, the provision of chemical signaling is a very well-explored concept in stem cell biology, in the context of both understanding adult stem cell niches and controlling (or not) the differentiation of pluripotent cells, but it is not the whole story. An increasing body of work now indicates that mechanobiological properties of the stem cell microenvironment – particularly the stiffness of the growth substratum – may be of fundamental importance in stem cell biology and regulation.

1.3 Substrate Stiffness in Cell Behavior

We saw earlier that certain cells have evolved to be able to detect externally applied force. However, virtually all mammalian cells need to apply force to their environment. This is seen perhaps most clearly in the "contact dependence" of most adult somatic cell types, where they must interact with a solid extracellular support in order to survive, grow, and divide. In the absence of such attachment, most cells – be they skin cells, muscle cells, brain cells, or otherwise – undergo a specialized form of controlled cell death called anoikis (Frisch and Screaton 2001). (Note that certain cell types, such as cells of the blood, do not share this feature, for obvious reasons.) So, what then is the signal that enables a cell to determine whether it is attached to a solid support? It all comes down to the cytoskeleton of the cell.

A cell's cytoskeleton is a complex arrangement of different polymer filaments that fulfil a number of vital functions – trafficking of organelles (such as endosomes and mitochondria), chromatid separation at mitosis, and motility. Cell motility depends on the interaction of a cell with its external environment, requiring the cell to move in relation to an external frame of reference. In the case of a contact-dependent cell type, this must be a solid support. By simple Newtonian mechanics, if a cell is to move in relation to such a solid support, it must exert a force on it. And if the cell is to gain any purchase on a material, the material must be able to accommodate and resist the force that the cell exerts. For this to occur, the cell must be able to generate tensile force within its cytoskeleton and do work.

1.3.1 A Historical Perspective on Stiffness Sensing

The notion of cells being able to probe the mechanical context of their environment has been appreciated for many years. Work in the 1970s showed that epithelial cells have markedly different morphologies and behaviors depending on whether they are grown on floating collagen gels or on hard growth substrata. Emerman et al. (1977) inferred that, aside from the different access to nutrients and different properties of gas exchange present in floating collagen substrates, the flexibility of the material could be affecting the shape of the cells by a postulated mechanical feedback. In later work (Shannon and Pitelka 1981), the same authors, quite directly, were able to conclude that stiffness (referred to as "flexibility" in their publications) was directly responsible for the functional phenotype of mammary cells on floating gels: while cells cultured as monolayers on floating collagen gels maintained a cuboidal secretory phenotype, cells cultured on the same collagen gels artificially stiffened by gluteraldehyde crosslinking appeared flattened and did not form the mature, secretory phenotype. Concurrently, other groups provided evidence for the accepted idea that cells exert force on the material on which they grow.

By developing a method of producing very thin membranes of silicone rubber, Harris et al. (1980) showed in striking visual images the degree to which silicone-adherent cells were able to deform the surface on which they grew. Most of these early studies did not further explore the biomechanical properties of such ECMs, but interpreted the key findings as being due to cell shape.

At around the same time, Folkman and Moscana (1978) were able to demonstrate experimentally (by reducing the adhesiveness of cell culture substrata) that there was a clear correlation between cell spreading and cell proliferation. This idea had been predicted by other researchers (e.g., Dulbecco 1970), who observed a higher mitotic index in cells given space to spread at the periphery of an artificially created *in vitro* "wound." Nevertheless, Folkman and Moscana (1978) were first to show direct evidence of a dependence of cell division on cell spreading, independent of the effects of (for example) cell–cell contact or cell density. These experiments were extended by Ingber and Jamieson (1985), who developed the idea of the "tensegrity" model of the cell's cytoskeleton – that is to say, that cell phenotype and tissue formation could be regulated by the mechanical phenomena occurring in the cytoskeleton. This led Inger and Folkman (1989) to show the importance of matrix "malleability" in the control of *in vitro*-simulated angiogenesis.

As techniques in bioengineering developed, other groups confirmed the dependence of cell shape and spreading on other cell functions besides division. For example, Watt et al. (1988) developed a method of depositing adhesive ECM islands of areas of between 500 and 2000 μm^2. Primary keratinocytes, plated on and confined to these islands, showed clear phenotypic differences depending on the degree to which they spread. In general, cells on larger islands (which had more space to spread out) synthesized more DNA than those on smaller islands, and the former remained undifferentiated while the latter did not. This idea was investigated several years later by Chen et al. (1997), who demonstrated via experiments based on the principle of depositing defined patterns of ECM on otherwise nonadhesive surfaces that cell spreading, rather than ECM contact area *per se*, influenced cell behavior, including apoptosis and cell proliferation.

Despite a great deal of evidence from the late 1970s and 1980s that the "malleability" or "flexibility" of ECMs could influence how cells behaved, including ideas about how intracellular tension might translate into biochemical signals, it was not until 1997 that the first formal test of how matrix stiffness affects cell behavior was conducted. Pelham and Wang (1997) employed a commonly used laboratory material – polyacrylamide – and varied the ratio of the monomer backbone of the polymer to its crosslinker to produce materials with a range of defined stiffnesses, which they measured simply by hanging weights from the polymer and measuring the extension (many will be familiar with the equivalent school-lab test of Hooke's law). Importantly, they attached thin films of these gels to a solid (glass support) and were able to covalently attach a matrix protein to the surface using polyacrylamide, converting the polymer into a material that could support the culture of a range of mammalian cells. Pelham and Wang were able to show that cells on stiffer substrates exhibited more stable focal adhesions than those on softer surfaces, which were more irregularly shaped and dynamic. The development of this (seemingly simple) technology was timely for those interested in cell traction dynamics, who had been inspired by Harris et al.'s (1980) work on substrate wrinkling. For example, Jacobson and colleagues had previously attempted to extend Harris' work to quantify the tractions that cells exerted on surfaces by using rubber substratum under tension

(Oliver et al. 1995; Dembo et al. 1996). However, these techniques were never optimized for use with mammalian cells. Subsequent to Pelham and Wang's publication, however, Wang teamed up with Micah Dembo to use the polyacrylamide method, combined with the introduction of fiduciary particles incorporated within the gels, to directly measure traction forces (Dembo and Wang 1999). This technique is now called "traction force microscopy" and is an established technique in a number of research fields, with more than 400 publications recorded in PubMed to date (e.g., Plotnikov et al. 2014). In addition, polyacrylamide surfaces also enabled the direct study of empirically defined ECM stiffnesses on a range of cell types. For example, in an echo of Inger and Folkman (1989), Deroanne et al. (2001) showed that a reduced substrate stiffness promoted tubulogenesis in endothelial cells, while Wang's group extended its earlier findings by showing that substrate stiffness could affect the motility of cells (Lo et al. 2000) and was a more important factor in the behavior of normal cells than were transformed cell lines (Wang et al. 2000).

Other groups began to take interest. In 2004, a group led by Dennis Discher showed that ECM stiffness was particularly important in the growth and differentiation of muscle cells (Engler et al. 2004). It demonstrated that while the formation of myotubes from myoblasts was unaffected by the stiffness of the ECM (though the subsequent phenotypic differentiation was affected), only those myotubes on ECMs with a stiffness corresponding to the stiffness of the tissue found *in vivo* formed striations. Together with the earlier observations, this study brought into sharp focus some of the disadvantages of the accepted methods of cultivating cells on rigid materials (glass or plastic). To date, most groups still work with rigid growth materials, but it is notable that there is a keen drive to provide more realistic methods of organ/tissue culture for drug testing (Feng et al. 2013), and several companies now make a business from selling growth substrata of defined stiffness (e.g., Matrigen, www.matrigen.com).

Subsequently, Discher's group highlighted the importance of mechanosensing in tissue cells (Discher et al. 2005), before publishing a seminal research paper showing that matrix elasticity alone can direct the differentiation of stem cells (Engler et al. 2006). The influence of this latter publication is reflected in the number of citations it has received (>5000) and the increase in the popularity of research on stem cell mechanotransduction.

1.4 Stem Cells and Substrate Stiffness

Discher et al. (2005) showed that a population of stem cells isolated from the bone marrow – mesenchymal stem cells (MSCs) (note that this term is somewhat controversial: the cells they studied may be more accurately referred to as "marrow stromal cells," a mixed population of primary cells likely to contain populations of stem and progenitor cells (Bianco et al. 2013)) – assumed different morphologies as a function of substrate stiffness. Moreover, over a period of several days, cells adherent to soft matrices (<1 kPa) began to express proteins specific to neuronal lineages, those on intermediate stiffnesses (~10 kPa) began to express markers of muscle differentiation, and those on stiffer surfaces began to express markers of bone cell differentiation (~30 kPa). This was tentatively shown not to be due merely to ECM surfaces preferentially selecting the adherence of one progenitor over another, as the authors could show transdifferentiation of cells over a period of time. These data reflect earlier work by McBeath et al. (2004), who showed adipogenic differentiation of MSCs confined to small islands and osteogenic differentiation on large islands (using a similar strategy to that employed by Watt et al. 1988). One might infer from these data that it is the stiffness-mediated change in cell shape that controls the phenotypic response, but Tee et al. (2011) have shown that when cell spreading is controlled and equalized on substrates of differing stiffnesses, cells remain able to modulate their cytoskeletal properties based on the stiffness, independent of the degree of spreading.

1.4.1 ESCs and Substrate Stiffness

What about other stem cells? Li et al. (2006) have shown that human ESCs can be maintained in an undifferentiated state on polymeric substrates with tunable stiffnesses. Later, Evans et al. (2009) showed that substrate elastic modulus can affect the initial differentiation behavior of murine ESCs, with stiffer substrates promoting mesendodermal differentiation and softer surfaces promoting ectodermal differentiation. This led to a greater differentiation of these stem cells to the osteogenic lineage. In the same study, collagen-functionalized polydimethylsiloxane (PDMS) was used as an ECM material, with stiffnesses ranging from 40 kPa to several megapascals, rather than the 0.1–50.0 kPa range that is investigated using polyacrylamide (Evans et al. 2009). In a series of papers later published by Ning Wang's group (Chowdhury et al. 2010a, 2010b; Poh et al. 2010), it was found that in contrast to many other mammalian cell types, murine ESCs are not sensitive to the modulus of their substrate and do not spread when in an undifferentiated state, even on stiff surfaces. In addition, cultivating cells on soft substrates could promote sustained self-renewal even in the absence of chemical factors normally required for self-renewal (leukemia inhibitory factor, LIF). Finally, mechanical stimulation of ESCs by exertion of torsional forces at the cell surface using arginylglycylaspartic acid (RGD)-conjugated beads could induce differentiation. This highly interesting work suggests that murine ESCs are an example of a cell type that does not have the ability to probe and sense ECM stiffness, but does however have the ability to detect applied force. Reflecting this, it has been shown that murine ESCs are unusual among mammalian cells in not being dependent on adherence to a surface for survival – they can be grown in suspension when cell aggregation is prevented. Recent work has shown that murine ESCs can be maintained in suspension in spinner flasks when an antibody against E-cadherin is added to the growth medium (Mohamet et al. 2010). This may also explain the requirement for the widespread use of gelatin coating as a substrate for murine ESC culture. Though one might expect gelatin to promote cell adhesion, in some cases it is used as an additive to prevent protein adsorption to surfaces and cell attachment (Milne et al. 2005). While this has never been tested, it may be speculated that gelatin facilitates self-renewal of murine ESC by allowing the growth of loosely adherent colonies while preventing the growth of more adherent, differentiated cells that arise spontaneously during cultivation. In an interesting discussion, Chowdhury et al. (2010b) speculated that early single-celled eukaryotes may have been subject to an evolutionary advantage that made them stiffer, enabling them to engage in mechanical functions such as invasion and crawling around the earth's primitive ocean floors, and that the mechanical state of ESCs is an echo of the early origins of multicellular life. It is therefore probable that in work investigating the effect of stiffness on murine ESC differentiation, the true effect of matrix stiffness is on the differentiation or selection of progenitor cells that arise stochastically during the very early stages of ESC commitment. Despite this, matrix rigidity or stiffness has been shown to affect the differentiation of murine ESCs into a range of different cell types, including cardiomyocytes (Shkumatov et al. 2014), pancreatic β cells (Candiello et al. 2013), endoderm (Jaramillo et al. 2012), and neurons (Keung et al. 2012).

Human ESCs are strikingly different from their murine counterparts. Whereas the latter form compact, sometimes multilayered, domelike colonies *in vitro*, the former grow as tightly packed epithelial sheets (Figure 1.3). In fact, the survival of human ESCs is linked to their cell–cell adhesive properties, and propagation efficiency decreases markedly on cell dissociation (in direct contrast to murine ESCs). It has been shown that this can be mitigated by the inclusion of a rho-associated protein kinase (ROCK) inhibitor (Y-27632) in the growth medium, which is thought to act by inhibiting cell contractility (Watanabe et al. 2007). An increase in the activity of the actin–myosin system is thought to be the reason for this apoptosis, which is usually prevented when the cells are adherent to one another and cytoskeleton tension is optimal (Ohgushi et al. 2010). Some have speculated that this reflects the embryonic origin of human ESCs as compared to murine ESCs. Human ESCs are similar to cells of the epiblast – a polarized epithelium that arises in the blastocyst – while murine ESCs are similar to the inner cell mass, which has no obvious polarity (Figure 1.3). Correct development of

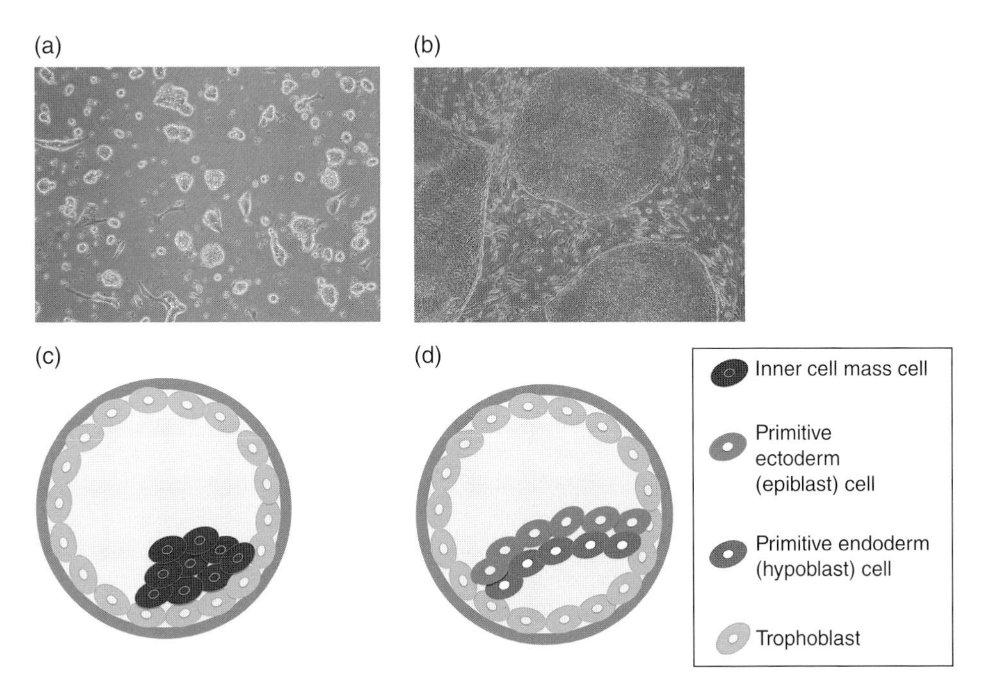

Figure 1.3 ESCs: differences in origin. (a) Murine ESCs form domelike, rounded colonies several cell layers thick, whereas (b) human ESCs form flattened, epithelial colonies. This may reflect differences in their origins. (c) Murine ESCs are thought to be analogous to cells of the inner cell mass of the embryo, which has no obvious polarity. (d) On the other hand, human ESCs (and murine EpiSCs) are likely to be more closely related to cells of the epiblast of the blastocyst. This structure is a polarized epithelium covering a basement membrane on the surface of the primitive endoderm (hypoblast). (*See insert for color representation of the figure.*)

the primitive ectoderm from the epiblast relies on appropriate patterning of cells, and it may be that cells that lose epithelial integrity and become detached from the epiblast sense the change in their mechanical microenvironment and are programmed to die by apoptosis (Ohgushi and Sasai 2011). Note that murine epiblast stem cells (EpiSCs) – which have many of the characteristics of human ESCs – can now be isolated from murine blastocysts at later time points compared to the original ESCs researched since Evans and Kaufman's 1981 paper (Brons et al. 2007), indicating that murine and human ESCs as commonly studied reflect mammalian tissues at two distinct developmental time points.

These data imply that matrix attachment and control of cell contractility in human ESCs within colonies may be more critical for the early differentiation of these cells than for murine ESCs. But in contrast to the large literature on murine ESCs and adult stem cells, research on the effect of matrix mechanical properties on human ESCs is poorly represented. Work from Healy's group demonstrated that human ESCs could be grown on materials with tunable stiffnesses (Li et al. 2006), but a PubMed search of "('human embryonic stem cell' OR 'human embryonic stem cells') AND ('stiffness' OR 'elasticity' OR 'rigidity')" at the time of writing yields fewer than 30 publications, many of which focus on the rheological properties of the cells themselves, rather than specific effects on their differentiation or self-renewal. In an example of the later, direct approach, Sun et al. (2012) investigated the effects of stiffness on human ESCs of using bendable PDMS pillar arrays, which they contended could be used to approximate "effective" stiffnesses of between ~2 and >1000 kPa, and measured self-renewal markers and E-cadherin expression in single cells and small aggregates of cells. They found higher expression levels of OCT4 in cells on matrices with higher effective stiffnesses, reflecting the fact that these cells are mechanosensitive and exhibit the correct phenotypic responses only when adherent to a surface with the optimal stiffness. Narayanan et al. (2014) produced growth substrates exhibiting a range of stiffnesses by decellularizing native ECMs and were able to show lineage-specific differentiation, though note here that because the chemical and physical properties of the ECMs were adjusted together with stiffness, it is not possible to judge any independent stiffness effect. Finally, Arshi et al. (2013) used a PDMS system similar to that of Evans et al. (2009) and found a preference for ESCs (initially differentiated in suspension culture as embryoid bodies) to differentiate into cardiomyocytes on surfaces of a higher stiffness.

One possible reason why the literature on the effect of matrix stiffness is limited in the case of human ESCs is their rather fickle growth conditions. Though the culture and isolation of human ESCs was first reported in 1998, it remains technically challenging and labor-intensive to grow these cells. Today, in most labs, human ESCs are grown on a feeder layer of murine embryonic fibroblasts. These cells provide a host of insoluble and soluble chemical cues to facilitate self-renewal. Otherwise, human ESCs are routinely grown on a propriety complex matrix preparation called Matrigel in the presence of medium conditioned by embryonic fibroblasts. Discovering a matrix that allows a more convenient method of propagating cells is currently a priority in the field, and matrices based on laminin isoforms are a particularly active area of research (Rodin et al. 2014).

To try and facilitate growth of human ESCs on polyacrylamide substrates, Weaver's group has published a methods paper demonstrating crosslinking of Matrigel to polyacrylamide surfaces (Lakins et al. 2012). This group fabricated polyacrylamide gels of ~100 μm depth and crosslinked Matrigel to the hydrogel via an ultraviolet (UV)-catalyzed conjugation of a protein-reactive N-succidimidyl ester to the gel surface. Interestingly, the group found clear morphological differences at different stiffnesses. Colonies on soft materials formed epithelial layers that were more columnar in nature than those on stiffer substrates, with a higher aspect ratio, basally displaced nuclei, and better developed E-cadherin staining at adherens junctions between cells. Very recently, the same group published another paper advancing this method and incorporating traction force microscopy to measure matrix deformations beneath colonies (Przybyla et al. 2016). These two papers are particularly exciting as they begin to address the often overlooked question of how groups of cells perceive stiffness, as compared to individual cells. In addition, they provide a methodological framework for probing the effect of the mechanical microenvironment in very early developmental events, by determining how local changes in, for instance, tension in epithelia map to changes in the phenotypic behavior of cells. This avenue of research is likely to yield some very interesting data over the coming years.

1.4.2 Collective Cell Behavior in Substrate Stiffness Sensing

One key question that is beginning to gain recognition is how cells might behave collectively to probe the stiffness properties of the extracellular environment that they inhabit. It is certainly true that many cells behave individually when probing the mechanical properties of their environment (e.g., mesenchymal cells, such as fibroblasts and macrophages). A far greater number of cell types, however, rely on cell–cell contact. This is particularly true in epithelia, where tight adherens junctions ensure barrier function and the integrity of the tissue as a whole (e.g., in skin and gut, and of course in embryonic tissues, as already discussed). In this way, the mechanobiology of the growth environment may be probed collectively at the tissue level rather than at the single cell level (which is the level most often studied in the literature). By acting together in this way, such cell collectives may be able to gain physical and mechanical information about their environment that would be unobtainable if they were to act as individual cells. As a rather crude analogy, consider a long line of people linked arm to arm. If a trap door were to open somewhere along the line, the force required to prevent the unfortunate people previously standing on it would have to be borne by others in the line. Someone standing a significant distance away from the trapdoor would understand that an event had happened, know that it had happened in a particular direction, and have to change their behavior according (perhaps by leaning to one side). There is plenty of evidence that cells behave in the same way, *in vivo* and *in vitro*. For example, groups led by Fredberg and Trepat have shown long-range force propagation in epithelia, in response to both pushing and pulling. When epithelial layers are disrupted, or when a gap in an epithelium is engineered by allowing cells to grow around a post and then removing it, cells migrate and divide to reoccupy the empty space (Tambe et al. 2011; Anon et al. 2012). Trepat and Fredberg (2011) use the analogy of a mosh pit to help explain how this happens. Cells are constantly migrating, but in epithelia they are closely packed and therefore constrained. Following the formation of a wound edge, cells make net movements into the space, dragging (and being constrained by) cells behind them, to which they are attached. This leads to the formation of more space in areas distal to the “wound,” allowing other cells the freedom to migrate in the direction of the space, or to stimulate their division.

These types of "long-range" force transmission may become particularly important in the mechanisms through which groups of cells probe substrate stiffness. In the early experiments on floating collagen gels, Emerman et al. (1977) found that mammary epithelial cells grown on unconstrained, floating soft collagen gels were able to contract the gels to around one-quarter of their original size, illustrating that groups of epithelial cells can exert significant force at their basal surfaces. Later, Trepat (2009) found that colonies of Madin–Darby canine kidney (MDCK) epithelial cells grown on basally adhered polyacrylamide gels (<100 μm in thickness) were insensitive to their substrate stiffness when grown as colonies but not when grown as single cells. These observations are important because they suggest, first, that groups of cells acting in concert are able to significant deform soft ECM materials and, second, that constraining a colony of cells may render the cells incapable of detecting the modulus of the material on which they grow. To understand why this might be the case, we must consider a number of reported studies on cells grown on substrates that vary in thickness.

Buxboim et al. (2010) developed a technique for casting polyacrylamide gels of various thicknesses adhered to an underlying glass support (in a manner similar to Pelham and Wang 1997). They found that even at very low elastic moduli (<1 kPa), at certain depths, cells started to behave as if they were on much stiffer gels (when the thickness was decreased to <10 μm). Lin et al. (2010) have provided a theoretical explanation for this phenomenon, which relates this "critical depth" to a value approximate to the lateral dimension of the adherent cell. In simple terms, the reason that cells can detect substrate depth in this way is because of the manner in which a cell probes the stiffness of its substrate. As a cell makes focal adhesions, it begins to contract, exerting a shear force on the ECM and detecting the stiffness of the material by monitoring the resistance to this force. However, in this case the force required to deform the surface of the ECM a given distance is dependent not only on the Young's modulus of the material, but also on the thickness. One can understand this more clearly by using an analogy: it is much easier to shear the surface of a deep plate of jello (referred to as jelly in the UK) than that of a very thin one, even though the Young's modulus of the material remains constant (see Evans and Gentleman 2014). As there is less gel, and because it is prevented from moving at its basal surface, a lateral shear deformation of a given magnitude will impart a much greater strain on the thin gel than on the thick one (Figure 1.4).

What are the consequences of this effect for groups of cells? Epithelial cells foster very tight intracellular junctions, and are able to exert significant force on ECMs, and therefore significant strains. Recent data from Zarkoob et al. (2015) support this. This group found average surface matrix deformations of ~4.2 μm for single keratinocytes, compared with 19.4 μm for groups of around eight cells, with some deformations reaching more than 100 μm. Similarly, Mertz et al. (2012) found that contractile forces scaled with the size of the colony. Though not tested, these deformations must penetrate significant depths into hydrogel substrates. Taking into account theoretical considerations of depth sensing (Lin et al. 2010; Edwards and Schwarz 2011; Banerjee and Marchetti 2012), this may provide an explanation for why Trepat (2009) found no effect of substrate on large colonies (millimeter-size) of MDCK cells: by acting collectively, cell-comprising colonies measure a greater stiffness on fixed elastic substrata than their Young's modulus would suggest. How deeply might colonies of cell sheets feel? This remains unknown, but based on the theoretical work and on Emerman's observations in the 1970s, it might

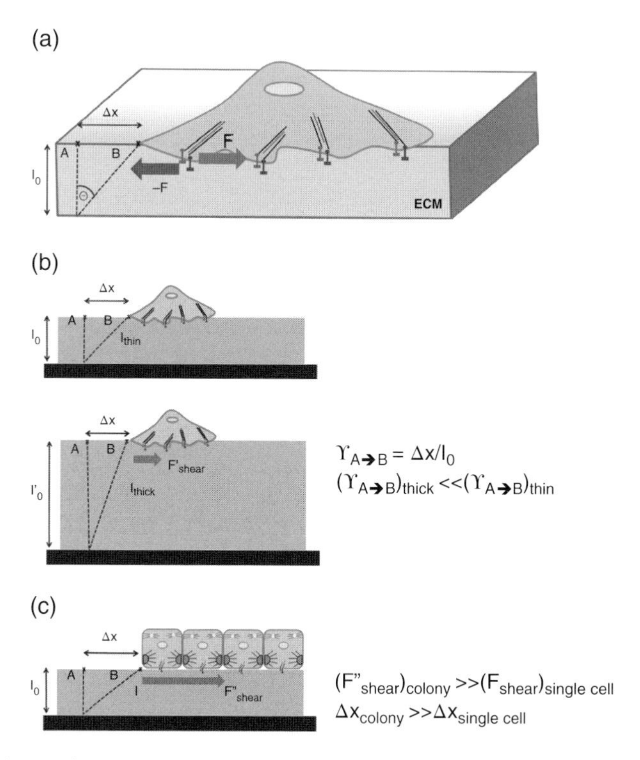

Figure 1.4 Cellular mechanosensing of substrate thickness. (a) In order to contract a gel from A to B for a distance Δx, a cell needs to form focal adhesions on a solid support, as shown in the cartoon schematic (the cones represent the integrin connections), and then exert a force (right arrow). During this process, a tensile force (skewed lines represent the actin–myosin dependent contraction) is generated in its cytoskeleton. The material has to be able to resist and accommodate to the force that the cell applies, in this case a shear force. (b) and (c) illustrate in a simplified way the difference in force that a cell must exert in order to contract a thick versus a thin gel of equal shear modulus. (b) The shear strain is measured as the ratio between the transverse displacement of the gel (Δx) and its initial length (l). An attached cell exerts a shear force on the gel (top left) and deforms it to a distance Δx. An equal deformation Δx in the direction A → B requires a greater shear strain ($\Upsilon_{A\to B}$) on thin gel compared with thick. Even if the shear modulus of the material is the same, the shear stress required to deform the thin gel is greater than that required for the thick gel. As a consequence, the tension generated in the cytoskeleton may reach a critical threshold on thin gels, causing the cell to spread more; on thick gels, the cell may be unable to generate the same tension, and thus remains rounded. (c) For colonies of cells, the transverse displacement may be greater than that for a single cell. This may be a collective behavior mediated by tight intracellular interactions. Note that this figure is for explanatory purposes only and ignores many variables.

be related to the size of the sheet: many hundreds of microns. Furthermore, epithelia would be able to detect not only uniform changes in substrate thickness, but also regional changes determined by stiff objects or heterogeneities deep within the hydrogel.

Collective stiffness sensing may have important implications in many areas of biology. One example is in skin wound-healing. To facilitate wound coverage and healing, a new layer of epithelium must migrate out over the surface of the underlying granulation tissue. The stiffness of this material and its heterogeneity may play an important role in

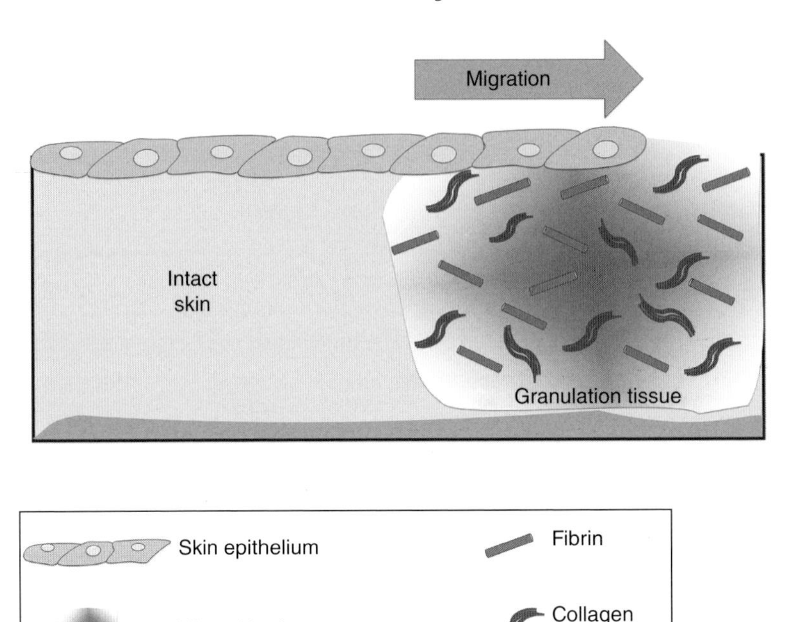

Figure 1.5 Mechanosensing of the stiffness and heterogeneity of the wound granulation tissue is important in wound-healing. A new layer of epithelium has to migrate from the intact skin over the granulation tissue. Granulation tissue is formed in the wound bed after skin injury. It is a heterogeneous material, manly formed of fibrin and type III collagen. During the healing process, its stiffness varies, and this might influence skin repair.

the rate of cell migration and in cell patterning and differentiation (Figure 1.5). Some data support this already (Wang et al. 2012). Further, embryonic development involves the rapid movement of a variety of different epithelia. Such epithelial tissues may be influenced not only by chemical signaling gradients, but also by the material characteristics of structures in other parts of the embryo, which they detect as mechanical gradients. Mechanically informed patterning may be a crucial but overlooked mechanism of mammalian development.

1.5 Material Structure and Future Perspectives in Stem Cell Mechanobiology

Artificial ECMs are useful for reductionist experiments aimed at determining the effect of a limited number of parameters on cell behavior, but they do not approach the complexity of ECMs *in vivo*. Those that have been investigated *in vitro* include (but are not limited to) synthetic materials such as polyacrylamide, PDMS, polyethylene glycol (PEG), and naturally derived materials such as alginate, collagen, and hyaluronic acid. These materials are discussed in a recent review (Evans and Gentleman 2014). Polyacrylamide has been the most popular, probably due to its ease of fabrication, linear elastic behavior, nontoxicity, and low cost. However, like all materials used in such experiments, it does have some limitations. First, though it is generally accepted that it behaves in an elastically linear fashion (Style et al. 2014), at greater deformations (such as those that might be exerted by groups of cells) this approximation does not hold (Boudou et al. 2009). Second, it is nonadsorbing, so matrix proteins must be chemically crossslinked to its surface to enable cell attachment. In principle, this could be any ECM protein with a reactive side group, but type I collagen is most often used. Third, and perhaps most importantly, it is difficult to modulate its stiffness without affecting the degree of crosslinking and porosity of the gel. This point was illustrated recently by Trappmann et al. (2012), who provided evidence to show it is the porosity of polyacrylamide, rather than the bulk stiffness, that is responsible for the cell behavior seen at low stiffnesses. They showed that porous gels that had been artificially stiffened had very similar effects to their unstiffened counterparts: cells remained rounded on both. They attributed this to the difference in ligand tethering on soft (large pores) versus stiff (small pores) gels: on the latter, the anchoring points between the collagen ECM are close together, while on the former they are wide apart. The authors suggested that large ligand spacing leads to loosely attached, floppy collagen fibers, which are perceived by the cell as a lower gel stiffness, regardless of the bulk modulus. It is somewhat difficult to reconcile these data with the large body of literature on traction force microscopy, however, which relies upon the displacement of fiduciary markers *within* the hydrogel for the calculation of cellular contractile forces. And if this theory were correct, it would predict that there are no depth-dependent effects on detection of substrate elasticity: surface ligand chemistry remains constant *regardless* of the thickness of the gel.

Despite this, most natural ECMs do not behave in a predictable, linear fashion. For example, collagen has a hierarchical structure, consisting of interlocking fibrils that may be elongated or coiled. It has been shown to exhibit viscoelastic behavior, its stiffness changing over time and with the rate and magnitude of applied force (Knapp 1997). These nonlinear effects can be envisaged by thinking of the collagen as a tangled ball of fibers: under a given applied force, some of the fibers will resist under tension, while others will be under shear or compression. Illustrating this, it was found in studies of cells growing on collagen substrata of various thicknesses that fibroblasts began to "depth-sense" the underlying glass support at much greater depths than they did on polyacrylamide (~65 versus ~5 μm) (Rudnicki et al. 2013). This was attributed to the effect of fiber alignment, which becomes more prominent at smaller scales. In effect, as the thickness decreases, cells begin to measure the stiffness of collagen fiber bundles held taut from their anchor point at the underlying glass surface to the surface of the gel at the cells' focal adhesions, rather than the stiffness of a randomly arranged network of interconnected fibers. It is likely that many other ECMs found *in vivo*, such as those composed of other collagens or of composites of collagens and other natural materials, will display similar behavior.

1.6 Conclusion

Stem cell mechanobiology is a fascinating topic that will continue to occupy the efforts of a diverse range of researchers over the years to come. This research will require the input and collaboration of experts from a great variety of backgrounds, from physicists and engineers to medical scientists and developmental biologists. Understanding how our bodies develop and respond to disease and infection is essential to human health and well-being, and mechanobiology is now appreciated as a key player in this field (e.g., in cancer (Butcher et al. 2009) and in scarring (Engler et al. 2008)). In biomedical engineering, materials and devices must be designed to interface closely with body tissues, and an appreciation of the mechanical effects of this will be important. In regenerative medicine, materials are designed to encourage the restoration of tissue function, which depends largely on the behavior of stem cells; answering fundamental questions about the response of stem cells to ECM mechanics, in parallel to the cohesive tissues in which they reside, will be key to the success of these strategies in the future.

References

Anon, E., X. Serra-Picamal, P. Hersen, N. C. Gauthier, M. P. Sheetz, X. Trepat, and B. Ladoux. 2012. "Cell crawling mediates collective cell migration to close undamaged epithelial gaps." *Proceedings of the National Academy of Sciences of the United States of America* **109**(27): 10891–96. doi:10.1073/pnas.1117814109.

Arshi, A., Y. Nakashima, H. Nakano, S. Eaimkhong, D. Evseenko, J. Reed, et al. 2013. "Rigid microenvironments promote cardiac differentiation of mouse and human embryonic stem cells." *Science and Technology of Advanced Materials* **14**(2). doi:10.1088/1468-6996/14/2/025003.

Banerjee, S. and M. Cristina Marchetti. 2012. "Contractile stresses in cohesive cell layers on finite-thickness substrates." *Physical Review Letters* **109**(10). doi:10.1103/PhysRevLett.109.108101.

Bianco, P., X. Cao, P. S. Frenette, J. J. Mao, P. G. Robey, P. J. Simmons, and C.-Y. Wang. 2013. "The meaning, the sense and the significance: translating the science of mesenchymal stem cells into medicine." *Nature Medicine* **19**(1): 35–42. doi:10.1038/nm.3028.

Boudou, T., J. Ohayon, C. Picart, R. I. Pettigrew, and P. Tracqui. 2009. "Nonlinear elastic properties of polyacrylamide gels: implications for quantification of cellular forces." *Biorheology*. doi:10.3233/BIR-2009-0540.

Brons, I.G., L. E Smithers, M. W. B. Trotter, P. Rugg-Gunn, B. Sun, S. M. Chuva de Sousa Lopes, et al. 2007. "Derivation of pluripotent epiblast stem cells from mammalian embryos." *Nature* **448**(7150): 191–95. doi:10.1038/nature05950.

Butcher, D. T., T. Alliston, and V. M. Weaver. 2009. "A tense situation: forcing tumour progression." *Nature Reviews. Cancer* **9**(2): 108–22. doi:10.1038/nrc2544.

Buxboim, A., K. Rajagopal, A. E. X. Brown, and D. E. Discher. 2010. "How deeply cells feel: methods for thin gels." *Journal of Physics. Condensed Matter: An Institute of Physics Journal* **22**(19). doi:10.1088/0953-8984/22/19/194116.

Candiello, J., S. S. Singh, K. Task, P. N. Kumta, and I. Banerjee. 2013. "Early differentiation patterning of mouse embryonic stem cells in response to variations in alginate substrate stiffness." *Journal of Biological Engineering* 7(1): 9. doi:10.1186/1754-1611-7-9.

Chen, C. S., M. Mrksich, S. Huang, G. M. Whitesides, and D. E. Ingber. 1997. "Geometric control of cell life and death." *Science (New York, N.Y.)* **276**(5317): 1425–28. doi:10.1126/science.276.5317.1425.

Chowdhury, F., Y. Li, Y. Chuin Poh, T. Yokohama-Tamaki, N. Wang, and T. S. Tanaka. 2010a. "Soft substrates promote homogeneous self-renewal of embryonic stem cells via downregulating cell-matrix tractions." *PloS One* **5**(12). doi:10.1371/journal.pone.0015655.

Chowdhury, F., S. Na, D. Li, Y.-C. Poh, T. S. Tanaka, F. Wang, and N. Wang. 2010b. "Material properties of the cell dictate stress-induced spreading and differentiation in embryonic stem cells." *Nature Materials* **9**(1): 82–8. doi:10.1038/nmat2563.

Damon, D. H., R. R. Lobb, P. A. D'Amore, and J. A. Wagner. 1989. "Heparin potentiates the action of acidic fibroblast growth factor by prolonging its biological half-life." *Journal of Cellular Physiology* **138**(2): 221–6. doi:10.1002/jcp.1041380202.

Dembo, M. and Y. L. Wang. 1999. "Stresses at the cell-to-substrate interface during locomotion of fibroblasts." *Biophysical Journal* **76**(4): 2307–16. doi:10.1016/S0006-3495(99)77386-8.

Dembo, M., T. Oliver, A. Ishihara, and K. Jacobson. 1996. "Imaging the traction stresses exerted by locomoting cells with the elastic substratum method." *Biophysical Journal* **70**(4): 2008–22. doi:10.1016/S0006-3495(96)79767-9.

Deroanne, C. F., C. M. Lapiere, and B. V. Nusgens. 2001. "In vitro tubulogenesis of endothelial cells by relaxation of the coupling extracellular matrix-cytoskeleton." *Cardiovascular Research* **49**(3): 647–58. PMID:11166278.

Discher, D. E., P. Janmey, and Y. L. Wang. 2005. "Tissue cells feel and respond to the stiffness of their substrate." *Science* **310**(5751): 1139–43. doi:10.1126/science.1116995.

Ducher, G., D. Courteix, S. Même, C. Magni, J. F. Viala, and C. L. Benhamou. 2005. "Bone geometry in response to long-term tennis playing and its relationship with muscle volume: a quantitative magnetic resonance imaging study in tennis players." *Bone* **37**(4): 457–66. doi:10.1016/j.bone.2005.05.014.

Dulbecco, R. 1970. "Topoinhibition and serum requirement of transformed and untransformed cells." *Nature* **227**: 802–6. PMID:4317354.

Edwards, C. M. and U. S. Schwarz. 2011. "Force localization in contracting cell layers." *Physical Review Letters* **107**(12). doi:10.1103/PhysRevLett.107.128101.

Emerman, J. T., J. Enami, D. R. Pitelka, and S. Nandi. 1977. "Hormonal effects on intracellular and secreted casein in cultures of mouse mammary epithelial cells on floating collagen membranes." *Proceedings of the National Academy of Sciences of the United States of America* **74**(10): 4466–70. doi:10.1073/pnas.74.10.4466.

Engler, A. J., M. A. Griffin, S. Sen, C. G. Bonnemann, H. L. Sweeney, and D. E. Discher. 2004. "Myotubes differentiate optimally on substrates with tissue-like stiffness: pathological implications for soft or stiff microenvironments." *Journal of Cell Biology* **166**(6): 877–87. doi:10.1083/jcb.200405004.

Engler, A. J., S. Sen, H. L. Sweeney, and D. E. Discher. 2006. "Matrix elasticity directs stem cell lineage specification." *Cell* **126**(4): 677–89. doi:10.1016/j.cell.2006.06.044.

Engler, A. J., C. Carag-Krieger, C. P. Johnson, M. Raab, H.-Y. Tang, D. W. Speicher, J. W. Sanger, J. M. Sanger, and D. E. Discher. 2008. "Embryonic cardiomyocytes beat best on a matrix with heart-like elasticity: scar-like rigidity inhibits beating." *Journal of Cell Science* **121**(Pt. 22): 3794–802. doi:10.1242/jcs.029678.

Evans, M. J. and M. H. Kaufman. 1981. "Establishment in culture of pluripotential cells from mouse embryos." *Nature* **292**(5819): 154–6. doi:10.1038/292154a0.

Evans, N. D. and E. Gentleman. 2014. "The role of material structure and mechanical properties in cell–matrix interactions". *Journal of Materials Chemistry B* **2**: 2345–56. doi:10.1039/C3TB21604G.

Evans, N. D., C. Minelli, E. Gentleman, V. LaPointe, S. N. Patankar, M. Kallivretaki, X. Chen, C. J. Roberts, and M. M. Stevens. 2009. "Substrate stiffness affects early differentiation events in embryonic stem cells." *European Cells & Materials* **18**: 1–4. PMID:19768669.

Feng, J., Y. Tang, Y. Xu, Q. Sun, F. Liao, and D. Han. 2013. "Substrate stiffness influences the outcome of antitumor drug screening in vitro." *Clinical Hemorheology and Microcirculation* **55**(1): 121–31. doi:10.3233/CH-131696.

Folkman, J. and A. Moscona. 1978. "Role of cell shape in growth control." *Nature* **273**(5661): 345–9. doi:10.1038/273345a0.
Frisch, S. M. and R. A. Screaton. 2001. "Anoikis mechanisms." *Current Opinion in Cell Biology* **13**(5): 555–62. doi:10.1016/S0955-0674(00)00251-9.
Harris, A. K., P. Wild, and D. Stopak. 1980. "Silicone rubber substrata: a new wrinkle in the study of cell locomotion." *Science (New York, N.Y.)* **208**(4440): 177–9. doi:10.1126/science.6987736.
Ingber, D. E. and J. Folkman. 1989. "Mechanochemical switching between growth and differentiation during fibroblast growth factor-stimulated angiogenesis in vitro: role of extracellular matrix." *Journal of Cell Biology* **109**(1): 317–30. PMID:2473081.
Ingber D. E. and J. D. Jamieson. 1985 "Cells as tensegrity structures: architectural regulation of histodifferentiation by physical forces tranduced over basement membrane." In: L. C. Andersson, C. G. Gahmberg and P. Ekblom (eds.). *Gene Expression During Normal and Malignant Differentiation.* Orlando, FL: Academic Press. pp. 13–32.
Jaramillo, M., S. S. Singh, S. Velankar, P. N. Kumta, and I. Banerjee. 2012. "Inducing endoderm differentiation by modulating mechanical properties of soft substrates." *Journal of Tissue Engineering and Regenerative Medicine* **9**(1): 1–12. doi:10.1002/term.1602.
Keung, A. J., P. Asuri, S. Kumar, and D. V. Schaffer. 2012. "Soft microenvironments promote the early neurogenic differentiation but not self-renewal of human pluripotent stem cells." *Integrative Biology* **4**(9): 1049–58. doi:10.1039/c2ib20083j.
King, N., C. T. Hittinger, and S. B. Carroll. 2003. "Evolution of key cell signaling and adhesion protein families predates animal origins." *Science (New York, NY)* **301**(5631): 361–3. doi:10.1126/science.1083853.
Klein-Nulend, J., A. van der Plas, C. M. Semeins, N. E. Ajubi, J. A. Frangos, P. J. Nijweide, and E. H. Burger. 1995. "Sensitivity of osteocytes to biomechanical stress in vitro." *The FASEB Journal: Official Publication of the Federation of American Societies for Experimental Biology* **9**(5): 441–5. PMID:7896017.
Knapp, D. M. 1997. "Rheology of reconstituted type I collagen gel in confined compression." *Journal of Rheology* **41**: 971. doi:10.1122/1.550817.
Krahl, H., U. Michaelis, H. G. Pieper, G. Quack, and M. Montag. 1994. "Stimulation of bone growth through sports: a radiologic investigation of the upper extremities in professional tennis players." *American Journal of Sports Medicine* **22**(6): 751–7. PMID:7856798.
Lakins, J. N., A. R. Chin, and V. M. Weaver. 2012. "Exploring the link between human embryonic stem cell organization and fate using tension-calibrated extracellular matrix functionalized polyacrylamide gels." *Methods in Molecular Biology* **916**: 317–50. doi:10.1007/978-1-61779-980-8-24.
Lavker R. N. and T. T. Sun. 1982. "Heterogeneity in epidermal basal keratinocytes: morphological and functional correlations." *Science* **215**: 1239–41. PMID:7058342.
Li, Y. J., E. H. Chung, R. T. Rodriguez, M. T. Firpo, and K. E. Healy. 2006. "Hydrogels as artificial matrices for human embryonic stem cell self-renewal." *Journal of Biomedical Materials Research A* **79**(1): 1–5. doi:10.1002/jbm.a.30732.
Lin, Y. C., D. T. Tambe, C. Y. Park, M. R. Wasserman, X. Trepat, R. Krishnan, et al. 2010. "Mechanosensing of substrate thickness." *Physical Review. E. Statistical, Nonlinear, and Soft Matter Physics* **82**(4 Pt. 1): 041918. PMID:21230324.

Lo, C. M., H. B. Wang, M. Dembo, and Y. L. Wang. 2000. "Cell movement is guided by the rigidity of the substrate." *Biophysics Journal* **79**(1): 144–52. doi:10.1016/S0006-3495(00)76279-5.

Lumpkin, E. A., K. L. Marshall, and A. M. Nelson. 2010. "The cell biology of touch." *Journal of Cell Biology* **191**(2): 237–48. doi:10.1083/jcb.201006074.

Maehle, A.-H. 2011. "Ambiguous cells: the emergence of the stem cell concept in the nineteenth and twentieth centuries." *Notes and Records of the Royal Society* **65**(4): 359–78. doi:10.1098/rsnr.2011.0023.

McBeath, R., D. M. Pirone, C. M. Nelson, K. Bhadriraju, and C. S Chen. 2004. "Cell shape, cytoskeletal tension, and RhoA regulate stem cell lineage commitment." *Developmental Cell* **6**(4): 483–95. PMID:15068789.

Mertz, A. F., S. Banerjee, Y. Che, G. K. German, Y. Xu, C. Hyland, et al. 2012. "Scaling of traction forces with the size of cohesive cell colonies." *Physical Review Letters* **108**(19): 198101. doi:10.1103/PhysRevLett.108.198101.

Milne, H. M., C. J. Burns, P. E. Squires, N. D. Evans, J. Pickup, P. M. Jones, and S. J. Persaud. 2005. "Uncoupling of nutrient metabolism from insulin secretion by overexpression of cytosolic phospholipase A2." *Diabetes* **54**(1): 116–24. PMID:15616018.

Mohamet, L., M. L. Lea, and C. M. Ward. 2010. "Abrogation of E-cadherin-mediated cellular aggregation allows proliferation of pluripotent mouse embryonic stem cells in shake flask bioreactors." *PloS One* **5**(9): e12921. doi:10.1371/journal.pone.0012921.

Narayanan, K., V. Y. Lim, J. Shen, Z. W. Tan, D. Rajendran, S.-C. Luo, et al. 2014. "Extracellular matrix-mediated differentiation of human embryonic stem cells: differentiation to insulin-secreting beta cells." *Tissue Engineering Part A* **20**(1–2): 424–33. doi:10.1089/ten.TEA.2013.0257.

Ohgushi, M. and Y. Sasai. 2011. "Lonely death dance of human pluripotent stem cells: ROCKing between metastable cell states." *Trends in Cell Biology* **21**(5): 274–82. doi:10.1016/j.tcb.2011.02.004.

Ohgushi, M., M. Matsumura, M. Eiraku, K. Murakami, T. Aramaki, A. Nishiyama, et al. 2010. "Molecular pathway and cell state responsible for dissociation-induced apoptosis in human pluripotent stem cells." *Cell Stem Cell* **7**(2): 225–39. doi:10.1016/j.stem.2010.06.018.

Oliver, T., M. Dembo, and K. Jacobson. 1995. "Traction forces in locomoting cells." *Cell Motility and the Cytoskeleton* **31**(3): 225–40. doi:10.1002/cm.970310306.

Pelham, R. J. Jr. and Y. Wang. 1997. "Cell locomotion and focal adhesions are regulated by substrate flexibility." *Proceedings of the National Academy of Sciences of the United States of America* **94**(25): 13661–5. PMID:9391082.

Plotnikov, S. V., B. Sabass, U. S. Schwarz, and C. M. Waterman. 2014. "High-resolution traction force microscopy." *Methods in Cell Biology* **123**: 367–94. doi:10.1016/B978-0-12-420138-5.00020-3.

Poh, Y. C., F. Chowdhury, T. S. Tanaka, and N. Wang. 2010. "Embryonic stem cells do not stiffen on rigid substrates." *Biophysical Journal* **99**(2): L19–21. doi:10.1016/j.bpj.2010.04.057.

Przybyla L., J. N. Lakins, R. Sunyer, X. Trepat, V. M. Weaver. 2016. "Monitoring developmental force distributions in reconstituted embryonic epithelia". *Methods* **94**: 101–13. PMID:26342256.

Ramalho-Santos, M. and H. Willenbring. 2007. "On the origin of the term 'stem cell.'" *Cell Stem Cell* **1**(1): 35–8. doi:10.1016/j.stem.2007.05.013.

Reya, T. and H. Clevers. 2005. "Wnt signalling in stem cells and cancer." *Nature* **434**(7035): 843–50. PMID:15829953.

Rodin, S., L. Antonsson, C. Niaudet, O. E. Simonson, E. Salmela, E. M. Hansson, et al. 2014. "Clonal culturing of human embryonic stem cells on laminin-521/E-cadherin matrix in defined and xeno-free environment." *Nature Communications* **5**: 3195. doi:10.1038/ncomms4195.

Rudnicki, M. S., H. A. Cirka, M. Aghvami, E. A. Sander, Q. Wen, and K. L. Billiar. 2013. "Nonlinear strain stiffening is not sufficient to explain how far cells can feel on fibrous protein gels." *Biophysical Journal* **105**(1): 11–20. doi:10.1016/j.bpj.2013.05.032.

Shannon, J. M. and D. R. Pitelka. 1981. "The influence of cell shape on the induction of functional differentiation in mouse mammary cells in vitro." *In Vitro* **17**(11): 1016–28. PMID:7033108.

Shkumatov, A., K. Baek, and H. Kong. 2014. "Matrix rigidity-modulated cardiovascular organoid formation from embryoid bodies." *PloS One* **9**(4): e94764. doi:10.1371/journal.pone.0094764.

Sibonga, J. D., H. J. Evans, H. G. Sung, E. R. Spector, T. F. Lang, V. S. Oganov, et al. 2007. "Recovery of spaceflight-induced bone loss: bone mineral density after long-duration missions as fitted with an exponential function." *Bone* **41**(6): 973–8. doi:10.1016/j.bone.2007.08.022.

Style, R. W., R. Boltyanskiy, G. K. German, C. Hyland, C. W. MacMinn, A. F. Mertz, et al. 2014. "Traction force microscopy in physics and biology." *Soft Matter* **10**(23): 4047–55. doi:10.1039/c4sm00264d.

Sun, Y., L. G. Villa-Diaz, R. H. W. Lam, W. Chen, P. H. Krebsbach, and J. Fu. 2012. "Mechanics regulates fate decisions of human embryonic stem cells." *PloS One* **7**(5): e37178. doi:10.1371/journal.pone.0037178.

Takahashi, K. and S. Yamanaka. 2006. "Induction of pluripotent stem cells from mouse embryonic and adult fibroblast cultures by defined factors." *Cell* **126**(4): 663–76. doi:10.1016/j.cell.2006.07.024.

Tambe, D. T., C. C. Hardin, T. E. Angelini, K. Rajendran, C. Y. Park, X. Serra-Picamal, et al. 2011. "Collective cell guidance by cooperative intercellular forces." *Nature Materials* **10**(6): 469–75. doi:10.1038/nmat3025.

Taylor, R.E., C. Zheng, R. P. Jackson, J. C. Doll, J. C. Chen, K. R. S. Holzbaur, et al. 2009. "The phenomenon of twisted growth: humeral torsion in dominant arms of high performance tennis players." *Computer Methods in Biomechanics and Biomedical Engineering.* **12**(1): 83–93. PMID:18654877.

Tee, S. Y., J. Fu, C. S. Chen, and P. A. Janmey. 2011. "Cell shape and substrate rigidity both regulate cell stiffness." *Biophysical Journal* **100**(5): L25–7. doi:10.1016/j.bpj.2010.12.3744.

Thomson, J. A. 1998. "Embryonic stem cell lines derived from human blastocysts." *Science* **282**(5391): 1145–7. doi:10.1126/science.282.5391.1145.

Trappmann, B., J. E. Gautrot, J. T. Connelly, D. G. Strange, Y. Li, M. L. Oyen, et al. 2012. "Extracellular-matrix tethering regulates stem-cell fate." *Nature Materials* **11**(7): 642–9. doi:10.1038/nmat3339.

Trepat, X. 2009. "Physical forces during collective cell migration." *Nature Physics* **5**: 426–30. doi:10.1038/nphys1269.

Trepat, X. and J. J. Fredberg. 2011. "Plithotaxis and emergent dynamics in collective cellular migration." *Trends in Cellular Biology* **21**(11): 638–46. doi:10.1016/j.tcb.2011.06.006.

Wang, H. B., M. Dembo, and Y. L. Wang. 2000. "Substrate flexibility regulates growth and apoptosis of normal but not transformed cells." *American Journal of Physiology. Cell Physiology* **279**(5): C1345–50. doi:11029281.

Wang, Y., G. Wang, X. Luo, J. Qiu, and C. Tang. 2012. "Substrate stiffness regulates the proliferation, migration, and differentiation of epidermal cells." *Burns* **38**(3): 414–20. doi:10.1016/j.burns.2011.09.002.

Watanabe, K., M. Ueno, D. Kamiya, A. Nishiyama, M. Matsumura, T. Wataya, et al. 2007. "A ROCK inhibitor permits survival of dissociated human embryonic stem cells." *Nature Biotechnology* **25**(6): 681–6. doi:10.1038/nbt1310.

Watt, F. M., P. W. Jordan, and C. H. O'Neill. 1988. "Cell shape controls terminal differentiation of human epidermal keratinocytes." *Proceedings of the National Academy of Sciences of the United States of America* **85**(15): 5576–80. PMCID:PMC281801.

Zarkoob H., S. Bodduluri, S. V. Ponnaluri, J. C. Selby, and E. A. Sander. 2015 "Substrate stiffness affects human keratinocyte colony formation." *Cellular and Molecular Bioengineering* **8**: 32–50. PMID:26019727.

2

Molecular Pathways of Mechanotransduction

From Extracellular Matrix to Nucleus

Hamish T. J. Gilbert and Joe Swift

Wellcome Trust Centre for Cell-Matrix Research, Faculty of Biology, Medicine and Health, University of Manchester, Manchester, UK

2.1 Introduction: Mechanically Influenced Cellular Behavior

A broad range of cellular phenomena are responsive to the mechanical properties of the local environment. These include alterations to: cell morphology (Pelham and Wang 1997) and contractility (Discher et al. 2005), often manifested in changes to cell spread area and the "focal adhesion" (FA) complexes that interface the cellular cytoskeleton and substrate; cell motility or "durotaxis" – movement directed by a gradient of matrix stiffness (Lo et al. 2000; Winer et al. 2009; Hadjipanayi et al. 2009b; Raab et al. 2012); cell proliferation rates (Klein et al. 2009; Hadjipanayi et al. 2009a); and apoptosis (Wang et al. 2000). One of the most exciting effects of environmental mechanics is on stem cell fate, leading them either to remain quiescent or to divide asymmetrically to facilitate commitment to lineage. The "stem cell niche" is broadly defined as the set of local environmental influences that can affect stem cell behavior, and combines chemical (e.g., soluble factor) and mechanical (e.g., substrate stiffness) inputs, both of which include contributions from the extracellular matrix (ECM) and surrounding cells (Schofield 1978). Mesenchymal stem cells (MSCs) have often been used in studies of mechanobiological processes, with early reports noting sensitivity to mechanical stimulation (Pittenger et al. 1999). More recent work has shown MSCs to have increased tendency toward soft-tissue lineages, such as fat, when cultured on soft substrate, and toward stiff-tissue lineages, such as bone, when on stiff substrate (Engler et al. 2006). The mechanical inputs that cells interpret are a combination of force and geometry over length scales of nano- to micrometers (Vogel and Sheetz 2006), but in all cases signals are eventually transduced through to changes at a molecular level. To give the required specificity of action, for example in turning a genetic program on or off, these molecular-scale signals must be regulated with exquisite spatial and temporal accuracy.

This chapter discusses each of the primary modes of molecular mechanosensing, starting outside the cell in the matrix and working into the nucleus (see Figure 2.1). Though a number of specific signaling modes are discussed and exemplified, spatial and temporal control of signaling is achieved through two recurring motifs: (i) *force-mediated regulation of activity through chemical modification*; and (ii) *force-mediated regulation*

Mechanobiology: Exploitation for Medical Benefit, First Edition. Edited by Simon C. F. Rawlinson.

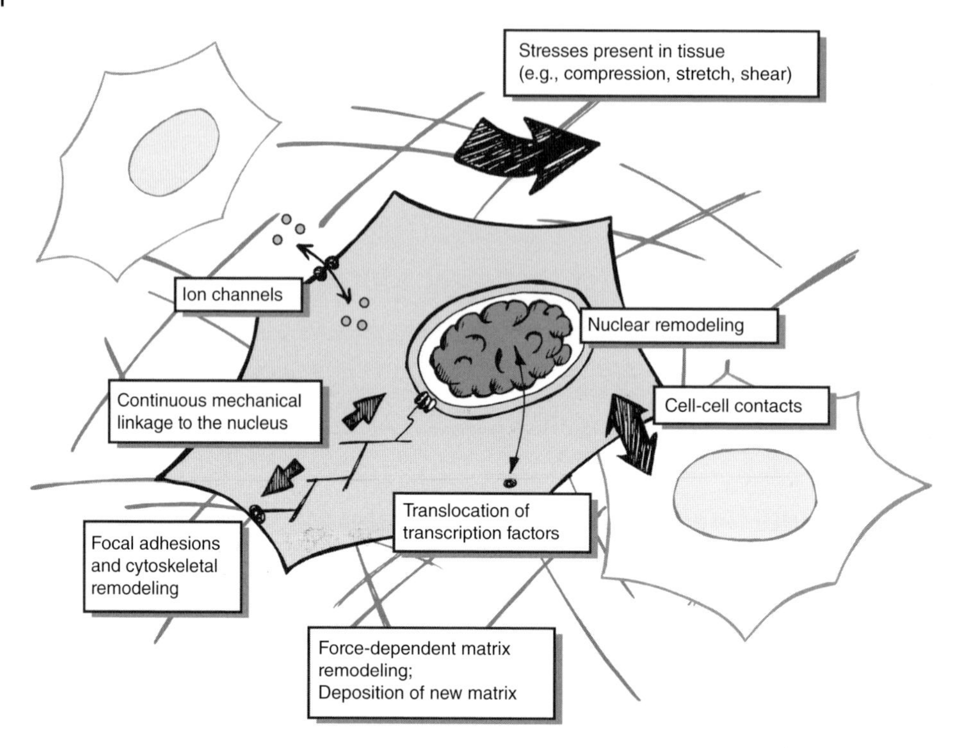

Figure 2.1 Overview of mechanotransduction pathways. Cells in tissue are subjected to stresses in the forms of compression, stretch, and shear; these perturbations may reach a cell through contact with matrix and extracellular fluids, or through cell–cell interfaces. Forces are passed through FAs at the cell surface, via a network of continuous mechanical linkages in the cytoskeleton, and into the nucleus. Structures within the matrix, cellular membrane, cytoskeleton, and nucleus are continuously remodeled in response to mechanical perturbations, and changes are transduced into molecular signaling pathways, such as through activation of ion channels or transcription factors (TFs). These signals are ultimately interpreted to affect cellular behavior.

of activity through change in the distribution, localization, or conformation of molecules. In many cases, these processes occur in concert; for example, a post-translational modification such as phosphorylation may alter the mobility of a protein, or a change in protein conformation may regulate its susceptibility to modification. We will also address some of the technological advances that have made the study of mechanotransduction pathways feasible, in the development of increasingly sophisticated *in vitro* models of tissue, as well as analytical methods that have allowed detailed study of mechanical properties, morphology, and composition.

2.2 Mechanosensitive Molecular Mechanisms

2.2.1 Continuous Mechanical Linkages from Outside the Cell to the Nucleus

The structure of a cell is maintained as it pulls against the matrix and cells that surround it, in a system of permanent stress caused by the cytoskeleton and myosin molecular motors (Ingber 2006). Cell–matrix contacts are mediated by membrane-spanning proteins called integrins, which have domains that can link matrix proteins, such as collagen, to intracellular FA complexes that interface with actin proteins, the major building blocks of the cytoskeleton (Puklin-Faucher and Sheetz 2009; Watt and Huck 2013). Contacts between cells link to the cytoskeleton through "tight" and "adherens" junction complexes that tether to actin, and through desmosome complexes that interact with proteins such as keratin and other cytoplasmic intermediate filaments (IFs; a family of structural proteins forming multimeric filaments of characteristic width and no directional polarity) (Jamora and Fuchs 2002). Tension in the cytoskeleton is maintained by feedback that regulates myosin and actin activity, causing cells in an incompliant environment to pull harder against their surroundings (Ingber 2003).

The nucleus, the stiffest organelle, is an integral part of the mechanical structure of the cell and is tethered to the cytoskeleton by the linker of nucleoskeleton and cytoskeleton (LINC) complex. Nesprin proteins tether the nuclear envelope to the actin network, to cytoplasmic IFs through plectin, and to the microtubule network (another multimeric, cytoplasmic structural component) through kinesin and dynein complexes. Within the nuclear envelope, nesprins bind to the SUN (Sad1p, UNC-84) domain-containing family of inner nuclear membrane proteins, and these in turn bind to IF lamin proteins that line the inside of the nuclear envelope. Lamins define the mechanical properties of the nucleus, but also interact with chromatin and a broad range of regulatory proteins. There is therefore a continuous mechanical linkage between the ECM of bulk tissue and the cell's regulatory center, the nucleus (Maniotis et al. 1997), facilitating rapid signaling (Li et al. 2007).

2.2.2 Force-Mediated Matrix Remodeling

The concentration, assembly state, and chemical modification of ECM proteins are major contributors to the mechanical properties of tissue. The ECM is classically represented as a rigid network that is populated by cells, but the reality is more dynamic. ECM molecules such as tenascin unfold elastically when subjected to force (Oberhauser et al. 1998) and the cytoskeletal tension generated by fibroblast cells has been shown to be sufficient to unfold domains within extracellular fibronectin (Baneyx et al. 2002). Further studies have demonstrated that this cell-induced conformational remodeling of fibronectin is part of the maturation process of newly deposited matrix, acting to align and stiffen the fibrils and modulate their interaction with other biomolecules (Antia et al. 2008). Changes in conformation induced by mechanical stressing can modulate the susceptibility of proteins to enzymatic degradation. When collagen-I fibrils were subjected to localized stretching, the stretched regions were protected against enzymatic proteolysis (Flynn et al. 2010). This mechanism forms the potential basis for a "stress strengthening" system, whereby matrix that is needed (as it bears a mechanical load) is maintained but unnecessary matrix can be turned over (Swift and Discher 2014). Deposition of fresh matrix is also a mechanically sensitive process. For example, chondrocytes – the cells responsible for forming protective cartilage – increase transcription of genes regulating the matrix components collagen-2, aggrecan, and transforming growth factor beta (TGF-β) in response to loading (reviewed in Grodzinsky et al. 2000).

Matrix is continually remodeled during development, as the mechanical demands placed upon tissues change. During chick development, all tissues were found to be uniformly soft in the early embryonic disc, but load-bearing tissues such as those in the heart were gradually stiffened through increased expression of collagen-1, while tissues in the brain remained soft (Majkut et al. 2013). The morphology of structures within the matrix, such as the diameter of fibril structures in articular cartilage, is also altered during maturation (Gannon et al. 2015). Many tissues, such as muscle and connective tissue, continue to stiffen into old age, in a process linked to increased chemical crosslinking in the matrix (Bailey 2001). This process of tissue stiffening may be both a consequence and a contributing factor in age-related disease. Living tissue is thus in a constant dynamic state of degradation, renewal, and remodeling (Xu et al. 2009).

2.2.3 Force-Sensitive Protein Unfolding in the Cytoskeleton and FA Complexes

In most proteins, the amino acid chain is collapsed down into a well-defined, folded, and functional state. Protein function can be lost if the protein is unfolded (or in many disease states, misfolded), a state brought about by a range of perturbations, such as increased temperature or the addition of chaotropic chemical agents. Seminal experiments have shown that protein unfolding can be reversible (as a function of the concentration of a chemical denaturant) and that folding itself is in many cases largely determined by the amino acid sequence (Anfinsen 1973), though in the cell, correct folding is often assisted by molecular chaperone proteins (Hartl et al. 2011). Proteins can also be unfolded by force, a phenomenon that can occur both in the matrix and within the cell itself. Protein unfolding under tension has been demonstrated for a range of intracellular proteins, such as cytoskeletal titin, where domains were sequentially and to some extent reversibly unfolded with an atomic force microscope (AFM) probe (Marszalek et al. 1999). While there is no direct equivalence between mechanical and chemical pathways of unfolding (Kumar and Li 2010), there are common features in the resulting unfolded states: proteins are distorted, potentially leading to impaired function and the exposure of previously buried, often hydrophobic amino acid residues, which may promote undesirable reaction, protein–protein interaction, and aggregation. This exposure of buried residues has allowed the study of force-induced protein unfolding *in situ* within the cell. For example, chemical tagging of cysteine residues followed by quantification by mass spectrometry (MS) has allowed unfolding studies of β-spectrin in red blood cells (RBCs) (Johnson et al. 2007a) and of lamin-A in the nuclei of adherent cells (Swift et al. 2013b). There is increasing evidence that alterations to protein interaction or reactivity following force-induced conformation change may be a common cellular regulatory mechanism. For example, the conformation of FA protein p130Cas is extended under tension, allowing phosphorylation at a tyrosine residue that modulates the activity of the enzyme Rap1 (Sawada et al. 2006). Regulation of kinase activity in FAs is a key mechanosensing motif that has been characterized in a range of experimental systems, including endothelial cells, osteoblasts, and fibroblasts (Mammoto et al. 2012). Substrate stiffness is also reflected in the phosphorylation of non-muscle myosin-IIa, controlling assembly of the myosin complex and thus having a regulatory role in cell polarization and durotaxis (Raab et al. 2012). Protein unfolding may also be undesirable (consequent to a failure to regulate stress), and protective chaperone proteins, such as HSP70, recognize client proteins via hydrophobic residues exposed during unfolding (Balaburski et al. 2013).

2.2.4 Nuclear Remodeling: Lamins and Chromatin

The nucleus is both the cell's "control room," as the location of DNA storage and transcription, and a significant mechanical component within the cell, as often the largest and stiffest organelle. The IF lamin proteins define the mechanical properties of the nucleus, but also mediate a broad range of molecular interactions (Dechat et al. 2010). The somatic cells of humans and many other vertebrates contain two main families of lamin protein: "A-type" (with A and C forms derived from alternative splicing of the *LMNA* gene transcript) and "B-type" (B1 and B2 forms derived from *LMNB1* and *LMNB2*, respectively). Characterizations of stress-induced nuclear deformation in nuclei with different lamina compositions show that A-type lamins primarily contribute to a liquid-like, viscous response, while B-type lamins promote elasticity (Shin et al. 2013; Swift et al. 2013b; Harada et al. 2014). Efforts to quantify lamins by immunostaining (Broers et al. 1997), and more recently by MS (Swift et al. 2013a,b), have shown that the composition of the nuclear lamina is a function of tissue stiffness and, by extension, the stresses present in the tissues, cells, and their nuclei. Thus, B-type lamins tend to be over-represented in the nuclei of cells in soft-tissue and hematopoietic systems, whereas nuclei in stiff and highly stressed tissues are dominated by A-type lamins, where the lamina may act as protective, shock-absorbing cage around the DNA. The tissue-specific composition and characteristics of the lamina may therefore be important in maintaining robustness to stress, and can also influence matrix-directed differentiation (Swift et al. 2013b) and cellular motility (Rowat et al. 2013; Harada et al. 2014). Responsive and dynamic regulation of the lamina is therefore key, and this may be achieved through specific post-translational modification of key proteins. For example, lamins are hyperphosphorylated during mitosis, but there is also evidence of interphase phosphorylation of lamin-A (Kochin et al. 2014); stress-dependent phosphorylation of lamin-A promotes its turnover and has been shown to form the basis of a mechanosensitive regulation of lamina composition (Swift et al. 2013b; Buxboim et al. 2014).

Lamins are a key component of the system of protein linkages that transmits mechanical signals into the nucleus (Simon and Wilson 2011). They are involved in a large variety of binding interactions within the nucleus (Wilson and Foisner 2010), including to lamin-B receptor (LBR) (Solovei et al. 2013), actin (Simon et al. 2010), emerin (which is involved in mechanically sensitive processes, both mediating nuclear stiffness (Guilluy et al. 2014) and binding to TF MKL1 (Ho et al. 2013)), and DNA (Shoeman and Traub 1990; Luderus et al. 1992; Stierle et al. 2003), particularly in the form of "lamina-associated domains" (LADs) of less transcriptionally active heterochromatin (Wagner and Krohne 2007). It is perhaps this multifaceted binding of lamin and its ability to tether DNA directly that have led to interest in the function of the lamina in chromatin regulation (Guelen et al. 2008; Kim et al. 2011; Zullo et al. 2012; Kind et al. 2013; Lund et al. 2013; Meuleman et al. 2013). However, it is not yet understood how the lamina contributes to the conserved spatial relationships within the nucleus, such as "chromosome territories" (Cremer and Cremer 2001; Iyer et al. 2012) and "transcriptional hotspots" (Fraser and Bickmore 2007), which are thought to influence transcriptional activity.

2.2.5 Force-Sensitive Ion Channels

The regulation of local concentrations of small, mobile soluble factors offers a rapid mechanism for cellular message transduction. This is often achieved via channels that allow selective permeability across membranes. Application of cyclic strain to endothelial cells allows a fast influx of calcium ions through the mechanosensitive TRPV4 channel, heading a signaling cascade that leads to cytoskeletal remodeling (Matthews et al. 2010). Recent work in chondrocytes has demonstrated TRPV4 activity-driven genetic regulation and subsequent matrix remodeling, and suggests the TRPV4 mechanosensing pathway as a potential target for osteoarthritis therapy (O'Conor et al. 2014). Small-molecule or ionic messengers such as calcium ions, inositol trisphosphate, diacylglycerol, and nitric oxide are central to a range of signaling pathways in a variety of tissues (Mammoto et al. 2012), but in many cases they have a nonspecific effect on cellular properties and functions. Subjecting chondrocytes to physiological changes in osmotic pressure has been found to cause pronounced alterations to chromatin packing (Irianto et al. 2013). However, transducing such broad physical changes into specific genetic programs, such as those that control matrix remodeling, is likely to require more targeted factors.

2.2.6 Transcription Factor Translocation

Though the preceding two subsections discussed mechanosensitive processes, it is perhaps not apparent in these cases how specific and targeted genetic programs can be affected. Additional precision may be provided by TF proteins that bind to particular DNA sequences, thus allowing blunt inputs – forces and perturbations acting without coherence – to be converted from mechanical to biochemical signals, activating individual genes at precise spatial locations within the nucleus. TF activity can be modulated in a number of ways, via changes in binding, local concentration, conformation, and modification of cofactors. Regulation of concentration in the nucleus by translocation across the nuclear membrane is emerging as a common motif, though the driving forces behind such movement – and how mechanosensitivity is achieved – are less apparent. A possible mechanism is changes to concentrations of binding sites. This is seen in interactions involving TF proteins MKL1, serum response factor (SRF) (Miralles et al. 2003), and emerin (Ho et al. 2013), and has roles in matrix-directed differentiation (Connelly et al. 2010). It is further supported by the fact that lamin-A level affects the translocation of retinoic acid-binding factors and thus regulates its own transcription (Swift et al. 2013b). Other mechanically influenced TF translocation processes include yes-associated protein 1 (YAP1), which is translocated to the nuclei of MSCs cultured on stiff substrates and has apparent roles in differentiation (Dupont et al. 2011) and matrix remodeling in cancer (Calvo et al. 2013); translocation of TF Nkx2.5, which causes matrix-dependent suppression of tensile cell phenotypes (Dingal et al. 2016); and the movement of TFs RelA and Oct1 between nucleus and cytoplasm during transient breakages in the nuclear envelope of fibroblasts with defective lamina caused by disease-associated mutations to lamin-A (De Vos et al. 2011).

2.2.7 Why Have Parallel Force-Sensing Mechanisms? The Need for Complexity in Biological Systems

This chapter has so far described a broad range of mechanisms, highlighting the likelihood that mechanical sensitivity arises from a diverse and complementary set of phenomena. The recent emergence of technologies that allow systems-level analysis has shown that in many cases regulation may be achieved through a complex set of feedback loops and complementary or competing processes. Indeed, not only is complexity likely given the diversity of transcripts, proteins, and small molecules present within the cell, but it may also be *necessary* to provide the nuanced responses required for life. Complexity may be required in order to provide sufficient sensitivity to input, and by extension insensitivity to stochastic noise. This sensitivity must often be delivered with specificity across a broad dynamic range, for example in time, space, or across a range of force inputs. Given that the response of any singular molecular process is not graded across a physiological range of tensions, it follows that multiple pathways are needed to deliver mechanosensitivity. Indeed, the full range of mechanosensing pathways may not yet have been elucidated: despite bulk tissue stiffness ranging between roughly 0.3 (brain) and 40 kPa (precalcified bone) (Discher et al. 2009), recent evidence suggests that the dynamic response of the key TFs YAP1 and RUNX2 – at least in terms of translocation within the cell – may be fully utilized at between 2 and 6 kPa (Yang et al. 2014).

2.3 Methods Enabling the Study of Mechanobiology

2.3.1 Culturing Monolayers of Cells in Two Dimensions

A range of *in vitro* model systems has been established to provide a mechanically defined microenvironment for cells, and this has laid the foundations for the study of mechanobiological responses and mechanotransduction pathways (see Figure 2.2). Such systems can be separated into those that stimulate cells in two- or three-dimensional (2D or 3D) model systems and those that do so within their native environment (e.g., tissue explants or *ex vivo* organ cultures).

The effects of cellular tension were observed in seminal studies of cells cultured on deformable silicone rubber substrates (Harris et al. 1980). The importance of substrate mechanical properties has since been investigated, for example by changing the stiffness of polyacrylamide gel substrates through alteration of their chemical composition (Pelham and Wang 1997). Substrates are typically rendered biocompatible by coating them with ECM molecules (e.g., collagen-I), and cells are able to "feel" the substrate to which they are adhered through the generation of myosin-dependent cytoskeletal tension (Huebsch et al. 2010). By culturing on thin layers of soft gel, cells have been shown to be sensitive to underlying stiffness several microns below the substrate surface (Buxboim et al. 2010). More recent work has developed substrates with mechanical properties that can be modified by exposure to light (Frey and Wang 2009), allowing studies of how cells respond to dynamic mechanical changes and "remember" earlier conditions (Yang et al. 2014).

In addition to their sensitivity to substrate stiffness, cells can respond to surface geometry and topology, and are able to detect surface patterns at the nanoscale (Stevens and George 2005). Recent findings indicate cells alter their cytoskeletal and nuclear shape when cultured on nanogrooved surfaces, leading to nuclear lamina and chromosome remodeling, and subsequent changes to the transcriptome and proteome (McNamara et al. 2012). Methods have also been developed whereby cells are cultured on arrays of silicone "microposts" – the deformability of the posts defines the effective stiffness experienced by the cells, while monitoring of the displacement of the posts allows the forces exerted by the cells to be calculated (Yang et al. 2011).

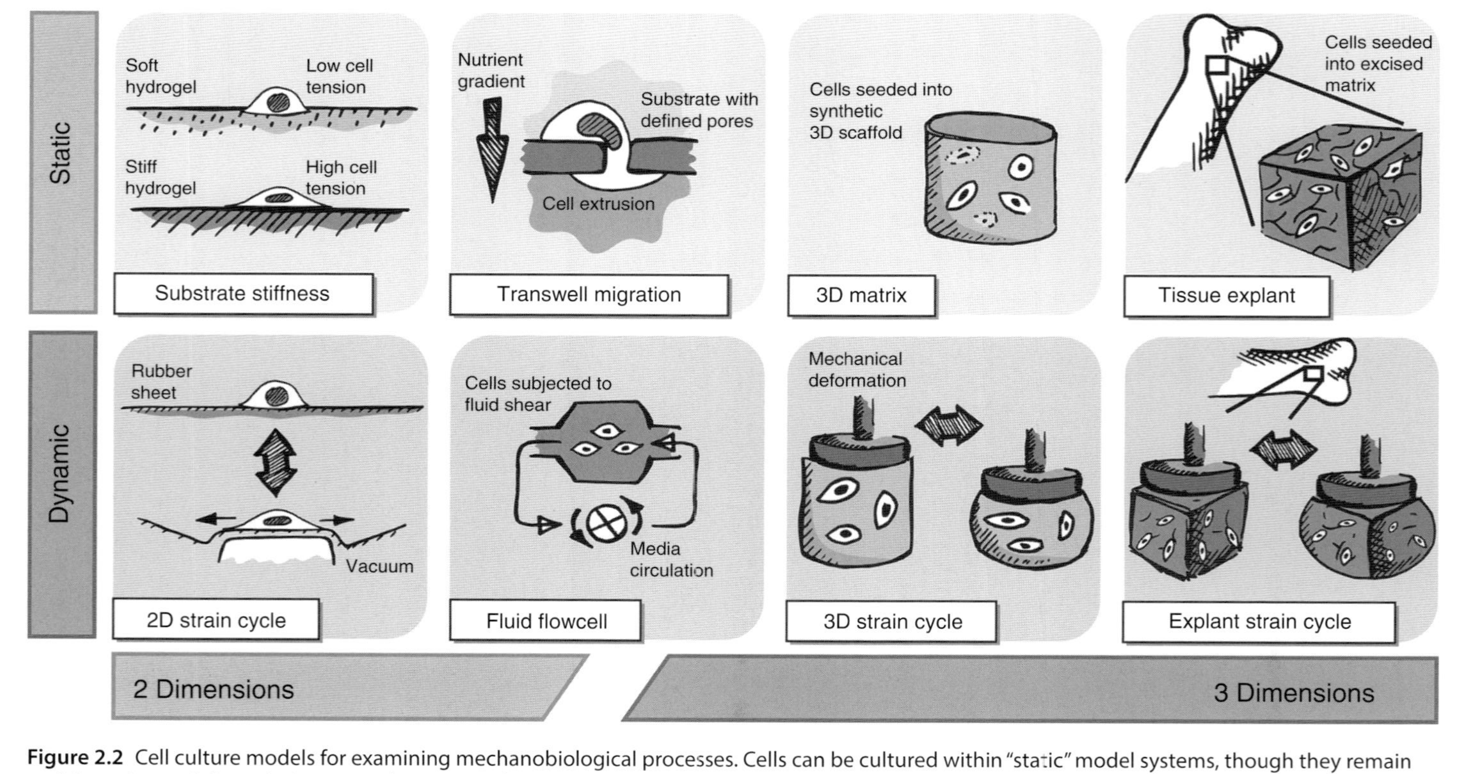

Figure 2.2 Cell culture models for examining mechanobiological processes. Cells can be cultured within "static" model systems, though they remain mobile and may deform their surroundings, or in "dynamic" systems that are subjected to externally imposed loading cycles to better reflect the stresses and strains in living tissue. The simplest 2D models allow for easy cell manipulation, imaging, and scale-up, but more sophisticated models with greater dimensionality offer a closer representation of tissue. The decellularization and repopulation of matrix extracted from tissue offers perhaps the closest model of the native chemical and mechanical environment short of carrying out experiments *in vivo*.

2.3.2 Three-Dimensional Culture Systems

With few exceptions, cells in tissue experience a 3D environment. *In vitro* culture systems are thus continually being developed in order to more accurately represent this reality. These systems typically present a 3D "scaffold" structure with which cells can interact. For example, cells can be encapsulated in hydrogels, allowing even dispersal throughout the substrate (reviewed in Tibbitt and Anseth 2009), or they can be seeded on to 3D porous materials and then allowed to migrate into the scaffold (e.g., Li et al. 2006). The scaffolds provide a cellular environment that can be made chemically complex (e.g., by addition of ECM proteins or specific ligands) and can be engineered with defined mechanical properties or deformability. Culturing of cells within a 3D environment has been shown to increase cell–matrix interactions through increased expression of integrins. Furthermore, this more physiologically relevant environment leads to fundamental changes in cell behaviors, including growth and survival (Cukierman et al. 2001). Like monolayer systems, 3D hydrogels with soft and stiff mechanical properties encourage MSC differentiation toward adipogenic and osteogenic lineages, respectively, though this occurs independently of both cell shape (cells remain spherical in both soft and stiff hydrogels) and actomyosin-induced cytoskeletal tension (Huebsch et al. 2010; Parekh et al. 2011).

Interstitial fluid flow is an important mechanical stimulus found in blood vessels, the lymphatic system, and bone. In both two and three dimensions, *in vitro* fluid flow systems have been used to generate mechanoresponses in cell types including endothelial cells, lymphocytes, fibroblasts, osteoblasts, and MSCs (reviewed in Polacheck et al. 2013). The use of 3D scaffolds in addition to fluid perfusion has been shown to lead to osteogenic differentiation of MSCs (Bancroft et al. 2002); however, it is difficult in this case to separate mechanical effects from the effects of shear-induced gradients in soluble signaling factors.

2.3.3 Models of Dynamic Tissue

Living tissue is not static, but is subjected to deformations and forces that may vary or be periodic in time. Systems have been developed, and are now commercially available and widely used, that allow cells to be strained in 2D culture. Straining can be static or dynamic (with a defined, periodic loading cycle), making these systems highly tunable and enabling mechanical stimulation using strains relevant to the cell type under investigation (for a review of cell-straining techniques, see Tondon et al. 2014). For example, the cyclic straining of MSCs adhered to type-I collagen-coated flexible membranes resulted in the increased expression of characteristic osteogenic markers, which appeared to be regulated through the involvement of stretch-activated ion channels (Kearney et al. 2010).

For a more accurate representation of the *in vivo* mechanical environment, for example in reflecting local heterogeneity, tissue explants or whole-organ culture systems can be mechanically stimulated. For example, it has been demonstrated that compression of bovine cartilage explants in the presence of inflammatory cytokines has an anticatabolic effect. Here, moderate compressive load reduced cytokine-induced catabolism, leading to reduced apoptosis and aggrecanase activity and increased anabolic gene expression (Li et al. 2013). Mechanical stimulation (e.g., compression) of tissue explants ensures that cells are loaded within their native ECM and results in many indirect mechanically induced effects, including hydrostatic pressures, fluid flow, and changes in osmotic pressures and charge densities. The benefit of these systems is that they are highly physiologically relevant, but the added complexity makes it difficult to attribute effects to specific molecular or cellular properties.

2.3.4 Microscopy in the Quantification of Mechanical Properties

Tissues and cells are complex, inhomogeneous structures and as such behave anisotropically when interacting with mechanical inputs. It is therefore beneficial to consider spatially resolved – in addition to global – mechanical properties of complex matrices and cells. A key tool, AFM, creates a high-resolution topographical image by recording the deflections of a cantilever attached to a sharp tip rastered across a sample. Imaging can be performed on samples in physiologically relevant environments (Radmacher 2002), and physical deformation of the sample by the tip allows a simultaneous recording of mechanical properties. As such, AFM has for many years been used to collect whole-cell topographical (e.g., Butt et al. 1990) and mechanical (e.g., Schneider et al. 1997) data.

Recent developments in AFM technology have allowed mechanical characterizations of subcellular features, such as measurement of the microrheological properties of cellular membranes (Gavara and Chadwick 2010). In one example, AFM of cell nuclei showed that measurements made *in situ* within whole cells gave a truer representation of endogenous stiffness (Liu et al. 2014); additionally, the authors reported a link between nuclear stiffness and the metastatic capacity of bladder cancer cell lines, with the more metastatic cells having softer nuclei. This work highlights the benefits of imaging under physiological conditions and demonstrates how a more comprehensive understanding of cellular mechanics can aid our understanding of pathologies and potentially lead to development of better diagnostic tools.

By combining AFM with additional microscopy methods (Gorelik et al. 2002), it is possible to identify specific molecular structures through fluorescent tagging. Hybrid optical/mechanical methods have been applied to problems in mechanobiology, for example in following the formation of actin filaments in response to the manipulation of FAs (Trache and Lim 2009). In this system, biotinylated fibronectin was attached to an AFM tip and placed proximal to a viable cell in order to enable a FA to form. A controlled force was then applied to the tip, administering a defined mechanical perturbation while a fluorescent confocal microscope visualized actin polymerization.

In situ mechanical stressing of the actin cytoskeleton has been measured in real time using a combination of confocal fluorescence microscopy and Förster resonance energy transfer (FRET) – a technique capable of assessing the proximity of protein domains. In this system, a FRET construct was inserted within structural proteins to yield an engineered molecular reporter of strain (Meng et al. 2008). In a similar study, a force-sensitive FRET probe was incorporated into the FA-localized protein vinculin (Grashoff et al. 2010) and confocal microscopy together with a sensitive force-calibrated probe was used to quantify the force applied to the protein within a stable FA.

2.3.5 -Omics Science

The mechanistic links between matrix, cell, and nucleus have been recognized for many years (see, for example, Slavkin and Greulich 1975). Some studies, often based on careful analysis of microscopy images, now seem remarkably prescient: for example, diagrams detailing a continuous linkage between fibrils in the matrix and envelope-tethered DNA in the nucleus were published long before the individual proteins had been identified (Bissell et al. 1982). The identities of the molecules contained within cells and tissues, though not necessarily their functions, have been elucidated over recent years by the introduction and adoption of new technologies, typically characterized by rich, untargeted datasets and the suffix, "-omics."

Genomic sequencing has allowed the DNA code of a range of organisms to be catalogued (Church and Gilbert 1984), including key systems of study ranging from *E. coli* (Blattner et al. 1997) to humans (Venter et al. 2001). How the genetic code is read is subject to another layer of environmentally specific regulation, broadly described by the term "epigenetics," and often influences the transcription of genes. Transcriptomic methods have allowed quantification of mRNA in cells and tissues (Lockhart and Winzeler 2000), utilizing a modified polymerase chain reaction (quantitative PCR) to interrogate extracted material. The resulting products of mRNA translation and the functional machines within cells – proteins – have been catalogued and quantified by proteomics (Aebersold and Mann 2003), driven by advances in MS technology. For example, liquid chromatography tandem mass spectrometry (LC-MS/MS) has allowed quantification of digested protein fragments, which are typically identified *in silico* by reference to the genomic code, while matrix-assisted laser desorption/ionization (MALDI) MS has allowed whole proteins, or even protein complexes, to be examined. Methods have also been developed to quantify the post-translational modification of proteins (Olsen and Mann 2013): these studies can be either untargeted (e.g., in establishing a broad "phosphoproteome"; Beausoleil et al. 2006) or targeted (by using immunoprecipitation (IP) to enrich a certain target system, focusing on modifications to a particular protein, e.g., lamin; Swift et al. 2013b). Targeted MS approaches have also been used to examine protein–protein interactions, for example in the identification of lamin-binding proteins (Kubben et al. 2011) or of how FA complex assembly is affected by myosin-II contractility (Schiller et al. 2011).

While -omics methods can provide a powerful global perspective on a process, a complete understanding of a pathway may necessitate temporal and spatial resolution. Here, -omics technologies can struggle, as they typically rely on measuring the averaged properties of a population. Because of stochastic differences in gene expression and protein levels, individual differences between even genetically identical cells can be significant (Niepel et al. 2009). This means that defining any cell population as "homogeneous" is inaccurate, making it even more difficult to determine how cells respond over a time course. One of the frontiers of -omics science is therefore to enable analysis with single-cell resolution (Wang and Bodovitz 2010).

Spatial resolution has been achieved in MS using MALDI imaging, in which a laser is rastered across a sample to give localized mass analysis. However, limitations in resolution mean that this method is currently better suited to analysis through cross-sections of tissue than at a subcellular level (Schwamborn and Caprioli 2010). Hybrid DNA sequencing methods have been developed to give some degree of spatial resolution within the nucleus: chromatin immunoprecipitation sequencing (ChIP-Seq) allows identification of all the gene loci bound by a particular TF (Johnson et al. 2007b), while chromosome conformation capture (3C), along with more recent derivative technologies, allows a snapshot of the 3D structure of chromatin to be captured by determining regions of DNA located adjacent to one another (Dekker et al. 2002).

Bigger datasets have necessitated new and more powerful methods for handling and interpreting data, and systems biology is a now well-established theoretical partner to -omics analysis. Recent reports have combined systems modeling with MS experiments to describe how populations of stem cells achieve robust fate responses following stimulation, for example (Ahrends et al. 2014). There have also been efforts recently to develop "systems mechanobiology" models in which common parameters for describing complex biological systems, such as concentrations and binding coefficients, are joined by mechanical elements, such as stress-dependent rates of protein turnover (Dingal and Discher 2014).

2.4 Conclusion

This chapter has established that mechanical inputs to cells are transduced to molecular-level signals by a complex range of processes, encompassing several mechanisms. These processes are likely targets of perturbation in disease and aging as the mechanical properties of tissue are altered. For example: liver tissue stiffens during fibrosis (Yin et al. 2007); arteries stiffen in response to damage (Klein et al. 2009); increased chemical crosslinking causes tissue to stiffen during aging (Bailey 2001); and mechanical wearing, consequent to overloading or abnormal load distributions, is associated with diseases of load-bearing tissues, including osteoporosis and osteoarthritis (Grodzinsky et al. 2000). In addition, ECM stiffness has been linked to a range of cancer processes (Kumar and Weaver 2009), with evidence connecting breast tissue density, stiffness, and the progression of breast cancer, for example (Mouw et al. 2014). An understanding of the factors that cause these mechanical changes and how they feed back into cellular behavior offers exciting potential for identifying new pathways to target for therapy. Target pathways may encompass compositional changes in ECM, molecular stress-sensors (including their conformational changes and downstream signaling cascades), and how TFs are post-translationally modified and translocated. Given the multifaceted and seemingly parallel mechanisms of mechanotransduction, there may also be an opportunity to develop combinatorial therapeutic approaches aimed at combating pathway overlap and redundancy. A molecular foundation to our understanding of mechanobiological processes therefore has potential to open a complementary approach to a broad range of medical challenges.

Acknowledgements

HTJG and JS are supported by ARUK and BBSRC David Phillips Fellowships, respectively. The authors thank Dr. Mark Jackson for comments and discussion.

References

Aebersold, R. and M. Mann. 2003. "Mass spectrometry-based proteomics." *Nature* **422**: 198–207. doi:papers://993FCEA4-B0CF-4173-A396-9AA8E807B7EE/Paper/p7887.

Ahrends, R., A. Ota, K. M. Kovary, T. Kudo, B. O. Park, and M. N. Teruel. 2014. "Controlling low rates of cell differentiation through noise and ultrahigh feedback." *Science* **344**(6190): 1384–9. doi:10.1126/science.1252079.

Anfinsen, C. B. 1973. "Principles that govern folding of protein chains." *Science* **181**(4096): 223–30. doi:10.1126/science.181.4096.223.

Antia, M., G. Baneyx, K. E. Kubow, and V. Vogel. 2008. "Fibronectin in aging extracellular matrix fibrils is progressively unfolded by cells and elicits an enhanced rigidity response." *Faraday Discussions* **139**: 229–49. doi:10.1039/b718714a.

Bailey, A. J. 2001. "Molecular mechanisms of ageing in connective tissues." *Mechanisms of Ageing and Development* **122**(7): 735–55. doi:10.1016/s0047-6374(01)00225-1.

Balaburski, G. M., J. I. J. Leu, N. Beeharry, S. Hayik, M. D. Andrake, G. Zhang, et al. 2013. "A modified HSP70 inhibitor shows broad activity as an anticancer agent." *Molecular Cancer Research* **11**(3): 219–29. doi:10.1158/1541-7786.mcr-12-0547-t.

Bancroft, G. N., V. I. Sikavitsast, J. van den Dolder, T. L. Sheffield, C. G. Ambrose, J. A. Jansen, and A. G. Mikos. 2002. "Fluid flow increases mineralized matrix deposition in 3D perfusion culture of marrow stromal osteloblasts in a dose-dependent manner." *Proceedings of the National Academy of Sciences of the United States of America* **99**(20): 12 600–5. doi:10.1073/pnas.202296599.

Baneyx, G., L. Baugh, and V. Vogel. 2002. "Fibronectin extension and unfolding within cell matrix fibrils controlled by cytoskeletal tension." *Proceedings of the National Academy of Sciences of the United States of America* **99**(8): 5139–43. doi:10.1073/pnas.072650799.

Beausoleil, S. A., J. Villen, S. A. Gerber, J. Rush, and S. P. Gygi. 2006. "A probability-based approach for high-throughput protein phosphorylation analysis and site localization." *Nature Biotechnology* **24**(10): 1285–92. doi:10.1038/nbt1240.

Bissell, M. J., H. G. Hall, and G. Parry. 1982. "How does the extracellular matrix direct gene-expression?" *Journal of Theoretical Biology* **99**(1): 31–68. doi:10.1016/0022-5193(82)90388-5.

Blattner, F. R., G. Plunkett, C. A. Bloch, N. T. Perna, V. Burland, M. Riley, et al. 1997. "The complete genome sequence of *Escherichia coli* K-12." *Science* **277**(5331): 1453–62. doi:10.1126/science.277.5331.1453.

Broers, J. L. V., B. M. Machiels, H. J. H. Kuijpers, F. Smedts, R. van den Kieboom, Y. Raymond, and F. C. S. Ramaekers. 1997. "A- and B-type lamins are differentially expressed in normal human tissues." *Histochemistry and Cell Biology* **107**: 505–17. doi:papers://993FCEA4-B0CF-4173-A396-9AA8E807B7EE/Paper/p12971.

Butt, H. J., E. K. Wolff, S. A. C. Gould, B. D. Northern, C. M. Peterson, and P. K. Hansma. 1990. "Imaging cells with the atomic force microscope." *Journal of Structural Biology* **105**(1–3): 54–61. doi:10.1016/1047-8477(90)90098-w.

Buxboim, A., K. Rajagopal, A. E. X. Brown, and D. E. Discher. 2010. "How deeply cells feel: methods for thin gels." *Journal of Physics-Condensed Matter* **22**(19): 194116. doi:10.1088/0953-8984/22/19/194116.

Buxboim, A., J. Swift, J. Irianto, A. Athirasala, Y.-R. C. Kao, K. R. Spinler, et al. 2014. "Matrix elasticity regulates lamin-A,C phosphorylation and turnover with feedback to actomyosin." *Current Biology* **24** (16): 1909–17. doi:10.1016/j.cub.2014.07.001.

Calvo, F., N. Ege, A. Grande-Garcia, S. Hooper, R. P. Jenkins, S. I. Chaudhry, et al. 2013. "Mechanotransduction and YAP-dependent matrix remodelling is required for the generation and maintenance of cancer-associated fibroblasts." *Nature Cell Biology* **15**(6): 637. doi:10.1038/ncb2756.

Church, G. M. and W. Gilbert. 1984. "Genomic sequencing." *Proceedings of the National Academy of Sciences of the United States of America* **81**(7): 1991–5. doi:10.1073/pnas.81.7.1991.

Connelly, J. T., J. E. Gautrot, B. Trappmann, D. W. M. Tan, G. Donati, W. T. S. Huck, and F. M. Watt. 2010. "Actin and serum response factor transduce physical cues from the microenvironment to regulate epidermal stem cell fate decisions." *Nature Cell Biology* **12**(7): 711–18. doi:10.1038/ncb2074.

Cremer, T. and C. Cremer. 2001. "Chromosome territories, nuclear architecture and gene regulation in mammalian cells." *Nature Reviews Genetics* **2**(4): 292–301. doi:10.1038/35066075.

Cukierman, E., R. Pankov, D. R. Stevens, and K. M. Yamada. 2001. "Taking cell-matrix adhesions to the third dimension." *Science* **294**(5547): 1708–12. doi:10.1126/science.1064829.

De Vos, W. H., F. Houben, M. Kamps, A. Malhas, F. Verheyen, J. Cox, et al. 2011. "Repetitive disruptions of the nuclear envelope invoke temporary loss of cellular compartmentalization in laminopathies." *Human Molecular Genetics* **20**(21): 4175–86. doi:10.1093/hmg/ddr344.

Dechat, T., S. A. Adam, P. Taimen, T. Shimi, and R. D. Goldman. 2010. "Nuclear lamins." *Cold Spring Harbor Perspectives in Biology* **2**(11): a000547. PMCID:PMC2964183.

Dekker, J., K. Rippe, M. Dekker, and N. Kleckner. 2002. "Capturing chromosome conformation." *Science* **295**(5558): 1306–11. doi:10.1126/science.1067799.

Dingal, P. C. D. P. and D. E. Discher. 2014. "Systems mechanobiology: tension-inhibited protein turnover is sufficient to physically control gene circuits." *Biophysical Journal* **107**(11): 2734–43. doi:10.1016/j.bpj.2014.10.042.

Dingal, P. C. D. P., D. Bradshaw, S. Cho, M. Raab, A. Buxbiom, J. Swift, and D. E. Discher. 2016. "Minimal matrix model of scars with fractal heterogeneity reveals a mechano-sensitive repressor of the rigid stem cell phenotype." *Nature Materials* submitted.

Discher, D. E., P. Janmey, and Y. L. Wang. 2005. "Tissue cells feel and respond to the stiffness of their substrate." *Science* **310**(5751): 1139–43. doi:10.1126/science.1116995.

Discher, D., C. Dong, J. J. Fredberg, F. Guilak, D. Ingber, P. Janmey, et al. 2009. "Biomechanics: cell research and applications for the next decade." *Annals of Biomedical Engineering* **37**(5): 847–59. doi:10.1007/s10439-009-9661-x.

Dupont, S., L. Morsut, M. Aragona, E. Enzo, S. Giulitti, M. Cordenonsi, et al. 2011. "Role of YAP/TAZ in mechanotransduction." *Nature* **474**(7350): 179–83. doi:10.1038/nature10137.

Engler, A. J., S. Sen, H. L. Sweeney, and D. E. Discher. 2006. "Matrix elasticity directs stem cell lineage specification." *Cell* **126**(4): 677–89. doi:10.1016/j.cell.2006.06.044.

Flynn, B. P., A. P. Bhole, N. Saeidi, M. Liles, C. A. DiMarzio, and J. W. Ruberti. 2010. "Mechanical strain stabilizes reconstituted collagen fibrils against enzymatic degradation by mammalian collagenase matrix metalloproteinase 8 (MMP-8)." *PloS One* **5**(8): e12337. doi:10.1371/journal.pone.0012337.

Fraser, P. and W. Bickmore. 2007. "Nuclear organization of the genome and the potential for gene regulation." *Nature* **447**(7143): 413–17. doi:10.1038/nature05916.

Frey, M. T. and Y. L. Wang. 2009. "A photo-modulatable material for probing cellular responses to substrate rigidity." *Soft Matter* **5**(9): 1918–24. doi:10.1039/b818104g.

Gannon, A. R., T. Nagel, A. P. Bell, N. C. Avery, and D. J. Kelly. 2015. "The changing role of the superficial region in determining the dynamic compressive properties of articular cartilage during postnatal development." *Osteoarthritis and Cartilage* **23**(6): 975–84. doi:10.1016/j.joca.2015.02.003.

Gavara, N. and R. S. Chadwick. 2010. “Noncontact microrheology at acoustic frequencies using frequency-modulated atomic force microscopy.” *Nature Methods* **7**(8): 650–4. doi:10.1038/nmeth.1474.

Gorelik, J., A. Shevchuk, M. Ramalho, M. Elliott, C. Lei, C. F. Higgins, et al. 2002. “Scanning surface confocal microscopy for simultaneous topographical and fluorescence imaging: application to single virus-like particle entry into a cell.” *Proceedings of the National Academy of Sciences of the United States of America* **99**(25): 16 018–23. doi:10.1073/pnas.252458399.

Grashoff, C., B. D. Hoffman, M. D. Brenner, R. B. Zhou, M. Parsons, M. T. Yang, et al. 2010. “Measuring mechanical tension across vinculin reveals regulation of focal adhesion dynamics.” *Nature* **466**(7303): 263–6. doi:10.1038/nature09198.

Grodzinsky, A. J., M. E. Levenston, M. Jin, and E. H. Frank. 2000. “Cartilage tissue remodeling in response to mechanical forces.” *Annual Review of Biomedical Engineering* **2**: 691–713. doi:10.1146/annurev.bioeng.2.1.691.

Guelen, L., L. Pagie, E. Brasset, W. Meuleman, M. B. Faza, W. Talhout, et al. 2008. “Domain organization of human chromosomes revealed by mapping of nuclear lamina interactions.” *Nature* **453**(7197): 948–51. doi:10.1038/nature06947.

Guilluy, C., L. D. Osborne, L. Van Landeghem, L. Sharek, R. Superfine, R. Garcia-Mata, and K. Burridge. 2014. “Isolated nuclei adapt to force and reveal a mechanotransduction pathway in the nucleus.” *Nature Cell Biology* **16**(4): 376–81. doi:10.1038/ncb2927.

Hadjipanayi, E., V. Mudera, and R. A. Brown. 2009a. “Close dependence of fibroblast proliferation on collagen scaffold matrix stiffness.” *Journal of Tissue Engineering and Regenerative Medicine* **3**(2): 77–84. doi:10.1002/term.136.

Hadjipanayi, E., V. Mudera, and R. A. Brown. 2009b. “Guiding cell migration in 3D: a collagen matrix with graded directional stiffness.” *Cell Motility and the Cytoskeleton* **66**(3): 121–8. doi:10.1002/cm.20331.

Harada, T., J. Swift, J. Irianto, J. W. Shin, K. R. Spinler, A. Athirasala, et al. 2014. “Nuclear lamin stiffness is a barrier to 3D migration, but softness can limit survival.” *Journal of Cell Biology* **204**(5): 669–82. doi:10.1083/jcb.201308029.

Harris, A. K., P. Wild, and D. Stopak. 1980. “Silicone-rubber substrata – new wrinkle in the study of cell locomotion.” *Science* **208**(4440): 177–9. doi:10.1126/science.6987736.

Hartl, F. U., A. Bracher, and M. Hayer-Hartl. 2011. “Molecular chaperones in protein folding and proteostasis.” *Nature* **475**(7356): 324–32. doi:10.1038/nature10317.

Ho, C. Y., D. E. Jaalouk, M. K. Vartiainen, and J. Lammerding. 2013. “Lamin A/C and emerin regulate MKL1-SRF activity by modulating actin dynamics.” *Nature* **497**(7450): 507–11. doi:10.1038/nature12105.

Huebsch, N., P. R. Arany, A. S. Mao, D. Shvartsman, O. A. Ali, S. A. Bencherif, et al. 2010. “Harnessing traction-mediated manipulation of the cell/matrix interface to control stem-cell fate.” *Nature Materials* **9**(6): 518–26. doi:10.1038/nmat2732.

Ingber, D. E. 2003. “Mechanobiology and diseases of mechanotransduction.” *Annals of Medicine* **35**(8): 564–77. doi:10.1080/07853890310016333.

Ingber, D. E. 2006. “Cellular mechanotransduction: putting all the pieces together again.” *FASEB Journal* **20**(7): 811–27. doi:10.1096/fj.05-5424rev.

Irianto, J., J. Swift, R. P. Martins, G. D. McPhail, M. M. Knight, D. E. Discher, and D. A. Lee. 2013. “Osmotic challenge drives rapid and reversible chromatin condensation in chondrocytes.” *Biophysical Journal* **104**(4): 759–69. doi:10.1016/j.bpj.2013.01.006.

Iyer, K. V., S. Maharana, S. Gupta, A. Libchaber, T. Tlusty, and G. V. Shivashankar. 2012. "Modeling and experimental methods to probe the link between global transcription and spatial organization of chromosomes." *PloS One* **7**(10): 14. doi:10.1371/journal.pone.0046628.

Jamora, C. and E. Fuchs. 2002. "Intercellular adhesion, signalling and the cytoskeleton." *Nature Cell Biology* **4**(4): e101–8. doi:10.1038/ncb0402-e101.

Johnson, C. P., H. Y. Tang, C. Carag, D. W. Speicher, and D. E. Discher. 2007a. "Forced unfolding of proteins within cells." *Science* **317**(5838): 663–6. doi: 10.1126/science.1139857.

Johnson, D. S., A. Mortazavi, R. M. Myers, and B. Wold. 2007b. "Genome-wide mapping of in vivo protein-DNA interactions." *Science* **316**(5830): 1497–502. doi:10.1126/science.1141319.

Kearney, E. M., E. Farrell, P. J. Prendergast, and V. A. Campbell. 2010. "Tensile strain as a regulator of mesenchymal stem cell osteogenesis." *Annals of Biomedical Engineering* **38**(5): 1767–79. doi:10.1007/s10439-010-9979-4.

Kim, Y., A. A. Sharov, K. McDole, M. Cheng, H. Hao, C. M. Fan, et al. 2011. "Mouse B-type lamins are required for proper organogenesis but not by embryonic stem cells." *Science* **334**(6063): 1706–10. doi:10.1126/science.1211222.

Kind, J., L. Pagie, H. Ortabozkoyun, S. Boyle, S. S. de Vries, H. Janssen, et al. 2013. "Single-cell dynamics of genome-nuclear lamina interactions." *Cell* **153**(1): 178–92. doi:10.1016/j.cell.2013.02.028.

Klein, E. A., L. Yin, D. Kothapalli, P. Castagnino, F. J. Byfield, T. Xu, et al. 2009. "Cell-cycle control by physiological matrix elasticity and in vivo tissue stiffening." *Current Biology* **19**(18): 1511–18. doi:10.1016/j.cub.2009.07.069.

Kochin, V., T. Shimi, E. Torvaldson, S. A. Adam, A. Goldman, C. G. Pack, et al. 2014. "Interphase phosphorylation of lamin A." *Journal of Cell Science* **127**(12): 2683–96. doi:10.1242/jcs.141820.

Kubben, N., J. W. Voncken, G. Konings, M. van Weeghel, M. M. G. van den Hoogenhof, M. Gijbels, et al. 2011. "Post-natal myogenic and adipogenic developmental defects and metabolic impairment upon loss of A-type lamins." *Nucleus* **2**(3): 195–207. doi:10.4161/nucl.2.3.15731.

Kumar, S. and M. S. Li. 2010. "Biomolecules under mechanical force." *Physics Reports – Review Section of Physics Letters* **486**(1–2): 1–74. doi:10.1016/j.physrep.2009.11.001.

Kumar, S. and V. Weaver. 2009. "Mechanics, malignancy, and metastasis: the force journey of a tumor cell." *Cancer and Metastasis Reviews* **28**(1–2): 113–27. doi:10.1007/s10555-008-9173-4.

Li, W. J., J. A. Cooper, R. L. Mauck, and R. S. Tuan. 2006. "Fabrication and characterization of six electrospun poly(alpha-hydroxy ester)-based fibrous scaffolds for tissue engineering applications." *Acta Biomaterialia* **2**(4): 377–85. doi:10.1016/j.actbio.2006.02.005.

Li, J., G. P. Chen, L. L. Zheng, S. J. Luo, and Z. H. Zhao. 2007. "Osteoblast cytoskeletal modulation in response to compressive stress at physiological levels." *Molecular and Cellular Biochemistry* **304**(1–2): 45–52. doi:10.1007/s11010-007-9484-8.

Li, Y., E. H. Frank, Y. Wang, S. Chubinskaya, H. H. Huang, and A. J. Grodzinsky. 2013. "Moderate dynamic compression inhibits pro-catabolic response of cartilage to mechanical injury, tumor necrosis factor-alpha and interleukin-6, but accentuates degradation above a strain threshold." *Osteoarthritis and Cartilage* **21**(12): 1933–41. doi:10.1016/j.joca.2013.08.021.

Liu, H. J., J. Wen, Y. Xiao, J. Liu, S. Hopyan, M. Radisic, et al. 2014. "In situ mechanical characterization of the cell nucleus by atomic force microscopy." *ACS Nano* **8**(4): 3821–8. doi:10.1021/nn500553z.

Lo, C. M., H. B. Wang, M. Dembo, and Y. L. Wang. 2000. "Cell movement is guided by the rigidity of the substrate." *Biophysical Journal* **79**(1): 144–52. PMID:10866943.

Lockhart, D. J. and E. A. Winzeler. 2000. "Genomics, gene expression and DNA arrays." *Nature* **405**(6788): 827–36. doi:10.1038/35015701.

Luderus, M. E. E., A. Degraaf, E. Mattia, J. L. Denblaauwen, M. A. Grande, L. Dejong, and R. Vandriel. 1992. "Binding of matrix attachment regions to lamin-B1." *Cell* **70**(6): 949–59. doi:10.1016/0092-8674(92)90245-8.

Lund, E., A. R. Oldenburg, E. Delbarre, C. T. Freberg, I. Duband-Goulet, R. Eskeland, et al. 2013. "Lamin A/C-promoter interactions specify chromatin state-dependent transcription outcomes." *Genome Research* **23**(10): 1580–9. doi:10.1101/gr.159400.113.

Majkut, S., T. Idema, J. Swift, C. Krieger, A. Liu, and D. E. Discher. 2013. "Heart-specific stiffening in early embryos parallels matrix and myosin expression to optimize beating." *Current Biology* **23**(23): 2434–9. doi:10.1016/j.cub.2013.10.057.

Mammoto, A., T. Mammoto, and D. E. Ingber. 2012. "Mechanosensitive mechanisms in transcriptional regulation." *Journal of Cell Science* **125**(13): 3061–73. doi:10.1242/jcs.093005.

Maniotis, A. J., C. S. Chen, and D. E. Ingber. 1997. "Demonstration of mechanical connections between integrins cytoskeletal filaments, and nucleoplasm that stabilize nuclear structure." *Proceedings of the National Academy of Sciences of the United States of America* **94**(3): 849–54. doi:10.1073/pnas.94.3.849.

Marszalek, P. E., H. Lu, H. B. Li, M. Carrion-Vazquez, A. F. Oberhauser, K. Schulten, and J. M. Fernandez. 1999. "Mechanical unfolding intermediates in titin modules." *Nature* **402**(6757): 100–3. PMID:10573426.

Matthews, B. D., C. K. Thodeti, J. D. Tytell, A. Mammoto, D. R. Overby, and D. E. Ingber. 2010. "Ultra-rapid activation of TRPV4 ion channels by mechanical forces applied to cell surface beta 1 integrins." *Integrative Biology* **2**(9): 435–42. doi:10.1039/c0ib00034e.

McNamara, L. E., R. Burchmore, M. O. Riehle, P. Herzyk, M. J. P. Biggs, C. D. W. Wilkinson, et al. 2012. "The role of microtopography in cellular mechanotransduction." *Biomaterials* **33**(10): 2835–47. doi:10.1016/j.biomaterials.2011.11.047.

Meng, F., T. M. Suchyna, and F. Sachs. 2008. "A fluorescence energy transfer-based mechanical stress sensor for specific proteins in situ." *FEBS Journal* **275**(12): 3072–87. doi:10.1111/j.1742-4658.2008.06461.x.

Meuleman, W., D. Peric-Hupkes, J. Kind, J. B. Beaudry, L. Pagie, M. Kellis, et al. 2013. "Constitutive nuclear lamina-genome interactions are highly conserved and associated with A/T-rich sequence." *Genome Research* **23**(2): 270–80. doi:10.1101/gr.141028.112.

Miralles, F., G. Posern, A. I. Zaromytidou, and R. Treisman. 2003. "Actin dynamics control SRF activity by regulation of its coactivator MAL." *Cell* **113**(3): 329–42. doi:10.1016/s0092-8674(03)00278-2.

Mouw, J. K., Y. Yui, L. Damiano, R. O. Bainer, J. N. Lakins, I. Acerbi, et al. 2014. "Tissue mechanics modulate microRNA-dependent PTEN expression to regulate malignant progression." *Nature Medicine* **20**(4): 360. doi:10.1038/nm.3497.

Niepel, M., S. L. Spencer, and P. K. Sorger. 2009. "Non-genetic cell-to-cell variability and the consequences for pharmacology." *Current Opinion in Chemical Biology* **13**(5–6): 556–61. doi:10.1016/j.cbpa.2009.09.015.

O'Conor, C. J., H. A. Leddy, H. C. Benefield, W. B. Liedtke, and F. Guilak. 2014. "TRPV4-mediated mechanotransduction regulates the metabolic response of chondrocytes to dynamic loading." *Proceedings of the National Academy of Sciences of the United States of America* **111**(4): 1316–21. doi:10.1073/pnas.1319569111.

Oberhauser, A. F., P. E. Marszalek, H. P. Erickson, and J. M. Fernandez. 1998. "The molecular elasticity of the extracellular matrix protein tenascin." *Nature* **393**(6681): 181–5. PMID:9603523.

Olsen, J. V. and M. Mann. 2013. "Status of large-scale analysis of post-translational modifications by mass spectrometry." *Molecular & Cellular Proteomics* **12**(12): 3444–52. PMID:24187339.

Parekh, S. H., K. Chatterjee, S. Lin-Gibson, N. M. Moore, M. T. Cicerone, M. F. Young, and C. G. Simon. 2011. "Modulus-driven differentiation of marrow stromal cells in 3D scaffolds that is independent of myosin-based cytoskeletal tension." *Biomaterials* **32**(9): 2256–64. doi:10.1016/j.biomaterials.2010.11.065.

Pelham, R. J. and Y. L. Wang. 1997. "Cell locomotion and focal adhesions are regulated by substrate flexibility." *Proceedings of the National Academy of Sciences of the United States of America* **94**(25): 13 661–5. doi:10.1073/pnas.94.25.13661.

Pittenger, M. F., A. M. Mackay, S. C. Beck, R. K. Jaiswal, R. Douglas, J. D. Mosca, et al. 1999. "Multilineage potential of adult human mesenchymal stem cells." *Science* **284**(5411): 143–7. doi:10.1126/science.284.5411.143.

Polacheck, W. J., R. Li, S. G. M. Uzel, and R. D. Kamm. 2013. "Microfluidic platforms for mechanobiology." *Lab on a Chip* **13**(12): 2252–67. doi:10.1039/c3lc41393d.

Puklin-Faucher, E. and M. P. Sheetz. 2009. "The mechanical integrin cycle." *Journal of Cell Science* **122**(2): 179–86. doi:10.1242/jcs.042127.

Raab, M., J. Swift, P. C. D. P. Dingal, P. Shah, J. W. Shin, and D. E. Discher. 2012. "Crawling from soft to stiff matrix polarizes the cytoskeleton and phosphoregulates myosin-II heavy chain." *Journal of Cell Biology* **199**(4): 669–83. doi:10.1083/jcb.201205056.

Radmacher, M. 2002. "Measuring the elastic properties of living cells by the atomic force microscope." In: L. Wilson, P. Matsudaira, B. Jena, and J. K. Horber (eds.). Atomic Force Microscopy in Cell Biology. Orlando, FL: Academic Press. pp. 67–90.

Rowat, A. C., D. E. Jaalouk, M. Zwerger, W. L. Ung, I. A. Eydelnant, D. E. Olins, et al. 2013. "Nuclear envelope composition determines the ability of neutrophil-type cells to passage through micron-scale constrictions." *Journal of Biological Chemistry* **288**(12): 8610–18. doi:10.1074/jbc.M112.441535.

Sawada, Y., M. Tamada, B. J. Dubin-Thaler, O. Cherniavskaya, R. Sakai, S. Tanaka, and M. P. Sheetz. 2006. "Force sensing by mechanical extension of the Src family kinase substrate p130Cas." *Cell* **127**(5): 1015–26. doi:10.1016/j.cell.2006.09.044.

Schiller, H. B., C. C. Friedel, C. Boulegue, and R. Fassler. 2011. "Quantitative proteomics of the integrin adhesome show a myosin II-dependent recruitment of LIM domain proteins." *Embo Reports* **12**(3): 259–66. doi:10.1038/embor.2011.5.

Schneider, S. W., Y. Yano, B. E. Sumpio, B. P. Jena, J. P. Geibel, M. Gekle, and H. Oberleithner. 1997. "Rapid aldosterone-induced cell volume increase of endothelial cells measured by the atomic force microscope." *Cell Biology International* **21**(11): 759–68. doi:10.1006/cbir.1997.0220.

Schofield, R. 1978. "The relationship between spleen colony-forming cell and the hematopoietic stem-cell." *Blood Cells* **4**(1–2): 7–25. PMID:747780.

Schwamborn, K. and R. M. Caprioli. 2010. "Molecular imaging by mass spectrometry – looking beyond classical histology." *Nature Reviews Cancer* **10**(9): 639–46. doi:10.1038/nrc2917.

Shin, J. W., K. R. Spinler, J. Swift, J. A. Chasis, N. Mohandas, and D. E. Discher. 2013. "Lamins regulate cell trafficking and lineage maturation of adult human hematopoietic cells." *Proceedings of the National Academy of Sciences of the United States of America* **110**(47): 18 892–7. doi:10.1073/pnas.1304996110.

Shoeman, R. L. and P. Traub. 1990. "The in vitro DNA-binding properties of purified nuclear lamin proteins and vimentin." *Journal of Biological Chemistry* **265**(16): 9055–61. PMID:2345165.

Simon, D. N. and K. L. Wilson. 2011. "The nucleoskeleton as a genome-associated dynamic 'network of networks'." *Nature Reviews Molecular Cell Biology* **12**(11): 695–708. doi:10.1038/nrm3207.

Simon, D. N., M. S. Zastrow, and K. L. Wilson. 2010. "Direct actin binding to A- and B-type lamin tails and actin filament bundling by the lamin A tail." *Nucleus* **1**(3): 264–72. PMID:21327074.

Slavkin, H. C. and R. C. Greulich. 1975. *Extracellular Matrix Influences on Gene Expression*. New York: Academic Press.

Solovei, I., A. S. Wang, K. Thanisch, C. S. Schmidt, S. Krebs, M. Zwerger, et al. 2013. "LBR and lamin A/C sequentially tether peripheral heterochromatin and inversely regulate differentiation." *Cell* **152**(3): 584–98. doi:10.1016/j.cell.2013.01.009.

Stevens, M. M. and J. H. George. 2005. "Exploring and engineering the cell surface interface." *Science* **310**(5751): 1135–8. doi:10.1126/science.1106587.

Stierle, V. N., J. L. Couprie, C. Ostlund, I. Krimm, S. Zinn-Justin, P. Hossenlopp, et al. 2003. "The carboxyl-terminal region common to lamins A and C contains a DNA binding domain." *Biochemistry* **42**(17): 4819–28. doi:10.1021/bi020704g.

Swift, J. and D. E. Discher. 2014. "The nuclear lamina is mechano-responsive to ECM elasticity in mature tissue." *Journal of Cell Science* **127**(14): 3005–15. doi:10.1242/jcs.149203.

Swift, J., T. Harada, A. Buxboim, J. W. Shin, H. Y. Tang, D. W. Speicher, and D. E. Discher. 2013a. "Label-free mass spectrometry exploits dozens of detected peptides to quantify lamins in wildtype and knockdown cells." *Nucleus – Austin* **4**(6): 450–9. doi:10.4161/nucl.27413.

Swift, J., I. L. Ivanovska, A. Buxboim, T. Harada, P. C. D. P. Dingal, J. Pinter, et al. 2013b. "Nuclear lamin-A scales with tissue stiffness and enhances matrix-directed differentiation." *Science* **341** (6149): 1240104. doi: 10.1126/science.1240104.

Tibbitt, M. W. and K. S. Anseth. 2009. "Hydrogels as extracellular matrix mimics for 3D cell culture." *Biotechnology and Bioengineering* **103**(4): 655–63. doi:10.1002/bit.22361.

Tondon, A., C. Haase, and R. Kaunas. 2014. "Mechanical stretch assays in cell culture systems." In: C. P. Neu and G. M. Genin (eds.). *Handbook of Imaging in Biological Mechanics*. Boca Raton, FL: CRC Press. pp. 313–22.

Trache, A. and S. M. Lim. 2009. "Integrated microscopy for real-time imaging of mechanotransduction studies in live cells." *Journal of Biomedical Optics* **14**(3): 13. doi:10.1117/1.3155517.

Venter, J. C., M. D. Adams, E. W. Myers, P. W. Li, R. J. Mural, G. G. Sutton, et al. 2001. "The sequence of the human genome." *Science* **291**(5507): 1304–51. doi:10.1126/science.1058040.

Vogel, V. and M. Sheetz. 2006. "Local force and geometry sensing regulate cell functions." *Nature Reviews Molecular Cell Biology* **7**(4): 265–75. doi:10.1038/nrm1890.

Wagner, N. and G. Krohne. 2007. "LEM-domain proteins: new insights into lamin-interacting proteins." *International Review of Cytology* **261**: 1–46. doi:10.1016/s0074-7696(07)61001-8.

Wang, D. J. and S. Bodovitz. 2010. "Single cell analysis: the new frontier in 'omics.'" *Trends in Biotechnology* **28**(6): 281–90. doi:10.1016/j.tibtech.2010.03.002.

Wang, H. B., M. Dembo, and Y. L. Wang. 2000. "Substrate flexibility regulates growth and apoptosis of normal but not transformed cells." *American Journal of Physiology – Cell Physiology* **279**(5): C1345–50. PMID:11029281.

Watt, F. M. and W. T. S. Huck. 2013. "Role of the extracellular matrix in regulating stem cell fate." *Nature Reviews Molecular Cell Biology* **14**(8): 467–73. doi:10.1038/nrm3620.

Wilson, K. L. and R. Foisner. 2010. "Lamin-binding proteins." *Cold Spring Harbor Perspectives in Biology* **2**(4): 17. doi:10.1101/cshperspect.a000554.

Winer, J. P., P. A. Janmey, M. E. McCormick, and M. Funaki. 2009. "Bone marrow-derived human mesenchymal stem cells become quiescent on soft substrates but remain responsive to chemical or mechanical stimuli." *Tissue Engineering* **15**(1): 147–54. doi:10.1089/ten.tea.2007.0388.

Xu, R., A. Boudreau, and M. Bissell. 2009. "Tissue architecture and function: dynamic reciprocity via extra- and intra-cellular matrices." *Cancer and Metastasis Reviews* **28**(1–2): 167–76. doi:10.1007/s10555-008-9178-z.

Yang, M. T., J. P. Fu, Y. K. Wang, R. A. Desai, and C. S. Chen. 2011. "Assaying stem cell mechanobiology on microfabricated elastomeric substrates with geometrically modulated rigidity." *Nature Protocols* **6**(2): 187–213. doi:10.1038/nprot.2010.189.

Yang, C., M. W. Tibbitt, L. Basta, and K. S. Anseth. 2014. "Mechanical memory and dosing influence stem cell fate." *Nature Materials* **13**(6): 645–52. doi:10.1038/nmat3889.

Yin, M., J. A. Talwalkar, K. J. Glaser, A. Manduca, R. C. Grimm, P. J. Rossman, et al. 2007. "Assessment of hepatic fibrosis with magnetic resonance elastography." *Clinical Gastroenterology and Hepatology* **5**(10): 1207–13. doi:10.1016/j.cgh.2007.06.012.

Zullo, J. M., I. A. Demarco, R. Pique-Regi, D. J. Gaffney, C. B. Epstein, C. J. Spooner, et al. 2012. "DNA sequence-dependent compartmentalization and silencing of chromatin at the nuclear lamina." *Cell* **149**(7): 1474–87. doi:10.1016/j.cell.2012.04.035.

3

Sugar-Coating the Cell

The Role of the Glycocalyx in Mechanobiology

Stefania Marcotti and Gwendolen C. Reilly

Department of Materials Science and Engineering, INSIGNEO Institute for In Silico Medicine, Sheffield, UK

3.1 What is the Glycocalyx?

The cell coat or glycocalyx is a proteoglycan-rich layer on the external surface of the cell membrane. Its thickness and proteoglycan composition vary according to the cell type and function. Glycocalyx means "sugar cup," and as well as long proteoglycan chains, it consists of small glycoproteins and glycosylated proteins, such that it contains a dense layer of glycosaminoglycan (GAG) chains (Tarbell and Pahakis 2006). This layer is also known as the pericellular matrix (PCM), though that term is more commonly used for the thick proteoglycan layer around chondrocytes forming the chondron, to distinguish it from the extracellular matrix (ECM), which forms the bulk cartilage. The terms "glycocalyx" and "PCM" are often used interchangeably, but the chondrocyte PCM is particularly thick and is characterized by specific composition and features; it is reviewed in detail elsewhere (Guilak et al. 2006; Wilusz et al. 2014). In this chapter, we will not consider the chondrocyte PCM in detail, but will focus on the glycocalyx of endothelial, bone, and muscle cells.

The glycocalyx components can be connected to the cell membrane via transmembrane proteoglycan-binding proteins or can span through the phospholipidic double layer. Glycoproteins and proteoglycans have a strong negative charge and attract water, so the glycocalyx is broadly very soft and water-saturated. Its gel-like characteristics modulate adhesion by providing resistance to certain protein–protein adhesions and enabling weak binding to specific molecules. To allow protein–protein binding, such as that between integrins and the ECM molecule fibronectin (Paszek et al. 2014), an energy barrier has to be overcome, with the proteoglycan molecules pushed aside or squashed to allow contact (Rutishauser et al. 1988; Soler et al. 1997, 1998; Sabri et al. 2000; Lipowsky 2012).

Even as a purely static structure that protects the cell membrane, the glycocalyx has important roles in health and disease. The endothelial glycocalyx is involved in the physiological functions of homeostasis, vessel permeability, regulation of leukocyte–endothelial cell interactions, regulation of clotting and complement cascades, growth factor binding, and lubrication, and is also responsible for the transduction of fluid flow-induced mechanical forces into biochemical intracellular signals (Tarbell and

Mechanobiology: Exploitation for Medical Benefit, First Edition. Edited by Simon C. F. Rawlinson.

Figure 3.1 Reconstructed confocal image from the Z-stacks of hyaluronic acid (HA) coats (biotinylated HA-binding protein) and primary cilia (antibody to acetylated alpha tubulin) on MLO-A5 cells (murine preosteocytes). The co-stain on the cilium indicates the presence of the glycocalyx on this fluid-detecting organelle.

Pahakis 2006; Reitsma et al. 2007; Weinbaum et al. 2007). In cancer, the glycocalyxes of circulating tumor cells play a fundamental role in the metastatic process. Effectively, by changing their composition and overexpressing specific proteoglycans or anchoring proteins, the glycocalyxes can increase the tumorigenicity of these cells. The glycocalyxes of both the circulating tumor cells and the target inflamed endothelial cells also control their interactions and affect adhesion processes and access by the potential tumor-suppressant drug (reviewed by Mitchell and King 2014). The immune system uses the glycocalyx as part of the pathogen recognition system and controls the inflammatory response by decoding the specific signature of the glycan/glycan-binding protein complex (reviewed by Schnaar 2015). In the bladder, the glycocalyx protects the endothelial bladder lining from the urine, and damage to the glycocalyx is a feature of cystitis and painful bladder syndrome (reviewed by Parsons 2007; Iacovelli et al. 2013).

As a modulator of cell binding and shape, the glycocalyx clearly has a role to play in interacting with important mechanotransduction structures discussed elsewhere in this book, such as integrins and membrane channel receptors. Intriguingly, we recently demonstrated that the glycocalyx is present on the fluid-detecting organelle that protrudes from the cell membrane – the primary cilia – as shown in Figure 3.1 (Delaine-Smith et al. 2014).

In this chapter, we will focus on the glycocalyx as a mechanosensor and force-modulation structure in its own right. We will briefly describe the physiology, morphology, and role in disease of the best characterized sugar coats. We will then describe some intriguing evidence for the role of the glycocalyx in the mechanobiology of muscle and bone and outline some other diseases in which it may be implicated.

3.2 Composition of the Glycocalyx

GAGs are sequences of disaccharide repeats that form chain oligosaccharides in which the two saccharides define the GAG. GAGs can be weakly or strongly negatively charged, depending on the sugars in the disaccharide. These GAG chains attach to core proteins to form proteoglycans. Proteoglycans have traditionally been difficult to characterize, due to their huge variety. Lectins are natural cell-membrane proteins that bind to specific carbohydrate groups. While lectins, usually extracted from plants, can have affinity for specific sugar units, identifying the predominant sugars present does not necessarily specify the proteoglycan molecule (Varki et al. 2009). Recently, however, specific proteoglycan antibodies and binding proteins have been identified, increasing the specificity of the detection of the components of the glycocalyx.

Two of the major components of the glycocalyx in different cell types are hyaluronic acid (HA) and the family of heparan sulfate proteoglycans (HSPGs). HA, or hyaluronan, is an unusual GAG in that it does not attach to a core protein but to another core HA chain, creating long bottle brush-like molecules. HA has been described as the structural proteoglycan of the glycocalyx in most cells (Scott and Heatley 1999; Reilly et al. 2003; Toole 2004; Evanko et al. 2007; Reitsma et al. 2007) and is weakly negatively charged. It is a synthesized in the plasma membrane by addition of sugar to the reducing polymer end through the action of HA synthases. The nonreducing end protrudes in the pericellular space. HA coats can be either anchored in the plasma membrane (if not released from its synthase enzyme) or bound to receptors (e.g., CD44 receptor) (Laurent and Fraser 1992; Banerji et al. 2007; Nijenhuis et al. 2009). Various functions in cell physiology appear to be related to HA (Nijenhuis et al. 2009). HA has been found to maintain the selective permeability permeation properties of the apical endothelial glycocalyx acting as a barrier to large macromolecules (Henry and Duling 1999) and was proposed to be able to reconstruct the endothelial glycocalyx after damage (Rubio-Gayosso et al. 2006; Singleton et al. 2006). In collaboration with the surface receptor CD44, HA is responsible for the assembly of the chondrocyte's PCM (Knudson et al. 1996). Mutations in HA synthases lead to diseases such as malignant progression in cancer (Toole et al. 2002; Toole 2004; Mitchell and King 2014) and cardiac and vascular defects (Toole et al. 2002; Zhu et al. 2014).

The HSPGs are the next most commonly discussed group of proteoglycans in the context of the glycocalyx. In endothelium, the most prominent GAGs are heparan sulfate (HS), at about 50–90% (in particular, syndecans, glypicans, and perlecans). Because they contain a sulfate group, HSPGs are highly negatively charged and bind a large range of proteins under physiological conditions. Various HSPGs exist, with different core proteins in specific cellular locations. One of their functions is sequestration within secretory vesicles of proteins bound within the ECM or to the cell surface. HSPGs can bind effectively to many ligands (e.g., cytokines, growth factors, ECM proteins), mainly through the HS chains, and are involved in diverse cellular processes, including cell–cell adhesion, cell–ECM adhesion, internalization, and clearance (Bernfield et al. 1999; Sarrazin et al. 2011). Moreover, HSs are critical in many biological features related to embryonic development, skeletogenesis and tissue homeostasis, matrix remodeling, and wound-healing (Kirn-Safran et al. 2009). When structural permutations occur, HSs are involved in a variety of pathologic conditions, including tumor angiogenesis, pathogen adhesion, and neurodegenerative disorders (Whitelock and Melrose 2011).

3.3 Morphology of the Glycocalyx

Visualizing and measuring the cell coat has proved extremely difficult in the past, because of its high water content and ability to collapse when dehydrated or in a non-physiological solution. Therefore, it has been less studied than other cell membrane

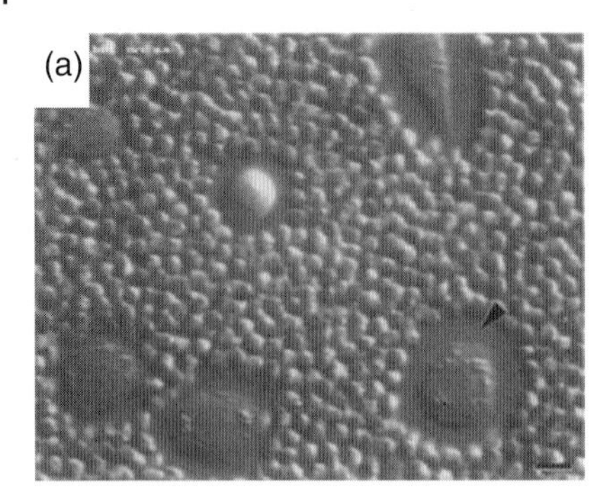

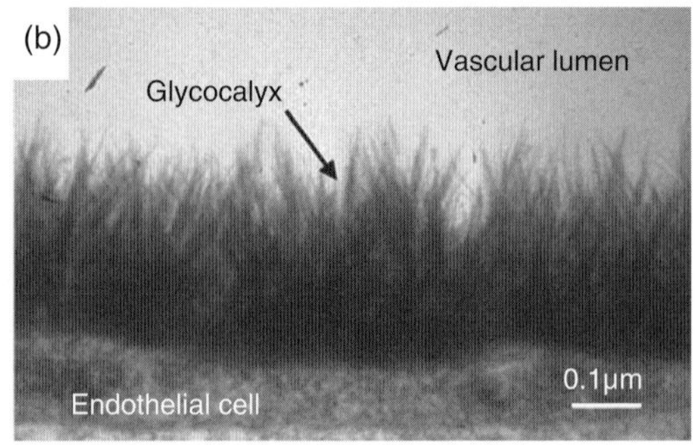

Figure 3.2 (a) RBC exclusion zone around a chondrocyte, indicated by the black arrow. This area corresponds to the PCM of the chondrocyte. Bar: 10 μm. *Source:* From Lee et al. (1993). (b) Electron microscopic views of the glycocalyx in isolated guinea pig hearts. *Source:* From Chappell et al. (2009).

components. A major difficulty in imaging proteoglycan pericellular layers is related to the fixation methods needed to maintain the physiological hydration of the mesh (Evanko et al. 2009; de la Motte and Drazba 2011). The glycocalyx was initially measured as an absence: a "red blood cell (RBC) exclusion zone" around the cell membrane (Figure 3.2a). It was noted that if small particles such as RBCs are added to cultured cells, there is a space between the RBCs and the cell membrane. The thickness of this "exclusion zone" can give a measurement of the thickness of the glycocalyx or PCM (Lee et al. 1993; Vink and Duling 2000; Reitsma et al. 2007; Kim et al. 2009; Alphonsus and Rodseth 2014). Direct-measurement methods were introduced with the advance of experimental techniques (Figure 3.2b).

These methods included transmission electron microscopy (TEM), intravital microscopy, confocal and two-photon laser scanning microscopy, sidestream dark-field imaging, and microparticle image velocimetry, coupled with different fixation methods and glycocalyx markers. However, regarding the thickness of the endothelial glycocalyx, *in vitro* and *in vivo* measurements lead to highly variable values, depending on the assessment method used. These values ranged from 0.02 to 8.90 μm *in vivo* or *ex vivo* and from 0.01 to 3.00 μm *in vitro* in one study (Ebong et al. 2011), calling into question the validity of *in vitro* models. Recently, a protocol was developed to image the endothelial glycocalyx using TEM coupled to rapid freezing/freeze-substitution fixation (Ebong et al. 2011). This technique allows for high-spatial resolution imaging of the glycocalyx in its hydrated and protein-rich configuration. It was tested on bovine aortic endothelial cells and rat fat-pad endothelial cells, with 11.35 ± 0.21 and 5.38 ± 1.13 μm glycocalyx measurements, respectively. These values are closer to those obtained *in vivo* (e.g., by fluorescence labeling of blood vessels) and confirm the validity of *in vitro* studies, highlighting the importance of the fixation method.

3.4 Mechanical Properties of the Glycocalyx

A parameter necessary to understand the mechanobiological role of glycocalyx is its mechanical properties. Different approaches have been used to study the endothelial glycocalyx, including reflectance interference contrast microscopy, microrheology, and atomic force microscopy (AFM).

In order to study the endothelial glycocalyx with reflectance interference contrast microscopy, the light interference pattern created between a bead and the coverslip reflections was evaluated as a measure of the stiffness of the GAG layer, and was found to be on the order of 0.05 Pa (Job et al. 2012). In passive microrheology, the thermal fluctuations of optically trapped (sub)micrometric colloidal particles are measured. By manipulating the particle position, it is possible to indent the PCM at different heights with respect to the membrane and to measure its shear elastic and viscous moduli. In one study, these moduli for metastatic prostate epithelial cancer cells were found to be about 0.35 and 14.0 Pa, respectively (Nijenhuis et al. 2012).

AFM has been widely used in the last 2 decades as a versatile platform for imaging and studying biological samples (Müller and Dufrêne 2011). Recently, researchers have also applied it to the study of the mechanical properties of the endothelial glycocalyx (Table 3.1). Oberleithner et al. (2011) measured the thickness and stiffness of the endothelial glycocalyx of split-open *ex vivo* human umbilical cord arteries. As a proof of concept of glycocalyx indentation, they treated bovine aortic endothelial cells with HS and measured the stiffness and thickness of the glycocalyx before and after treatment by means of a coupled AFM/fluorescence microscopy set-up. Their measurements resulted in a glycocalyx thickness of 400 nm and stiffness of 0.25 mPa. The same group (Wiesinger et al. 2013) recently applied a similar protocol to *ex vivo* and *in vitro* analyses of the mechanical properties change when the endothelial glycocalyx is subjected to enzymatic degraders or sepsis mediators. Both treatments resulted in the softening (i.e., reduced stiffness) reduction in thickness of the glycocalyx.

O'Callaghan et al. (2011) tested bovine lung microvasculature endothelial cells (BLMVECs) by measuring the elastic properties of the glycocalyx after selective enzymatic degradation of HS and HA. Moreover, they treated cells with a cytoskeleton disruptor to differentiate between the cell and glycocalyx contributions. The glycocalyx elastic modulus gave ~250 Pa for an expected glycocalyx thickness of 200 nm. Enzymatic degradation affected the rate of stiffness increase in relation to indentation depth, while disruption of the cytoskeleton inhibited the modulus increase with the indentation.

Bai and Wang (2012) investigated the spatial and temporal distribution and the mechanical properties of human umbilical vein endothelial cells (HUVECs). They analyzed cells with intact or degraded glycocalycies (selective removal of sialic acid or HS) at different time points and found a developed glycocalyx after day 14, with a thickness of 300–1000 nm and a Young's modulus of 390 Pa.

Marsh and Waugh (2013) investigated the endothelial glycocalyx of HUVECs cultured under fluid flow. By using a two-layer model derived from polymer analysis and fitting 20 repeated indentations over 25 different cells, they obtained a glycocalyx thickness of 380 ± 50 nm and a stiffness of 700 ± 500 Pa.

Sokolov et al. (2013) proposed a protocol to quantitatively measure the elastic modulus of cells by taking into account the contribution of the surface brush. To validate the protocol, human cervical epithelial cells were evaluated, obtaining a thickness of the brush of about 1.4 μm. However, the stiffness of the glycocalyx cannot be extrapolated from the customized model.

As is clear from Table 3.1, there is no agreement concerning endothelial glycocalyx stiffness. This may be due to the different approaches employed, or to the physiological variability between cell lines. An important point relates to the actual sensing of the glycocalyx: due to its meshlike structure, it is possible that pyramidal tips or too-small

Table 3.1 Comparison of AFM indentation experiments measuring the mechanical properties of the glycocalyx.

	Oberleithner et al. (2011)	**O'Callaghan et al.** (2011)	**Bai and Wang** (2012)	**Marsh and Waugh** (2013)	**Sokolov et al.** (2013)
Tip	Polystyrene beads (1 μm diameter)	Silica bead (18 μm diameter)	Pyramidal (20 nm radius)	Beads (2.4 μm diameter)	Silica beads (5 μm diameter)
Cantilever	Triangular 0.01 N/m	Rectangular 0.03 N/m	Rectangular 0.38 N/m	Arrow 0.03 N/m	V-shaped 0.25 N/m
Sample	*Ex vivo* human arteries in HEPES buffer + 1% FBS	Monolayer of BLMVECs in medium	Monolayer of HUVECs in medium	Monolayer of HUVECs	Low-density human cervical epithelial cells in inorganic buffer
Number of measurements	–	>80	>25	>500	10–30
Location of measurements	–	Nuclei and cell junctions (*a posterior*)	Nucleus, middle, and edge	Within 2 μm of the nucleus	$50 \times 50\,\mu m^2$ grid over the nucleus
Glycocalyx stiffness	0.000 25 Pa	250 ± 30 Pa	390 Pa	700 ± 500 Pa	–
Glycocalyx thickness	400 nm	200 nm	300 nm to 1 μm	380 ± 50 nm	1.4 μm

HEPES, 4-(2-hydroxyethyl)-1-piperazineethanesulfonic acid; FBS, fetal bovine syndrome; BLMVEC, bovine lung microvasculature endothelial cell; HUVEC, human umbilical vein endothelial cell.

colloidal ones penetrate this layer without compressing it, or that the sensitivity of the probe is too weak to detect the actual point of contact with the glycocalyx.

3.5 Mechanobiology of the Endothelial Glycocalyx

The endothelial glycocalyx covers the endothelial cells and represents the contact surface with the bloodstream. Under physiological conditions, an extended endothelial surface layer is created by the association of the glycocalyx macromolecules with blood-borne components through ionic interactions or specific bindings (for a review, see Tarbell and Pahakis 2006). Among other functions, the endothelial glycocalyx has a mechanosensor role in the transduction of fluid flow signal. A "wind in the trees" model has been proposed to explain the fluid flow mechanotransduction: the fluid flow (wind) is sensed by the GAGs (branches) and transmitted to the cell membrane or the cytoskeleton (ground) through the core protein (tree trunk). This simple conceptual model has been shown to be plausible by structural glycocalyx observations, since the core proteins are sufficiently stiff to act as transmitters without significant deflection (Squire et al. 2001; Weinbaum et al. 2003).

Various experiments have been carried out to demonstrate that mechanoresponsive signals can be transmitted by the endothelial glycocalyx (Tarbell 2010; Curry and Adamson 2012). For example, by exposing bovine aortic endothelial cells to shear stress after selective removal of HS, chondroitin sulfate (CS), and HA, Pahakis et al. (2007) demonstrated the involvement of HS and HA, but not CS, in nitric oxide (NO) release. Moreover, it has been demonstrated that the very composition of the endothelial glycocalyx can change when fluid flow is applied *in vitro*, suggesting a direct mechanosensitive response (Gouverneur et al. 2006).

As the glycocalyx controls the permeability and plays a vasculoprotective role in healthy vessels, these functions fail when disrupted, leading to pathological situations. Recently, some evidence was collected which highlights the involvement of endothelial glycocalyx in various diseases (reviewed by Nieuwdorp et al. 2005; Reitsma et al. 2007; Becker et al. 2010; Henrich et al. 2010; Alphonsus and Rodseth 2014; Kolářová et al. 2014). However, further investigations are required to decide whether this involvement acts as a cause or consequence of vascular impairment (van den Berg and Vink 2006). Diabetes is characterized by both micro- and macrovasculature disorders affecting vessel permeability and dilation through mechanisms that have not yet been fully elucidated. It has been proved that hyperglycemic acute and chronic conditions damage the glycocalyx, suggesting that the role of the glycocalyx in the permeability of macromolecules (e.g., albumin) could mediate vascular impairment (Nieuwdorp et al. 2006a,b).

Microvasculature dysfunction is also involved in the ischemia/reperfusion process. Endothelial cells undergo damage caused by oxidative stress, resulting in leukocyte response and increased vascular permeability. Mulivor and Lipowsky (2004) first showed that the thickness of the glycocalyx is reduced by shedding of GAGs consequent to ischemia/reperfusion events. Similar evidence was collected *in vivo* in patients undergoing vascular surgery (Rehm et al. 2007; Bruegger et al. 2009). The glycocalyx shedding hints at a possible involvement in endothelial tissue damage; however, its relative contribution and its use as a target for therapeutic actions are yet to be fully understood. Similar shedding events are observed in the systemic inflammatory response syndrome, where the degradation is initiated by inflammatory mediators such as tumor necrosis factor alpha (TNF-α) (Henry and Duling 2000; Chappell et al. 2009) and leads to detection of glycocalyx components in the blood of patients in septic shock (Nelson et al. 2008). Moreover, variations in the mechanical properties of the endothelial glycocalyx when exposed to sepsis mediators have been shown, with the average softening of the proteoglycan layer occurring alongside the reduction in thickness (Wiesinger et al. 2013). This observation highlights the importance of a mechanobiological approach to the study of glycocalyx physiopathology.

The endothelial glycocalyx may also be involved in atherosclerosis. Experimental data show a reduction in vasculoprotective capacity at sites with higher atherogenic risks (van den Berg et al. 2006) and highlight the oxidized low-density lipoprotein (LDL)-induced degradation of the glycocalyx with consequent local platelet adhesion (Vink et al. 2000). Sites with high atherogenic risk are characterized by low and oscillatory wall shear stress (reviewed by Peiffer et al. 2013); therefore, it seems reasonable to hypothesize a correlation with the ability of the glycocalyx to sense the shear stress stimuli. Hence, the glycocalyx is involved in many pathological conditions characterized by vasculature impairment and dysfunction, and some therapeutic strategies targeting the glycocalyx are beginning to be investigated (for review, see Becker et al. 2010; Kolářová et al. 2014).

3.6 Does the Glycocalyx Play a Mechanobiological Role in Bone?

In bone, osteocytes are encased in the dense mineralized ECM and connect through tiny channels (canaliculi) via long processes (dendrites). There is a nonmineralized proteoglycan-rich pericellular space between the osteocyte membrane and the wall of the mineralized tissue of 0.1–1.0 μm. This space has been proposed to contain the glycocalyx and play a role in mechanosensation (Weinbaum et al. 1994; Cowin et al. 1995). Initially, Cowin's focus was on the negative charges of the proteoglycans and their ability to facilitate load-induced streaming potentials (Cowin et al. 1995), but based on the emerging literature on the mechanobiology of endothelial cells, a force-transmission role for the osteocyte glycocalyx was postulated and modeled. Weinbaum et al. (1994) proposed an idealized mathematical model of the osteocyte dendrites. This model represents an individual canaliculus, with its central cell process as two concentric cylinders connected through a series of filamentous tethering elements. These elements are thought to be the proteoglycan chains of the glycocalyx, which will experience drag as the fluid flows through the pericellular spaces. This drag action would be transmitted to the cell membrane as circumferential strains, amplifying the otherwise small strains seen in bone loading (You et al. 2001). A schematic of the bone glycocalyx is shown in Figure 3.3.

This hypothesis is supported by our own *in vitro* studies, which characterized the cultured bone cell glycocalyx as including a substantial HA component (Reilly et al. 2003; Burra et al. 2010; Morris et al. 2010), plus small amounts of other proteoglycans, such as CS in the form of decorin (Ingram et al. 1993) and biglycan (Takagi et al. 1991). Given the role of HA in the mechanobiology of the endothelium (Pahakis et al. 2007; Kumagai et al. 2009) and its function in cell attachment (Zimmerman et al. 2002), we

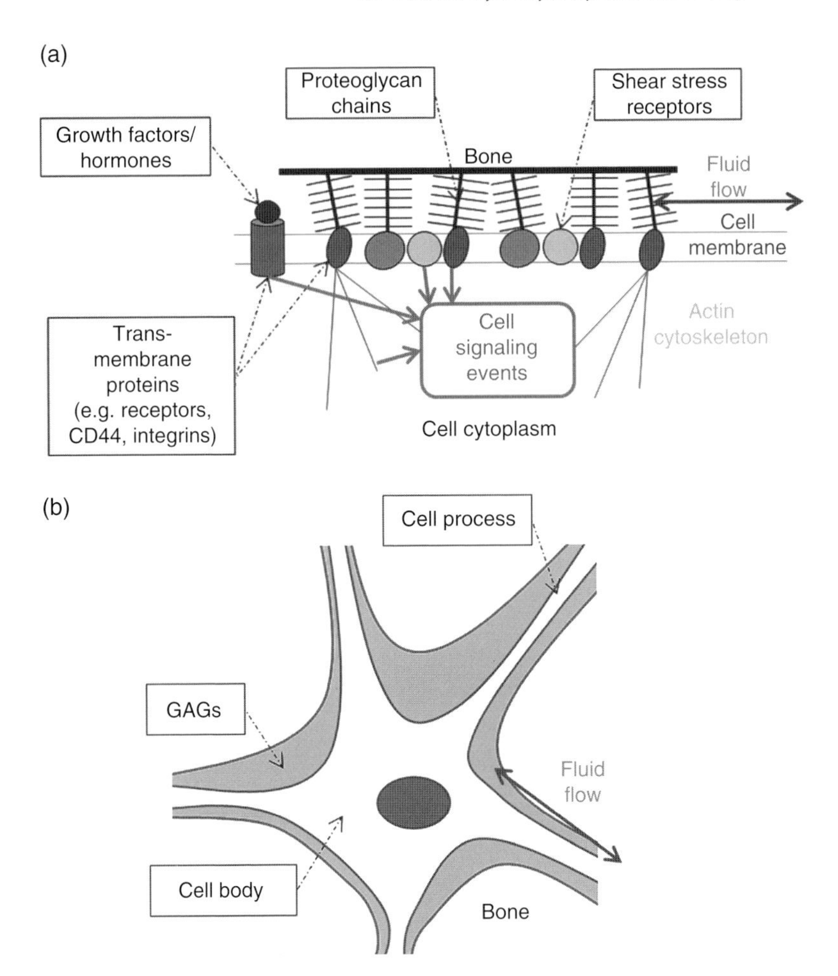

Figure 3.3 Schematic of the bone glycocalyx. (a) Osteocyte lying in the lacunocanalicular network, highlighting the disposition of the bone glycocalyx. *Source:* From Morris et al. (2010). (b) Potential mechanism of fluid flow sensing. The fluid flow passes through the proteoglycan chains, causing cell signaling events, which affect the cytoskeleton. *Source:* From Reilly et al. (2003).

hypothesized that an HA-rich glycocalyx in bone would be important in mechanotransduction. Moreover, the CD44 transmembrane protein can bind to the actin cytoskeleton with a switch active/inactive mechanism (Thorne et al. 2004), suggesting a possible means of force transmission from the outer to the inner cell compartment (Singleton et al. 2006).

However since the initial study on HA, more evidence has emerged of the important role of integrins, with ultrastructural studies by McNamara et al. (2009) indicating that the osteocyte dendrite contains small hillocks that directly contact the bone wall via protein–protein binding. Initially, McNamara et al. (2009) postulated the integrin attachments were more important for mechanosensation than the proteoglycan "tethering elements" in the spaces in between them. In parallel, studies showed HSPGs (i.e., perlecan) to be present in the osteocyte pericellular space and to be critical in maintaining its size (Thompson et al. 2011). Moreover, experiments in mice demonstrated an increased solute transport and a decrease in fiber density in perlecans-deficient animals when compared to controls (Wang et al. 2014). Putting this work together, it appears that HSPGs, possibly supported by HA, are the tethering elements stabilizing the integrin attachments and that all three components are key players in the structure of the osteocyte–bone interface, and therefore in bone mechanotransduction.

The bone glycocalyx has been less investigated than the endothelial glycocalyx, so its role in pathological conditions is not yet established. However, there is some evidence of its involvement in mechanotransduction based on *in vitro* experiments. In our experiments, the degradation of the osteocyte's glycocalyx reduced prostaglandin E_2 (PGE_2) release, suggesting a contribution of the glycocalyx in bone mechanotransduction related to bone remodelling, with PGE_2 as a key mediator of the bone resorbing cells, the osteoclasts. It was later confirmed that not only PGE_2 but also collagen matrix deposition induced by flow can be inhibited by glycocalyx removal or blocking of the HA-binding protein CD44, providing support for the hypothesis that CD44/HA binding has an important role to play in bone formation (Morris et al. 2010). Selective hyaluronidase-induced degradation of the dendritic processes of osteocyte glycocalyx led to the abolishment of opening of Cx43 hemichannels when mechanical stimulation of the dendrites was applied. Moreover, the hyaluronidase treatment negatively affected the integrin attachment formation between the dendrites and the plate. These findings suggest that the dendritic processes are responsible for mechanotransduction and that the glycocalyx plays an important role in their mechanosensitivity (Burra et al. 2010, 2011).

Interestingly, there are several skeletal defects associated with proteoglycan gene mutations, especially of HSPGs (Bishop et al. 2007). One example is a skeletal disorder called hereditary multiple exostoses, which is characterized by the growth of multiple bony protuberances (osteochondromas), usually at the epiphyseal growth plate of long bones. This disorder is caused by mutations in the EXT genes, which are enzymes required to build the HSPG chains (Stickens et al. 2005; Zak et al. 2011). Another example relates to the mutation in the HSPG Glypican 3, which causes Simpson–Golabi–Behmel syndrome in humans: an overgrowth syndrome clinically characterized by multiple congenital abnormalities (Tenorio et al. 2014). It has been shown that this genetic mutation can cause the reduction of calcified trabecular bone and a delay in the appearance of osteoclasts (bone-resorbing cells) (Viviano et al. 2005). While these diseases seem to be associated with skeletal development, it is yet to be elucidated whether there is an effect on the mechanobiology of the system which contributes to the dysfunctional skeletal cell behavior.

3.7 Glycocalyx in Muscle

Recently, the glycocalyx of other cell types has begun to be investigated, in order to better understand its role in mechanotransduction. In particular, Juffer et al. (2014) explored the glycocalyx in muscle fibers in order to study the fluid shear stress response of myoblasts *in vitro*. They found that an intact glycocalyx is necessary for shear stress-stimulated NO production, suggesting its active role in the transmission of mechanical stimuli to the mechanosensor complexes of the cell. This, together with the available information on bone and endothelium, highlights the necessity of further investigations to unveil the importance of the glycocalyx in physiological and pathological conditions. In the future, as the initial clinical evidence for vascular impairment treatment suggests (Becker et al. 2010), the glycocalyx could be targeted for specific drug therapies and tissue-engineered approaches.

3.8 How Can the Glycocalyx be Exploited for Medical Benefit?

While our understanding of the role of glycobiology in general and the glycocalyx in particular in the human body is still rather rudimentary in comparison to other cellular components, we already have a vast array of tools that could potentially manipulate proteoglycans in the body. For example, anti-inflammatories that protect from atherosclerosis may be doing so through a mechanism that protects the endothelial glycocalyx from damage by cytokines (Wheeler-Jones et al. 2010). The most obvious proteoglycan supplements are over-the-counter proteoglycan-based "nutraceuticals" such as HA or CS. These are sold as being beneficial to skin and cartilage, and particularly as potentially alleviating the symptoms of osteoarthritis. There is controversy within the scientific community over these products: the Osteoarthritis Research Society International (OARSI) (McAlindon et al. 2014) classified the use of CS as not having clear benefits for symptom relief and as not appropriate for disease modification. On the other hand, some positive clinical evidence has been shown of a low risk score, a moderate to high effect size, and a high risk/benefit score (Henrotin et al. 2014). These chemical compounds are classified as dietary supplements in many countries (including the United States), meaning there is no need for medical prescription. However, clinical trials using different-quality products could have led to different results: there may be a gap between the results of clinical trials conducted in a well-characterized osteoarthritis patient set and the advice given to the general population (Henrotin et al. 2014; Cutolo et al. 2015).

Local delivery (injection or infusion) of HA, for example to the bladder (Iavazzo et al. 2007) or a joint (Brandt et al. 2000; Cutolo et al. 2015), may have had more success in alleviating symptoms than proteoglycan pills, though the mechanism and significance of the effects is still rather unclear. Additionally, many anticancer drugs are under development which target either proteoglycans or their catabolic and anabolic enzymes (Adamia et al. 2005; Raman et al. 2010); these could have additional mechanobiology modulatory effects. Once we have a better understanding of how circulating proteoglycans are processed and incorporated into the glycocalyx, techniques that specifically target a cell's mechanobiology apparatus and cause minimal side effects can be envisaged. Alternatively, as our understanding of the genetics of the proteoglycan synthases progresses, there may be ways to improve a cell's ability to elaborate its glycocalyx and modulate its response to mechanical forces.

3.9 Conclusion

It is clear that the glycocalyx is an important mechanotransduction mechanism in the mechanosensitive tissue of the body: the cardiovascular and musculoskeletal systems. The glycocalyx has also been demonstrated to be a significant factor in the progression of many diseases, in particular atherosclerosis, osteoarthritis, and cancer. However, the link between the mechanobiological role of the glycocalyx and its role in disease states is still unclear. Many proteoglycan-modulating therapeutics under current investigation have the potential to be used to further explore the mechanobiological role of the glycocalyx in health and disease.

References

Adamia, S., C. Maxwell, and L. Pilarski. 2005. "Hyaluronan and hyaluronan synthases: potential therapeutic targets in cancer." *Current Drug Target – Cardiovascular & Hematological Disorders* **5**(1): 3–14. doi:10.2174/1568006053005056.

Alphonsus, C. S. and R. N. Rodseth. 2014. "The endothelial glycocalyx: a review of the vascular barrier." *Anaesthesia* **69**(7): 777–84. doi:10.1111/anae.12661.

Bai, K. and W. Wang. 2012. "Spatio-temporal development of the endothelial glycocalyx layer and its mechanical property in vitro." *Journal of the Royal Society, Interface/the Royal Society* **9**(74): 2290–8. doi:10.1098/rsif.2011.0901.

Banerji, S., A. J. Wright, M. Noble, D. J. Mahoney, I. D. Campbell, A. J. Day, and D. G Jackson. 2007. "Structures of the Cd44-hyaluronan complex provide insight into a fundamental carbohydrate-protein interaction." *Nature Structural & Molecular Biology* **14**(3): 234–9. doi:10.1038/nsmb1201.

Becker, B. F., D. Chappell, D. Bruegger, T. Annecke, and M. Jacob. 2010. "Therapeutic strategies targeting the endothelial glycocalyx: acute deficits, but great potential." *Cardiovascular Research* **87**(2): 300–10. doi:10.1093/cvr/cvq137.

Bernfield, M., M. Gotte, P. Woo Park, O. Reizes, M. L. Fitzgerald, J. Lincecum, and M. Zako. 1999. "Functions of cell surface heparan sulfate proteoglycans." *Annual Review of Biochemistry* **68**: 729–77. doi:10.1146/annurev.genet.37.061103.090226.

Bishop, J. R., M. Schuksz, and J. D. Esko. 2007. "Heparan sulphate proteoglycans fine-tune mammalian physiology." *Nature* **446**(7139): 1030–7. doi:10.1038/nature05817.

Brandt, K. D., G. N. Smith Jr., and L. S. Simon. 2000. "Intraarticular injection of hyaluronan as treatment for knee osteoarthritis: what is the evidence?" *Arthritis and Rheumatism* **43**(6): 1192–203. PMID:10857778.

Bruegger, D., M. Rehm, J. Abicht, J. Oliver Paul, M. Stoeckelhuber, M. Pfirrmann, et al. 2009. "Shedding of the endothelial glycocalyx during cardiac surgery: on-pump versus off-pump coronary artery bypass graft surgery." *Journal of Thoracic and Cardiovascular Surgery* **138** (6): 1445–7. doi:10.1016/j.jtcvs.2008.07.063.

Burra, S., D. P. Nicolella, W. Loren Francis, C. J. Freitas, N. J. Mueschke, K. Poole, and J. X. Jiang. 2010. "Dendritic processes of osteocytes are mechanotransducers that induce the opening of hemichannels." *Proceedings of the National Academy of Sciences of the United States of America* **107**(31): 13 648–53. doi:10.1073/pnas.1009382107.

Burra, S., D. P. Nicolella, and J. X. Jiang. 2011. "Dark horse in osteocyte biology: glycocalyx around the dendrites is critical for osteocyte mechanosensing." *Communicative & Integrative Biology* **4**(1): 48–50. doi:10.4161/cib.4.1.13646.

Chappell, D., K. Hofmann-Kiefer, M. Jacob, M. Rehm, J. Briegel, U. Welsch, et al. 2009. "TNF-α induced shedding of the endothelial glycocalyx is prevented by hydrocortisone and antithrombin." *Basic Research in Cardiology* **104**(1): 78–89. doi:10.1007/s00395-008-0749-5.

Cowin, S. C., S. Weinbaum, and Y. Zeng. 1995. "A case for bone canaliculi generated as the anatomical potential." *Journal of Biomechanics* **28**(11): 1281–97. PMID:8522542.

Curry, F. E. and R. H. Adamson. 2012. "Endothelial glycocalyx: permeability barrier and mechanosensor." *Annals of Biomedical Engineering* **40**(4): 828–39. doi:10.1007/s10439-011-0429-8.

Cutolo, M., F. Berenbaum, M. Hochberg, L. Punzi, and J. Reginster. 2015. "Commentary on recent therapeutic guidelines for osteoarthritis." *Seminars in Arthritis and Rheumatism* **44**: 611–17. doi:10.1016/j.semarthrit.2014.12.003.

De la Motte, C. A., and J. A. Drazba. 2011. "Viewing hyaluronan: imaging contributes to imagining new roles for this amazing matrix polymer." *Journal of Histochemistry and Cytochemistry* **59**(3): 252–7. doi:10.1369/0022155410397760.

Delaine-Smith, R. M., A. Sittichokechaiwut, and G. C. Reilly. 2014. "Primary cilia respond to fluid shear stress and mediate flow-induced calcium deposition in osteoblasts." *FASEB Journal* **28**(1): 430–9. doi:10.1096/fj.13-231894.

Ebong, E. E., F. P. Macaluso, D. C. Spray, and J. M. Tarbell. 2011. "Imaging the endothelial glycocalyx in vitro by rapid freezing/freeze substitution transmission electron microscopy." *Arteriosclerosis, Thrombosis, and Vascular Biology* **31**(8): 1908–15. doi:10.1161/ATVBAHA.111.225268.

Evanko, S. P., M. I. Tammi, R. H. Tammi, and T. N. Wight. 2007. "Hyaluronan-dependent pericellular matrix." *Advanced Drug Delivery Reviews* **59**(13): 1351–65. doi:10.1016/j.addr.2007.08.008.

Evanko, S. P., S. Potter-Perigo, P. Y. Johnson, and T. N. Wight. 2009. "Organization of hyaluronan and versican in the extracellular matrix of human fibroblasts treated with the viral mimetic poly I:C." *Journal of Histochemistry and Cytochemistry* **57**(11): 1041–60. doi:10.1369/jhc.2009.953802.

Gouverneur, M., J. A. E. Spaan, H. Pannekoek, R. D. Fontijn, and H. Vink. 2006. "Fluid shear stress stimulates incorporation of hyaluronan into endothelial cell glycocalyx." *American Journal of Physiology – Heart and Circulatory Physiology* **290**: H458–2. doi:10.1152/ajpheart.00592.2005.

Guilak, F., L. G. Alexopoulos, M. L. Upton, I. Youn, J. Bong Choi, L. Cao, et al. 2006. "The pericellular matrix as a transducer of biomechanical and biochemical signals in articular cartilage." *Annals of the New York Academy of Sciences* **1068**(1): 498–512. doi:10.1196/annals.1346.011.

Henrich, M., M. Gruss, and M. A. Weigand. 2010. "Sepsis-induced degradation of endothelial glycocalix." *Scientific World Journal* **10**: 917–23. doi:10.1100/tsw.2010.88.

Henrotin, Y., M. Marty, and A. Mobasheri. 2014. "What is the current status of chondroitin sulfate and glucosamine for the treatment of knee osteoarthritis?" *Maturitas* **78**(3): 184–7. doi:10.1016/j.maturitas.2014.04.015.

Henry, C. B. and B. R. Duling. 1999. "Permeation of the luminal capillary glycocalyx is determined by hyaluronan." *American Journal of Physiology* **277**(2 Pt. 2): H508–14. PMID:10444475.

Henry, C. B. and B. R. Duling. 2000. "TNF-alpha increases entry of macromolecules into luminal endothelial cell glycocalyx." *American Journal of Physiology – Heart and Circulatory Physiology* **279**: H2815–23. PMID:11087236.

Iacovelli, V., L. Topazio, G. Gaziev, P. Bove, G. Vespasiani, and E. Finazzi Agrò. 2013. "Intravesical glycosaminoglycans in the management of chronic cystitis." *Minerva Urologica e Nefrologica* **65**(4): 249–62. PMID:24091478.

Iavazzo, C., S. Athanasiou, E. Pitsouni, and M. E. Falagas. 2007. "Hyaluronic acid: an effective alternative treatment of interstitial cystitis, recurrent urinary tract infections, and hemorrhagic cystitis?" *European Urology* **51**(6): 1534–41. doi:10.1016/j.eururo.2007.03.020.

Ingram, R. T., B. L. Clarke, L. W. Fisher, and L. A. Fitzpatrick. 1993. "Distribution of noncollagenous proteins in the matrix of adult human bone: evidence of anatomic and functional heterogeneity." *Journal of Bone and Mineral Research* **8**: 1019–29. doi:10.1002/jbmr.5650080902.

Job, K. M., R. O. Dull, and V. Hlady. 2012. "Use of reflectance interference contrast microscopy to characterize the endothelial glycocalyx stiffness." *American Journal of Physiology – Lung Cellular and Molecular Physiology* **302**(12): L1242–9. doi:10.1152/ajplung.00341.2011.

Juffer, P., A. D. Bakker, J. Klein-Nulend, and R. T. Jaspers. 2014. "Mechanical loading by fluid shear stress of myotube glycocalyx stimulates growth factor expression and nitric oxide production." *Cell Biochemistry and Biophysics* **69**(3): 411–19. doi:10.1007/s12013-013-9812-4.

Kim, S., P. Kai Ong, O. Yalcin, M. Intaglietta, and P. C. Johnson. 2009. "The cell-free layer in microvascular blood flow." *Biorheology* **46**(3): 181–9. doi:10.3233/BIR-2009-0530.

Kirn-Safran, C., M. C. Farach-Carson, and D. D. Carson. 2009. "Multifunctionality of extracellular and cell surface heparan sulfate proteoglycans." *Cellular and Molecular Life Sciences* **66**(21): 3421–34. doi:10.1007/s00018-009-0096-1.

Knudson, W., D. J. Aguiar, Q. Hua, and C. B. Knudson. 1996. "CD44-anchored hyaluronan-rich pericellular matrices: an ultrastructural and biochemical analysis." *Experimental Cell Research* **228**(2): 216–28. PMID:8912714.

Kolářová, H., B. Ambrůzová, L. Sviválková Šindlerová, A. Klinke, and L. Kubala. 2014. "Modulation of endothelial glycocalyx structure under inflammatory conditions." *Mediators of Inflammation* **2014**: 694312. doi:10.1155/2014/694312.

Kumagai, R., X. Lu, and G. S. Kassab. 2009. "Role of glycocalyx in flow-induced production of nitric oxide and reactive oxygen species." *Free Radical Biology and Medicine* **47**(5): 600–7. doi:10.1016/j.freeradbiomed.2009.05.034.

Laurent, T. C. and R. E. Fraser. 1992. "Hyaluronan." *FASEB Journal* **6**: 2397–404. PMID:1563592.

Lee, G. M., B. Johnstone, K. Jacobson, and B. Caterson. 1993. "The dynamic structure of the pericellular matrix on living cells." *Journal of Cell Biology* **123**(6): 1899–907. doi:10.1083/jcb.123.6.1899.

Lipowsky, H. H. 2012. "The endothelial glycocalyx as a barrier to leukocyte adhesion and its mediation by extracellular proteases." *Annals of Biomedical Engineering* **40**(4): 840–8. doi:10.1007/s10439-011-0427-x.

Marsh, G. and R. E. Waugh. 2013. "Quantifying the mechanical properties of the endothelial glycocalyx with atomic force microscopy." *Journal of Visualized Experiments* **72**: e50163. doi:10.3791/50163.

McAlindon, T. E., R. R. Bannuru, M. C. Sullivan, N. K. Arden, F. Berenbaum, S. M. Bierma-Zeinstra, et al. 2014. "OARSI guidelines for the non-surgical management of knee osteoarthritis." *Osteoarthritis and Cartilage* **22**(3): 363–88. doi:10.1016/j.joca.2014.01.003.

McNamara, L. M., R. J. Majeska, S. Weinbaum, V. Friedrich, and M. B. Schaffler. 2009. "Attachment of osteocyte cell processes to the bone matrix." *Anatomical Record* **292**: 355–63. doi:10.1002/ar.20869.

Mitchell, M. J. and M. R. King. 2014. "Physical biology in cancer. 3. The role of cell glycocalyx in vascular transport of circulating tumor cells." *American Journal of Physiology – Cell Physiology* **306**(2): C89–97. doi:10.1152/ajpcell.00285.2013.

Morris, H. L., C. I. Reed, J. W. Haycock, and G. C. Reilly. 2010. "Mechanisms of fluid-flow-induced matrix production in bone tissue engineering." *Proceedings of the Institution of Mechanical Engineers. Part H, Journal of Engineering in Medicine* **224**(12): 1509–21. doi: 10.1243/09544119JEIM751.

Mulivor, A. W. and H. H. Lipowsky. 2004. "Inflammation- and ischemia-induced shedding of venular glycocalyx." *American Journal of Physiology – Heart and Circulatory Physiology* **286**(5): H1672–80. doi:10.1152/ajpheart.00832.2003.

Müller, D. J. and Y. F. Dufrêne. 2011. "Atomic force microscopy: a nanoscopic window on the cell surface." *Trends in Cell Biology* **21**(8): 461–9. doi:10.1016/j.tcb.2011.04.008.

Nelson, A., I. Berkestedt, A. Schmidtchen, L. Ljunggren, and M. Bodelsson. 2008. "Increased levels of glycosaminoglycans during septic shock: relation to mortality and the antibacterial actions of plasma." *Shock (Augusta, Ga.)* **30**(6): 623–7. doi:10.1097/SHK.0b013e3181777da3.

Nieuwdorp, M., M. C. Meuwese, H. Vink, J. B. L. Hoekstra, J. J. P. Kastelein, and E. S. G. Stroes. 2005. "The endothelial glycocalyx: a potential barrier between health and vascular disease." *Current Opinion in Lipidology* **16**: 507–11. PMID:16148534.

Nieuwdorp, M., H. L. Mooij, J. Kroon, B. Atasever, J. A. E. Spaan, C. Ince, et al. 2006a. "Endothelial glycocalyx damage coincides with microalbuminuria in type 1 diabetes." *Diabetes* **55**: 1127–32. PMID:16567538.

Nieuwdorp, M., T. W. van Haeften, M. C. L. G. Gouverneur, H. L. Mooij, M. H. P. van Lieshout, M. Levi, et al. 2006b. "Loss of endothelial glycocalyx during acute hyperglycemia coincides with endothelial dysfunction and coagulation activation in vivo." *Diabetes* **55**: 480–6. PMID:16443784.

Nijenhuis, N., D. Mizuno, J. A. E. Spaan, and C. F. Schmidt. 2009. "Viscoelastic response of a model endothelial glycocalyx." *Physical Biology* **6**: 025014. doi:10.1088/1478-3975/6/2/025014.

Nijenhuis, N., D. Mizuno, J. A. E. Spaan, and C. F. Schmidt. 2012. "High-resolution microrheology in the pericellular matrix of prostate cancer cells." *Journal of the Royal Society Interface* **9**(73): 1733–44. doi:10.1098/rsif.2011.0825.

O'Callaghan, R., K. M. Job, R. O. Dull, and V. Hlady. 2011. "Stiffness and heterogeneity of the pulmonary endothelial glycocalyx measured by atomic force microscopy." *American Journal of Physiology – Lung Cellular and Molecular Physiology* **301**(3): L353–60. doi:10.1152/ajplung.00342.2010.

Oberleithner, H., W. Peters, K. Kusche-Vihrog, S. Korte, H. Schillers, K. Kliche, and K. Oberleithner. 2011. "Salt overload damages the glycocalyx sodium barrier of vascular endothelium." *Pflügers Archiv: European Journal of Physiology* **462**: 519–28. doi:10.1007/s00424-011-0999-1.

Pahakis, M. Y., J. R. Kosky, R. O. Dull, and J. M. Tarbell. 2007. "The role of endothelial glycocalyx components in mechanotransduction of fluid shear stress." *Biochemical and Biophysical Research Communications* **355**: 228–33. doi:10.1016/j.bbrc.2007.01.137.

Parsons, C. L. 2007. "The role of the urinary epithelium in the pathogenesis of interstitial cystitis/prostatitis/urethritis." *Urology* **69**(4 Suppl.): 9–16. doi:10.1016/j.urology.2006.03.084.

Paszek, M. J., C. C. DuFort, O. Rossier, R. Bainer, J. K. Mouw, K. Godula, et al. 2014. "The cancer glycocalyx mechanically primes integrin-mediated growth and survival." *Nature* **511**(7509): 319–25. doi:10.1038/nature13535.

Peiffer, V., S. J. Sherwin, and P. D. Weinberg. 2013. "Does low and oscillatory wall shear stress correlate spatially with early atherosclerosis? A systematic review." *Cardiovascular Research* **99**(2): 242–50. doi:10.1093/cvr/cvt044.

Raman, K. and B. Kuberan. 2010. "Chemical tumor biology of heparan sulfate proteoglycans." *Current Chemical Biology* **4**(1): 20–31. doi:10.2174/187231310790226206.Chemical.

Rehm, M., D. Bruegger, F. Christ, P. Conzen, M. Thiel, M. Jacob, et al. 2007. "Shedding of the endothelial glycocalyx in patients undergoing major vascular surgery with global and regional ischemia." *Circulation* **116**(17): 1896–906. doi:10.1161/CIRCULATIONAHA.106.684852.

Reilly, G. C., T. R. Haut, C. E Yellowley, H. J. Donahue, and C. R. Jacobs. 2003. "Fluid flow induced PGE 2 release by bone cells is reduced by glycocalyx degradation whereas calcium signals are not." *Biorheology* **40**: 591–603. PMID:14610310.

Reitsma, S., D. W. Slaaf, H. Vink, M. A. M. J. Van Zandvoort, and M. G. A. Oude Egbrink. 2007. "The endothelial glycocalyx: composition, functions, and visualization." *Pflügers Archiv: European Journal of Physiology* **454**(3): 345–59. doi:10.1007/s00424-007-0212-8.

Rubio-Gayosso, I., S. H. Platts, and B. R. Duling. 2006. "Reactive oxygen species mediate modification of glycocalyx during ischemia-reperfusion injury." *American Journal of Physiology – Heart and Circulatory Physiology* **290**: H2247–56. doi:10.1152/ajpheart.00796.2005.

Rutishauser, U., A. Acheson, A. K. Hall, D. M. Mann, and J. Sunshine. 1988. "The neural cell adhesion molecule (NCAM) as a regulator of cell-cell interactions." *Science* **240**(4848): 53–7. PMID:3281256.

Sabri, S., M. Soler, C. Foa, A. Pierres, A. Benoliel, and P. Bongrand. 2000. "Glycocalyx modulation is a physiological means of regulating cell adhesion." *Journal of Cell Science* **113**(Pt. 9): 1589–600. PMID:10751150.

Sarrazin, S., W. C. Lamanna, and J. D. Esko. 2011. "Heparan sulfate proteoglycans." *Cold Spring Harbor Perspectives in Biology* **3**(7): 1–33. doi:10.1101/cshperspect.a004952.

Schnaar, R. L. 2015. "Glycans and glycan-binding proteins in immune regulation: a concise introduction to glycobiology for the allergist." *Journal of Allergy and Clinical Immunology* **135** (3): 609–15. doi:10.1016/j.jaci.2014.10.057.

Scott, J. E., and F. Heatley. 1999. "Hyaluronan forms specific stable tertiary structures in aqueous solution: a 13C NMR study." *Proceedings of the National Academy of Sciences of the United States of America* **96**(9): 4850–55. doi:10.1073/pnas.96.9.4850.

Singleton, P. A., S. M. Dudek, S. Fan Ma, and J. G. N. Garcia. 2006. "Transactivation of sphingosine 1-phosphate receptors is essential for vascular barrier regulation: novel role for hyaluronan and CD44 receptor family." *Journal of Biological Chemistry* **281**(45): 34 381–93. doi:10.1074/jbc.M603680200.

Sokolov, I., M. E. Dokukin, and N. V. Guz. 2013. "Method for quantitative measurements of the elastic modulus of biological cells in AFM indentation experiments." *Methods (San Diego, Calif.)* **60**(2): 202–13. doi:10.1016/j.ymeth.2013.03.037.

Soler, M., C. Merant, C. Servant, M. Fraterno, C. Allasia, J. C. Lissitzky, et al. 1997. "Leukosialin (CD43) behavior during adhesion of human monocytic THP-1 cells to red blood cells." *Journal of Leukocyte Biology* **61**(5): 609–18. PMID:9129210.

Soler, M., S. Desplat-Jego, B. Vacher, L. Ponsonnet, M. Fraterno, P. Bongrand, et al. 1998. "Adhesion-related glycocalyx study: quantitative approach with imaging-spectrum in the

energy filtering transmission electron microscope (EFTEM)." *FEBS Letters* **429**(1): 89–94. doi:10.1016/S0014-5793(98)00570-5.
Squire, J. M., M. Chew, G. Nneji, C. Neal, J. Barry, and C. Michel. 2001. "Quasi-periodic substructure in the microvessel endothelial glycocalyx: a possible explanation for molecular filtering?" *Journal of Structural Biology* **136**(3): 239–55. doi:10.1006/jsbi.2002.4441.
Stickens, D., B. M. Zak, N. Rougier, J. D. Esko, and Z. Werb. 2005. "Mice deficient in Ext2 lack heparan sulfate and develop exostoses." *Development* **132**(22): 5055–68. doi:10.1242/dev.02088.Mice.
Takagi, M., M. Maeno, A. Kagami, Y. Takahashi, and K. Otsuka. 1991. "Biochemical and immunocytochemical characterization of mineral binding proteoglycans in rat bone." *Journal of Histochemistry and Cytochemistry: Official Journal of the Histochemistry Society* **39**(1): 41–50. doi:10.1177/39.1.1898498.
Tarbell, J. M. 2010. "Shear stress and the endothelial transport barrier." *Cardiovascular Research* **87**: 320–30. doi:10.1093/cvr/cvq146.
Tarbell, J. M. and M. Y. Pahakis. 2006. "Mechanotransduction and the glycocalyx." *Journal of Internal Medicine* **259**(4): 339–50. doi:10.1111/j.1365-2796.2006.01620.x.
Tenorio, J., P. Arias, V. Martínez-Glez, F. Santos, S. García-Miñaur, J. Nevado, and P. Lapunzina. 2014. "Simpson-Golabi-Behmel syndrome types I and II." *Orphanet Journal of Rare Diseases* **9**(1): 138. doi:10.1186/s13023-014-0138-0.
Thompson, W. R., S. Modla, B. J. Grindel, K. J. Czymmek, C. B. Kirn-Safran, L. Wang, et al. 2011. "Perlecan/Hspg2 deficiency alters the pericellular space of the lacunocanalicular system surrounding osteocytic processes in cortical bone." *Journal of Bone and Mineral Research* **26**(3): 618–29. doi:10.1002/jbmr.236.
Thorne, R. F., J. W. Legg, and C. M. Isacke. 2004. "The role of the CD44 transmembrane and cytoplasmic domains in co-ordinating adhesive and signalling events." *Journal of Cell Science* **117**(Pt. 3): 373–80. doi:10.1242/jcs.00954.
Toole, B. P. 2004. "Hyaluronan: from extracellular glue to pericellular cue." *Nature Reviews. Cancer* **4**(7): 528–39. doi:10.1038/nrc1391.
Toole, B. P., T. N. Wight, and M. I. Tammi. 2002. "Hyaluronan-cell interactions in cancer and vascular disease." *Journal of Biological Chemistry* **277**(7): 4593–6. doi:10.1074/jbc.R100039200.
Van den Berg, B., and H. Vink. 2006. "Glycocalyx perturbation: cause or consequence of damage to the vasculature?" *American Journal of Physiology – Heart and Circulatory Physiology* **290**: H2174–5. doi:10.1152/ajpheart.00197.2006.
Van den Berg, B. M., J. A. E. Spaan, T. M. Rolf, and H. Vink. 2006. "Atherogenic region and diet diminish glycocalyx dimension and increase intima-to-media ratios at murine carotid artery bifurcation." *American Journal of Physiology – Heart and Circulatory Physiology* **290**(2): H915–20. doi:10.1152/ajpheart.00051.2005.
Varki, A., R.D. Cummings, J.D. Esko, H.H. Freeze, P. Stanley, C.R. Bertozzi, et al. (eds.). 2009. *Essentials of Glycobiology*, 2nd edn. Cold Spring Harbor, NY: Cold Spring Harbor Laboratory Press.
Vink, H. and B. R. Duling. 2000. "Capillary endothelial surface layer selectively reduces plasma solute distribution volume." *American Journal of Physiology – Heart and Circulatory Physiology* **278**: H285–9. doi:10.1007/s003390201284.
Vink, H., A. A. Constantinescu, and J. A. E. Spaan. 2000. "Oxidized lipoproteins degrade the endothelial surface layer implications for platelet-endothelial cell adhesion." *Circulation* **101**: 1500–2. doi:10.1161/01.CIR.0000061910.39145.F0.

Viviano, B. L., L. Silverstein, C. Pflederer, S. Paine-Saunders, K. Mills, and S. Saunders. 2005. "Altered hematopoiesis in glypican-3-deficient mice results in decreased osteoclast differentiation and a delay in endochondral ossification." *Developmental Biology* **282**(1): 152–62. doi:10.1016/j.ydbio.2005.03.003.

Wang, B., X. Lai, C. Price, W. R. Thompson, W. Li, T. R. Quabili, et al. 2014. "Perlecan-containing pericellular matrix regulates solute transport and mechanosensing within the osteocyte lacunar-canalicular system." *Journal of Bone and Mineral Research* **29**(4): 878–91. doi:10.1002/jbmr.2105.

Weinbaum, S., S. C. Cowin, and Y. Zeng. 1994. "A model for the excitation of osteocytes by mechanical loading-induced bone fluid shear stresses." *Journal of Biomechanics* **27**(3): 339–60. PMID:8051194.

Weinbaum, S., X. Zhang, Y. Han, H. Vink, and S. C. Cowin. 2003. "Mechanotransduction and flow across the endothelial glycocalyx." *Proceedings of the National Academy of Sciences of the United States of America* **100**(13): 7988–95. PMID:12810946.

Weinbaum, S., J. M. Tarbell, and E. R. Damiano. 2007. "The structure and function of the endothelial glycocalyx layer." *Annual Review of Biomedical Engineering* **9**: 121–67. doi:10.1146/annurev.bioeng.9.060906.151959.

Wheeler-Jones, C. P. D., Charlotte E. F., and A. A. Pitsillides. 2010. "Targeting hyaluronan of the endothelial glycocalyx for therapeutic intervention." *Current Opinion in Investigational Drugs* **11**(9): 997–1006. PMID:20730694.

Whitelock, J. and J. Melrose. 2011. "Heparan sulfate proteoglycans in healthy and diseased systems." *Wiley Interdisciplinary Reviews: Systems Biology and Medicine* **3**(6): 739–51. doi:10.1002/wsbm.149.

Wiesinger, A., W. Peters, D. Chappell, D. Kentrup, S. Reuter, H. Pavenstädt, et al. 2013. "Nanomechanics of the endothelial glycocalyx in experimental sepsis." *PloS One* **8**(11): e80905. doi:10.1371/journal.pone.0080905.

Wilusz, R. E., J. Sanchez-Adams, and F. Guilak. 2014. "The structure and function of the pericellular matrix of articular cartilage." *Matrix Biology* **39**: 25–32. doi:10.1016/j.matbio.2014.08.009.

You, L., S. C. Cowin, M. B. Schaffler, and S. Weinbaum. 2001. "A model for strain amplification in the actin cytoskeleton of osteocytes due to fluid drag on pericellular matrix." *Journal of Biomechanics* **34**: 1375–86. doi:10.1016/S0021-9290(01)00107-5.

Zak, B. M., M. Schuksz, E. Koyama, C. Mundy, D. E. Wells, Y. Yamaguchi, et al. 2011. "Compound heterozygous loss of Ext1 and Ext2 is sufficient for formation of multiple exostoses in mouse ribs and long bones." *Bone* **48**(5): 979–87. doi:10.1016/j.bone.2011.02.001.

Zhu, X., X. Deng, G. Huang, J. Wang, J. Yang, S. Chen, et al. 2014. "A novel mutation of hyaluronan synthase 2 gene in chinese children with ventricular septal defect." *PloS One* **9**(2): 1–5. doi:10.1371/journal.pone.0087437.

Zimmerman, E., B. Geiger, and L. Addadi. 2002. "Initial stages of cell-matrix adhesion can be mediated and modulated by cell-surface hyaluronan." *Biophysical Journal* **82**: 1848–57. doi:10.1016/S0006-3495(02)75535-5.

4

The Role of the Primary Cilium in Cellular Mechanotransduction

An Emerging Therapeutic Target

Kian F. Eichholz and David A. Hoey

Trinity Centre for Bioengineering, Trinity Biomedical Sciences Institute and Department of Mechanical and Manufacturing Engineering, School of Engineering, Trinity College Dublin, Dublin, Ireland
Department of Mechanical, Aeronautical and Biomedical Engineering, Centre for Applied Biomedical Engineering Research, Materials and Surface Science Institute, University of Limerick, Limerick, Ireland

4.1 Introduction

Our bodies are constantly exposed to physical forces that contribute significantly to tissue development and physiology, and unfortunately in certain circumstances to disease. The way in which the cells within our tissues sense these physical forces and transduce them into a biochemical cellular response is known as "mechanotransduction," and is a poorly understood phenomenon. Defective or abnormal mechanotransduction may result in a vast range of pathologies, including diseases in cardiology, neurology, nephrology, oncology, and orthopedics, among many other areas (Ingber 2003). There are several ways in which mechanotransduction may be abnormal, including alterations in cell mechanical properties, the structure of the extracellular matrix (ECM), and the molecular mechanisms used in mechanotransduction. Understanding the fundamental mechanisms that mediate this mechanotransduction may reveal novel targets for therapeutic development. The concept of mechanotherapy encompasses a range of treatments that may be used to mediate mechanotransduction in order to obtain a desired therapeutic result. It has been formally defined by Huang et al. (2013) as "therapeutic interventions that reduce and reverse injury to damaged tissues or promote the homeostasis of healthy tissues by mechanical means at the molecular, cellular, or tissue level."

The primary cilium is a solitary, immotile, "antenna-like" structure approximately 0.2 μm in diameter and extending several microns from the cell surface into the extracellular environment (Figure 4.1). It forms a distinct cellular compartment in which receptors, ion channels, and signaling molecules specifically localize, allowing key biochemical and biophysical pathways to be activated, amplified, and regulated (Berbari et al. 2009; Nachury 2014). It is therefore not surprising that defects in ciliary proteins associated with structure and/or signaling within the cilium have been linked to a

Mechanobiology: Exploitation for Medical Benefit, First Edition. Edited by Simon C. F. Rawlinson.

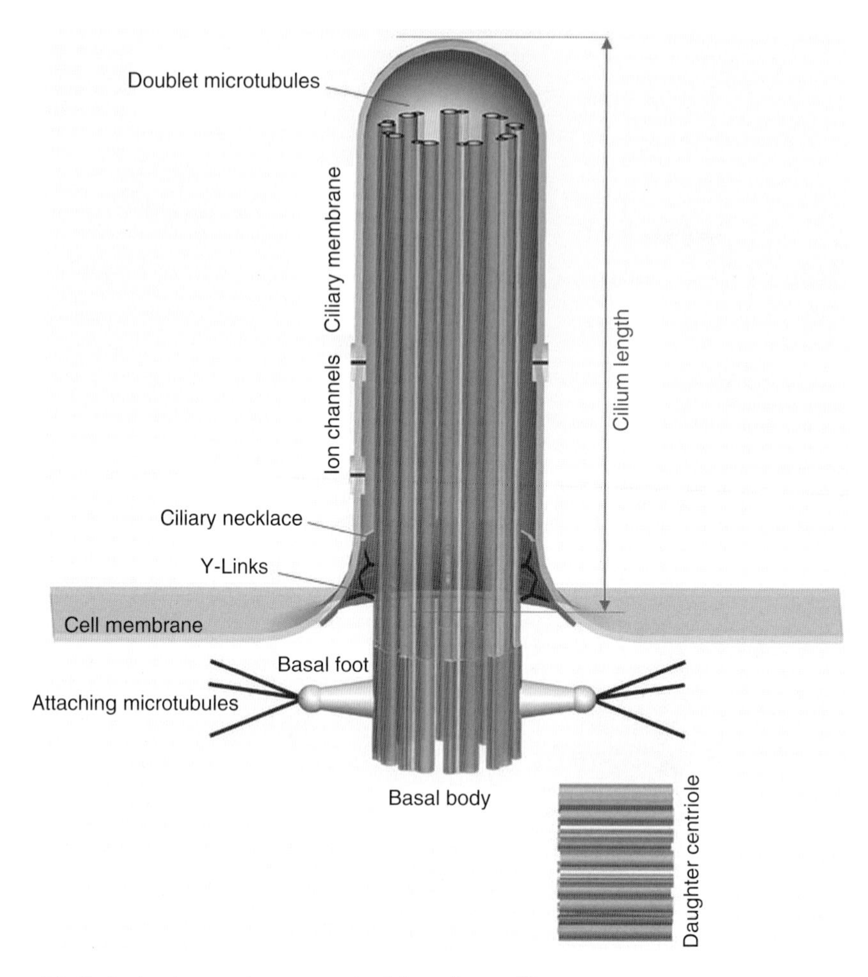

Figure 4.1 Basic structure and components of the primary cilium.

myriad of pathologies, known as "ciliopathies" (Tobin and Beales 2009; Hildebrandt et al. 2011). Interestingly, given the interaction of organelles with the extracellular environment, the primary cilium has been implicated as an important structure in sensing and relaying mechanical signals such as fluid shear, compression, vibration, and pressure in many tissues, as shown in Table 4.1. Therefore, understanding how this organelle contributes to cellular mechanotransduction may provide novel targets for mechanotherapy development.

In this chapter, we introduce the primary cilium and its role in mechanotransduction (Section 4.2.1); discuss the structure of the cilium and how this contributes to cilium mechanics and cell mechanosensitivity (Section 4.2.2 and 4.2.3); explore the molecular mechanisms of cilia-mediated mechanotransduction (Section 4.2.4); and examine potential avenues by which the cilium may be targeted therapeutically to treat disease (Section 4.3).

Table 4.1 Tissues in which primary cilia play a role in mechanotransduction.

Location in body	Cell type	Source
Kidney	Epithelial cell	Praetorius and Spring (2001)
Liver	Cholangiocyte	Masyuk et al. (2006)
Cartilage	Chondrocyte	Wann et al. (2012b)
Endothelium	Endothelial cell	Nauli et al. (2008)
Vasculature	Vascular smooth-muscle cells	Lu et al. (2008)
Bone	Osteocyte	Malone et al. (2007)
Bone marrow	Mesenchymal stem cell (MSC)	Hoey et al. (2012b)

4.2 The Primary Cilium

4.2.1 Primary Cilium Mechanobiology

Early reports of primary cilia-associated mechanobiology were predominately in the kidney. Primary cilia are present on the epithelium, where they extend out from the cell surface into the tubule lumen of the nephron (Yoder 2007). Schwartz et al. (1997) were the first to demonstrate and characterize bending of the kidney epithelial cell primary cilium under physiological flow conditions, and these authors postulated that this may be one mechanism by which kidney cells can sense the surrounding mechanical environment. Further work by Praetorius and Spring (2001) revealed that this flow-induced bending of the kidney primary cilium resulted in an extracellular calcium-dependent increase in intracellular calcium. Interestingly, defects in the cilium or alterations in urine production and flow result in a diminished calcium signal, leading to polycystic kidney disease (PKD), where cell proliferation is unregulated and cyst formation may occur (Singla and Reiter 2006). Other tissues in which primary cilia have demonstrated an important role in flow mechanosensing similar to that seen in the kidney include the liver and the endothelium. In the liver, cholangiocytes (epithelial lining cells of the intrahepatic bile duct) possess a primary cilium that deflects under bile flow; this is required for flow-mediated increases in calcium signaling and decreases in cyclic adenosine monophosphate (cAMP) signaling. Defects in the cilium have been linked to abnormalities in absorption, secretion, cell–matrix interactions, and hyperproliferation of cells (Masyuk et al. 2008), with end consequences leading to cyst formation and hepatic fibrosis. In blood vessels, the endothelium plays a role in controlling and regulating blood flow via the sensing of blood flow characteristics. Endothelial cells are highly responsive to shear stresses, and the primary cilium in particular may play a role in sensitizing endothelial cells to shear stress. For example, endothelial cells with absent cilia demonstrate a reduced responsiveness to fluid forces, as was confirmed by detecting and studying the shear-responsive protein KLF2 (Hierck et al. 2008).

The primary cilium has also demonstrated a role in mechanotransduction where the dominant mechanical stimulus is not fluid flow-induced shear. This is very much evident in the musculoskeletal system. For example, cartilage consists of a densely packed collagenous ECM in which mechanosensitive chondrocytes are embedded. These cells extend a cilium that is directly attached to the matrix via integrins (McGlashan et al. 2006).

Wann et al. (2012a) demonstrated that primary cilia mediate mechanotransduction in chondrocytes under compression, while Irianto et al. (2014) found that depletion of cilia results in thicker articular cartilage with altered mechanical properties, indicating an important function for the cilium in maintaining healthy cartilage. In bone tissue, the osteocyte is believed to be the main mechanosensory cell, with essential functions in controlling bone remodeling. Osteocytes are embedded within the lacuna–canalicular (LC) network of bone and are subjected to strain, pressure, and fluid shear as a result of daily ambulation. Osteocytes possess a primary cilium and mediate mechanically induced increases in osteogenic gene expression (Malone et al. 2007). Furthermore, transgenic mice in which the primary cilia have been deleted in osteoblasts and osteocytes have an inhibited ability to form bone in response to physical loading (Temiyasathit et al. 2012).

As the number of tissues in which the primary cilium is involved in mechanotransduction is increasing, it is becoming very evident that this cellular organelle has an integral role in maintaining tissue homeostasis and that it may be a suitable therapeutic target for when homeostasis is lost, such as in disease.

4.2.2 Primary Cilium Structure

The primary cilium is a complex structure consisting of several components that form a distinct cellular microdomain extending into the extracellular space, capable of sensing and transducing key signals. The structural components are discussed in this section and illustrated in Figure 4.1.

The core structure of the cilium, which protrudes beyond the surface of the cell, consists of an internal framework of nine doublet microtubules arranged circumferentially around the long axis of the organelle. It is called the axoneme. The hollow doublet microtubules extend from the triplet microtubule arrangement of the mother centriole (basal body) within the cell. The manner in which they are positioned circumferentially reveals a larger tubular structure, with a hollow central section within the cilium. This arrangement provides for a more efficient placement of mass, with the tubular characteristics causing the mass to be distributed farther from the central axis, thus increasing the second moment of area and also the ciliary stiffness. Surrounding the axoneme is the ciliary membrane. While this is continuous with the cell membrane, there are differences between the two in terms of membrane morphology and behavior, with the ciliary membrane being shown to have different ion-binding and osmotic properties (Satir and Gilula 1970). The ciliary membrane is decorated with transmembrane molecules such as ion channels and receptors, which play important roles in cilia-mediated mechanotransduction.

Separating the ciliary compartment from the cytoplasm is the transition zone. This is located at the proximal end of the doublet microtubules, and comprises the ciliary necklace (Farbman 1992) and structures known as Y-links, which connect the ciliary necklace to the outer microtubule doublets and the plasma membrane (Szymanska and Johnson 2012). Located between these Y-links are pores, which are believed to regulate active transport to and from the ciliary compartment, in a similar manner to that witnessed with nuclear transport (Kee et al. 2012). There are also several subciliary components within the cytoplasm, including the mother centriole (basal body) and daughter centriole (which make up the centrosome), the basal feet, and attaching microtubules. The mother centriole transitions into the basal body after docking with the cell membrane (Ke and Yang 2014). The basal body is located just below the transition zone, between the axoneme and the cell cytoplasm, and acts as a template for ciliogenesis (Hoey et al. 2012a). It is directly connected to the ciliary membrane via distal appendages called transitional fibers or alar sheets. The basal feet extend radially around the basal body, and provide the cilium with a foundation for secure attachment to the cell cytoskeleton via attaching microtubules.

4.2.3 Primary Cilium Mechanics

The mechanical characteristics and behaviors of the primary cilium are fundamental to the modes by which it carries out sensing and transducing functions, as they dictate the ability of the cilium to extend into the extracellular space and the degree to which the cilium deforms under load. In order to study and characterize the mechanical properties of the cilium, a representative model must first be chosen to fit the bending profile of the organelle under a known applied load. One of the most basic models is Euler–Bernoulli beam theory, which was the first published model of primary cilia mechanics (Schwartz et al. 1997). This model assumes that the cilium behaves as a miniature cantilevered beam protruding from the cell surface. Schwartz used it to accurately predict the bending profile of an epithelial cilium witnessed experimentally under several flow regimens, and as a consequence was able to discern the mechanical properties (flexural rigidity, EI) of the organelle to be approximately $3.1 \times 10^{-23}\ N/m^2$. Several more advanced models have been presented in recent years, incorporating a more complex ciliary structure, rotation at the ciliary base, and fluid-structure interactions (Rydholm et al. 2010; Young et al. 2012; Battle et al. 2014; Downs et al. 2014; Khayyeri et al. 2015; Resnick 2015).

For the purpose of this chapter, Euler–Bernoulli beam theory will be used to model the primary cilium as a cylindrical cantilevered beam (see Box 4.1). This model, along with advanced techniques for experimentally visualizing the deformation of the primary cilium under loading, can be combined to give a simplified view of how the cilium deforms and contributes to cellular mechanotransduction. It must be noted that the model in Box 4.1 is simplified in comparison to that used by Schwartz, and is only valid for small deformations. However, by studying classical beam theory, the importance of engineering mechanics can be appreciated in determining the effect of structural variables in regulating ciliary mechanics. It can be seen that the length of the beam (which in this case refers to the primary cilium) and the flexural rigidity are the most influential parameters in determining the bending profile. Equation 4.1 may be rearranged to show that the deflection of the cilium is proportional to its length to the power of 4, illustrating how cilia length may drastically alter the deformation of the cilium and consequentially the mechanosensitive response of the cell under a consistent loading condition. One mechanism by which the primary cilium may be utilized by the cell as a mechanosensor is via the opening of stretch-activated ion channels along the ciliary axoneme (Hoey et al. 2012a), as illustrated in Figure 4.2. Given this, along with the knowledge of how beam theory can be applied to the primary cilium, it can be seen that a longer cilium will result in a greater drag force, greater deflections, and the potential for a greater degree of ion channel activation. Thus, the cell may use alteration of the length of the primary cilium as a mechanism for regulating the degree of mechanotransduction; there is

Box 4.1 Studying the Mechanics of the Primary Cilium

The mechanical characteristics and behavior of the primary cilium are fundamental to the modes by which it carries out its sensing and transducing functions. Euler–Bernoulli beam theory can be used an introduction to primary cilium bending, but it is only valid for small deflections The deflection (y) at any point (x) along a cilium of length (L), elastic modulus (E), and second moment of area (I) for a uniformly distributed load (F) can be given by the following:

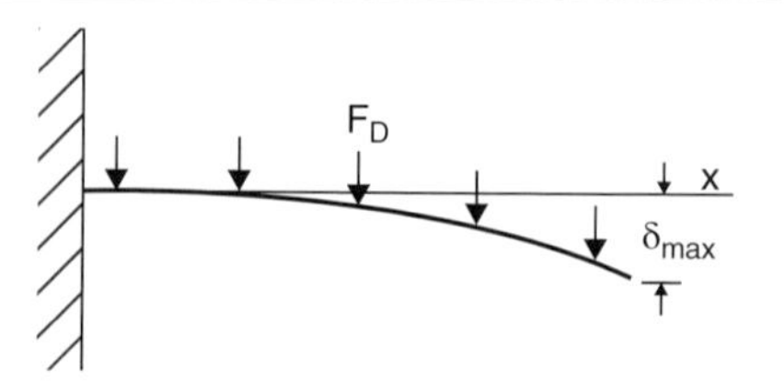

Free body diagram of primary cilium subjected to fluid flow

$$y = \frac{Fx^4}{24EI}\left(x^2 + 6l^2 - 4lx\right) \tag{4.1}$$

The maximum deflection (δ_{max}) at the tip of the cilium can be determined by the following equation, where (L) is substituted for (x). It is seen that deflection is proportional to the length to the power of four, indicating the importance of length in deflection and therefore mechanosensitivity.

$$\delta_{max} = \frac{FL^4}{8EI} \tag{4.2}$$

The force on the cilium due to loading may be approximated by computing the drag force per unit length (F_D) of a fluid of density (ρ) and velocity (v) over an infinity long cylinder while considering the projected area seen by the cilium due to flow, which is unit length multiplied by diameter (d):

$$F_D = \frac{1}{2}\rho v^2 C_D d \tag{4.3}$$

Where the coefficient of drag (C_D) for a given Reynolds number (Re) may be approximated as follows (Tritton 1988):

$$C_D = \frac{8\pi}{Re(2.002 - lnRe)} \tag{4.4}$$

This gives a final expression for the force exerted on the cilium per unit length due to loading-induced fluid flow. This may be multiplied by cilium length to obtain the total drag force:

$$F_D = \frac{4\pi\rho v^2 d}{Re(2.002 - ln Re)} \tag{4.5}$$

The second moment of area of the cilium may be approximated using the formula for a circle, or by considering more representative cases using circles or annuli arranged radially around the centre of the cilium. However, it is common for the elastic modulus and second moment of area to be combined, giving the flexural rigidity (EI):

$$EI = \frac{FL^4}{8\delta_{max}} \quad (4.6)$$

The force can be entered as the drag force due to fluid flow, while the length and maximum deflection can be determined using real-time microscopy. The flexural rigidity of the primary cilium gives an overall indication of its resistance to bending.

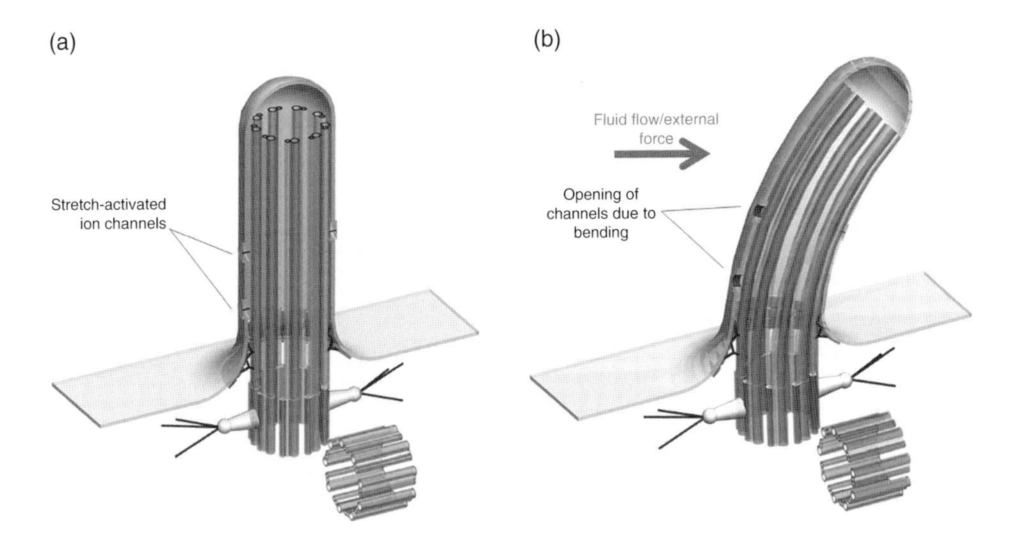

Figure 4.2 Schematic of the primary cilium, illustrating (a) stretch-activated ion channels under static conditions and (b) the opening of these channels in the presence of ciliary bending.

evidence to suggest this, with an alteration in cilium length being seen in the presence of various modes of mechanical stimulation in renal (Resnick and Hopfer 2007), epithelial, mesenchymal (Besschetnova et al. 2010), and tendon (Gardner et al. 2011) cells. Moreover, the material properties of the cilium itself are also highly influential for how it deforms in a loaded environment. As can be intuitively understood from an engineering perspective, stiffness has an inverse relationship to deflection, as can be seen by considering Equation 4.1 again: stiffness is the second variable, along with the second moment of area, and when these are multiplied together to get their product, the value for flexural rigidity is obtained. For a given cilium length, the degree of bending under a known load will depend on the flexural rigidity. Interestingly, as with cilium length, there is evidence to suggest that the primary cilium may alter its stiffness in response to mechanical loading (Hoey et al. 2012a); this would allow for similar regulation of cilium-mediated mechanosensation for different force intensities and exposure times.

4.2.4 Primary Cilium-Mediated Mechanotransduction

Deformation of the primary cilium under physical load is believed to trigger a biochemical response that activates downstream cellular signaling, as illustrated in Figure 4.2 for the case of channel opening due to bending. The molecular mechanism behind this phenomenon is an area of great interest, as like ciliary mechanics, it may be targeted therapeutically to treat disease.

One of the most common cellular biochemical signals elicited through physical perturbation is a flux of intracellular calcium, and interestingly it has been shown that the primary cilium is required for this response in numerous cell types (Praetorius and Spring 2001; Masyuk et al. 2006). In fact, given the development of advanced techniques such as patch clamp methods (Decaen et al. 2013) and cilia-localized fluorescence resonance energy transfer (FRET) sensors (Su et al. 2013; Lee et al. 2015), it has been demonstrated that the cilium acts a distinct calcium signaling compartment, whereby calcium enters the cilium via ion channels along the axoneme (Delling et al. 2013). Numerous mechanosensitive ion channels, such as transient receptor potential vanilloid 4 (TRPV4), polycystin 2 (PC2) and piezo-type mechanosensitive ion channel component 1 (Piezo1), have been shown to localize to the cilium and are required for loading-induced activation of calcium signaling in many tissues, including bone, cartilage, and the kidney (Hoey et al. 2012a). They therefore represent targets that may be activated via agonists to mimic the effect of loading.

Like calcium, cAMP is an important biochemical signal that is manipulated by physical perturbation. The application of fluid shear has been shown to decrease the levels of cAMP in liver and bone cells in a primary cilium-dependent manner (Masyuk et al. 2006; Kwon et al. 2010). Furthermore, this decrease in cAMP is mediated by a cilia-localized enzyme called adenylyl cyclase 6 (AC6). Animals deficient in AC6 demonstrate an inhibited ability to form bone in response to loading, indicating that AC6 is required for mechanotransduction. Given that AC6 is calcium ion-inhibitable, it has been hypothesized that calcium enters the ciliary compartment via channels identified earlier, inhibiting AC6, reducing cAMP levels, and so mediating downstream cellular signaling.

4.3 Cilia-Targeted Therapeutic Strategies

There are several strategies in which the primary cilium is targeted in an effort to overcome the wide range of problems associated with ciliopathies and/or diseases of mechanotransduction. It may be targeted in terms of its structure or mechanics (introduced in Sections 4.2.2 and 4.2.3), which influences its ability to act as a mechanosensitive organelle for the cell, or it may be targeted by therapeutically activating or suppressing the molecular mechanisms of cilia-mediated mechanotransduction (introduced in Section 4.2.4).

4.3.1 Targeting Ciliary Structure and Mechanics

As can be seen from Section 4.2.3, there are several key parameters which may dictate the degree of cilia-mediated mechanotransduction, including cilia length and flexural rigidity. Fundamental to these is the actual existence of a cilium, as several ciliopathies are characterized by a defect in ciliogenesis.

Mutations in at least 89 genes resulting in defects in cilia formation, structure, or function have been linked to 23 recognized human syndromes. Therefore, cilium-focused gene therapy strategies are now being tested for the restoration of gene transcription and cilium formation (McIntyre et al. 2013). Though it does not involve the primary cilia, but instead the closely related motile cilia, primary cilia dyskinesia (PCD) arises due to a defect in ciliary structure that inhibits the motile cilium's ability to beat and clear the airways. This results in pulmonary infections, which eventually destroy the lungs. Recent *ex vivo* cultures of epithelial cells from a PCD patient were given a healthy copy of the gene DNAI1, introduced via lentiviral transduction, which restored gene transcription and ciliary structure and rescued cilia motility (Chhin et al. 2009). To date, cilium-focused gene therapy strategies have been successful in several cell types, including retinal photoreceptors and olfactory sensory neurons (McIntyre et al. 2013), and they hold great promise for future therapeutic development.

As cilia length and flexural rigidity have been linked to ciliary deformation and therefore, potentially, to mechanotransduction, targeting these parameters may be a way of tuning the mechanosensitivity of a cell. A recent study by Spasic and Jacobs (2015) utilized fenoldopam and lithium chloride to increase the length of the primary cilium in osteocytes (bone cells). Interestingly, osteocytes with longer cilia responded to fluid shear with an enhanced osteogenic response, suggesting that these cells may be more mechanoresponsive. Therefore, it may be possible to utilize them in enhancing cellular mechanosensitivity, which could be beneficial in treating diseases of load-bearing tissues, such as osteoporosis (bone-loss disease). Furthermore, cell substrate topography has recently been shown to control cilia length, demonstrating that biophysical cues can also dictate cilia structure and potentially cellular mechanosensitivity (McMurray et al. 2013).

4.3.2 Targeting the Molecular Mechanism of Cilia-Mediated Mechanotransduction

In addition to targeting cilia structure/mechanics as a way of altering cellular mechanosensitivity, we can also target the molecular mechanisms of cilia-mediated mechanotransduction in order to mimic the effect of applied mechanical load and cilia deformation.

Bending of the cilium under load is believed to trigger stretch-activated channels such as TRPV4, PC2, and Piezo1 along the axoneme that initiates a calcium-dependent cell-signaling mechanism. Therefore, activation of said channels through pharmacological means may represent an approach to mimicking physical stimulation of the cilium and activating downstream "mechanoresponsive" signaling independent of cilia deformation. For example, TRPV4 localizes to the primary cilium in several tissues and has been shown to play important roles in cilia-mediated mechanotransduction (Gradilone et al. 2007; Luo et al. 2014; Lee et al. 2015). In particular, a recent study has demonstrated that cilia-localized TRPV4 is required to sense and regulate intraocular pressure, with defects in this mechanotransduction mechanism leading to glaucoma (Luo et al. 2014). Interestingly, a US patent has been issued which describes a technique for targeting TRPV4 utilizing channel-specific agonists in order to activate this mechanism and treat glaucoma (Sun 2015). Similar approaches have been proposed in other tissues, such as bone (Lee et al. 2015), cartilage (O'Conor et al. 2014), and kidney

(Pochynyuk et al. 2013). Key to the development of such therapeutics is a sound knowledge of the molecular mechanisms of cilia-mediated mechanotransduction, which may be cell- and tissue-dependent.

4.4 Conclusion

The primary cilium has emerged in recent years as a nexus of extracellular signal sensing and has firmly established a role for itself in regulating the mechanobiology of several tissues. Deciphering the structural and molecular mechanisms by which this organelle mediates mechanotransduction will yield novel therapeutics that target the cilium in order to treat disease.

Acknowledgements

The authors would like to acknowledge funding from the European Research Council under the European Community's Seventh Framework Programme (FP7/2007-2013) under ERC grant agreement no.336882. We also gratefully acknowledge Science Foundation Ireland for their support (SFI/13/ERC/L2864), and, finally, the Irish Research Council.

References

Battle, C., C. M. Ott, D. T. Burnette, J. Lippincott-Schwartz, and C. F. Schmidt. 2014. "Intracellular and extracellular forces drive primary cilia movement." *Proceedings of the National Academy of Sciences* **122**(5): 1410–15. doi:10.1073/pnas.1421845112.

Berbari, N. F., A. K. O'Connor, C. J. Haycraft, and B. K. Yoder. 2009. "The primary cilium as a complex signaling center." *Current Biology* **19**(13): R526–35. doi:10.1016/j.cub.2009.05.025.

Besschetnova, T. Y., E. Kolpakova-Hart, Y. Guan, J. Zhou, B. R. Olsen, and J. V. Shah. 2010. "Identification of signaling pathways regulating primary cilium length and flow-mediated adaptation." *Current Biology* **20**(2): 182–7. doi:10.1016/j.cub.2009.11.072.

Chhin, B., D. Negre, O. Merrot, J. Pham, Y. Tourneur, D. Ressnikoff, et al. 2009. "Ciliary beating recovery in deficient human airway epithelial cells after lentivirus ex vivo gene therapy." *PLoS Genetics* **5**(3): e1000422. doi:10.1371/journal.pgen.1000422.

Decaen, P. G., M. Delling, T. N. Vien, and D. E. Clapham. 2013. "Direct recording and molecular identification of the calcium channel of primary cilia." *Nature* **504**(7479): 315–18. doi:10.1038/nature12832.

Delling, M., P. G. Decaen, J. F. Doerner, S. Febvay, and D. E. Clapham. 2013. "Primary cilia are specialized calcium signalling organelles." *Nature* **504**(7479): 311–14. doi:10.1038/nature12833.

Downs, M. E., A. M. Nguyen, F. A. Herzog, D. A. Hoey, and C. R. Jacobs. 2014. "An experimental and computational analysis of primary cilia deflection under fluid flow." *Computer Methods in Biomechanics and Biomedical Enginerring* **17**(1): 2–10. doi: 10.1080/10255842.2011.653784.

Farbman, A. I. 1992. *Cell Biology of Olfaction*. Cambridge: Cambridge University Press.
Gardner, K., S. P. Arnoczky, and M. Lavagnino. 2011. "Effect of in vitro stress-deprivation and cyclic loading on the length of tendon cell cilia in situ." *Journal of Orthopaedic Research* **29**(4): 582–7. doi:10.1002/jor.21271.
Gradilone, S. A., A. I. Masyuk, P. L. Splinter, J. M. Banales, B. Q. Huang, P. S. Tietz, et al. 2007. "Cholangiocyte cilia express TRPV4 and detect changes in luminal tonicity inducing bicarbonate secretion." *Proceedings of the National Academy of Sciences of the United States of America* **104**(48): 19 138–43. doi:10.1073/pnas.0705964104.
Hierck, B. P., K. Van Der Heiden, F. E. Alkemade, S. Van De Pas, J. V. Van Thienen, B. C. W. Groenendijk, et al. 2008. "Primary cilia sensitize endothelial cells for fluid shear stress." *Developmental Dynamics* **237**(3): 725–35. doi:10.1002/dvdy.21472.
Hildebrandt, F., T. Benzing, and N. Katsanis. 2011. "Mechanisms of disease: ciliopathies." *New England Journal of Medicine* **364**(16): 1533–43.
Hoey, D. A., M. E. Downs, and C. R. Jacobs. 2012a. "The mechanics of the primary cilium: an intricate structure with complex function." *Journal of Biomechanics* **45**(1): 17–26. doi:10.1016/j.jbiomech.2011.08.008.
Hoey, D. A., S. Tormey, S. Ramcharan, F. J. O'Brien, and C. R. Jacobs. 2012b. "Primary cilia-mediated mechanotransduction in human mesenchymal stem cells." *Stem Cells* **30**(11): 2561–70. doi:10.1002/stem.1235.
Huang, C., J. Holfeld, W. Schaden, D. Orgill, and R. Ogawa. 2013. "Mechanotherapy: revisiting physical therapy and recruiting mechanobiology for a new era in medicine." *Trends in Molecular Medicine* **19**(9): 555–64. doi:10.1016/j.molmed.2013.05.005.
Ingber, D. E. 2003. "Mechanobiology and diseases of mechanotransduction." *Annals of Medicine* **35**(8): 564–77. doi:10.1080/07853890310016333.
Irianto, J., G. Ramaswamy, R. Serra, and M. M. Knight. 2014. "Depletion of chondrocyte primary cilia reduces the compressive modulus of articular cartilage." *Journal of Biomechanics* **47**(2): 579–82. doi:10.1016/j.jbiomech.2013.11.040.
Ke, Y. N. and W. X. Yang. 2014. "Primary cilium: an elaborate structure that blocks cell division?" *Gene* **547**(2): 175–85. doi:10.1016/j.gene.2014.06.050.
Kee, H. L., J. F. Dishinger, T. L. Blasius, C. J. Liu, B. Margolis, and K. J. Verhey. 2012. "A size-exclusion permeability barrier and nucleoporins characterize a ciliary pore complex that regulates transport into cilia." *Nature Cell Biology* **14**(4): 431–7. doi:10.1038/ncb2450.
Khayyeri, H., S. Barreto, and D. Lacroix. 2015. "Primary cilia mechanics affects cell mechanosensation: a computational study." *Journal of Theorical Biology* **379**: 38–46. doi:10.1016/j.jtbi.2015.04.034.
Kwon, R. Y., S. Temiyasathit, P. Tummala, C. C. Quah, and C. R. Jacobs. 2010. "Primary cilium-dependent mechanosensing is mediated by adenylyl cyclase 6 and cyclic AMP in bone cells." *FASEB Journal* **24**(8): 2859–868. doi:10.1096/fj.09-148007.
Lee, K. L., M. D. Guevarra, A. M. Nguyen, M. C. Chua, Y. Wang, and C. R. Jacobs. 2015. "The primary cilium functions as a mechanical and calcium signaling nexus." *Cilia* **4**: 7. doi:10.1186/s13630-015-0016-y.
Lu, C. J., H. Du, J. Wu, D. A. Jansen, K. L. Jordan, N. Xu, et al. 2008. "Non-random distribution and sensory functions of primary cilia in vascular smooth muscle cells." *Kidney and Blood Pressure Research* **31**(3): 171–84. doi:10.1159/000132462.

Luo, N., M. D. Conwell, X. J. Chen, C. I. Kettenhofen, C. J. Westlake, L. B. Cantor, et al. 2014. "Primary cilia signaling mediates intraocular pressure sensation." *Proceedings of the National Academy of Sciences of the United States of America* **111**(35): 12 871–6. doi:10.1073/pnas.1323292111.

Malone, A. M., C. T. Anderson, P. Tummala, R. Y. Kwon, T. R. Johnston, T. Stearns, and C. R. Jacobs. 2007. "Primary cilia mediate mechanosensing in bone cells by a calcium-independent mechanism." *Proceedings of the National Academy of Sciences of the United States of America* **104**(33): 13 325–30. doi:10.1073/pnas.0700636104.

Masyuk, A. I., T. V. Masyuk, P. L. Splinter, B. Q. Huang, A. J. Stroope, and N. F. LaRusso. 2006. "Cholangiocyte cilia detect changes in luminal fluid flow and transmit them into intracellular Ca2+ and cAMP signaling." *Gastroenterology* **131**(3): 911–20. doi:10.1053/j.gastro.2006.07.003.

Masyuk, A. I., T. V. Masyuk, and N. F. LaRusso. 2008. "Cholangiocyte primary cilia in liver health and disease." *Developmental Dynamics* **237**(8): 2007–12. doi:10.1002/dvdy.21530.

McGlashan, S. R., C. G. Jensen, and C. A. Poole. 2006. "Localization of extracellular matrix receptors on the chondrocyte primary cilium." *Journal of Histochemistry & Cytochemistry* **54**(9): 1005–14. doi:10.1369/jhc.5A6866.2006.

McIntyre, J. C., C. L. Williams, and J. R. Martens. 2013. "Smelling the roses and seeing the light: gene therapy for ciliopathies." *Trends in Biotechnology* **31**(6): 355–63. doi:10.1016/j.tibtech.2013.03.005.

McMurray, R. J., A. K. Wann, C. L. Thompson, J. T. Connelly, and M. M. Knight. 2013. "Surface topography regulates wnt signaling through control of primary cilia structure in mesenchymal stem cells." *Scientific Reports* **3**: 3545. doi:10.1038/srep03545.

Nachury, M. V. 2014. "How do cilia organize signalling cascades?" *Philosophical Transactions of the Royal Society of London. Series B, Biological Sciences* **369**(1650). doi:10.1098/rstb.2013.0465.

Nauli, S. M., Y. Kawanabe, J. J. Kaminski, W. J. Pearce, D. E. Ingber, and J. Zhou. 2008. "Endothelial cilia are fluid shear sensors that regulate calcium signaling and nitric oxide production through polycystin-1." *Circulation* **117**(9): 1161–71. doi:10.1161/CIRCULATIONAHA.107.710111.

O'Conor, C. J., H. A. Leddy, H. C. Benefield, W. B. Liedtke, and F. Guilak. 2014. "TRPV4-mediated mechanotransduction regulates the metabolic response of chondrocytes to dynamic loading." *Proceedings of the National Academy of Sciences of the United States of America* **111**(4): 1316–21. doi:10.1073/pnas.1319569111.

Pochynyuk, O., O. Zaika, R. G. O'Neil, and M. Mamenko. 2013. "Novel insights into TRPV4 function in the kidney." *Pflugers Archiv European Journal of Physiology* **465**(2): 177–86. doi:10.1007/s00424-012-1190-z.

Praetorius, H. A. and K. R. Spring. 2001. "Bending the MDCK cell primary cilium increases intracellular calcium." *Journal of Membrane Biology* **184**(1): 71–79. PMID:11687880.

Resnick, A. 2015. "Mechanical properties of a primary cilium as measured by resonant oscillation." *Biophysical Journal* **109**(1): 18–25. doi:10.1016/j.bpj.2015.05.031.

Resnick, A. and U. Hopfer. 2007. "Force-response considerations in ciliary mechanosensation." *Biophysical Journal* **93**(4): 1380–90. doi:10.1529/biophysj.107.105007.

Rydholm, S., G. Zwartz, J. M. Kowalewski, P. Kamali-Zare, T. Frisk, and H. Brismar. 2010. "Mechanical properties of primary cilia regulate the response to fluid flow." *American Journal of Physiology – Renal Physiology* **298**(5): F1096–102. doi:10.1152/ajprenal.00657.2009.

Satir, P. and N. B. Gilula. 1970. "Cell junction in a lamellibranch gill ciliated epithelium – localization of pyroantimonate precipitate." *Journal of Cell Biology* **47**(2): 468–87. doi:10.1083/jcb.47.2.468.

Schwartz, E. A., M. L. Leonard, R. Bizios, and S. S. Bowser. 1997. "Analysis and modeling of the primary cilium bending response to fluid shear." *American Journal of Physiology* **272**(1 Pt. 2): F132–8. PMID:9039059.

Singla, V. and J. F. Reiter. 2006. "The primary cilium as the cell's antenna: signaling at a sensory organelle." *Science* **313**(5787): 629–33. doi:10.1126/science.1124534.

Spasic, M. and C. Jacobs. 2015. "Pharmacologically increasing primary cilia length enhances osteocyte mechanotransduction." Paper read at Orthopaedic Research Society, at Las Vegas, NV.

Su, S., S. C. Phua, R. DeRose, S. Chiba, K. Narita, P. N. Kalugin, et al. 2013. "Genetically encoded calcium indicator illuminates calcium dynamics in primary cilia." *Nature Methods* **10**(11): 1105–7. doi:10.1038/nmeth.2647.

Sun, Y. 2015. Targeting primary cilia to treat glaucoma. Patent application number: 20150231148.

Szymanska, K. and C. A. Johnson. 2012. "The transition zone: an essential functional compartment of cilia." *Cilia* **1**(1): 10. doi:10.1186/2046-2530-1-10.

Temiyasathit, S., W. J. Tang, P. Leucht, C. T. Anderson, S. D. Monica, A. B. Castillo, et al. 2012. "Mechanosensing by the primary cilium: deletion of Kif3A reduces bone formation due to loading." *PLoS One* **7**(3): e33368. doi:10.1371/journal.pone.0033368.

Tobin, J. L. and P. L. Beales. 2009. "The nonmotile ciliopathies." *Genetics in Medicine* **11**(6): 386–402. doi:10.1097/GIM.0b013e3181a02882.

Tritton, D. J. 1988. *Physical Fluid Dynamics*, 2nd edn. Oxford: Oxford University Press.

Wann, A. K. T., N. Zuo, C. J. Haycraft, C. G. Jensen, C. A. Poole, S. R. McGlashan, and M. M. Knight. 2012a. "Primary cilia mediate mechanotransduction through control of ATP-induced Ca2+ signaling in compressed chondrocytes." *FASEB Journal* **26**(4): 1663–71. doi:10.1096/fj.11-193649.

Wann, A. K., N. Zuo, C. J. Haycraft, C. G. Jenson, C. A. Poole, S. R. McGlashan, and M. M. Knight. 2012b. "The primary cilium orchestrates chondrocyte mechanotransduction." *Osteoarthritis and Cartilage* **20**: S241–2. doi:10.1016/j.joca.2012.02.399.

Yoder, B. K. 2007. "Role of primary cilia in the pathogenesis of polycystic kidney disease." *Journal of the American Society of Nephrology* **18**(5): 1381–8. doi:10.1681/ASN.2006111215.

Young, Y. N., M. Downs, and C. R. Jacobs. 2012. "Dynamics of the primary cilium in shear flow." *Biophysical Journal* **103**(4): 629–39. doi:10.1016/j.bpj.2012.07.009.

5

Mechanosensory and Chemosensory Primary Cilia in Ciliopathy and Ciliotherapy

Surya M. Nauli[1], Rinzhin T. Sherpa[1], Caretta J. Reese[1], and Andromeda M. Nauli[2]

[1] *Department of Biomedical & Pharmaceutical Sciences, Chapman University, Irvine, CA, USA*
[2] *Department of Pharmaceutical Sciences, Marshall B. Ketchum University, Fullerton, CA, USA*

5.1 Introduction

Over the past few decades, we have been able to integrate mechanobiology into the molecular basis of disease. Mechanical forces in the cellular microenvironment have been recognized as critical regulators of molecular responses, biochemical reactions, gene expression, and tissue development. Recent insights into cellular mechanotransduction point to the primary cilium as an important cellular organelle responsible for sensing mechanical fluid-shear stress and eliciting downstream effects.

As mechanosensory organelles, primary cilia are usually classified as a “9 + 0” type, based on their structural microtubule arrangement (Figure 5.1). Based on the motility characteristic, primary cilia are generally classified as nonmotile organelles. Like any other organelle within a cell, cilia have many important and specialized cellular functions. Classification of cilia can thus provide a broad spectrum of understanding of their role in cellular function.

Structural or functional abnormalities of primary cilia can result in a spectrum of clinical diseases that are associated with gross anatomical changes in tissue and/or organ structure (Hildebrandt et al. 2011; Waters and Beales 2011). It is therefore not hard to understand that a single cell is required to sense and respond to mechanical signals mediated from the extracellular microenvironment and to translate those stimuli into intracellular signaling events. Furthermore, this mechanotransduction process is crucial for maintaining a healthy cellular structure and function, and cannot be disrupted. With this understanding, medical intervention focused on restoring proper ciliary function stands out as a promising approach to remedy mechano-associated diseases.

Mechanobiology: Exploitation for Medical Benefit, First Edition. Edited by Simon C. F. Rawlinson.

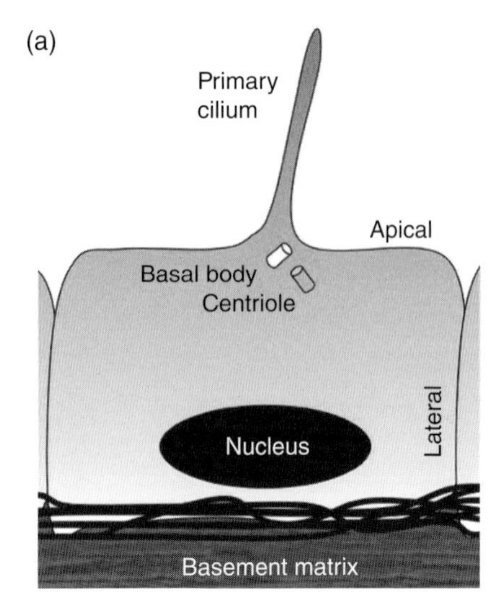

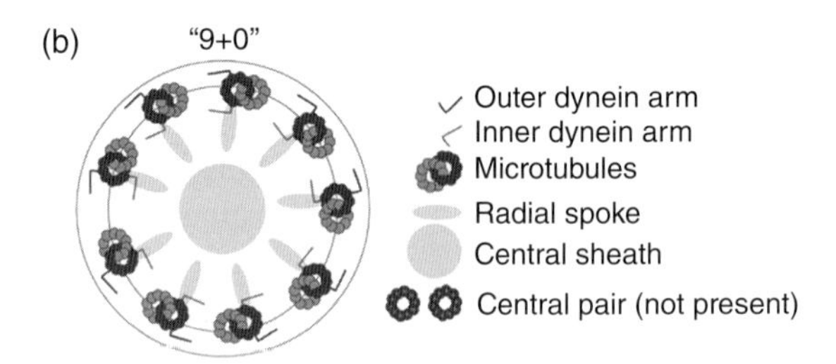

Figure 5.1 Primary cilium as a sensory organelle. (a) Side view of the primary cilium, which acts as a cellular organelle. A primary cilium is projected at the apical membrane of many cell types. The cilium is extended from a mother centriole, also known as a basal body. (b) Based on the central pair of microtubules in the axoneme seen from the cross-section, a primary cilium is generally categorized into a "9 + 0" structure. It was once thought that a cilium with "9 + 0" axoneme was always immotile, but some are now known to be motile, making classification more complex.

5.2 Mechanobiology and Diseases

Though genetic mutations or abnormal functions of various proteins can be traced back to mechanotransduction signaling, the etiology of mechanopathophysiology has been difficult to analyze, primarily due to two main challenges. The first, described by Nauli et al. (2011), illustrates that mechanical forces are very difficult to analyze and differentiate in a complex physiological system *in vivo*. Though different mechanical forces are known to be distinct from one another in cell culture or *ex vivo* studies (Nauli et al. 2013; Prasad et al. 2014a), the biophysical properties that these forces possess are extremely complex and can alter one another's properties *in vivo*. For example, there are at least five different mechanical forces that a blood vessel can encounter (Table 5.1): if pressure (force) is terminated through occlusion in one segment of an artery, the other four forces in the same artery will be altered. In other words, if pressure were prevented from occurring in an artery, the artery would not have the ability to experience stretch, strain, compression, or shear stress forces as a consequence.

Table 5.1 Mechanical forces within a blood vessel. *Source:* http://creativecommons.org/licenses/by/3.0/http://www.hindawi.com/journals/ijvm/2011/376281/.

Types of forces	Differentiations of forces
Stretch	Distention force by surrounding muscle
Cyclic strain	Pulsatile force by turbulent flow of blood
Compression	Contractile force by differential pressure in the vessel
Pressure	Systolic force on intima surface by kinetic flow of blood
Shear stress	Drag force along intima surface by kinetic flow of blood

Table 5.2 Mechanotransduction diseases.

Disease	Branch of medicine	Reference
Atherosclerosis	Cardiology	Baeyens et al. (2014)
Ehlers–Danlos syndrome	Dermatology	Ogawa and Hsu (2013)
Heartburn	Gastroenterology	Dusenkova et al. (2014)
Glomerulosclerosis	Nephrology	Wilson and Dryer (2014)
Migrane	Neurology	Strassman and Levy (2006)
Metastasis	Oncology	Polacheck et al. (2014)
Glaucoma	Ophthalmology	Lei et al. (2014)
Rheumatoid arthritis	Orthopedics	Sato et al. (2014)
Congenital deafness	Pediatrics	Zou et al. (2014)
Emphysema	Pulmonary medicine	Suki et al. (2012)
Pre-eclampsia	Reproductive medicine	Kohler et al. (1998)
Urinary incontinence	Urology	Kanasaki et al. (2013)

The second challenge, proposed by Ingber (2003), is that mechanotransduction does not necessarily contain a classic "stimulus–response" coupling. Any external mechanical forces will need to impose on the pre-existing force balance. In other words, the pre-existing force, coupled with additional force stimuli applied to the system, governs the overall cellular response. Thus, the existing forces have already complicated our studies on mechanotransduction. If these forces are not assessed, their impact on the surrounding microcellular environment will be overlooked.

Nonetheless, abnormalities in the mechanotransduction cascade have long been implicated in clinical diseases (Ingber 2003). Though diseases associated with mechanotransduction abnormalities involve most branches of medicine (Table 5.2), it is important to note that their mechanism or etiology is difficult to envisage: studying these diseases in an attempt to determine whether they are caused by changes in cell mechanics, alterations in tissue structures, or deregulation of mechanochemical conversions has been extremely challenging.

Primary cilia have been identified as sensory organelles by Nauli and others, making the association between a mechanical property and disease much easier to study. Genetic identification, proteomic discovery, and the localization of many proteins to cilia have demonstrated the mechanosensory fluid role of primary cilia in many vestibular organs. Various organs depend on the mechanosensory characteristics of cilia to sense and transmit extracellular signals into intracellular biochemical responses. Cilia possess the ability to sense a variety of fluid movements in the body, including blood in the vasculature (Nauli et al. 2008; AbouAlaiwi et al. 2009), urine in kidney nephrons (Nauli et al. 2003, 2006), interstitial fluid in the bone matrix (Whitfield 2008), bile in the hepatic biliary system (Masyuk et al. 2006), pancreatic juice in the pancreatic duct (Rydholm et al. 2010), cerebrospinal fluid (CSF) in the neuronal tube (Narita et al. 2010), and fluid pressure in the inner ears (Kim et al. 2003; Lepelletier et al. 2013). The inability to sense fluid shear stress in these vestibular organs can contribute to multiple-organ pathogenesis (e.g., hypertension to hydrocephalus or deafness to cystic organ formation). Consequently, abnormal primary cilia function and/or ciliary proteins are now linked to various developmental disorders, known as "ciliopathies." These include left–right asymmetry defect, nephronophthisis, Bardet–Biedl syndrome (BBS), oral facial syndrome, polycystic kidney disease (PKD), obesity, hypertension, and aneurysm.

5.3 Primary Cilia as Biomechanics

The primary cilium has garnered much interest in the last few years, though it was once thought to be a dormant vestigial organelle with no known function. It is a microtubule-based, antenna-like structure found in a single copy on the apical surface of fully differentiated mammalian cells (Figure 5.1). The diameter of a cilium is approximately 0.25 μm, and its length can vary from 2 to 50 μm.

Mechanosensory studies on primary cilia in different organ systems have confirmed that cilia are responsive to fluid shear stress. Activation of cilia can be accomplished by bending with either suction through a micropipette (Praetorius and Spring 2001), apical fluid perfusion through a change in flow rate (Prasad et al. 2014b), or twisting using magnetic beads (Nauli et al. 2013). Cilia act as microsensory compartments, and their role depends on mechanoproteins such as polycystin-1 (Nauli et al. 2003, 2008), polycystin-2 (AbouAlaiwi et al. 2009), fibrocystin (Wang et al. 2007), or transient receptor potential-4 (Kottgen et al. 2008), as well as many others that have recently been discovered (Table 5.3). Thus, the overall functions of the sensory cilia compartments depend on the proper localization of the functional proteins in the cilium (Figure 5.2).

Cells that no longer possess functional cilia show a loss in response to fluid-flow induced intracellular calcium influx. The primary cilium is able to respond to bending through the use of calcium entry via mechanically sensitive channels (Jin et al. 2014b). The initial calcium influx into the cilia results in the gradual development of calcium-induced calcium-release mechanisms in intracellular stores (Jin et al. 2014a). Large increases in the calcium levels of a cell may activate calcium-sensitive channels or calcium-dependent processes, ranging from cell proliferation to cell death. Without structural cilia, a cellular response to fluid flow could not be detected, though all sensory machineries were still present (Aboualaiwi et al. 2014).

Table 5.3 Ciliary proteins, by subcellular localization.

Ciliary tip	**References**
EB1	Pedersen et al. (2003), Schroder et al. (2007)
Gli	Haycraft et al. (2005), Liem et al. (2009)
KIF7	Endoh-Yamagami et al. (2009), Liem et al. (2009)
Smo	Corbit et al. (2005), Haycraft et al. (2005), Liem et al. (2009)
Sufu	Jia et al. (2009)
Ciliary axoneme	**References**
DNAH11	Bartoloni et al. (2002)
DNAH5	Ibanez-Tallon et al. (2004)
DNAH7	Zhang et al. (2002)
DNAI1	Pennarun et al. (1999)
Dyf-1	Ou et al. (2005a), Dave et al. (2009)
Dyf-3	Murayama et al. (2005), Ou et al. (2005b)
DYNC2H1	May et al. (2005)
DYNC2LI1	Rana et al. (2004)
Hydin	Davy and Robinson (2003), Pazour et al. (2005)
IFT140	Tsujikawa and Malicki (2004)
IFT172	Huangfu et al. (2003), Sun et al. (2004), Pedersen et al. (2005), Gorivodsky et al. (2009), Lunt et al. (2009)
IFT20	Follit et al. (2006), Jonassen et al. (2008)
IFT46	Gouttenoire et al. (2007)
IFT52	Liu et al. (2005), Tsujikawa and Malicki (2004)
IFT57/curly	Krock and Perkins (2008), Lunt et al. (2009)
IFT57/hippi	Houde et al. (2006), Tsujikawa and Malicki (2004)
IFT80	Beales et al. (2007)
IFT81	Sun et al. (2004), Lucker et al. (2005)
IFT88	Murcia et al. (2000), Pazour et al. (2000), Haycraft et al. (2001), Qin et al. (2001), Yoder et al. (2002b)
Kif17	Jenkins et al. (2006)
Kif3A/B	Marszalek et al. (1999, 2000), Takeda et al. (1999), Lin et al. (2003)
MDHC7	Neesen et al. (2001), Vernon et al. (2005)
PACRG	Lorenzetti et al. (2004), Dawe et al. (2005)
PF13	Omran et al. (2008)
PF16	Sapiro et al. (2002), Zhang et al. (2005)
PF2	Rupp and Porter (2003)
PF20	Zhang et al. (2004)
Tektin	Tanaka et al. (2004)

(*Continued*)

Table 5.3 (Continued)

Ciliary base (centrosome)	References
ALMS1	Hearn et al. (2005), Graser et al. (2007), Li et al. (2007), Mikule et al. (2007)
BBS1	Oliveira and Goodell (2003), Davis et al. (2007), Oeffner et al. (2008)
BBS2	Nishimura et al. (2004a), Nachury et al. (2007), Oeffner et al. (2008)
BBS3	Fan et al. (2004a)
BBS4	Kim et al. (2004), Gerdes et al. (2007), Oeffner et al. (2008)
BBS5	Li et al. (2004), Yen et al. (2006)
BBS6	Kim et al. (2005)
BBS7	Oliveira and Goodell (2003), Blacque et al. (2004)
BBS8	Ansley et al. (2003), Blacque et al. (2004)
CC2D2A	Gorden et al. (2008)
Cep164	Graser et al. (2007)
CEP290	Gorden et al. (2008), Kim et al. (2008)
EBI	Askham et al. (2002), Piehl et al. (2004), Schroder et al. (2007)
Fa2p	Mahjoub et al. (2004)
FAPP2	Vieira et al. (2006)
Fin1	Grallert and Hagan (2002)
Fleer	Pathak et al. (2007)
Jouberin	Eley et al. (2008)
MKS-1	Kyttala et al. (2006), Dawe et al. (2007),
MKS-3	Smith et al. (2006), Dawe et al. (2007), Tammachote et al. (2009)
Nek1	Mahjoub et al. (2005), White and Quarmby (2008)
Nek2	Bahe et al. (2005)
Nek7	Yissachar et al. (2006), Kim et al. (2007)
Nek8	Mahjoub et al. (2005), Otto et al. (2008)
NPHP-1	Otto et al. (2003), Winkelbauer et al. (2005)
NPHP-2	Otto et al. (2003)
NPHP-3	Olbrich et al. (2003), Bergmann et al. (2008)
NPHP-4	Mollet et al. (2005), Winkelbauer et al. (2005)
NPHP-5	Otto et al. (2005)
NPHP-6	Sayer et al. (2006)
ODF2	Donkor et al. (2004), Ishikawa et al. (2005)
OFD1	Romio et al. (2004), Ferrante et al. (2006)
p-150	Askham et al. (2002)
PCM-1	Kim et al. (2004, 2008), Graser et al. (2007), Mikule et al. (2007)

Pericentrin	Jurczyk et al. (2004), Graser et al. (2007), Mikule et al. (2007)
POC12/MKS1	Kyttala et al. (2006), Dawe et al. (2007), Weatherbee et al. (2009)
Rab8	Kim et al. (2008)
Rootletin	Yang et al. (2002, 2005), Bahe et al. (2005)
RPGR	Shu et al. (2005)
Seahorse	Morgan et al. (2005), Kishimoto et al. (2008)
UNC	Baker et al. (2004)
Ciliary membrane	**References**
EGFR	Ma et al. (2005)
Fibrocystin	Ward et al. (2003), Wang et al. (2007)
Mchr1	Berbari et al. (2008)
PDGFRα	Schneider et al. (2005)
Polycystin-1	Barr and Sternberg (1999), Yoder et al. (2002a)
Polycystin-2	Barr and Sternberg (1999), Pazour et al. (2002), Yoder et al. (2002a)
Somatostatin-3 receptor	Schulz et al. (2000)
Serotonin-6 receptor	Brailov et al. (2000)
Tie-1,Tie-2 receptors	Teilmann and Christensen (2005)
TRPN1	Kim et al. (2003), Shin et al. (2005)
TRPV4	Qin et al. (2005), Teilmann et al. (2005)
Ciliary soluble compartment	**References**
14-3-3	Fan et al. (2004b)
Adenylyl cyclase	Menco (2005), Bishop et al. (2007)
Arl13b	Cantagrel et al. (2008), Hori et al. (2008)
Arl2l1	Sun et al. (2004)
ATP synthase	Hu and Barr (2005)
β-arrestin-2	Menco (2005)
CaM kinase II	Menco (2005)
CAML	Nagano et al. (2005)
CRB1	Fan et al. (2004b)
CRB3	Fan et al. (2004b, 2007), Omori and Malicki (2006)
Cystin	Hou et al. (2002), Yoder et al. (2002a)
GRK3	Menco (2005)
GSK3β	Etienne-Manneville and Hall (2003), Thoma et al. (2007)
Importin	Fan et al. (2007)
Mek1/2	Schneider et al. (2005)
OSEG family	Avidor-Reiss et al. (2004)
Par3	Fan et al. (2004b), Nishimura et al. (2004b), Sfakianos et al. (2007)

(Continued)

Table 5.3 (Continued)

Par6	Fan et al. (2004b)
Phosphodiesterase	Menco (2005)
PKC	Etienne-Manneville and Hall (2003), Fan et al. (2004b)
pVHL	Okuda et al. (1999), Lolkema et al. (2007), Thoma et al. (2007)
STAT6	Low et al. (2006)
Tubby	Mukhopadhyay et al. (2005), Mak et al. (2006)
TULP2	Stolc et al. (2005)

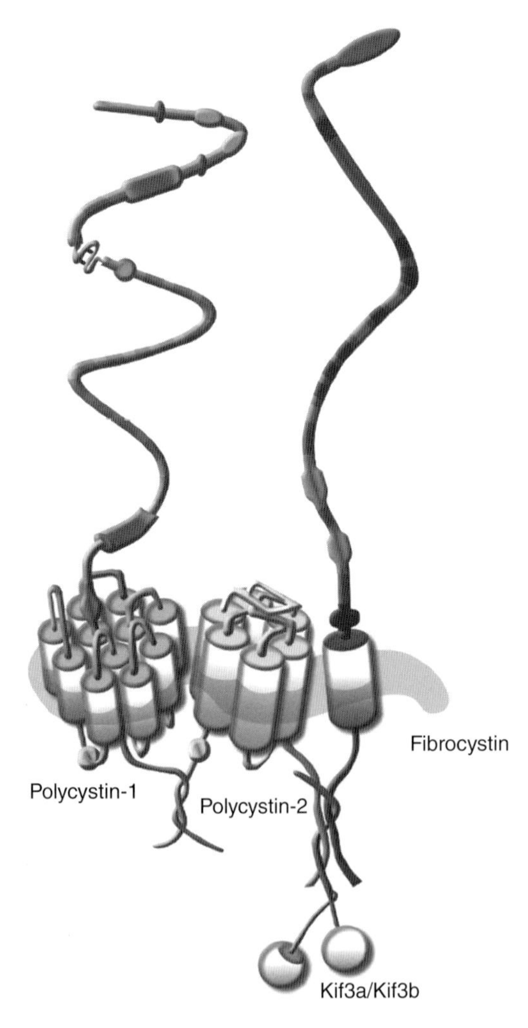

Figure 5.2 Mechanosensory primary cilia are dependent on functional sensory proteins. Polycystin-1, polycystin-2, and fibrocystin form a mechanosensory complex protein in the cilium to sense fluid shear stress. Polycystin-1 and polycystin-2 interact with each other at their COOH termini, forming a polycystin complex. It is predicted that fibrocystin interacts with this complex through polycystin-2, with kif acting as an adaptor protein.

5.4 Modulating Mechanobiology Pathways

Studies of cilia biology have shown that primary cilia act as coordinators in signaling pathways during development and tissue homeostasis. Cilia are composed of receptors, ion channels, and various transporter proteins. This composition enables primary cilia to play a critical role in several transduction pathways, including hedgehog (Hh), Wnt, planar cell polarity, and platelet-derived growth factor (PDGF) (Corbit et al. 2005; Schneider et al. 2005; Zilber et al. 2013; Muntean et al. 2014). Mechanosensory pathways have also been probed, and any abnormalities in these pathways can result in hypertension and/or aneurysm formation (Figure 5.3).

5.4.1 Potential Intervention for Ciliotherapy

Both ciliary length and ciliary function are tightly regulated (Abdul-Majeed et al. 2012). Longer cilia tend to have a greater sensitivity to fluid shear stress (Upadhyay et al. 2014). Activation of the ciliary dopamine receptor will increase cilia length. More specifically,

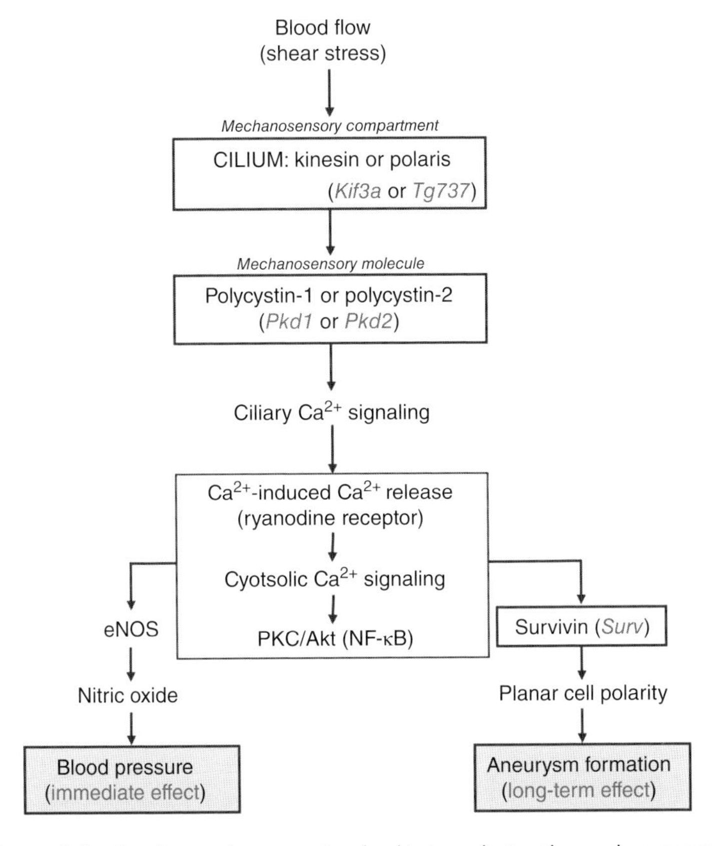

Figure 5.3 Intracellular signaling pathways are involved in transducing the mechanosensory function of primary cilia. Mechanistic divergence pathways initiated from primary cilia are responsible for blood pressure maintenance and aneurysm formation. Abnormal primary cilia induce high blood pressure earlier than aneurysm formation. However, abnormal survivin function is sufficient to form an aneurysm without altering blood pressure.

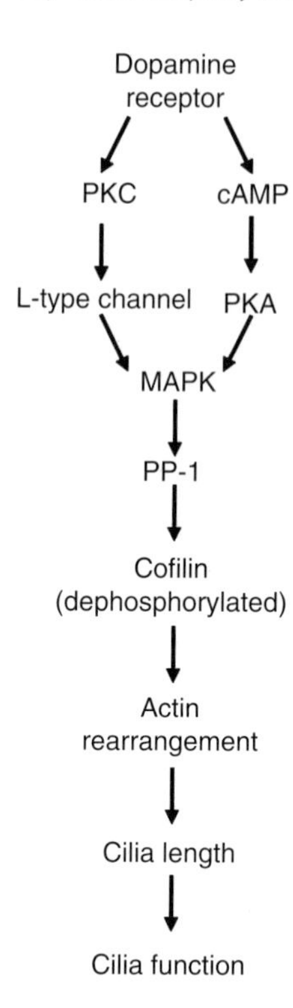

Figure 5.4 Ciliary dopamine receptor can regulate cilia length and function through a complex cellular pathway. Both calcium- and cAMP-dependent protein kinases (PKC and PKA) are involved in regulating cilia length through MAP kinase (MAPK) and protein phosphatase 1 (PP-1). PP-1 plays an important role in actin rearrangement, which is a requirement for cilia length regulation. As cilia length optimally increases, the cilia function will become more sensitive in response to fluid shear stress.

dopamine receptor type 5 (DR5) is localized to primary cilia. As such, DR5-specific agonist is among the few stimuli that require cilia for ciliary and intracellular signal transductions (Abdul-Majeed and Nauli 2011). DR5 activation increases cilia length through cofilin and actin polymerization (Figure 5.4).

The idea that pharmacological DR5 activation could be used as ciliotherapy is evident from *in vitro* studies involving the termination of mechanociliary function through silencing of DR5 expression (Abdul-Majeed and Nauli 2011). DR5 activation also restores cilia function in the mechanoinsensitive cells. Because the chemosensory function of cilia via DR5 can alter the mechanosensory function through changes in sensitivity to fluid shear stress, it has been proposed that DR5 has functional chemo- and mechanosensory roles in primary cilia (Abdul-Majeed and Nauli 2011).

5.4.2 Potential Mechanotherapy

Patients with PKD suffer from uncontrolled hypertension. It has been shown that the vascular endothelia in PKD patients are mechanically compensated with abnormal primary cilia function (AbouAlaiwi et al. 2009). Activating DR5 can be used as a potential mechanotherapy, by altering the mechanosensory function of primary cilia. This type of therapy is also known as ciliotherapy (Kathem et al. 2014). Initial drug screening indicated that activation of ciliary DR5 in addition to the DR5-specific agonist (fenoldopam) increases nitric oxide (NO) biosynthesis in response to fluid shear stress in vascular endothelia. DR5 activation increases cilia length, and also rescues the mechanosensitivity of PKD endothelial cells from fluid shear stress. This, in turn, decreases the overall blood pressure in the PKD mouse model (Kathem et al. 2014). In a clinical study, hypertensive PKD patients had a significantly lower baseline level of NO than did hypertensive-only patients. DR5 activation decreased blood pressure in PKD patients (Kathem et al. 2014).

The baseline level of asymmetric dimethylarginine (ADMA), an endogenous inhibitor of eNOS and a marker for endothelial dysfunction, was significantly higher in the PKD group than in the hypertensive-only group (Kathem et al. 2014). ADMA is a physiological inhibitor of NO biosynthesis and is commonly used as a marker for assessing endothelial function in the clinical setting. Consistent with this idea, plasma ADMA levels are highly correlated with the severity of endothelial dysfunction, and high ADMA levels further impair blood flow and accelerate endothelial dysfunction in PKD patients. Compared to the hypertensive-only patients, the PKD patients had an abnormality in regulating NO biosynthesis. This is consistent with a previous study, which indicated that the vascular-lining endothelia of patients with PKD were dysfunctional due to their nonsensitivity to flow-induced NO biosynthesis (AbouAlaiwi et al. 2009).

Results from a less complex *in vivo* rodent system with endothelial cilia dysfunction also support the idea that a DR5–cilia–NO axis plays an important role in regulating blood pressure in PKD (Kathem et al. 2014). In a more complex clinical setting, dopaminergic receptor activation showed a potential therapeutic benefit on overall arterial blood pressure. Together, these studies serve as a proof of principle for targeted clinical therapy on primary cilia as a novel mechanism for modulating the progression of ciliopathy- and biomechanics-related diseases in general.

Though it was previously proposed that peripheral dopaminergic activation increases renal blood flow (Olsen 1998), we postulate that this vasodilation effect of dopamine on renal arteries acts by sensitizing primary cilia function. Without a doubt, a specific targeted therapy is more suitable for therapeutic management of different mechanical diseases. Future studies are warranted. Nonetheless, recent clinical studies suggest the possibility of using cilia-targeted therapy in PKD patients and hypertensive patients with mechanical-sensing dysfunction.

5.5 Conclusion

Our knowledge of mechanotransduction has advanced in the past several decades. This includes the recognition of primary cilia, which function as mechanosensory organelles. The importance of sensory cilia in different organ systems has also been confirmed, and many cilia-related diseases are still to be identified. There is no doubt that the biomedical approach to target mechanosensory primary cilia will continue to be debated in the years to come.

References

Abdul-Majeed, S. and S. M. Nauli. 2011. "Dopamine receptor type 5 in the primary cilia has dual chemo- and mechano-sensory roles." *Hypertension* **58**(2): 325–31. doi:10.1161/HYPERTENSIONAHA.111.172080.

Abdul-Majeed, S., B. C. Moloney, and S. M. Nauli. 2012. "Mechanisms regulating cilia growth and cilia function in endothelial cells." *Cellular and Molecular Life Sciences* **69**(1): 165–73. doi:10.1007/s00018-011-0744-0.

Aboualaiwi, W. A., B. S. Muntean, S. Ratnam, B. Joe, L. Liu, R. L. Booth, et al. 2014. "Survivin-induced abnormal ploidy contributes to cystic kidney and aneurysm formation." *Circulation* **129**(6): 660–72. doi:10.1161/CIRCULATIONAHA.113.005746.

AbouAlaiwi, W. A., M. Takahashi, B. R. Mell, T. J. Jones, S. Ratnam, R. J. Kolb, and S. M. Nauli. 2009. "Ciliary polycystin-2 is a mechanosensitive calcium channel involved in nitric oxide signaling cascades." *Circulation Research* **104**(7): 860–9. doi:10.1161/CIRCRESAHA.108.192765.

Ansley, S. J., J. L. Badano, O. E. Blacque, J. Hill, B. E. Hoskins, C. C. Leitch, et al. 2003. "Basal body dysfunction is a likely cause of pleiotropic Bardet-Biedl syndrome." *Nature* **425**(6958): 628–33. PMID:14520415.

Askham, J. M., K. T. Vaughan, H. V. Goodson, and E. E. Morrison. 2002. "Evidence that an interaction between EB1 and p150(Glued) is required for the formation and maintenance of a radial microtubule array anchored at the centrosome." *Molecular Biology of the Cell* **13**(10): 3627–45. PMID:12388762.

Avidor-Reiss, T., A. M. Maer, E. Koundakjian, A. Polyanovsky, T. Keil, S. Subramaniam, and C. S. Zuker. 2004. "Decoding cilia function: defining specialized genes required for compartmentalized cilia biogenesis." *Cell* **117**(4): 527–39. PMID:15137945.

Baeyens, N., M. J. Mulligan-Kehoe, F. Corti, D. D. Simon, T. D. Ross, J. M. Rhodes, et al. 2014. "Syndecan 4 is required for endothelial alignment in flow and atheroprotective signaling." *Proceedings of the National Academy of Sciences of the United States of America* **111**(48): 17 308–13. doi:10.1073/pnas.1413725111.

Bahe, S., Y. D. Stierhof, C. J. Wilkinson, F. Leiss, and E. A. Nigg. 2005. "Rootletin forms centriole-associated filaments and functions in centrosome cohesion." *Journal of Cell Biology* **171**(1): 27–33. PMID:16203858.

Baker, J. D., S. Adhikarakunnathu, and M. J. Kernan. 2004. "Mechanosensory-defective, male-sterile unc mutants identify a novel basal body protein required for ciliogenesis in Drosophila." *Development* **131**(14): 3411–22. PMID:15226257.

Barr, M. M. and P. W. Sternberg. 1999. "A polycystic kidney-disease gene homologue required for male mating behaviour in *C. elegans*." *Nature* **401**(6751): 386–9. PMID:10517638.

Bartoloni, L., J. L. Blouin, Y. Pan, C. Gehrig, A. K. Maiti, N. Scamuffa, et al. 2002. "Mutations in the DNAH11 (axonemal heavy chain dynein type 11) gene cause one form of situs inversus totalis and most likely primary ciliary dyskinesia." *Proceedings of the National Academy of Sciences of the United States of America* **99**(16): 10 282–6. PMID:12142464.

Beales, P. L., E. Bland, J. L. Tobin, C. Bacchelli, B. Tuysuz, J. Hill, et al. 2007. "IFT80, which encodes a conserved intraflagellar transport protein, is mutated in Jeune asphyxiating thoracic dystrophy." *Nature Genetics* **39**(6): 727–9. PMID:17468754.

Berbari, N. F., J. S. Lewis, G. A. Bishop, C. C. Askwith, and K. Mykytyn. 2008. "Bardet-Biedl syndrome proteins are required for the localization of G protein-coupled receptors to

primary cilia." *Proceedings of the National Academy of Sciences of the United States of America* **105**(11): 4242–6. doi:10.1073/pnas.0711027105.
Bergmann, C., M. Fliegauf, N. O. Bruchle, V. Frank, H. Olbrich, J. Kirschner, et al. 2008. "Loss of nephrocystin-3 function can cause embryonic lethality, Meckel-Gruber-like syndrome, situs inversus, and renal-hepatic-pancreatic dysplasia." *American Journal of Human Genetics* **82**(4): 959–70. doi:10.1016/j.ajhg.2008.02.017.
Bishop, G. A., N. F. Berbari, J. Lewis, and K. Mykytyn. 2007. "Type III adenylyl cyclase localizes to primary cilia throughout the adult mouse brain." *Journal of Comparative Neurology* **505**(5): 562–71. PMID:17924533.
Blacque, O. E., M. J. Reardon, C. Li, J. McCarthy, M. R. Mahjoub, S. J. Ansley, et al. 2004. "Loss of *C. elegans* BBS-7 and BBS-8 protein function results in cilia defects and compromised intraflagellar transport." *Genes & Development* **18**(13): 1630–42. PMID:15231740.
Brailov, I., M. Bancila, M. J. Brisorgueil, M. C. Miquel, M. Hamon, and D. Verge. 2000. "Localization of 5-HT(6) receptors at the plasma membrane of neuronal cilia in the rat brain." *Brain Research* **872**(1–2): 271–5. PMID:10924708.
Cantagrel, V., J. L. Silhavy, S. L. Bielas, D. Swistun, S. E. Marsh, J. Y. Bertrand, et al. 2008. "Mutations in the cilia gene ARL13B lead to the classical form of Joubert syndrome." *American Journal of Human Genetics* **83**(2): 170–9. doi:10.1016/j.ajhg.2008.06.023.
Corbit, K. C., P. Aanstad, V. Singla, A. R. Norman, D. Y. Stainier, and J. F. Reiter. 2005. "Vertebrate Smoothened functions at the primary cilium." *Nature* **437**(7061): 1018–21. doi:10.1038/nature04117.
Dave, D., D. Wloga, N. Sharma, and J. Gaertig. 2009. "DYF-1 is required for assembly of the axoneme in Tetrahymena thermophila." *Eukaryotic Cell* **8**(9): 1397–406. doi:10.1128/EC.00378-08.
Davis, R. E., R. E. Swiderski, K. Rahmouni, D. Y. Nishimura, R. F. Mullins, K. Agassandian, et al. 2007. "A knockin mouse model of the Bardet-Biedl syndrome 1 M390R mutation has cilia defects, ventriculomegaly, retinopathy, and obesity." *Proceedings of the National Academy of Sciences of the United States of America* **104**(49): 19 422–7. PMID:18032602.
Davy, B. E. and M. L. Robinson. 2003. "Congenital hydrocephalus in hy3 mice is caused by a frameshift mutation in Hydin, a large novel gene." *Human Molecular Genetics* **12**(10): 1163–70. PMID:12719380.
Dawe, H. R., H. Farr, N. Portman, M. K. Shaw, and K. Gull. 2005. "The Parkin co-regulated gene product, PACRG, is an evolutionarily conserved axonemal protein that functions in outer-doublet microtubule morphogenesis." *Journal of Cell Science* **118**(Pt. 23): 5421–30. PMID:16278296.
Dawe, H. R., U. M. Smith, A. R. Cullinane, D. Gerrelli, P. Cox, J. L. Badano, et al. 2007. "The Meckel-Gruber syndrome proteins MKS1 and meckelin interact and are required for primary cilium formation." *Human Molecular Genetics* **16**(2): 173–86. PMID:17185389.
Donkor, F. F., M. Monnich, E. Czirr, T. Hollemann, and S. Hoyer-Fender. 2004. "Outer dense fibre protein 2 (ODF2) is a self-interacting centrosomal protein with affinity for microtubules." *Journal of Cell Science* **117**(Pt. 20): 4643–51. PMID:15340007.
Dusenkova, S., F. Ru, L. Surdenikova, C. Nassenstein, J. Hatok, R. Dusenka, et al. 2014. "The expression profile of acid-sensing ion channel (ASIC) subunits ASIC1a, ASIC1b, ASIC2a, ASIC2b, and ASIC3 in the esophageal vagal afferent nerve subtypes." *American Journal of Physiology. Gastrointestinal and Liver Physiology* **307**(9): G922–30. doi:10.1152/ajpgi.00129.2014.

Eley, L., C. Gabrielides, M. Adams, C. A. Johnson, F. Hildebrandt, and J. A. Sayer. 2008. "Jouberin localizes to collecting ducts and interacts with nephrocystin-1." *Kidney International* **74**(9): 1139–49. doi:10.1038/ki.2008.377.

Endoh-Yamagami, S., M. Evangelista, D. Wilson, X. Wen, J. W. Theunissen, K. Phamluong, et al. 2009. "The mammalian Cos2 homolog Kif7 plays an essential role in modulating Hh signal transduction during development." *Current Biology* **19**(15): 1320–6. doi:10.1016/j.cub.2009.06.046.

Etienne-Manneville, S. and A. Hall. 2003. "Cdc42 regulates GSK-3beta and adenomatous polyposis coli to control cell polarity." *Nature* **421**(6924): 753–6. PMID:12610628.

Fan, Y., M. A. Esmail, S. J. Ansley, O. E. Blacque, K. Boroevich, A. J. Ross, et al. 2004a. "Mutations in a member of the Ras superfamily of small GTP-binding proteins causes Bardet-Biedl syndrome." *Nature Genetics* **36**(9): 989–93. PMID:15314642.

Fan, S., T. W. Hurd, C. J. Liu, S. W. Straight, T. Weimbs, E. A. Hurd, et al. 2004b. "Polarity proteins control ciliogenesis via kinesin motor interactions." *Current Biology* **14**(16): 1451–61. PMID:15324661.

Fan, S., V. Fogg, Q. Wang, X. W. Chen, C. J. Liu, and B. Margolis. 2007. "A novel Crumbs3 isoform regulates cell division and ciliogenesis via importin beta interactions." *Journal of Cell Biology* **178**(3): 387–98. PMID:17646395.

Ferrante, M. I., A. Zullo, A. Barra, S. Bimonte, N. Messaddeq, M. Studer, et al. 2006. "Oral-facial-digital type I protein is required for primary cilia formation and left-right axis specification." *Nature Genetics* **38**(1): 112–17. PMID:16311594.

Follit, J. A., R. A. Tuft, K. E. Fogarty, and G. J. Pazour. 2006. "The intraflagellar transport protein IFT20 is associated with the Golgi complex and is required for cilia assembly." *Molecular Biology of the Cell* **17**(9): 3781–92. PMID:16775004.

Gerdes, J. M., Y. Liu, N. A. Zaghloul, C. C. Leitch, S. S. Lawson, M. Kato, et al. 2007. "Disruption of the basal body compromises proteasomal function and perturbs intracellular Wnt response." *Nature Genetics* **39**(11): 1350–60. PMID:17906624.

Gorden, N. T., H. H. Arts, M. A. Parisi, K. L. Coene, S. J. Letteboer, S. E. van Beersum, et al. 2008. "CC2D2A is mutated in Joubert syndrome and interacts with the ciliopathy-associated basal body protein CEP290." *American Journal of Human Genetics* **83**(5): 559–71. doi:10.1016/j.ajhg.2008.10.002.

Gorivodsky, M., M. Mukhopadhyay, M. Wilsch-Braeuninger, M. Phillips, A. Teufel, C. Kim, et al. 2009. "Intraflagellar transport protein 172 is essential for primary cilia formation and plays a vital role in patterning the mammalian brain." *Developmental Biology* **325**(1): 24–32. doi:10.1016/j.ydbio.2008.09.019.

Gouttenoire, J., U. Valcourt, C. Bougault, E. Aubert-Foucher, E. Arnaud, L. Giraud, and F. Mallein-Gerin. 2007. "Knockdown of the intraflagellar transport protein IFT46 stimulates selective gene expression in mouse chondrocytes and affects early development in zebrafish." *Journal of Biological Chemistry* **282**(42): 30 960–73. PMID:17720815.

Grallert, A. and I. M. Hagan. 2002. "Schizosaccharomyces pombe NIMA-related kinase, Fin1, regulates spindle formation and an affinity of Polo for the SPB." *EMBO Journal* **21**(12): 3096–107. PMID:12065422.

Graser, S., Y. D. Stierhof, S. B. Lavoie, O. S. Gassner, S. Lamla, M. Le Clech, and E. A. Nigg. 2007. "Cep164, a novel centriole appendage protein required for primary cilium formation." *Journal of Cell Biology* **179**(2): 321–30. PMID:17954613.

Haycraft, C. J., P. Swoboda, P. D. Taulman, J. H. Thomas, and B. K. Yoder. 2001. "The *C. elegans* homolog of the murine cystic kidney disease gene Tg737 functions in a ciliogenic pathway and is disrupted in osm-5 mutant worms." *Development* **128**(9): 1493–505. PMID:11290289.

Haycraft, C. J., B. Banizs, Y. Aydin-Son, Q. Zhang, E. J. Michaud, and B. K. Yoder. 2005. "Gli2 and Gli3 localize to cilia and require the intraflagellar transport protein polaris for processing and function." *PLoS Genetics* **1**(4): e53. PMID:16254602.

Hearn, T., C. Spalluto, V. J. Phillips, G. L. Renforth, N. Copin, N. A. Hanley, and D. I. Wilson. 2005. "Subcellular localization of ALMS1 supports involvement of centrosome and basal body dysfunction in the pathogenesis of obesity, insulin resistance, and type 2 diabetes." *Diabetes* **54**(5): 1581–7. PMID:15855349.

Hildebrandt, F., T. Benzing, and N. Katsanis. 2011. "Ciliopathies." *New England Journal of Medicine* **364**(16): 1533–43. doi:10.1056/NEJMra1010172.

Hori, Y., T. Kobayashi, Y. Kikko, K. Kontani, and T. Katada. 2008. "Domain architecture of the atypical Arf-family GTPase Arl13b involved in cilia formation." *Biochemical and Biophysical Research Communications* **373**(1): 119–24. doi:10.1016/j.bbrc.2008.06.001.

Hou, X., M. Mrug, B. K. Yoder, E. J. Lefkowitz, G. Kremmidiotis, P. D'Eustachio, et al. 2002. "Cystin, a novel cilia-associated protein, is disrupted in the cpk mouse model of polycystic kidney disease." *Journal of Clinical Investigation* **109**(4): 533–40. PMID:11854326.

Houde, C., R. J. Dickinson, V. M. Houtzager, R. Cullum, R. Montpetit, M. Metzler, et al. 2006. "Hippi is essential for node cilia assembly and Sonic hedgehog signaling." *Developmental Biology* **300**(2): 523–33. PMID:17027958.

Hu, J. and M. M. Barr. 2005. "ATP-2 interacts with the PLAT domain of LOV-1 and is involved in Caenorhabditis elegans polycystin signaling." *Molecular Biology of the Cell* **16**(2): 458–69. PMID:15563610.

Huangfu, D., A. Liu, A. S. Rakeman, N. S. Murcia, L. Niswander, and K. V. Anderson. 2003. "Hedgehog signalling in the mouse requires intraflagellar transport proteins." *Nature* **426**(6962): 83–7. PMID:14603322.

Ibanez-Tallon, I., A. Pagenstecher, M. Fliegauf, H. Olbrich, A. Kispert, U. P. Ketelsen, et al. 2004. "Dysfunction of axonemal dynein heavy chain Mdnah5 inhibits ependymal flow and reveals a novel mechanism for hydrocephalus formation." *Human Molecular Genetics* **13**(18): 2133–41. PMID:15269178.

Ingber, D. E. 2003. "Mechanobiology and diseases of mechanotransduction." *Annals of Medicine* **35**(8): 564–77. PMID:14708967.

Ishikawa, H., A. Kubo, S. Tsukita, and S. Tsukita. 2005. "Odf2-deficient mother centrioles lack distal/subdistal appendages and the ability to generate primary cilia." *Nature Cell Biology* **7**(5): 517–24. PMID:15852003.

Jenkins, P. M., T. W. Hurd, L. Zhang, D. P. McEwen, R. L. Brown, B. Margolis, et al. 2006. "Ciliary targeting of olfactory CNG channels requires the CNGB1b subunit and the kinesin-2 motor protein, KIF17." *Current Biology* **16**(12): 1211–16. PMID:16782012.

Jia, J., A. Kolterud, H. Zeng, A. Hoover, S. Teglund, R. Toftgard, and A. Liu. 2009. "Suppressor of Fused inhibits mammalian Hedgehog signaling in the absence of cilia." *Developmental Biology* **330**(2): 452–60. doi:10.1016/j.ydbio.2009.04.009.

Jin, X., A. M. Mohieldin, B. S. Muntean, J. A. Green, J. V. Shah, K. Mykytyn, and S. M. Nauli. 2014a. "Cilioplasm is a cellular compartment for calcium signaling in response to mechanical and chemical stimuli." *Cellular and Molecular Life Sciences* **71**(11): 2165–78. doi:10.1007/s00018-013-1483-1.

Jin, X., B. S. Muntean, M. S. Aal-Aaboda, Q. Duan, J. Zhou, and S. M. Nauli. 2014b. "L-type calcium channel modulates cystic kidney phenotype." *Biochimica et Biophysica Acta* **1842**(9): 1518–26. doi:10.1016/j.bbadis.2014.06.001.

Jonassen, J. A., J. San Agustin, J. A. Follit, and G. J. Pazour. 2008. "Deletion of IFT20 in the mouse kidney causes misorientation of the mitotic spindle and cystic kidney disease." *Journal of Cell Biology* **183**(3): 377–84. doi:10.1083/jcb.200808137.

Jurczyk, A., A. Gromley, S. Redick, J. San Agustin, G. Witman, G. J. Pazour, et al. 2004. "Pericentrin forms a complex with intraflagellar transport proteins and polycystin-2 and is required for primary cilia assembly." *Journal of Cell Biology* **166**(5): 637–43. PMID:15337773.

Kanasaki, K., W. Yu, M. von Bodungen, J. D. Larigakis, M. Kanasaki, F. Ayala de la Pena, R. Kalluri, and W. G. Hill. 2013. "Loss of beta1-integrin from urothelium results in overactive bladder and incontinence in mice: a mechanosensory rather than structural phenotype." *FASEB Journal* **27**(5): 1950–61. doi:10.1096/fj.12-223404.

Kathem, S. H., A. M. Mohieldin, S. Abdul-Majeed, S. H. Ismail, Q. H. Altaei, I. K. Alshimmari, et al. 2014. "Ciliotherapy: a novel intervention in polycystic kidney disease." *Journal of Geriatric Cardiology* **11**(1): 63–73. doi:10.3969/j.issn.1671-5411.2014.01.001.

Kim, J., Y. D. Chung, D. Y. Park, S. Choi, D. W. Shin, H. Soh, et al. 2003. "A TRPV family ion channel required for hearing in Drosophila." *Nature* **424**(6944): 81–4. PMID:12819662.

Kim, J. C., J. L. Badano, S. Sibold, M. A. Esmail, J. Hill, B. E. Hoskins, et al. 2004. "The Bardet-Biedl protein BBS4 targets cargo to the pericentriolar region and is required for microtubule anchoring and cell cycle progression." *Nature Genetics* **36**(5): 462–70. PMID:15107855.

Kim, J. C., Y. Y. Ou, J. L. Badano, M. A. Esmail, C. C. Leitch, E. Fiedrich, et al. 2005. "MKKS/BBS6, a divergent chaperonin-like protein linked to the obesity disorder Bardet-Biedl syndrome, is a novel centrosomal component required for cytokinesis." *Journal of Cell Science* **118**(Pt. 5): 1007–20. PMID:15731008.

Kim, S., K. Lee, and K. Rhee. 2007. "NEK7 is a centrosomal kinase critical for microtubule nucleation." *Biochemical and Biophysical Research Communications* **360**(1): 56–62. PMID:17586473.

Kim, J., S. R. Krishnaswami, and J. G. Gleeson. 2008. "CEP290 interacts with the centriolar satellite component PCM-1 and is required for Rab8 localization to the primary cilium." *Human Molecular Genetics* **17**(23): 3796–805. doi:10.1093/hmg/ddn277.

Kishimoto, N., Y. Cao, A. Park, and Z. Sun. 2008. "Cystic kidney gene seahorse regulates cilia-mediated processes and Wnt pathways." *Developmental Cell* **14**(6): 954–61. doi:10.1016/j.devcel.2008.03.010.

Kohler, R., G. Schonfelder, H. Hopp, A. Distler, and J. Hoyer. 1998. "Stretch-activated cation channel in human umbilical vein endothelium in normal pregnancy and in preeclampsia." *Journal of Hypertension* **16**(8): 1149–56. PMID:9794719.

Kottgen, M., B. Buchholz, M. A. Garcia-Gonzalez, F. Kotsis, X. Fu, M. Doerken, et al. 2008. "TRPP2 and TRPV4 form a polymodal sensory channel complex." *Journal of Cell Biology* **182**(3): 437–47. doi:10.1083/jcb.200805124.

Krock, B. L. and B. D. Perkins. 2008. "The intraflagellar transport protein IFT57 is required for cilia maintenance and regulates IFT-particle-kinesin-II dissociation in vertebrate photoreceptors." *Journal of Cell Science* **121**(Pt. 11): 1907–15. doi:10.1242/jcs.029397.

Kyttala, M., J. Tallila, R. Salonen, O. Kopra, N. Kohlschmidt, P. Paavola-Sakki, et al. 2006. "MKS1, encoding a component of the flagellar apparatus basal body proteome, is mutated in Meckel syndrome." *Nature Genetics* **38**(2): 155–7. PMID:16415886.

Lei, Y., W. D. Stamer, J. Wu, and X. Sun. 2014. "Endothelial nitric oxide synthase-related mechanotransduction changes in aged porcine angular aqueous plexus cells." *Investigative Ophthalmology & Visual Science* **55**(12): 8402–8. doi:10.1167/iovs.14-14992.

Lepelletier, L., J. B. de Monvel, J. Buisson, C. Desdouets, and C. Petit. 2013. "Auditory hair cell centrioles undergo confined Brownian motion throughout the developmental migration of the kinocilium." *Biophysical Journal* **105**(1): 48–58. doi:10.1016/j.bpj.2013.05.009.

Li, J. B., J. M. Gerdes, C. J. Haycraft, Y. Fan, T. M. Teslovich, H. May-Simera, et al. 2004. "Comparative genomics identifies a flagellar and basal body proteome that includes the BBS5 human disease gene." *Cell* **117**(4): 541–52. PMID:15137946.

Li, G., R. Vega, K. Nelms, N. Gekakis, C. Goodnow, P. McNamara, et al. 2007. "A role for Alstrom syndrome protein, alms1, in kidney ciliogenesis and cellular quiescence." *PLoS Genetics* **3**(1): e8. PMID:17206865.

Liem, K. F. Jr., M. He, P. J. Ocbina, and K. V. Anderson. 2009. "Mouse Kif7/Costal2 is a cilia-associated protein that regulates Sonic hedgehog signaling." *Proceedings of the National Academy of Sciences of the United States of America* **106**(32): 13 377–82. doi:10.1073/pnas.0906944106.

Lin, F., T. Hiesberger, K. Cordes, A. M. Sinclair, L. S. Goldstein, S. Somlo, and P. Igarashi. 2003. "Kidney-specific inactivation of the KIF3A subunit of kinesin-II inhibits renal ciliogenesis and produces polycystic kidney disease." *Proceedings of the National Academy of Sciences of the United States of America* **100**(9): 5286–91. PMID:12672950.

Liu, A., B. Wang, and L. A. Niswander. 2005. "Mouse intraflagellar transport proteins regulate both the activator and repressor functions of Gli transcription factors." *Development* **132**(13): 3103–11. PMID:15930098.

Lolkema, M. P., D. A. Mans, C. M. Snijckers, M. van Noort, M. van Beest, E. E. Voest, and R. H. Giles. 2007. "The von Hippel-Lindau tumour suppressor interacts with microtubules through kinesin-2." *FEBS Letters* **581**(24): 4571–6. PMID:17825299.

Lorenzetti, D., C. E. Bishop, and M. J. Justice. 2004. "Deletion of the Parkin coregulated gene causes male sterility in the quaking(viable) mouse mutant." *Proceedings of the National Academy of Sciences of the United States of America* **101**(22): 8402–7. PMID:15148410.

Low, S. H., S. Vasanth, C. H. Larson, S. Mukherjee, N. Sharma, M. T. Kinter, et al. 2006. "Polycystin-1, STAT6, and P100 function in a pathway that transduces ciliary mechanosensation and is activated in polycystic kidney disease." *Developmental Cell* **10**(1): 57–69. PMID:16399078.

Lucker, B. F., R. H. Behal, H. Qin, L. C. Siron, W. D. Taggart, J. L. Rosenbaum, and D. G. Cole. 2005. "Characterization of the intraflagellar transport complex B core: direct interaction of the IFT81 and IFT74/72 subunits." *Journal of Biological Chemistry* **280**(30): 27 688–96. PMID:15955805.

Lunt, S. C., T. Haynes, and B. D. Perkins. 2009. "Zebrafish ift57, ift88, and ift172 intraflagellar transport mutants disrupt cilia but do not affect hedgehog signaling." *Developmental Dynamics* **238**(7): 1744–59. doi:10.1002/dvdy.21999.

Ma, R., W. P. Li, D. Rundle, J. Kong, H. I. Akbarali, and L. Tsiokas. 2005. "PKD2 functions as an epidermal growth factor-activated plasma membrane channel." *Molecular Cell Biology* **25**(18): 8285–98. PMID:16135816.

Mahjoub, M. R., M. Qasim Rasi, and L. M. Quarmby. 2004. "A NIMA-related kinase, Fa2p, localizes to a novel site in the proximal cilia of Chlamydomonas and mouse kidney cells." *Molecular Biology of the Cell* **15**(11): 5172–86. PMID:15371535.

Mahjoub, M. R., M. L. Trapp, and L. M. Quarmby. 2005. "NIMA-related kinases defective in murine models of polycystic kidney diseases localize to primary cilia and centrosomes." *Journal of the American Society of Nephrology* **16**(12): 3485–9. PMID:16267153.

Mak, H. Y., L. S. Nelson, M. Basson, C. D. Johnson, and G. Ruvkun. 2006. "Polygenic control of Caenorhabditis elegans fat storage." *Nature Genetics* **38**(3): 363–8. PMID:16462744.

Marszalek, J. R., P. Ruiz-Lozano, E. Roberts, K. R. Chien, and L. S. Goldstein. 1999. "Situs inversus and embryonic ciliary morphogenesis defects in mouse mutants lacking the KIF3A subunit of kinesin-II." *Proceedings of the National Academy of Sciences of the United States of America* **96**(9): 5043–8. PMID:10220415.

Marszalek, J. R., X. Liu, E. A. Roberts, D. Chui, J. D. Marth, D. S. Williams, and L. S. Goldstein. 2000. "Genetic evidence for selective transport of opsin and arrestin by kinesin-II in mammalian photoreceptors." *Cell* **102**(2): 175–87. PMID:10943838.

Masyuk, A. I., T. V. Masyuk, P. L. Splinter, B. Q. Huang, A. J. Stroope, and N. F. LaRusso. 2006. "Cholangiocyte cilia detect changes in luminal fluid flow and transmit them into intracellular Ca^{2+} and cAMP signaling." *Gastroenterology* **131**(3): 911–20. doi:10.1053/j.gastro.2006.07.003.

May, S. R., A. M. Ashique, M. Karlen, B. Wang, Y. Shen, K. Zarbalis, et al. 2005. "Loss of the retrograde motor for IFT disrupts localization of Smo to cilia and prevents the expression of both activator and repressor functions of Gli." *Developmental Biology* **287**(2): 378–89. PMID:16229832.

Menco, B. P. 2005. "The fine-structural distribution of G-protein receptor kinase 3, beta-arrestin-2, Ca2+/calmodulin-dependent protein kinase II and phosphodiesterase PDE1C2, and a Cl(-)-cotransporter in rodent olfactory epithelia." *Journal of Neurocytology* **34**(1–2): 11–36. PMID:16374707.

Mikule, K., B. Delaval, P. Kaldis, A. Jurcyzk, P. Hergert, and S. Doxsey. 2007. "Loss of centrosome integrity induces p38-p53-p21-dependent G1-S arrest." *Nature Cell Biology* **9**(2): 160–70. PMID:17330329.

Mollet, G., F. Silbermann, M. Delous, R. Salomon, C. Antignac, and S. Saunier. 2005. "Characterization of the nephrocystin/nephrocystin-4 complex and subcellular localization of nephrocystin-4 to primary cilia and centrosomes." *Human Molecular Genetics* **14**(5): 645–56. PMID:15661758.

Morgan, G. W., P. W. Denny, S. Vaughan, D. Goulding, T. R. Jeffries, D. F. Smith, et al. 2005. "An evolutionarily conserved coiled-coil protein implicated in polycystic kidney disease is involved in basal body duplication and flagellar biogenesis in Trypanosoma brucei." *Molecular Cell Biology* **25**(9): 3774–83. PMID:15831481.

Mukhopadhyay, A., B. Deplancke, A. J. Walhout, and H. A. Tissenbaum. 2005. "C. elegans tubby regulates life span and fat storage by two independent mechanisms." *Cell Metabolism* **2**(1): 35–42. PMID:16054097.

Muntean, B. S., X. Jin, F. E. Williams, and S. M. Nauli. 2014. "Primary cilium regulates CaV1.2 expression through Wnt signaling." *Journal of Cell Physiology* **229**(12): 1926–34. doi:10.1002/jcp.24642.

Murayama, T., Y. Toh, Y. Ohshima, and M. Koga. 2005. "The dyf-3 gene encodes a novel protein required for sensory cilium formation in Caenorhabditis elegans." *Journal of Molecular Biology* **346**(3): 677–87. PMID:15713455.

Murcia, N. S., W. G. Richards, B. K. Yoder, M. L. Mucenski, J. R. Dunlap, and R. P. Woychik. 2000. "The Oak Ridge Polycystic Kidney (orpk) disease gene is required for left-right axis determination." *Development* **127**(11): 2347–55. PMID:10804177.

Nachury, M. V., A. V. Loktev, Q. Zhang, C. J. Westlake, J. Peranen, A. Merdes, et al. 2007. "A core complex of BBS proteins cooperates with the GTPase Rab8 to promote ciliary membrane biogenesis." *Cell* **129**(6): 1201–13. PMID:17574030.

Nagano, J., K. Kitamura, K. M. Hujer, C. J. Ward, R. J. Bram, U. Hopfer, et al. 2005. "Fibrocystin interacts with CAML, a protein involved in Ca2+ signaling." *Biochemical and Biophysical Research Communications* **338**(2): 880–9. PMID:16243292.

Narita, K., T. Kawate, N. Kakinuma, and S. Takeda. 2010. "Multiple primary cilia modulate the fluid transcytosis in choroid plexus epithelium." *Traffic* **11**(2): 287–301. doi:10.1111/j.1600-0854.2009.01016.x.

Nauli, S. M., F. J. Alenghat, Y. Luo, E. Williams, P. Vassilev, X. Li, et al. 2003. "Polycystins 1 and 2 mediate mechanosensation in the primary cilium of kidney cells." *Nature Genetics* **33**(2): 129–37. doi:10.1038/ng1076.

Nauli, S. M., S. Rossetti, R. J. Kolb, F. J. Alenghat, M. B. Consugar, P. C. Harris, et al. 2006. "Loss of polycystin-1 in human cyst-lining epithelia leads to ciliary dysfunction." *Journal of the American Society of Nephrology* **17**(4): 1015–25. doi:10.1681/ASN.2005080830.

Nauli, S. M., Y. Kawanabe, J. J. Kaminski, W. J. Pearce, D. E. Ingber, and J. Zhou. 2008. "Endothelial cilia are fluid shear sensors that regulate calcium signaling and nitric oxide production through polycystin-1." *Circulation* **117**(9): 1161–71. doi:10.1161/CIRCULATIONAHA.107.710111.

Nauli, S. M., X. Jin, and B. P. Hierck. 2011. "The mechanosensory role of primary cilia in vascular hypertension." *International Journal of Vascular Medicine* **2011**: 376281. doi:10.1155/2011/376281.

Nauli, S. M., X. Jin, W. A. AbouAlaiwi, W. El-Jouni, X. Su, and J. Zhou. 2013. "Non-motile primary cilia as fluid shear stress mechanosensors." *Methods in Enzymology* **525**: 1-20. doi:10.1016/B978-0-12-397944-5.00001-8.

Neesen, J., R. Kirschner, M. Ochs, A. Schmiedl, B. Habermann, C. Mueller, et al. 2001. "Disruption of an inner arm dynein heavy chain gene results in asthenozoospermia and reduced ciliary beat frequency." *Human Molecular Genetics* **10**(11): 1117–28. PMID:11371505.

Nishimura, D. Y., M. Fath, R. F. Mullins, C. Searby, M. Andrews, R. Davis, et al. 2004a. "Bbs2-null mice have neurosensory deficits, a defect in social dominance, and retinopathy associated with mislocalization of rhodopsin." *Proceedings of the National Academy of Sciences of the United States of America* **101**(47): 16 588–93. PMID:15539463.

Nishimura, T., K. Kato, T. Yamaguchi, Y. Fukata, S. Ohno, and K. Kaibuchi. 2004b. "Role of the PAR-3-KIF3 complex in the establishment of neuronal polarity." *Nature Cell Biology* **6**(4): 328–34. PMID:15048131.

Oeffner, F., C. Moch, A. Neundorf, J. Hofmann, M. Koch, and K. H. Grzeschik. 2008. "Novel interaction partners of Bardet-Biedl syndrome proteins." *Cell Motility and the Cytoskeleton* **65**(2): 143–55. PMID:18000879.

Ogawa, R. and C. K. Hsu. 2013. "Mechanobiological dysregulation of the epidermis and dermis in skin disorders and in degeneration." *Journal of Cellular and Molecular Medicine* **17**(7): 817–22. doi:10.1111/jcmm.12060.

Okuda, H., S. Hirai, Y. Takaki, M. Kamada, M. Baba, N. Sakai, et al. 1999. "Direct interaction of the beta-domain of VHL tumor suppressor protein with the regulatory domain of atypical PKC isotypes." *Biochemical and Biophysical Research Communications* **263**(2): 491–7. PMID:10491320.

Olbrich, H., M. Fliegauf, J. Hoefele, A. Kispert, E. Otto, A. Volz, et al. 2003. "Mutations in a novel gene, NPHP3, cause adolescent nephronophthisis, tapeto-retinal degeneration and hepatic fibrosis." *Nature Genetics* **34**(4): 455–9. PMID:12872122.

Oliveira, D. M. and M. A. Goodell. 2003. "Transient RNA interference in hematopoietic progenitors with functional consequences." *Genesis* **36**(4): 203–8. PMID:12929091.

Olsen, N. V. 1998. "Effects of dopamine on renal haemodynamics tubular function and sodium excretion in normal humans." *Danish Medical Bulletin* **45**(3): 282–97. PMID:9675540.

Omori, Y. and J. Malicki. 2006. "oko meduzy and related crumbs genes are determinants of apical cell features in the vertebrate embryo." *Current Biology* **16**(10): 945–57. PMID:16713951.

Omran, H., D. Kobayashi, H. Olbrich, T. Tsukahara, N. T. Loges, H. Hagiwara, et al. 2008. "Ktu/PF13 is required for cytoplasmic pre-assembly of axonemal dyneins." *Nature* **456**(7222): 611–16. doi:10.1038/nature07471.

Otto, E. A., B. Schermer, T. Obara, J. F. O'Toole, K. S. Hiller, A. M. Mueller, et al. 2003. "Mutations in INVS encoding inversin cause nephronophthisis type 2, linking renal cystic disease to the function of primary cilia and left-right axis determination." *Nature Genetics* **34**(4): 413–20. PMID:12872123.

Otto, E. A., B. Loeys, H. Khanna, J. Hellemans, R. Sudbrak, S. Fan, et al. 2005. "Nephrocystin-5, a ciliary IQ domain protein, is mutated in Senior-Loken syndrome and interacts with RPGR and calmodulin." *Nature Genetics* **37**(3): 282–8. PMID:15723066.

Otto, E. A., M. L. Trapp, U. T. Schultheiss, J. Helou, L. M. Quarmby, and F. Hildebrandt. 2008. "NEK8 mutations affect ciliary and centrosomal localization and may cause nephronophthisis." *Journal of the American Society of Nephrology* **19**(3): 587–92. doi:10.1681/ASN.2007040490.

Ou, G., O. E. Blacque, J. J. Snow, M. R. Leroux, and J. M. Scholey. 2005a. "Functional coordination of intraflagellar transport motors." *Nature* **436**(7050): 583–7. PMID:16049494.

Ou, G., H. Qin, J. L. Rosenbaum, and J. M. Scholey. 2005b. "The PKD protein qilin undergoes intraflagellar transport." *Current Biology* **15**(11): R410–11. PMID:15936258.

Pathak, N., T. Obara, S. Mangos, Y. Liu, and I. A. Drummond. 2007. "The zebrafish fleer gene encodes an essential regulator of cilia tubulin polyglutamylation." *Molecular Biology of the Cell* **18**(11): 4353–64. PMID:17761526.

Pazour, G. J., B. L. Dickert, Y. Vucica, E. S. Seeley, J. L. Rosenbaum, G. B. Witman, and D. G. Cole. 2000. "Chlamydomonas IFT88 and its mouse homologue, polycystic kidney disease gene tg737, are required for assembly of cilia and flagella." *Journal of Cell Biology* **151**(3): 709–18. PMID:11062270.

Pazour, G. J., J. T. San Agustin, J. A. Follit, J. L. Rosenbaum, and G. B. Witman. 2002. "Polycystin-2 localizes to kidney cilia and the ciliary level is elevated in orpk mice with polycystic kidney disease." *Current Biology* **12**(11): R378–80. PMID:12062067.

Pazour, G. J., N. Agrin, J. Leszyk, and G. B. Witman. 2005. "Proteomic analysis of a eukaryotic cilium." *Journal of Cell Biology* **170**(1): 103–13. PMID:15998802.

Pedersen, L. B., S. Geimer, R. D. Sloboda, and J. L. Rosenbaum. 2003. "The Microtubule plus end-tracking protein EB1 is localized to the flagellar tip and basal bodies in Chlamydomonas reinhardtii." *Current Biology* **13**(22): 1969–74. PMID:14614822.

Pedersen, L. B., M. S. Miller, S. Geimer, J. M. Leitch, J. L. Rosenbaum, and D. G. Cole. 2005. "Chlamydomonas IFT172 is encoded by FLA11, interacts with CrEB1, and regulates IFT at the flagellar tip." *Current Biology* **15**(3): 262–6. PMID:15694311.

Pennarun, G., E. Escudier, C. Chapelin, A. M. Bridoux, V. Cacheux, G. Roger, et al. 1999. "Loss-of-function mutations in a human gene related to Chlamydomonas reinhardtii dynein IC78 result in primary ciliary dyskinesia." *American Journal of Human Genetics* **65**(6): 1508–19. PMID:10577904.

Piehl, M., U. S. Tulu, P. Wadsworth, and L. Cassimeris. 2004. "Centrosome maturation: measurement of microtubule nucleation throughout the cell cycle by using GFP-tagged EB1." *Proceedings of the National Academy of Sciences of the United States of America* **101**(6): 1584–8. PMID:14747658.

Polacheck, W. J., A. E. German, A. Mammoto, D. E. Ingber, and R. D. Kamm. 2014. "Mechanotransduction of fluid stresses governs 3D cell migration." *Proceedings of the National Academy of Sciences of the United States of America* **111**(7): 2447–52. doi:10.1073/pnas.1316848111.

Praetorius, H. A. and K. R. Spring. 2001. "Bending the MDCK cell primary cilium increases intracellular calcium." *Journal of Membrane Biology* **184**(1): 71–9. PMID:11687880.

Prasad, R. M., X. Jin, W. A. Aboualaiwi, and S. M. Nauli. 2014a. "Real-time vascular mechanosensation through ex vivo artery perfusion." *Biological Procedures Online* **16**(1): 6. doi:10.1186/1480-9222-16-6.

Prasad, R. M., X. Jin, and S. M. Nauli. 2014b. "Sensing a sensor: identifying the mechanosensory function of primary cilia." *Biosensors (Basel)* **4**(1): 47–62. doi:10.3390/bios4010047.

Qin, H., J. L. Rosenbaum, and M. M. Barr. 2001. "An autosomal recessive polycystic kidney disease gene homolog is involved in intraflagellar transport in C. elegans ciliated sensory neurons." *Current Biology* **11**(6): 457–61. PMID:11301258.

Qin, H., D. T. Burnette, Y. K. Bae, P. Forscher, M. M. Barr, and J. L. Rosenbaum. 2005. "Intraflagellar transport is required for the vectorial movement of TRPV channels in the ciliary membrane." *Current Biology* **15**(18): 1695–9. PMID:16169494.

Rana, A. A., J. P. Barbera, T. A. Rodriguez, D. Lynch, E. Hirst, J. C. Smith, and R. S. Beddington. 2004. "Targeted deletion of the novel cytoplasmic dynein mD2LIC disrupts the embryonic organiser, formation of the body axes and specification of ventral cell fates." *Development* **131**(20): 4999–5007. PMID:15371312.

Romio, L., A. M. Fry, P. J. Winyard, S. Malcolm, A. S. Woolf, and S. A. Feather. 2004. "OFD1 is a centrosomal/basal body protein expressed during mesenchymal-epithelial transition in human nephrogenesis." *Journal of the American Society of Nephrology* **15**(10): 2556–68. PMID:15466260.

Rupp, G. and M. E. Porter. 2003. "A subunit of the dynein regulatory complex in Chlamydomonas is a homologue of a growth arrest-specific gene product." *Journal of Cell Biology* **162**(1): 47–57. PMID:12847082.

Rydholm, S., G. Zwartz, J. M. Kowalewski, P. Kamali-Zare, T. Frisk, and H. Brismar. 2010. "Mechanical properties of primary cilia regulate the response to fluid flow." *American Journal of Physiology. Renal Physiology* **298**(5): F1096–102. doi:10.1152/ajprenal.00657.2009.

Sapiro, R., I. Kostetskii, P. Olds-Clarke, G. L. Gerton, G. L. Radice, and J. F. Strauss III. 2002. "Male infertility, impaired sperm motility, and hydrocephalus in mice deficient in sperm-associated antigen 6." *Molecular Cell Biology* **22**(17): 6298–305. PMID:12167721.

Sato, M., K. Nagata, S. Kuroda, S. Horiuchi, T. Nakamura, M. Karima, et al. 2014. "Low-intensity pulsed ultrasound activates integrin-mediated mechanotransduction pathway in synovial cells." *Annals of Biomedical Engineering* **42**(10): 2156–63. doi:10.1007/s10439-014-1081-x.

Sayer, J. A., E. A. Otto, J. F. O'Toole, G. Nurnberg, M. A. Kennedy, C. Becker, et al. 2006. "The centrosomal protein nephrocystin-6 is mutated in Joubert syndrome and activates transcription factor ATF4." *Nature Genetics* **38**(6): 674–81. PMID:16682973.

Schneider, L., C. A. Clement, S. C. Teilmann, G. J. Pazour, E. K. Hoffmann, P. Satir, and S. T. Christensen. 2005. "PDGFRalphaalpha signaling is regulated through the primary cilium in fibroblasts." *Current Biology* **15**(20): 1861–6. doi:10.1016/j.cub.2005.09.012.

Schroder, J. M., L. Schneider, S. T. Christensen, and L. B. Pedersen. 2007. "EB1 is required for primary cilia assembly in fibroblasts." *Current Biology* **17**(13): 1134–9. PMID:17600711.

Schulz, S., M. Handel, M. Schreff, H. Schmidt, and V. Hollt. 2000. "Localization of five somatostatin receptors in the rat central nervous system using subtype-specific antibodies." *Journal of Physiology, Paris* **94**(3–4): 259–64. PMID:11088003.

Sfakianos, J., A. Togawa, S. Maday, M. Hull, M. Pypaert, L. Cantley, et al. 2007. "Par3 functions in the biogenesis of the primary cilium in polarized epithelial cells." *Journal of Cell Biology* **179**(6): 1133–40. PMID:18070914.

Shin, J. B., D. Adams, M. Paukert, M. Siba, S. Sidi, M. Levin, et al. 2005. "Xenopus TRPN1 (NOMPC) localizes to microtubule-based cilia in epithelial cells, including inner-ear hair cells." *Proceedings of the National Academy of Sciences of the United States of America* **102**(35): 12 572–7. PMID:16116094.

Shu, X., A. M. Fry, B. Tulloch, F. D. Manson, J. W. Crabb, H. Khanna, et al. 2005. "RPGR ORF15 isoform co-localizes with RPGRIP1 at centrioles and basal bodies and interacts with nucleophosmin." *Human Molecular Genetics* **14**(9): 1183–97. PMID:15772089.

Smith, U. M., M. Consugar, L. J. Tee, B. M. McKee, E. N. Maina, S. Whelan, et al. 2006. "The transmembrane protein meckelin (MKS3) is mutated in Meckel-Gruber syndrome and the wpk rat." *Nature Genetics* **38**(2): 191–6. PMID:16415887.

Stolc, V., M. P. Samanta, W. Tongprasit, and W. F. Marshall. 2005. "Genome-wide transcriptional analysis of flagellar regeneration in Chlamydomonas reinhardtii identifies orthologs of ciliary disease genes." *Proceedings of the National Academy of Sciences of the United States of America* **102**(10): 3703–7. PMID:15738400.

Strassman, A. M. and D. Levy. 2006. "Response properties of dural nociceptors in relation to headache." *Journal of Neurophysiology* **95**(3): 1298–306. doi:10.1152/jn.01293.2005.

Suki, B., R. Jesudason, S. Sato, H. Parameswaran, A. D. Araujo, A. Majumdar, et al. 2012. "Mechanical failure, stress redistribution, elastase activity and binding site availability on elastin during the progression of emphysema." *Pulmonary Pharmacology & Therapeutics* **25**(4): 268–75. doi:10.1016/j.pupt.2011.04.027.

Sun, Z., A. Amsterdam, G. J. Pazour, D. G. Cole, M. S. Miller, and N. Hopkins. 2004. "A genetic screen in zebrafish identifies cilia genes as a principal cause of cystic kidney." *Development* **131**(16): 4085–93. PMID:15269167.

Takeda, S., Y. Yonekawa, Y. Tanaka, Y. Okada, S. Nonaka, and N. Hirokawa. 1999. "Left-right asymmetry and kinesin superfamily protein KIF3A: new insights in determination of laterality and mesoderm induction by kif3A-/- mice analysis." *Journal of Cell Biology* **145**(4): 825–36. PMID:10330409.

Tammachote, R., C. J. Hommerding, R. M. Sinders, C. A. Miller, P. G. Czarnecki, A. C. Leightner, et al. 2009. "Ciliary and centrosomal defects associated with mutation and depletion of the Meckel syndrome genes MKS1 and MKS3." *Human Molecular Genetics* **18**(17): 3311–23. doi: 10.1093/hmg/ddp272.

Tanaka, H., N. Iguchi, Y. Toyama, K. Kitamura, T. Takahashi, K. Kaseda, et al. 2004. "Mice deficient in the axonemal protein Tektin-t exhibit male infertility and immotile-cilium syndrome due to impaired inner arm dynein function." *Molecular Cell Biology* **24**(18): 7958–64. PMID:15340058.

Teilmann, S. C. and S. T. Christensen. 2005. "Localization of the angiopoietin receptors Tie-1 and Tie-2 on the primary cilia in the female reproductive organs." *Cell Biology International* **29**(5): 340–6. PMID:15893943.

Teilmann, S. C., A. G. Byskov, P. A. Pedersen, D. N. Wheatley, G. J. Pazour, and S. T. Christensen. 2005. "Localization of transient receptor potential ion channels in primary and motile cilia of the female murine reproductive organs." *Molecular Reproduction and Development* **71**(4): 444–52. PMID:15858826.

Thoma, C. R., I. J. Frew, C. R. Hoerner, M. Montani, H. Moch, and W. Krek. 2007. "pVHL and GSK3beta are components of a primary cilium-maintenance signalling network." *Nature Cell Biology* **9**(5): 588–95. PMID:17450132.

Tsujikawa, M. and J. Malicki. 2004. "Intraflagellar transport genes are essential for differentiation and survival of vertebrate sensory neurons." *Neuron* **42**(5): 703–16. PMID:15182712.

Upadhyay, V. S., B. S. Muntean, S. H. Kathem, J. J. Hwang, W. A. Aboualaiwi, and S. M. Nauli. 2014. "Roles of dopamine receptor on chemosensory and mechanosensory primary cilia in renal epithelial cells." *Frontiers in Physiology* **5**: 72. doi:10.3389/fphys.2014.00072.

Vernon, G. G., J. Neesen, and D. M. Woolley. 2005. "Further studies on knockout mice lacking a functional dynein heavy chain (MDHC7). 1. Evidence for a structural deficit in the axoneme." *Cell Motility and the Cytoskeleton* **61**(2): 65–73. PMID:15838838.

Vieira, O. V., K. Gaus, P. Verkade, J. Fullekrug, W. L. Vaz, and K. Simons. 2006. "FAPP2, cilium formation, and compartmentalization of the apical membrane in polarized Madin-Darby canine kidney (MDCK) cells." *Proceedings of the National Academy of Sciences of the United States of America* **103**(49): 18 556–61. PMID:17116893.

Wang, S., J. Zhang, S. M. Nauli, X. Li, P. G. Starremans, Y. Luo, et al. 2007. "Fibrocystin/polyductin, found in the same protein complex with polycystin-2, regulates calcium responses in kidney epithelia." *Molecular Cell Biology* **27**(8): 3241–52. doi:10.1128/MCB.00072-07.

Ward, C. J., D. Yuan, T. V. Masyuk, X. Wang, R. Punyashthiti, S. Whelan, et al. 2003. "Cellular and subcellular localization of the ARPKD protein; fibrocystin is expressed on primary cilia." *Human Molecular Genetics* **12**(20): 2703–10. PMID:12925574.

Waters, A. M. and P. L. Beales. 2011. "Ciliopathies: an expanding disease spectrum." *Pediatric Nephrology* **26**(7): 1039–56. doi:10.1007/s00467-010-1731-7.

Weatherbee, S. D., L. A. Niswander, and K. V. Anderson. 2009. "A mouse model for Meckel Syndrome reveals Mks1 is required for ciliogenesis and Hedgehog signaling." *Human Molecular Genetics* **18**(23): 4565–75. doi:10.1093/hmg/ddp422.

White, M. C. and L. M. Quarmby. 2008. "The NIMA-family kinase, Nek1 affects the stability of centrosomes and ciliogenesis." *BMC Cell Biol* **9**: 29. doi:10.1186/1471-2121-9-29.

Whitfield, J. F. 2008. "The solitary (primary) cilium – a mechanosensory toggle switch in bone and cartilage cells." *Cellular Signalling* **20**(6): 1019–24. doi:10.1016/j.cellsig.2007.12.001.

Wilson, C. and S. E. Dryer. 2014. "A mutation in TRPC6 channels abolishes their activation by hypoosmotic stretch but does not affect activation by diacylglycerol or G protein signaling cascades." *American Journal of Physiology. Renal Physiology* **306**(9): F1018–25. doi:10.1152/ajprenal.00662.2013.

Winkelbauer, M. E., J. C. Schafer, C. J. Haycraft, P. Swoboda, and B. K. Yoder. 2005. "The C. elegans homologs of nephrocystin-1 and nephrocystin-4 are cilia transition zone proteins involved in chemosensory perception." *Journal of Cell Science* **118**(Pt. 23): 5575–87. PMID:16291722.

Yang, J., X. Liu, G. Yue, M. Adamian, O. Bulgakov, and T. Li. 2002. "Rootletin, a novel coiled-coil protein, is a structural component of the ciliary rootlet." *Journal of Cell Biology* **159**(3): 431–40. PMID:12427867.

Yang, J., J. Gao, M. Adamian, X. H. Wen, B. Pawlyk, L. Zhang, et al. 2005. "The ciliary rootlet maintains long-term stability of sensory cilia." *Molecular Cell Biology* **25**(10): 4129–37. PMID:15870283.

Yen, H. J., M. K. Tayeh, R. F. Mullins, E. M. Stone, V. C. Sheffield, and D. C. Slusarski. 2006. "Bardet-Biedl syndrome genes are important in retrograde intracellular trafficking and Kupffer's vesicle cilia function." *Human Molecular Genetics* **15**(5): 667–77. PMID:16399798.

Yissachar, N., H. Salem, T. Tennenbaum, and B. Motro. 2006. "Nek7 kinase is enriched at the centrosome, and is required for proper spindle assembly and mitotic progression." *FEBS Letters* **580**(27): 6489–95. PMID:17101132.

Yoder, B. K., X. Hou, and L. M. Guay-Woodford. 2002a. "The polycystic kidney disease proteins, polycystin-1, polycystin-2, polaris, and cystin, are co-localized in renal cilia." *Journal of the American Society of Nephrology* **13**(10): 2508–16. PMID:12239239.

Yoder, B. K., A. Tousson, L. Millican, J. H. Wu, C. E. Bugg Jr., J. A. Schafer, and D. F. Balkovetz. 2002b. "Polaris, a protein disrupted in orpk mutant mice, is required for assembly of renal cilium." *American Journal of Physiology. Renal Physiology* **282**(3): F541–52. PMID:11832437.

Zhang, Y. J., W. K. O'Neal, S. H. Randell, K. Blackburn, M. B. Moyer, R. C. Boucher, and L. E. Ostrowski. 2002. "Identification of dynein heavy chain 7 as an inner arm component of human cilia that is synthesized but not assembled in a case of primary ciliary dyskinesia." *Journal of Biological Chemistry* **277**(20): 17 906–15. PMID:11877439.

Zhang, Z., I. Kostetskii, S. B. Moss, B. H. Jones, C. Ho, H. Wang, et al. 2004. "Haploinsufficiency for the murine orthologue of Chlamydomonas PF20 disrupts spermatogenesis." *Proceedings of the National Academy of Sciences of the United States of America* **101**(35): 12 946–51. PMID:15328412.

Zhang, Z., B. H. Jones, W. Tang, S. B. Moss, Z. Wei, C. Ho, et al. 2005. "Dissecting the axoneme interactome: the mammalian orthologue of Chlamydomonas PF6 interacts with sperm-associated antigen 6, the mammalian orthologue of Chlamydomonas PF16." *Molecular & Cellular Proteomics* **4**(7): 914–23. PMID:15827353.

Zilber, Y., S. Babayeva, J. H. Seo, J. J. Liu, S. Mootin, and E. Torban. 2013. "The PCP effector Fuzzy controls cilial assembly and signaling by recruiting Rab8 and Dishevelled to the primary cilium." *Molecular Biology of the Cell* **24**(5): 555–65. doi:10.1091/mbc.E12-06-0437.

Zou, J., T. Zheng, C. Ren, C. Askew, X. P. Liu, B. Pan, et al. 2014. "Deletion of PDZD7 disrupts the Usher syndrome type 2 protein complex in cochlear hair cells and causes hearing loss in mice." *Human Molecular Genetics* **23**(9): 2374–90. doi:10.1093/hmg/ddt629.

6

Mechanobiology of Embryonic Skeletal Development

Lessons for Osteoarthritis

Andrea S. Pollard and Andrew A. Pitsillides

Comparative Biomedical Sciences, The Royal Veterinary College, London, UK

6.1 Introduction

Establishing limb skeletal form during embryonic development is a complex problem that involves integration of genetic, vascular, hormonal, and mechanical influences upon cartilage and bone. The skeleton is an organ which performs structural and mechanical roles, and it is therefore unsurprising that, like its postnatal growth, its development is influenced by mechanical loading. Developing skeletal elements are subject to mechanical forces from several different origins, including forces generated at a cellular level by the cytoskeleton, forces produced within tissues by growth, and loading produced by embryonic muscle contraction. In this chapter, we consider the response of the developing skeleton to embryo movement-related loading.

The skeleton experiences mechanical stimuli engendered by embryo movement for much of its development; this starts at least as early as day 4.5 in the chicken, day 12.5 in the mouse, and week 7.5–8.0 in humans (Hamburger and Balaban 1963; Bekoff 1981; de Vries et al. 1982; Carry et al. 1983; Hanson and Landmesser 2003). To date, it has not been possible to experimentally monitor the strain produced in developing bones by embryonic muscle contraction. However, a number of strategies exist for examining the impact of removal or stimulation of embryo movement on developmental processes, and these allow the role of mechanics in their regulation to be inferred. These include pharmacological manipulation of embryo movement and the use of mutant mouse models in which embryonic paralysis is an outcome.

These methods have been used to reveal the critical role played by movement-related mechanical stimuli in cartilage and bone development; processes involved in normal endochondral bone growth and joint formation are at least partially dependent on mechanical regulation. This chapter provides an overview of the role of mechanical stimuli in regulating joint cavitation and longitudinal bone growth, and also highlights the implications that mechanical regulation of developmental processes have for the study of osteoarthritis. Osteoarthritis is a disease of the whole joint, characterized by articular cartilage degeneration, subchondral bone thickening, and the formation of bony projections at the joint margins, called osteophytes. Osteoarthritis results in pain and disability, and it is a major worldwide healthcare burden. This chapter will inspect

Mechanobiology: Exploitation for Medical Benefit, First Edition. Edited by Simon C. F. Rawlinson.

how some characteristic changes in osteoarthritic joints should be revisited in the context of our new understanding of the mechanoregulation of joint formation and endochondral growth of embryonic cartilaginous skeletal elements.

6.2 An Overview of Embryonic Skeletal Development

In early embryogenesis, the appendicular skeleton, including the pectoral girdle, the pelvis, and the limbs, initially arises from the lateral plate mesoderm. These structures, like much of the axial skeleton, including the vertebrae and ribs, form through endochondral ossification. The limb bud mesenchyme condenses to form cartilage models of the future limb bones, which will eventually ossify. These condensations initially contain mesenchymal limb bud cells, which produce extracellular matrix (ECM) and begin to express factors associated with chondrogenesis, including SOX9, cadherin-2, neural cell adhesion molecule 1, and tenascin-C (Pitsillides and Beier 2011). The composition of the ECM plays an important role in mediating cell–cell contacts and differentiating mesenchymal stem cells into chondrocytes. In contrast, the flat bones of the skull and mandible, and part of the clavicle, develop through intramembranous ossification, during which mesenchymal cells differentiate directly into osteoprogenitors, without the formation of a cartilage model. There is evidence to suggest that the growth of bones with endochondral origins, but not of those derived by intramembranous ossification, is regulated by mechanical stimuli (Rawlinson et al. 1995).

Events during early limb development are controlled by gradients of expression of signaling molecules. The apical ectodermal ridge and zone of polarizing activity control proximodistal and anterior–posterior patterning events in the limb bud via signaling pathways such as hedgehog (sonic hedgehog, Shh and Indian hedgehog, Ihh) and WNT/β-catenin. Fibroblast growth factors, transforming growth factor beta (TGF-β) family members, and regulators of transcription, including Hox and Pax genes, also contribute to limb patterning. In this way, the cartilage models of the stylopod, zeugopod, and autopod elements (corresponding to the humerus, radius/ulna, and metacarpals/digits in the forelimb and femur, tibia/fibula, and metatarsals/digits in the hindlimb, respectively) emerge in a proximodistal sequence (Summerbell et al. 1973; Rowe and Fallon 1982; Tickle 1995, 2003; Tickle and Münsterberg 2001; Wolpert 2010). Once early patterning events have defined each skeletal element and each presumptive joint at the interzone (the region separating skeletal elements, where the future joint will form), several processes take place. Joint progenitor cells at the interzone do not undergo chondrogenesis like the adjacent cartilage elements. Rather, cavitation – the formation of a joint cavity – and formation of the synovium and associated joint structures occur. The chondrocytes in each individual cartilage model also undergo a highly regulated process of proliferation, maturation, and hypertrophy, in order to expand the length of each of the elements before they are eventually replaced by bone (Kronenberg 2003).

A number of studies utilizing animal models of embryonic paralysis have demonstrated that, while early patterning events appear to occur independently of mechanical regulation, both endochondral growth and joint cavitation require embryo movement (Figure 6.1).

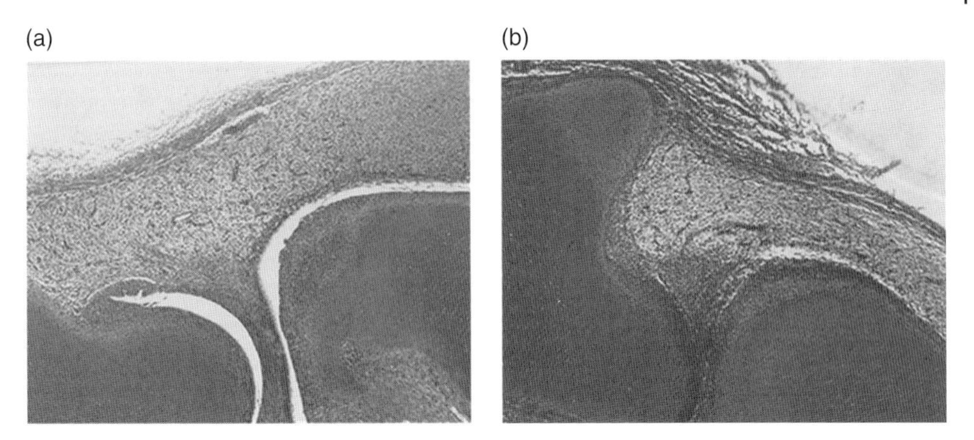

Figure 6.1 Joint cavitation is dependent upon embryo movement. Knee joints of embryonic chickens at 11 days into development. The distal femur and proximal tibia are visible in the sagittal plane. (a) A fully formed joint cavity in a normal embryo. (b) Failure of joint cavitation in response to pharmacological immobilization. (*See insert for color representation of the figure.*)

6.3 Regulation of Joint Formation

The initial phase of joint development involves demarcation of the interzone. Interzones first appear in developing limbs as densely cellular, homogenous regions between the growing cartilage elements, the latter showing initial expression of collagen type IIA in the ECM followed by a switch to collagen type IIB expression later in development. The interzones can be differentiated from these growing cartilage elements by their lack of collagen-IIB and Sox9 expression, and the expression of several "interzone markers," including GDF5, WNT9A, and versican, and activation of the MEK-ERK-1/2 (pERK-1/2) pathway (Thorogood and Hinchliffe 1975; Archer et al. 1994; Kavanagh et al. 2006; Ray et al. 2015).

The second phase of joint formation involves formation of the joint cavity. A number of previous studies have suggested that this process is achieved by coordinated cell death, but this has been contested by more recent studies, which observed no apoptotic cells and only occasional cell necrosis at the joint line during cavitation in chicken metatarsophalangeal joints, and found that cells previously thought to be "degenerating" at the joint line prior to cavitation in fact express GDF-5 and later form the surface layer of articular cartilage in the cavitated joint. Current paradigms describing joint formation dictate that, rather than cell death, coordinated changes in cell–cell adhesion and matrix composition are responsible for producing a cleft at the developing joint line between each future limb element. The primary event in joint cavity formation appears to be altered synthesis of hyaluronan (HA), a glycosaminoglycan expressed in high quantities in the synovial fluid of fully formed joints by cells immediately adjacent to the forming cavity. This differentiation of cells at the joint line to a phenotype which produces markedly higher levels of HA requires activation of ERK and occurs coincident with joint cavitation (Pitsillides et al. 1995; Bastow et al. 2005; Ito and Kida 2000).

Interactions between extracellular HA and cell surface HA-binding proteins are likely to influence tissue integrity at the forming joint. At low HA concentrations, HA and HA-binding protein interactions promote cell–cell adhesion, while at high HA concentrations, cell separation occurs. One such HA-binding protein is CD44, which is upregulated in expression at the cavitating joint by the interzone marker Wnt9a (Dowthwaite et al. 1999). There is substantial evidence that HA production and the interaction between CD44 and HA mediate the switch from tissue cohesion to separation, which occurs at the interzone, and that this loss of cohesion results in the formation of a cavity between developing cartilage elements (Dowthwaite et al. 1999).

This joint formation process appears to be dependent upon mechanical stimuli resulting from embryo movement. This was first established in the 1920s and 30s by the experiments of Murray (Murray 1926; Murray and Selby 1930) and Fell and Canti (1934). These experiments used explanted limbs from chicken embryos to show that early patterning events can take place in the absence of the limb's normally contiguous structures, including skeletal muscle, implying that precise specification of the joint's position takes place independently of movement – though this was not explicitly monitored. In stark contrast to this, however, these studies also established that embryonic muscle contraction is necessary for later joint cavitation. Drachman and Sokoloff (1966) demonstrated that the induction of paralysis in embryonic chicks *in ovo* with the neuromuscular blocking agents decamethonium bromide (DMB) or type A botulinum toxin results in a failure of cavitation at the knee, ankle, and toe joints (other joints were not examined), leading to cartilaginous fusion of opposing limb elements across the presumptive joint region. Similar failures in joint cavitation and secondary fusion of previously cavitated joints have been observed in more recent embryonic chick studies and in "muscleless" mutant mice (Hall and Herring 1990; Hosseini and Hogg 1991a,b; Ward et al. 1999; Nowlan et al. 2010a,b), in which the limb musculature fails to develop, resulting in embryo paralysis.

In the absence of embryo movement, interzone cells do not maintain their designated fate and expression of interzone markers is lost. This has been observed in the elbow joint of muscleless mice (Kahn et al. 2009). ERK-1/2 activity at the joint line is also downregulated in response to immobilization of embryonic chick, and this results in a failure in the increased HA synthesis, which normally leads to separation of cells at the forming joint cavity (Dowthwaite et al. 1999; Bastow et al. 2005; Kavanagh et al. 2006). Expression of UGDH (an enzyme necessary for HA production) and the HA-binding protein CD44 is partially or fully lost at the joint line in response to immobilization (Pitsillides and Ashhurst 2008). Our studies have revealed that constitutively active $p38^{MAPK}$ is also expressed at the forming joint line, where it appears to exert a mechanomodulatory effect on both local MEK-ERK pathway activation and HA production, as well as binding by surface cells of the developing articular cartilage (Lewthwaite et al. 2006).

Most recently, our unpublished findings have shown that developing joint line cells express high levels of cyclooxygenase-2 (COX-2) both *in vivo* and *in vitro*, and that COX-2-derived prostaglandin products drive HA synthesis and binding (Wheeler et al. pers. comm.). In paralyzed limbs, the joint progenitor cells at the interzone express factors associated with chondrogenesis, including Sox9 and Col2a1, and differentiate into chondrocytes, resulting in cartilaginous fusion of limb elements (Kahn et al. 2009). It is important to note that mechanosensitivity is not *always* a feature of joint formation; early interzone specification occurs independently of mechanical cues even in pharmacologically immobilized embryos, but the second phase of joint formation when cavitation occurs requires embryo movement. Taken together, these studies suggest that the mechanisms regulating joint cavitation, which are responsive to mechanical stimuli, are intimately linked with inflammation pathways (constitutively active MEK-ERK and $p38^{MAPK}$ and COX-2). This is unusual, as it implies that normal movement-dependent developmental joint formation processes exploit pathways thought "classically" to be involved in lymphocyte activation and the promotion of proinflammatory cytokine production.

6.4 Regulation of Endochondral Ossification

Joint formation is not the only event to show a dependence upon embryo movement. The critical role of movement in regulating endochondral ossification has also been investigated primarily in the context of limb development in pharmacologically immobilized chickens and mouse models of embryo paralysis. Longitudinal bone growth, both in the embryo after early patterning and postnatally, occurs at the growth plate, which is separated into distinct zones, where chondrocytes arranged in columns parallel to the long axis of the bone proliferate, mature, and differentiate while producing matrix (Olsen et al. 2000; Kronenberg 2003). The growth plate is a highly *dynamic* structure in which chondrocytes maintain their spatially fixed locations; longitudinal expansion is achieved via the addition of new cells and hypertrophic expansion (Hunziker et al. 1987). Immature chondrocytes undergo mitosis in the proliferative zone and produce collagen-II and proteoglycan-rich matrix. These cells intercalate into organized longitudinal columns, wherein chondrocytes progress from the proliferative zone to become prehypertrophic (identified by Ihh expression) and eventually undergo hypertrophy (identifiable by expression of collagen-X). In this hypertrophic zone, chondrocytes enlarge via three distinct phases: (i) "true hypertrophy," with a proportionate increase in dry mass and fluid uptake to achieve a 3× increase in volume; (ii) cellular swelling alone, with a 4× increase in volume without any increase in dry mass; and (iii) a repeat of the first phase, with further hypertrophy achieved by a proportional increase in dry mass and fluid uptake. It is interesting to note, parenthetically, that this final phase of hypertrophy is the primary contributor to differential elongation of limb elements between species that exhibit varied limb proportions (Cooper et al. 2013).

These events are regulated by paracrine signals, including gradients of parathyroid hormone related-peptide (PTHrP) and Ihh produced by the perichondrium (Chung et al. 2001; Kronenberg 2006). PTHrP is secreted by perichondrial cells and by cells at the ends of the long bones, and is expressed in a gradient extending toward the metaphysis. It acts on receptors on proliferative zone chondrocytes to stimulate proliferation and delay Ihh production and cell differentiation. When the source of PTHrP production is sufficiently distant, as in the prehypertrophic zone, Ihh can be produced. Ihh acts – in a feedback loop that is currently not fully understood – to stimulate the production of PTHrP at the ends of the bones, and stimulates perichondral cells to become bone-producing osteoblasts (Kronenberg 2003).

In ovo immobilization of chick embryos leads to significant reductions in limb bone length. This was first reported by Drachman and Sokoloff (1966) in the lower limbs of chicks immobilized with a neuromuscular blocking agent, decamethonium. More recent studies (Hall and Herring 1990; Hosseini and Hogg 1991a,b; Osborne et al. 2002) have built on this by demonstrating that induction of paralysis at relatively early time points does not significantly impact upon longitudinal bone growth. This provides further evidence that limb-patterning events and early growth occur independently of mechanical input. A requirement for embryo movement to drive endochondral bone growth appears to be acquired around the beginning of the final third of gestation in the chicken embryo. There is also evidence that a differential impact of *in ovo* immobilization on the growth of different limb regions may result in altered limb proportions in response to altered embryo movement (Lamb et al. 2003; Pitsillides 2006).

Investigation into the cellular basis of these changes in longitudinal growth indicates that embryo motility acts to influence skeletal growth via several mechanisms. The transition of cartilage into bone is slowed by immobilization of embryonic chickens, and the rate of calcification is increased in response to mechanical loading. In embryonic growth cartilage, reduced proliferation has been reported in chicks immobilized with DMB and in mouse models of paralysis (Roddy et al. 2011). A regulatory effect of mechanical loading on chondrocyte proliferation *in vitro* has also been observed. This suggests that mechanical stimuli are necessary for the recruitment and proliferation of immature chondrocytes in the growth plate. However, in immobilized zebrafish, which show a significant reduction in the size of all pharyngeal cartilage elements, and in muscle-deficient mouse embryos, chondrocyte number does not appear to be altered by the absence of muscle contraction (Shwartz et al. 2012). Chondrocyte intercalation into columns in the proliferative zone of the growth plate is, however, abnormal in these models, indicating that cell polarity is likely also dependent on mechanical stimuli. ECM production by growth-plate chondrocytes is also influenced by mechanical loading, with implications for the promotion of chondrocyte differentiation and the structural and mechanical integrity of the developing skeletal elements.

We have now found that immobilization of embryo chick limbs very rapidly results in slowing of cell cycle progression with failure of cells to progress through the affected growth plates (Figure 6.2); it also reduces hypertrophy levels, but this only becomes evident later (Pollard et al. submitted). Our array-based screening for genes associated with this growth plate sensitivity to movement-related mechanical stimuli show very strong links to a coincident dampening of mTOR pathway signaling (Pollard et al. submitted). Dissection of the most rapid effects of immobilization suggests that the primary target of movement is cell cycle progression via an mTOR-mediated mechanism in order to control the polarity required for efficient intercalation, which, in turn, accelerates proliferation and ECM production to hasten cartilage–bone transition and growth.

Though the molecular mechanisms governing mechanical regulation of growth plate dynamics have not been fully characterized, they are likely to involve Ihh/PTHrP signaling. The expression of these regulators of cartilage proliferation and differentiation has been shown to be mechanosensitive both *in vitro* and *in vivo*, and thus may allow for coordination of genetic and mechanical regulation in establishing limb skeletal form.

6.5 An Overview of Relevant Osteoarthritic Joint Changes

Osteoarthritis is a degenerative joint disease characterized by compromised articular cartilage integrity. The pathogenesis of osteoarthritis frequently involves switching from an articular to an endochondral growth plate chondrocyte phenotype (Staines et al. 2014). Articular chondrocytes usually display much less “dynamism” than those of

(a)

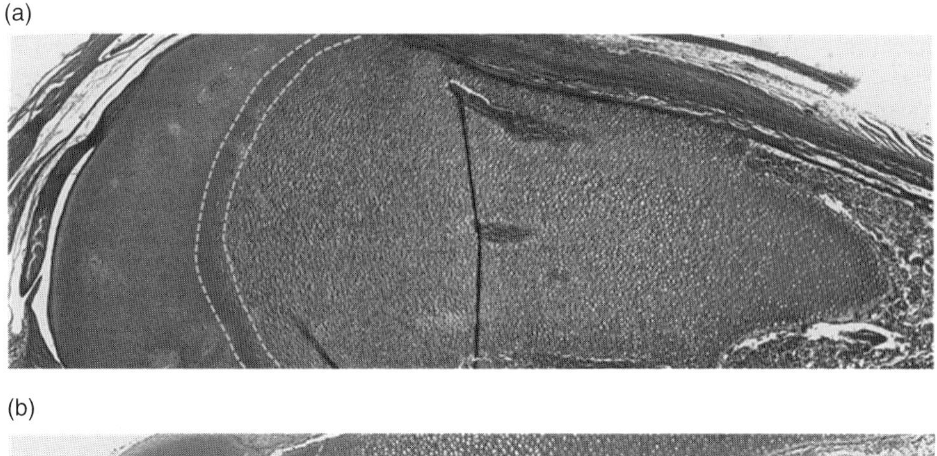

(b)

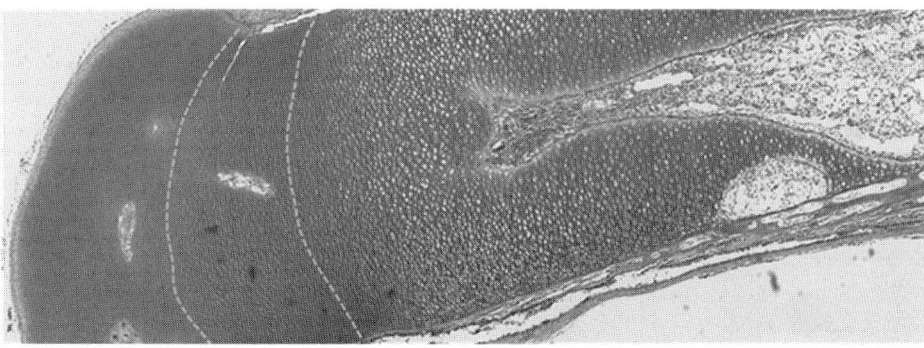

Figure 6.2 Growth plate cartilage from embryonic chickens at 18 days' incubation. The "proliferative zone" where cells express proliferative markers such as PCNA, which is expressed in the S-phase of the cell cycle, is indicated by dotted lines. (a) Growth plate of a normal embryo. (b) Growth plate of a pharmacologically immobilized embryo, demonstrating an expanded proliferative zone, resulting from the failure of cells to complete the cell cycle and progress through the growth plate. (*See insert for color representation of the figure.*)

the growth plate. The former exhibit a stable phenotype in healthy joints and rarely, if ever, undergo the processes seen in the growth plate, such as rapid proliferation and subsequent hypertrophy and apoptosis. This difference in behavior reflects the contrasting fates of these cartilage structures: the transience of the growth plate, with its resident chondrocytes ultimately being replaced by bone, could hardly differ more from the articular cartilage, wherein chondrocytes are required to be permanent, ideally lasting unblemished for the entire lifetime of the individual.

Many processes linked to chondrocyte hypertrophy contribute to the pathophysiology of osteoarthritis. These include diminished collagen-II, aggrecan, and SOX9 expression, elevated MMP13 expression, apoptosis, and, vitally, ECM mineralization, with associated blood vessel and osteoclast recruitment. Initiation of this growth phenotype by resident articular chondrocytes probably contributes to osteoarthritis degeneration. This direct input to cartilage destruction is accompanied by an indirect hypertrophic chondrocyte contribution to cartilage loss in osteoarthritis, reiterating its developmental paracrine role in orchestrating cartilage replacement with bone via regulation of endothelial cell, osteoblast, and osteoclast behavior. Cartilage growth and ossification is central to osteophyte formation and may contribute to subchondral bone sclerosis, articular cartilage thinning, and joint space narrowing in the osteoarthritic joint. Exposure of the underpinning pathways will improve our understanding, diagnosis, and treatment of osteoarthritis.

Detection of collagen-X (a distinctive hypertrophic chondrocyte marker) and raised Ihh, osteocalcin, CD36, and alkaline phosphatase levels point strongly to ectopic hypertrophic differentiation in osteoarthritic cartilage. Broadly similar genome-wide expression patterns in hypertrophic growth plate chondrocytes and experimental osteoarthritis endorse this view. One prominent common marker is MMP13, which seems to have a major role in cartilage destruction. We acknowledge that this view – informed mostly by these molecular signatures – must be tempered by an apparent lack of "true" hypertrophy, namely large cell volume, in osteoarthritic cartilage, implying central differences in the cell behavior. Links between hypertrophy and osteoarthritis have been bolstered, however, by a study in osteoarthritic patients showing that disease grade and collagen-X expression both correlate with cartilage calcification. This linkage might not be a general feature of osteoarthritis but it nonetheless supports a clinical relevance in human osteoarthritis. *This has led researchers to question to what extent these links between osteoarthritic chondrocyte characteristics and developmental pathways rely upon the mechanical loading function of these joint tissues for their interaction?*

6.6 Lessons for Osteoarthritis from Joint Formation

Disruption of cartilage homeostasis by aging, genetic predisposition, trauma, or metabolic disorder in osteoarthritis induces profound phenotypic chondrocyte modifications. These, in turn, promote the synthesis of a subset of factors that induce cartilage damage and target other joint tissues. Chondrocyte-derived inflammatory factors such as cytokines (e.g., IL-1/tumor necrosis factor alpha (TNFα)), chemokines, alarmins, prostanoids (e.g., PGE_2), adipokines, and nitric oxide (NO), and expression of their cell-surface receptors, clearly contribute to at least the latter stages of osteoarthritis. These chondrocyte responses in osteoarthritic joints are regulated to achieve complex modulation of catabolic and anabolic pathways (Goldring and Berenbaum 2004; Houard et al. 2013).

Intriguingly, while these inflammatory markers are linked to osteoarthritis, so is postnatal expression of the factors that are mechanodependently regulated during joint formation. Expression of CD44 by chondrocytes allows their interaction with the ECM and the transduction of mechanical signals to maintain articular cartilage integrity (Ostergaard et al. 1997). CD44 upregulation in articular cartilage is associated with osteoarthritis severity in human joints (Zhang et al. 2013). MEK-ERK1/2 pathway activation also correlates with elevated expression of degradative enzymes, including a disintegrin and metalloproteinase with thrombospondin motifs (ADAMTS) and matrix metalloproteinases (MMPs) with strong links to osteoarthritis development (Appleton et al. 2010; Pitsillides and Beier 2011). It is possible that activation of both CD44 and MEK-ERK is connected in developing embryo joints and in articular cartilage of osteoarthritic joints by common mechanoregulatory mechanisms. Studies are required to address this, since proof of such a connection would significantly alter our strategies for preventing, treating, and even reversing the pathobiology of osteoarthritis.

Constitutive activation of the MEK-ERK, p38^{MAPK}, and COX-2 pathways exploits signaling centered on nuclear factor κB (NF-κB). Regulation of these pathways – conventionally linked with inflammation – by movement in normal joint development and in osteoarthritic chondrocytes suggests that the cellular features linked to the onset and progression of osteoarthritis might be the product of local mechanical challenge. Recent findings indicate that the natural osteoarthritis in STR/Ort mice is also intimately linked to an "inflammatory" articular chondrocyte gene signature, dominated by NF-κB pathway signaling, even prior to osteoarthritis onset, suggesting that mechanodependent NF-κB activation predisposes cartilage to osteoarthritis (Poulet et al. 2012).

NF-κB family members are indeed known to orchestrate mechanical, inflammatory, and stress-activated processes. Two pivotal kinases, IκB kinase (IKK)α and IKKβ, activate NF-κB dimers to regulate expression of specific target genes involved in ECM remodeling and terminal chondrocyte differentiation. Indeed, IKKα functions *in vivo* to control hypertrophic differentiation and collagenase activity, and thus represents a potential therapeutic target in osteoarthritis (Olivotto et al. 2015). Other pathways that may shift the phenotype of normally quiescent articular chondrocytes, disrupting homeostasis and causing aberrant expression of proinflammatory and catabolic genes, include receptors such as discoidin domain receptor 2 and syndecan-4 and transcription factors such as NF-κB, C/EBPβ, ETS, Runx2, and hypoxia-inducible factor 2α (HIF-2α). Low-grade inflammation detected by proteomic/transcriptomic analyses of synovial fluids/membranes from osteoarthritic patients strengthens this view by identifying the membrane attack complex-mediated arm of complement – and its co-localization with MMP13 and activated ERK around chondrocytes – as being crucial in the pathogenesis of osteoarthritis (Wang et al. 2011). Discerning the contributions of these pathways to the initiation and progression of osteoarthritis is therefore likely pivotal (Goldring and Otero 2011). Indeed, the link we propose – between changes seen in osteoarthritic cartilage chondrocytes and those contributing to embryonic joint formation that exhibit sensitivity to limb movement – is likely best explored by comparing loaded and unloaded cartilage zones, where vulnerability to osteoarthritis likely diverges on the basis of local mechanical triggers.

6.7 Lessons for Osteoarthritis from Endochondral Ossification

As in joint formation, several key pathways in skeletal development known to control longitudinal bone growth are subject to regulation by the mechanical environment, and there is evidence that these pathways may be recapitulated in osteoarthritis. Markers formerly thought to be expressed only in the growth plate have been found in osteoarthritic articular cartilage. These factors are linked both to the ectopic proliferation that occurs in osteoarthritic cartilage and to chondrocyte hypertrophy and the ECM mineralization that follows, including MMP13, collagen-X, and Ihh. *Does the mechanical environment in the joint impact on the proposed role of developmental pathways in osteoarthritis, revealing new strategies for limiting its progression?*

Interlinks between the apparently discordant "transient" and inherently "stable" articular cartilage chondrocyte phenotypes are therefore crucial, and whether their switching contributes to osteoarthritis is not yet fully established (Thorogood and

Hinchliffe 1975; Archer et al. 1994; Kronenberg 2003). Indeed, recent evidence suggesting uncommon origins for growth-plate and articular cartilage chondrocytes only makes this hypothesis more controversial (Ito and Kida 2000; Bastow et al. 2005; Kavanagh et al. 2006). Our recent data reveal changes in the articular cartilage of STR/Ort mouse knee joints consistent with aberrant deployment of endochondral processes prior to osteoarthritis onset (Poulet et al. 2012). These data indicate, at least in the spontaneous human-like osteoarthritis in STR/Ort mice, that growth-related endochondral ossification abnormalities may forecast mechanisms of osteoarthritis development in articular cartilage. A meta-analysis of transcriptional profiles revealed elevation in functions linked with endochondral ossification, increased MMP13 and collagen type-X expression, and differential expression of known mineralization regulators in STR/Ort joint articular cartilage (Poulet et al. 2012).

It is vital that these observations were made in joints – albeit osteoarthritic ones – that were serving their "normal" mechanical load-bearing function, and that the changes exhibited were shared with those driven in the growth plate by movement of embryonic limbs. Indeed, osteoarthritis is often characterized by frequent new rounds of proliferation in individual chondrocyte lacunae and by newly identified roles for mTOR in osteoarthritis processes. Raised ECM production rates are also evident in early osteoarthritis and in the hastening of cartilage–bone transition seen at the cartilage–subchondral bone interface. It is therefore tempting to speculate that activation of these transient growth/hypertrophic chondrocyte behaviors in osteoarthritic joints is connected to those driven in developing embryo growth plates by movement, and that they exploit common mechanoregulatory mechanisms.

6.8 Conclusion

The requirement for movement in the formation of embryonic joints and in accelerating endochondral ossification for long bone growth is linked to osteoarthritis in many ways. Most obviously, they are linked by the emergence of osteoarthritis characteristics in articular chondrocytes, which suggest that they deploy these mechanodependent embryonic processes postnatally. Whether their deployment is beneficial or detrimental in osteoarthritic joints remains to be surmised; perhaps osteoarthritis is so prevalent because it is the sole response that choncdrocytes can make in the mechanically challenging environment of the joint. Recent genome-wide association studies (GWASs) seeking to identify genetic polymorphisms associated with osteoarthritis, where most of the genes and pathways identified are implicated in either joint or cartilage development (e.g., GDF5, SMAD3 FRZB, and DIO2), strengthen this link. Joint movement and osteoarthritis are also linked by developmental abnormalities, such as chondrodysplasias, that lead to osteoarthritis via altered joint geometry and loading. Tenuous links exist in the proposed developmental origins of osteoarthritis, which likely impact later growth and physiological joint function. In secondary osteoarthritis, the initial failure of the mechanodependent regulation of the developmental pathways involved in joint formation and endochondral growth may predispose to the development of osteoarthritis later in life. Finally, in some forms of osteoarthritis, linkage is evident in the switch from articular "permanent" chondrocytes to phenotypes characteristic of "transient" developing cartilage; ultimately, this is linked to chondrocyte hypertrophy, ECM degradation and loss of joint integrity. Ectopic activation in articular chondrocytes of developmental processes has long been suspected to contribute to osteoarthritis. This chapter indicates that this link may rely upon the mechanics engendered by movement.

Acknowledgements

This work was supported by the Anatomical Society. The authors are indebted to Christine Yu for the images of joints in normal and immobilized chick limbs.

References

Appleton, C. T., G. Shirine, E. Usmani, J. S. Mort, and F. Beier. 2010. "Rho/ROCK and MEK/ERK activation by transforming growth factor-α induces articular cartilage degradation." *Laboratory Investigation* **90**(1): 20–30. doi:10.1038/labinvest.2009.111.

Archer, C. W., H. Morrison, and A. A. Pitsillides. 1994. "Cellular aspects of the development of diarthrodial joints and articular cartilage." *Journal of Anatomy* **184**(Pt. 3): 447–56. PMID:7928634.

Bastow, E. R., K. J. Lamb, J. C. Lewthwaite, A. C. Osborne, E. Kavanagh, C. P. Wheeler-Jones, and A. A. Pitsillides. 2005. "Selective activation of the MEK-ERK pathway is regulated by mechanical stimuli in forming joints and promotes pericellular matrix formation." *Journal of Biological Chemistry* **280**(12): 11 749–58. PMID:15647286.

Bekoff, A. 1981. "Embryonic development of chick motor behaviour." *Trends in Neurosciences* **4**: 181–4. doi:http://dx.doi.org/10.1016/0166-2236(81)90059-X.

Carry, M. R., M. Morita, and H. O. Nornes. 1983. "Morphogenesis of motor endplates along the proximodistal axis of the mouse hindlimb." *Anatomical Record* **207**(3): 473–85. PMID:6650877.

Chung, U. I., E. Schipani, A. P. McMahon, and H. M. Kronenberg. 2001. "Indian hedgehog couples chondrogenesis to osteogenesis in endochondral bone development." *Journal of Clinical Investigation* **107**(3): 295–304. doi: 10.1172/jci11706.

Cooper, K. L., S. Oh, Y. Sung, R. R. Dasari, M. W. Kirschner, and C. J. Tabin. 2013. "Multiple phases of chondrocyte enlargement underlie differences in skeletal proportions." *Nature* **495**(7441): 375–8. doi:10.1038/nature11940.

de Vries, J. I. P., G. H. A. Visser, and H. F. R. Prechtl. 1982. "The emergence of fetal behaviour: I. Qualitative aspects." *Early Human Development* 7: 301–22. PMID:7169027.

Dowthwaite, G. P., A. C. Ward, J. Flannely, R. F. Suswillo, C. R. Flannery, C. W. Archer, and A. A. Pitsillides. 1999. "The effect of mechanical strain on hyaluronan metabolism in embryonic fibrocartilage cells." *Matrix Biology* **18**(6): 523–32. PMID:10607914.

Drachman, D. B. and Sokoloff, L. 1966. "The role of movement in embryonic joint development." *Developmental Biology* **14**: 401–20.

Fell, H. B. and R. G. Canti. 1934. "Experiments on the development in vitro of the avian knee-joint." *Proceedings of the Royal Society of London. Series B, Biological Sciences* **116**(799): 316–51. doi:10.1098/rspb.1934.0076.

Goldring, M. B. and F. Berenbaum. 2004. "The regulation of chondrocyte function by proinflammatory mediators: prostaglandins and nitric oxide." *Clinical Orthopaedics and Related Research* **427**(Suppl.): S37–46. PMID:15480072.

Goldring, M. B. and M. Otero. 2011. "Inflammation in osteoarthritis." *Current Opinion in Rheumatology* **23**(5): 471–8. doi:10.1097/BOR.0b013e328349c2b1.

Hall, B. K. and S. W. Herring. 1990. "Paralysis and growth of the musculoskeletal system in the embryonic chick." *Journal of Morphology* **206**(1): 45–56. doi: 10.1002/jmor.1052060105.

Hamburger, V. and M. Balaban. 1963. "Observations and experiments on spontaneous rhythmical behavior in the chick embryo." *Developmental Biology* **7**: 533–45. PMID:13952299.

Hanson, M. G. and L. T. Landmesser. 2003. "Characterization of the circuits that generate spontaneous episodes of activity in the early embryonic mouse spinal cord." *Journal of Neuroscience* **23**(2): 587–600. PMID:12533619.

Hosseini, A. and D. A. Hogg. 1991a. "The effects of paralysis on skeletal development in the chick embryo. I. General effects." *Journal of Anatomy* **177**: 159. PMID:1769890.

Hosseini, A. and D. A. Hogg. 1991b. "The effects of paralysis on skeletal development in the chick embryo. II. Effects on histogenesis of the tibia." *Journal of Anatomy* **177**: 169–78. PMID:1769891.

Houard, X., M. B. Goldring, and F. Berenbaum. 2013. "Homeostatic mechanisms in articular cartilage and role of inflammation in osteoarthritis." *Current Rheumatology Reports* **15**(11): 375. doi:10.1007/s11926-013-0375-6.

Hunziker, E. B., R. K. Schenk, and L. M. Cruz-Orive. 1987. "Quantitation of chondrocyte performance in growth-plate cartilage during longitudinal bone growth." *Journal of Bone and Joint Surgery. American Volume* **69**(2): 162–73. PMID:3543020.

Ito, M. M. and M. Y. Kida. 2000. "Morphological and biochemical re-evaluation of the process of cavitation in the rat knee joint: cellular and cell strata alterations in the interzone." *Journal of Anatomy* **197**(4): 659–79. doi: 10.1046/j.1469-7580.2000.19740659.x.

Kahn, J., Y. Shwartz, E. Blitz, S. Krief, A. Sharir, D. A. Breitel, et al. 2009. "Muscle contraction is necessary to maintain joint progenitor cell fate." *Developmental Cell* **16**(5): 734–43. doi: 10.1016/j.devcel.2009.04.013.

Kavanagh, E., V. L. Church, A. C. Osborne, K. J. Lamb, C. W. Archer, P. H. Francis-West, and A. A. Pitsillides. 2006. "Differential regulation of GDF-5 and FGF-2/4 by immobilisation in ovo exposes distinct roles in joint formation." *Developmental Dynamics* **235**(3): 826–34. doi: 10.1002/dvdy.20679.

Kronenberg, H. M. 2003. "Developmental regulation of the growth plate." *Nature* **423**(6937): 332–6. doi: 10.1038/nature01657.

Kronenberg, H. M. 2006. "PTHrP and skeletal development." *Annals of the New York Academy of Sciences* **1068**: 1–13. doi: 10.1196/annals.1346.002.

Lamb, K. J., J. C. Lewthwaite, J. P. Lin, D. Simon, E. Kavanagh, C. P. Wheeler-Jones, and A. A. Pitsillides. 2003. "Diverse range of fixed positional deformities and bone growth restraint provoked by flaccid paralysis in embryonic chicks." *International Journal of Experimental Pathology* **84**(4): 191–9.

Lewthwaite, J. C., E. R. Bastow, K. J. Lamb, J. Blenis, C. P. Wheeler-Jones, and A. A. Pitsillides. 2006. "A specific mechanomodulatory role for p38 MAPK in embryonic joint articular surface cell MEK-ERK pathway regulation." *Journal of Biological Chemistry* **281**(16): 11 011–18. doi: 10.1074/jbc.M510680200.

Murray, D. F. M. 1926. "An experimental study of the development of the limbs of the chick." *Proceedings of the Linnean Society of New South Wales* **51**: 179–263.

Murray, P. D. F. and D. Selby. 1930. "Intrinsic and extrinsic factors in the primary development of the skeleton." *Development Genes and Evolution* **122**(3): 629–62. doi: 10.1007/bf00573594.

Nowlan, N. C., C. Bourdon, G. Dumas, S. Tajbakhsh, P. J. Prendergast, and P. Murphy. 2010a. "Developing bones are differentially affected by compromised skeletal muscle formation." *Bone* **46**(5): 1275–85. doi:10.1016/j.bone.2009.11.026.

Nowlan, N. C., J. Sharpe, K. A. Roddy, P. J. Prendergast, and P. Murphy. 2010b. "Mechanobiology of embryonic skeletal development: Insights from animal models." *Birth Defects Research Part C: Embryo Today: Reviews* **90**(3): 203–13. doi: 10.1002/bdrc.20184.

Olivotto, E., M. Otero, K. B. Marcu, and M. B. Goldring. 2015. "Pathophysiology of osteoarthritis: canonical NF-κB/IKKβ-dependent and kinase-independent effects of IKKα in cartilage degradation and chondrocyte differentiation." *RMD Open* **1**(Suppl. 1): e000061. doi: 10.1136/rmdopen-2015-000061.

Olsen, B. R., A. M. Reginato, and W. Wang. 2000. "Bone development." *Annual Review of Cell Development Biology* **16**: 191–220. doi: 10.1146/annurev.cellbio.16.1.191.

Osborne, A. C., K. J. Lamb, J. C. Lewthwaite, G. P. Dowthwaite, and A. A. Pitsillides. 2002. "Short-term rigid and flaccid paralyses diminish growth of embryonic chick limbs and abrogate joint cavity formation but differentially preserve pre-cavitated joints." *Journal of Musculoskeletal & Neuronal Interaction* **2**(5): 448–56. PMID:15758413.

Ostergaard, K., D. M. Salter, C. B. Andersen, J. Petersen, and K. Bendtzen. 1997. "CD44 expression is up-regulated in the deep zone of osteoarthritic cartilage from human femoral heads." *Histopathology* **31**(5): 451–9. PMID:9416486.

Pitsillides, A. A. 2006. "Early effects of embryonic movement: 'a shot out of the dark.'" *Journal of Anatomy* **208**(4): 417–31. doi: 10.1111/j.1469-7580.2006.00556.x.

Pitsillides, A. A. and D. E. Ashhurst. 2008. "A critical evaluation of specific aspects of joint development." *Developmental Dynamics* **237**(9): 2284–94. doi:10.1002/dvdy.21654.

Pitsillides, A. A. and F. Beier. 2011. "Cartilage biology in osteoarthritis – lessons from developmental biology." *Nature Reviews. Rheumatology* **7**(11): 654–63. doi: 10.1038/nrrheum.2011.129.

Pitsillides, A. A., C. W. Archer, P. Prehm, M. T. Bayliss, and J. C. Edwards. 1995. "Alterations in hyaluronan synthesis during developing joint cavitation." *Journal of Histochemistry and Cytochemistry* **43**(3): 263–73. PMID:7868856.

Poulet, B., V. Ulici, T. C. Stone, M. Pead, V. Gburcik, E. Constantinou, et al. 2012. "Time-series transcriptional profiling yields new perspectives on susceptibility to murine osteoarthritis." *Arthritis and Rheumatism* **64**(10): 3256–66. doi: 10.1002/art.34572.

Ray, A., P. N. Singh, M. L. Sohaskey, R. M. Harland, and A. Bandyopadhyay. 2015. "Precise spatial restriction of BMP signaling is essential for articular cartilage differentiation." *Development* **142**(6): 1169–79. doi:10.1242/dev.110940.

Roddy, K. A., G. M. Kelly, M. H. van Es, P. Murphy, and P. J. Prendergast. 2011. "Dynamic patterns of mechanical stimulation co-localise with growth and cell proliferation during morphogenesis in the avian embryonic knee joint." *Journal of Biomechanics* **44**(1): 143–9. doi:10.1016/j.jbiomech.2010.08.039.

Rowe, D. A. and J. F. Fallon. 1982. "The proximodistal determination of skeletal parts in the developing chick leg." *Journal of Embryology and Experimental Morphology* **68**(1): 1–7. PMID:7108415.

Shwartz, Y., Z. Farkas, T. Stern, A. Aszodi, and E. Zelzer. 2012. "Muscle contraction controls skeletal morphogenesis through regulation of chondrocyte convergent extension." *Developmental Biology* **370**(1): 154–63. doi:10.1016/j.ydbio.2012.07.026.

Summerbell, D., J. H. Lewis, and L. Wolpert. 1973. "Positional information in chick limb morphogenesis." *Nature* **244**(5417): 492–6. PMID:4621272.

Thorogood, P. V. and J. R. Hinchliffe. 1975. "An analysis of the condensation process during chondrogenesis in the embryonic chick hind limb." *Journal of Embryology and Experimental Morphology* **33**(3): 581–606. PMID:1176861.

Tickle, C. 1995. "Vertebrate limb development." *Current Opinion in Genetics & Development* **5**(4): 478–484. doi:http://dx.doi.org/10.1016/0959-437X(95)90052-I.

Tickle, C. 2003. "Patterning systems – from one end of the limb to the other." *Developmental Cell* **4**(4): 449–58. doi:10.1016/s1534-5807(03)00095-9.

Tickle, C. and A. Münsterberg. 2001. "Vertebrate limb development – the early stages in chick and mouse." *Current Opinion in Genetics & Development* **11**(4): 476–81. PMID:11448636.

Wang, Q., A. L. Rozelle, C. M. Lepus, C. R. Scanzello, J. J. Song, D. M. Larsen, et al. 2011. "Identification of a central role for complement in osteoarthritis." *Nature Medicine* **17**(12): 1674–9. doi:10.1038/nm.2543.

Ward, A. C., G. P. Dowthwaite, A. A. Pitsillides. Hyaluronan in joint cavitation. *Biochem Soc Trans*. **27**(2):128–35. PMID: 10093721.

Wolpert, L. 2010. "Arms and the man: the problem of symmetric growth." *PLoS Biology* **8**(9): e1000477. doi:10.1371/journal.pbio.1000477.

Zhang, F.-J., W. Luo, S.-G. Gao, D.-Z. Su, Y.-S. Li, C. Zeng, and G.-H. Lei. 2013. "Expression of CD44 in articular cartilage is associated with disease severity in knee osteoarthritis." *Modern Rheumatology* **23**(6): 1186–91. doi:10.1007/s10165-012-0818-3.

7

Modulating Skeletal Responses to Mechanical Loading by Targeting Estrogen Receptor Signaling

Gabriel L. Galea[1] *and Lee B. Meakin*[2]

[1] *Developmental Biology of Birth Defects, UCL Great Ormond Street Institute of Child Health, London, UK*
[2] *School of Veterinary Sciences, University of Bristol, Bristol, UK*

7.1 Introduction

Mechanical loading is the primary functional determinant of bone mass and architecture. Withstanding loading-engendered strains (defined as the percentage change in dimension) is both the primary purpose of the skeleton and the stimulus which governs the (re)modeling activity of the cells responsible for determining bone structure. Increases in mechanical strain generated by increased load-bearing result in increased bone formation by osteoblast cells, increasing bone mass and improving bone architecture. This reduces strains back to a target level. Conversely, reductions in the bone strain environment, such as during disuse, result in increased activity of the osteoclast cells responsible for bone resorption, thus reducing bone mass and increasing strain. This homeostatic feedback loop, commonly referred to as the mechanostat (Figure 7.1), is active locally in each site of load-bearing bones, matching the forming and resorbing activities occurring in each region of each bone with the loading requirements imposed on it (Frost 1987; Skerry 2006).

The mechanostat appears to fail in later life, results in remodeling as in a state of disuse, such that resorption predominates despite ongoing loading (Skerry and Lanyon 2001; Meakin et al. 2014a). Consequently, bone mass is lost and architecture deteriorates, leading to fragility and an increased incidence of fractures at habitual levels of loading. These "fragility fractures" are characteristic of osteoporosis. Osteoporotic fractures are expected to occur in 50% of women and 30% of men over 50 years of age in their remaining lifetime (van Staa et al. 2001). The higher incidence in women is associated with a lower peak bone mass attained at the end of puberty and a more rapid decline in bone mass following the menopause. This period of rapid bone loss correlates with postmenopausal estrogen withdrawal. A reduction in circulating estrogens has also been correlated with aging-related bone loss in men (Ohlsson and Vandenput 2009). Hormone replacement therapy (HRT) to supplement endogenous estrogens is an effective treatment for postmenopausal osteoporosis, but its association with adverse health effects, including increased risk of various cancers, limits its clinical use (Beral 2003; Beral et al. 2007).

Mechanobiology: Exploitation for Medical Benefit, First Edition. Edited by Simon C. F. Rawlinson.

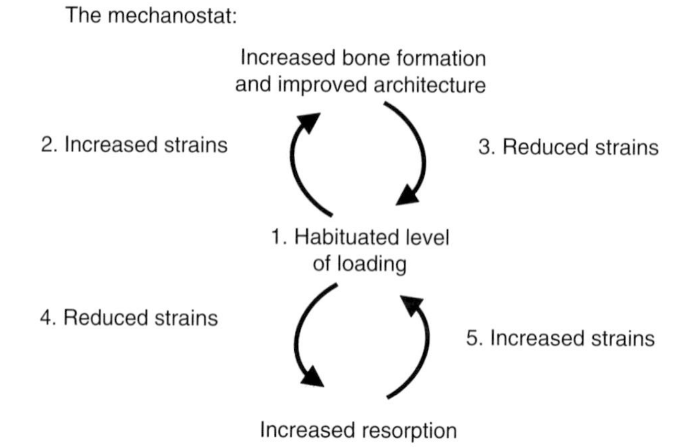

Figure 7.1 Schematic illustration of the mechanostat. (1) Bone remodels toward a habituated steady state in which customary levels of loading engender an acceptable mechanical strain stimulus. (2) When loading increases, such as during exercise, increased strain-related stimuli activate osteoblasts, which are responsible for bone formation, leading to increased bone mass and improved architecture. (3) Consequently, the improvements in bone structure return strains to an acceptable level at the new level of loading. (4) Conversely, when loading is reduced, as occurs during bed rest, strains decrease, such that the resorptive activity of osteoclasts predominates. (5) This reduces bone mass and increases strain levels toward those experienced in the habituated state.

In place of HRT, numerous selective estrogen receptor modulators (SERMs) have been developed in an attempt to selectively mimic the effects of estrogens on bone and other target organs. The effects of estrogens such as 17β-estradiol (E2) on bone are both osteogenic and antiresorptive. They increase proliferation and differentiation of osteoblasts from their precursors, leading to an increased number and activity of the cells responsible for bone formation. Simultaneously, estrogens reduce the differentiation of osteoclasts and increase their rate of apoptosis (programmed cell death), reducing resorption. Loss of circulating estrogens can therefore contribute to the predominance of resorption following the menopause and in later life. However, this raises a question: Why does the increase in mechanical strain consequent to reduced bone mass in an estrogen-deficient state not result in compensatory activation of the mechanisms that underlie the mechanostat? The discovery in 1998 that estrogen receptors expressed locally within mechanoresponsive osteoblastic cells contribute to these cells' responses to mechanical strain offers a potential answer to this question (Damien et al. 1998).

7.2 Biomechanical Activation of Estrogen Receptor Signaling: *In Vitro* Studies

Estrogen signaling is classically activated when systemically circulating E2 diffuses across cell membranes to bind its intracellular receptors ERα and the more recently discovered ERβ. Various novel estrogen receptors, including the transmembrane G-protein-coupled estrogen receptor 1 (GPER1) and estrogen-related receptors (ERRs), are now recognized and are known to influence bone cell function, but the mechanisms by which they exert their effects are not clearly understood. The classical estrogen

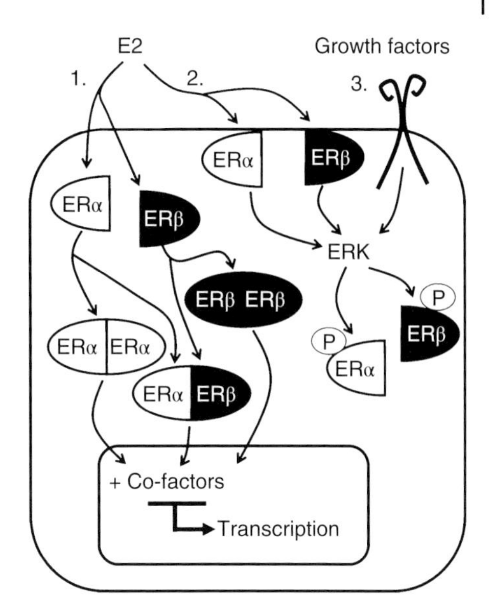

Figure 7.2 Simplified representation of the estrogen receptor signaling cascade. (1) Genomic ligand-dependent estrogen receptor signaling is initiated by estrogens such as 17β-estradiol (E2) diffusing across cell membranes to bind the estrogen receptors, primarily ERα and ERβ. These estrogen receptors then homo- or heterodimerize and translocate to the nucleus, where they interact with various cofactors to alter gene expression. (2) Nongenomic ligand-dependent estrogen receptor signaling is initiated when E2 binds the estrogen receptors typically at the cell membrane to trigger activation of protein kinases, including ERK. (3) Ligand-independent estrogen receptor signaling can be initiated by various growth factors binding cell-surface receptors that activate protein kinases, again including ERK, which are able to phosphorylate (P) the estrogen receptors and thus activate them in the absence of E2.

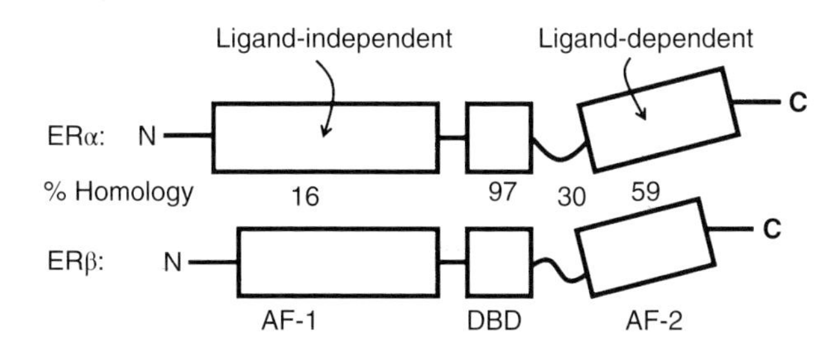

Figure 7.3 Schematic representation of the structure of ERα and ERβ. Both estrogen receptors have an AF-1 domain (which mediates interactions with other proteins), a short linker region, a DNA-binding domain (DBD), a hinge region, and a ligand-binding AF-2 domain. Per cent homologies between the two receptors are based on those previously reported by Dey et al. (2013). Structural differences between the receptors are exploited in the pharmacological development of SERMs.

receptors, ERα and ERβ, are steroid receptors that, upon activation by their ligand, homo- or heterodimerize (i.e., ERα binds to ERβ) and translocate to the nucleus, where they direct gene expression (Figure 7.2). In addition to this "genomic" mode of action, E2 can activate nongenomic signaling pathways, whereby ERα and ERβ interact with kinases such as the extracellular signal-regulated kinase (ERK), which forms part of the mitogen-activated protein kinase (MAPK) cascade. In turn, ERK can phosphorylate the estrogen receptors, activating them in the absence of their ligand (Figure 7.2). This interaction between the estrogen receptors and other proteins occurs in distinct structural domains from those to which E2 binds (Arnal et al. 2013).

The general structures of ERα and ERβ are similar: both have an N-terminal activation function 1 (AF-1) domain, a DNA-binding domain, a short hinge region (which conveys molecular flexibility), and a C-terminal AF-2 domain (Figure 7.3). The AF-1 domain largely mediates interactions with other proteins. For example, ERK phosphorylates ERα's AF-1 domain in osteoblastic cells (Jessop et al. 2001). Various AF-1-dependent interactions are specific to each estrogen receptor, because the AF-1 domain is highly divergent between ERα and ERβ, with only approximately 16% homology (Dey et al. 2013). In contrast, the estrogen receptors' DNA-binding domain is highly homologous, allowing binding to the same estrogen response elements (EREs). EREs are DNA sequences that are recognized and bound by the estrogen receptor's DNA-binding domains, altering gene expression. Binding of E2 to the estrogen receptors occurs at the AF-2 domain. AF-2 is approximately 60% homologous between ERα and ERβ, producing structural differences exploited in the development of SERMs to convey selectivity for either estrogen receptor isoform (Dey et al. 2013). Thus, activation of the estrogen receptors by binding of their endogenous ligand E2 to AF-2 results in "genomic" signaling, which directly alters gene expression, and "nongenomic" signaling through kinase cascades, including MAPK; however, the estrogen receptors can also be activated independently of E2 through phosphorylation of their AF-1 domain, again resulting in both "genomic" and "nongenomic" signaling.

Mechanical strain is one of the stimuli able to ligand-independently activate estrogen receptor signaling. This has been demonstrated using *in vitro* models of mechanical stimulation, in which osteoblastic cells are exposed to predefined levels of dynamic mechanical strain which, if applied to bone *in vivo*, would be expected to be osteogenic. One is the four-point bending model, in which osteoblastic cells are cultured on flexible plastic slides, which are then cyclically deformed between prongs at the extreme ends of each slide, causing them to bend upwards in an arc (Galea and Price 2015). The top surface of this arc is stretched (i.e., experiences tensile strain) such that cells cultured on it are exposed to strain through their substrate. In this four-point bending model, strain activates numerous cascades, including ERK and protein kinase A (PKA). Both ERK and PKA are able to phosphorylate the AF-1 domain of ERα, activating it (Jessop et al. 2001). This kinase-mediated activation of ERα occurs despite purposeful removal of steroid hormones from the cells' culture medium, indicating that strain activates ERα independently of its ligands.

Ligand-independent ERα activation in osteoblastic cells subjected to strain was initially reported to result in genomic signaling (Zaman et al. 2000). This was demonstrated by transfecting osteoblastic cells with exogenous DNA vectors containing two ERE sequences followed by a chloramphenicol acetyl transferase (CAT) gene. In these transfected cells, changes in ERE activation alter CAT expression. When these cells were treated with E2 as a positive control, their CAT activity increased as expected. Similarly, when they were subjected to strain by four-point bending in the absence of E2, their CAT activity also increased. In agreement with this, other studies have observed that exposure of similar osteoblastic cells to strain increases nuclear translocation of ERα. In order to determine the temporal pattern of ERα activation following strain, osteoblastic cells were harvested at various time points following strain and their level of ERα phosphorylation at specific AF-1 sites was compared to that of static control cells by Western blotting. This study revealed that ligand-independent ERα activation is among the first responses of osteoblastic cells to strain, increasing within 5 minutes of a brief episode of dynamic strain (Jessop et al. 2001).

The same studies that observed ERα activation and translocation to the nucleus following strain also identified increased levels of ERα on the cell membrane. Membrane-localized ERα is better able to activate nongenomic signaling pathways. Many such pathways are now recognized to be facilitated by ERα following strain. Among the most important of these is the canonical Wnt signaling pathway. Conventionally, canonical Wnt signaling is activated when Wnt glycoprotein ligands bind the co-receptors low-density lipoprotein receptor-related protein (LRP) and Frizzled at the cell membrane, leading to increased levels of the transcription factor β-catenin, which in the absence of Wnt signals is normally phosphorylated by glycogen synthase kinase (GSK)-3β and targeted for degradation. Wnt signaling increases β-catenin levels, promoting its translocation to the nucleus, where it acts as a transcription factor, increasing the expression of genes involved in proliferation and osteoblast differentiation. Exposure of primary osteoblasts harvested from the long bones of mice to mechanical strain *in vitro* rapidly (within 30 minutes) increases the nuclear translocation of β-catenin. However, nuclear translocation of β-catenin does not occur in cells from ERα knockout mice, suggesting that the presence of ERα in osteoblasts facilitates the increase in Wnt signaling (Armstrong et al. 2007).

One of the mechanisms by which this is achieved involves a function of ERα at the cell membrane, where it is able to interact with the pro-proliferative insulin-like growth factor (IGF) receptor (IGFR) (Sunters et al. 2009). Binding of ERα to the IGFR increases the sensitivity of IGFR to available IGF ligands, increasing IGF signaling. IGF signaling indirectly inhibits GSK-3β, thereby resulting in increased β-catenin levels. This pathway is not activated by strain in osteoblasts from ERα knockout mice, indicating that the presence of ERα facilitates β-catenin stabilization downstream of IGF signaling.

Wnt, IGF, and ERα signaling are all well-established pro-proliferative cascades in various cell types, including osteoblasts. It is therefore not surprising that one of the recognized outcomes of strain-related ERα signaling in osteoblastic cells is to increase proliferation. This is supported by various lines of evidence. Whereas osteoblasts derived from the long bones of wild-type mice increase their rate of proliferation following strain, those derived from ERα knockout mice do not (Lee et al. 2003). Pharmacological inhibition of ERα with antagonizing SERMs prevents the increase in osteoblastic cell proliferation normally observed following strain. Furthermore, transfecting additional ERα into osteoblastic cells before subjecting them to strain increases their proliferative response (Zaman et al. 2000). The proliferative response to strain appears to require canonical Wnt/β-catenin signaling, as inhibition of this signaling pathway by an endogenous, bone-specific antagonist called sclerostin prevents the increase in osteoblastic cell proliferation observed following strain (Galea et al. 2013a).

Sclerostin is a glycoprotein secreted by osteocytes: terminally differentiated osteoblasts embedded in the bone matrix, where they are ideally located to sense changes in mechanical strain. Strain decreases expression of *Sost* RNA, which codes for sclerostin; similarly, activation of the estrogen receptors by E2 also downregulates *Sost*. Surprisingly, however, a recent study found that selectively inhibiting ERβ, not ERα, prevents downregulation of *Sost* following strain, suggesting that ERβ expression in mature osteoblast cells expressing *Sost* contributes to strain-related activation of Wnt signaling, which is also known to be facilitated by ERα (Galea et al. 2013a). However, ERβ appears to oppose ERα by suppressing osteoblastic proliferation (Galea et al. 2013a). Consistent with this, long bone-derived osteoblasts from ERβ knockout mice show an enhanced proliferative response to strain compared with similarly derived cells from wild-type mice (Jessop et al. 2004).

In contrast to ERα, little is known about the mechanisms by which ERβ is activated following strain, or the signaling pathways in which it is involved. One study demonstrated that both ERα and ERβ contribute to ERK activation following strain, such that siRNA-mediated knockdown of either estrogen receptor blunted the ability of strain to increase levels of phosphorylated ERK (the active form of ERK) (Aguirre et al. 2007). This finding has been replicated in a recent study, in which primary osteoblasts from wild-type or ERβ knockout mice were subjected to oscillating fluid flow shear stress (FFSS) as an alternative *in vitro* model of mechanical stimulation (shear being a form of strain). FFSS increased levels of phosphorylated ERK in cells from wild-type but not from ERβ knockout mice (Castillo et al. 2014). However, while ERβ and ERα appear to cooperate to activate ERK, in other contexts they oppose each other's activity, as in their influence on osteoblastic cell proliferation. ERβ has been suggested to be the "dominant" estrogen receptor, as it is able to inhibit expression of various target genes upregulated by ERα homodimers (Lindberg et al. 2003). In addition, inhibition of ERα by ERβ may be partly a result of ERβ reducing ERα expression, as primary osteoblasts from ERβ knockout cells express higher levels of ERα than similarly harvested cells from wild-type mice (Castillo et al. 2014).

Taken together, these various *in vitro* studies demonstrate that ligand-independent functions of ERα and ERβ are components of the early strain-related cellular responses underlying the mechanostat. In the case of ERα, these functions include facilitation of canonical Wnt signaling, IGF signaling, and, ultimately, increased proliferation of osteoblastic cells. Osteocytes are terminally differentiated and therefore do not proliferate, but express the Wnt antagonist *Sost*/sclerostin, which is normally downregulated following mechanical loading. *In vitro*, downregulation of *Sost* by strain appears to be ERβ-mediated, potentially facilitating pro-proliferative Wnt activation in nearby osteoblasts. However, though ERα and ERβ can cooperate in processes such as ERK activation, within individual cells they can also oppose each other's activities, including alteration of gene expression. Some of these functions of the estrogen receptors within different members of the osteoblast lineage that are responsive to strain and estradiol are illustrated in Figure 7.4.

7.3 Skeletal Consequences of Altered Estrogen Receptor Signaling: *In Vivo* Mouse Studies

Antagonistic and compensatory interactions between ERα and ERβ have confounded efforts to delineate their roles in bone. One approach to the delineation of any gene's function is to knock it out using transgenic techniques to remove a part or all of its known coding regions from a mouse's genome. When knockout mice were first developed, it was only possible to delete the gene of interest in the germline, such that all cells at all stages of development lacked the target protein. This approach, generating "global" knockout mice, was applied to delete either ERα, ERβ, or both. However, the continued expression of truncated estrogen receptors in early knockout models rendered them incomplete. Sufficient expression of the ERα ligand-binding AF-2 domain was present in double ERα/ERβ knockout mice to enable estradiol to stimulate similar increases in their cortical, but not trabecular, bone mass to those seen in wild-type mice (Lindberg et al. 2002). This early finding is consistent with a much more recent study, in which, using purposeful targeting techniques, only ERα's AF-2 or AF-1 domain was selectively deleted. These studies also demonstrated that the AF-2 domain of ERα mediates E2's osteogenic effects in cortical bone, while the entire ERα is required in trabecular bone (Börjesson et al. 2011).

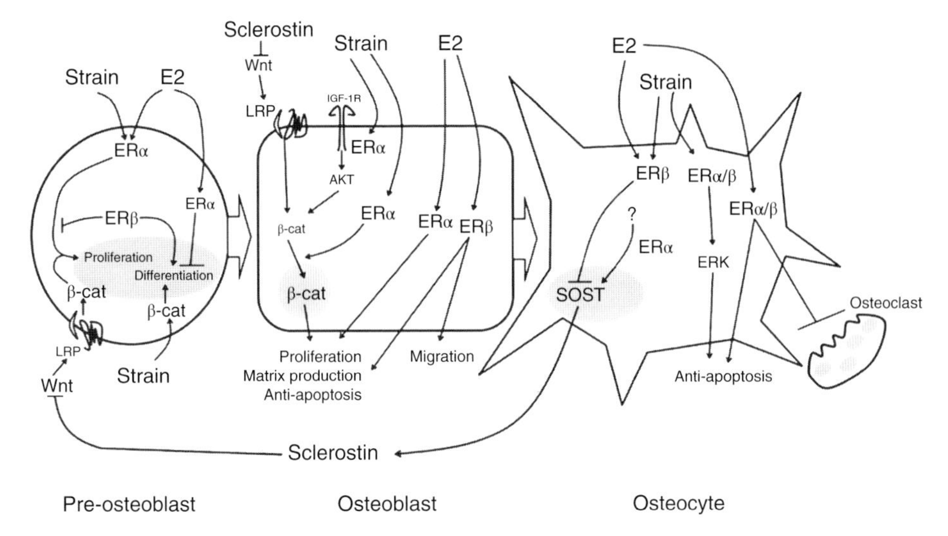

Figure 7.4 Schematic representation of the actions of estrogen receptor in different stages of the osteoblast lineage. Functions of the estrogen receptors illustrated here are inferred from mechanistic studies of osteoblastic cell types used to model osteoblastic cells in different stages of differentiation. Early osteoblasts can proliferate or differentiate, and though ERα promotes their proliferation and suppresses differentiation, there is evidence that ERβ promotes differentiation while inhibiting proliferation. In mature osteoblasts, ERα promotes proliferation and ERβ reduces proliferation, but both reduce apoptosis. ERβ and ERα both contribute to these cells' bone-forming functions. In response to mechanical strain, ERα facilitates osteogenic signaling pathways, including IGF and Wnt/β-catenin. Both receptors regulate *Sost* expression: ERβ mediates its acute downregulation via strain and estradiol, and ERα maintains its basal expression. Both estrogen receptors also reduce osteoclast recruitment by osteoblastic cells. β-cat, β-catenin. *Source:* Figure adapted with permission from Galea et al. 2013a.

Another limitation of studying global knockout mice is that compensatory adaptations can come into play over the course of the animal's lifetime (as recently reviewed in Galea et al. 2013b). This is the case in estrogen receptor knockout mice, as female ERα, but not ERβ, knockout mice have very high levels of circulating estrogens. Conversely, whereas IGF1 is reduced in ERα knockout female mice, it is elevated in those lacking ERβ, potentially exerting estrogen-independent osteogenic effects, including the role of IGF1 in the mechanostat. Furthermore, ERα expression is increased in the bone of ERβ knockout mice. Notwithstanding these limitations, various estrogen receptor knockout mice have been used to clarify the roles of the estrogen receptors in the mechanostat through *in vivo* loading studies.

In vivo loading of rodent bones has been achieved with many different model systems, discussion of which is beyond the scope of this chapter and can be found elsewhere (e.g., Meakin et al. 2014b). The model of greatest relevance to the studies described here is the noninvasive mouse axial tibial loading model (De Souza et al. 2005), in which anesthetized wild-type or knockout mice are placed in custom-made devices with their knee inside a loading cup attached to a mobile lever arm and their ankle held stable in a similar cup directly below the knee. As the upper arm displaces to apply a predefined force, this force is axially imposed through the tibia. Critically, because the tibia contains both cortical and trabecular bone (unlike the mouse ulna, which lacks meaningful trabecular bone), the osteogenic responses to loading can be studied in both these

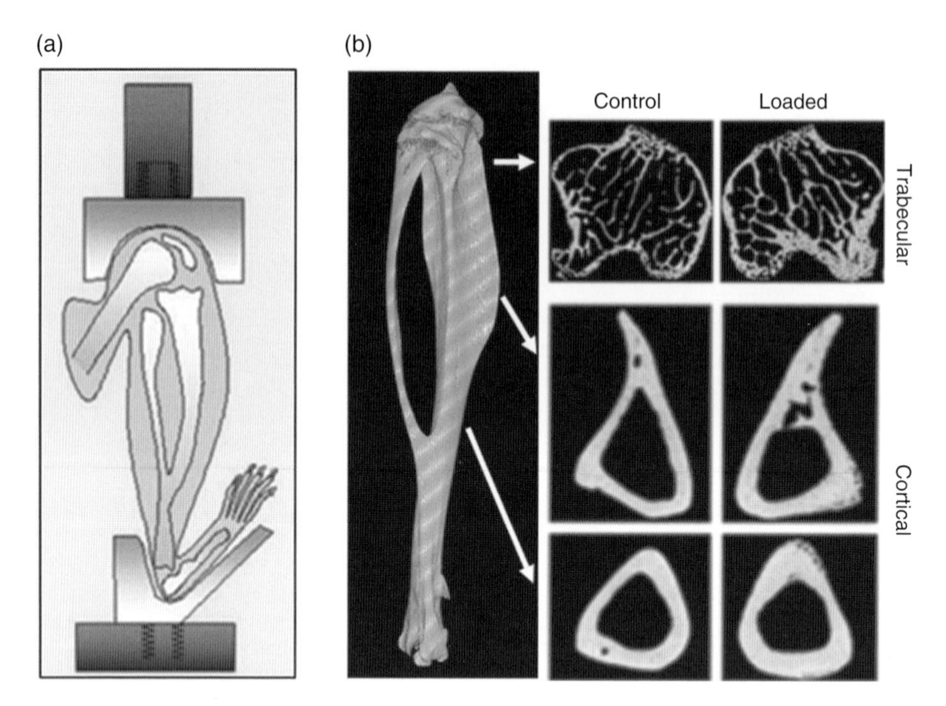

Figure 7.5 Illustration of the mouse axial tibial loading model used to investigate influences on the mechanostat. (a) Schematic diagram of the noninvasive mouse axial tibial loading model. The flexed knee is placed in a cup attached to the actuator arm of an electromagnetic materials testing machine, while the flexed ankle is placed in a cup attached to a load cell, which measures forces applied. (b) Micro-computed tomography (CT) images demonstrating the dramatic increase in both trabecular and cortical bone following 40 cycles of loading three times per week for 2 weeks. *Source:* Figure adapted from Sugiyama et al. (2008), with permission.

compartments (Figure 7.5). Subjecting adult female mice to 40 cycles of axial tibial loading at a peak strain magnitude of 2500 microstrain (με, effectively 0.25% change in total length) every other day for 2 weeks increases cortical bone area by over 20% and trabecular bone volume per tissue volume by over 50% (Sugiyama et al. 2011). The components of the strain-application waveform are important determinants of the ultimate osteogenic response, as discussed elsewhere (Meakin et al. 2014b). It is particularly important to match peak strain magnitudes engendered by loading when comparing wild-type and knockout mice with alterations in bone mass or architecture, as observed in ERα knockouts. In brief, this is achieved by attaching strain gauges to the tibiae of representative mice *ex vivo* and measuring the strains engendered by different magnitudes of load in order to identify loads that should be used to engender the desired peak strain stimulus in both genotypes.

In vivo loading studies comparing the osteogenic responses to loading between wild-type and ERα knockout mice have been reported by various authors and were recently reviewed by Galea et al. (2013b). These studies reproducibly demonstrate that ERα contributes to the osteogenic response to loading in the cortical bone of female mice such that, in the absence of ERα, loading at matched peak strain magnitudes is less able to cause bone formation (Lee et al. 2003) (Figure 7.6). These findings corroborate the various *in vitro* findings suggesting ERα facilitates the cellular mechanisms involved in

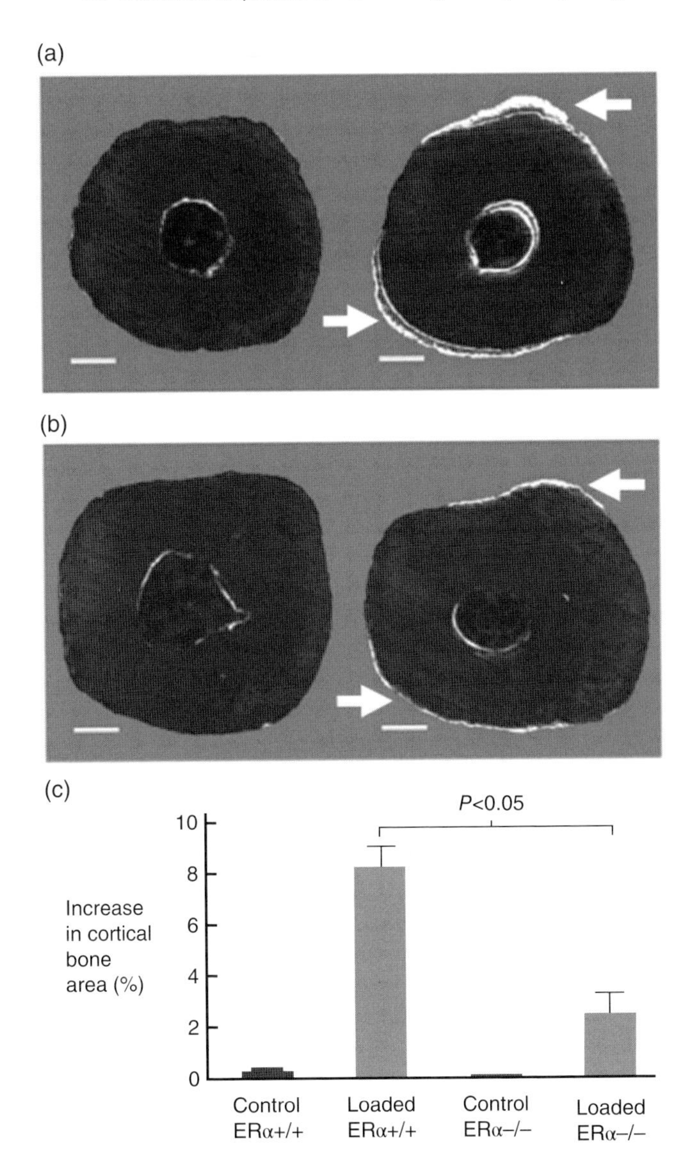

Figure 7.6 Deletion of ERα impairs the osteogenic response to loading in the cortical bone of female mice. Sections taken from the distal ulnae of mice to demonstrate cortical bone formation in (a) wild type and (b) ERα global knockout mice following 2 weeks of mechanical loading. Bones are labeled with fluorescent fluorochromes administered on the first and last days of loading. The distance between labels indicates new bone formation (arrows). (c) Quantification of new bone formation, demonstrating a blunted response to mechanical loading in ERα global knockout mice. *Source:* Figure adapted from Lee et al. (2003), with permission.

the mechanostat. Also, consistent with *in vitro* studies demonstrating that strain activates ERα through its ligand-independent AF-1 domain, selective deletion of ERα's AF-1, but not AF-2, domain is sufficient to blunt the osteogenic response to loading in the tibial cortical bone of female mice (Windahl et al. 2013a). In addition, removal of circulating estrogens through ovariectomy, paralleling *in vitro* steroid-depletion experiments, reinforces the ligand independence of the estrogen receptors' involvement in the mechanostat, as ovariectomy does not alter the osteogenic response to loading (Windahl et al. 2013a). However, despite these contributions of ERα to the mechanostat, its deletion does not reduce trabecular bone gain in female mice subjected to *in vivo* loading (Saxon et al. 2012).

Interpretation of *in vivo* ERα global knockout loading studies is limited by systemic changes, including increased estrogen levels. To account for this, modern transgenic techniques have been employed by several groups to selectively delete ERα in members of the osteoblast lineage, forming "targeted" knockouts. The technology used to generate these knockouts is beyond the scope of this chapter. In brief, it involves the expression of an "eraser" enzyme, Cre, under the control of the promoters of genes selectively expressed in the cells of interest, such as osteoblast differentiation factors. Cre excises DNA between "marker" sequences, referred to as "loxP sites." Thus, when ERα is flanked by loxP sites, it is normally expressed in most cells, but in osteoblasts induced to selectively express Cre, it is deleted. These approaches have been used to delete ERα at various stages of the osteoblast lineage, as discussed elsewhere (Galea et al. 2013b). Taken together, these studies suggest that ERα contributes to cortical bone mass in female mice primarily through its actions in early stages of the lineage. In fact, selective deletion of ERα in terminally differentiated osteocytes does not alter the osteogenic response to loading in female mice (Windahl et al. 2013b). These findings are consistent with the *in vitro* finding that ERα contributes to proliferative responses in osteoblastic cells, given that ERα expression in osteocytes, which are unable to proliferate, does not account for its contribution to the osteogenic response in the cortical bone of female mice following loading.

As with *in vitro* studies, *in vivo* studies investigating the effects of ERβ on the mechanostat have lagged far behind those on ERα. The first study to investigate ERβ's role in the mechanostat reported that female mice with partial ablation of ERβ expression show a smaller increase in cortical bone formation following noninvasive axial loading of the ulna, suggesting ERβ facilitates adaptation to loading, much like ERα (Lee et al. 2004). However, subsequent reports in mice with more complete ERβ ablation have shown an enhanced cortical osteogenic response to loading, leading to the conclusion that ERβ's effect is to inhibit the osteogenic response to loading (Saxon et al. 2007, 2012). It is only possible to speculate on the inconsistencies between these studies, because insights into the role(s) of ERβ gained from knockout models must be interpreted with caution given the potentially compensatory upregulation of ERα (Windahl et al. 2001; Castillo et al. 2014) and changes in ERβ's effects on gene expression with versus without ERα (Lindberg et al. 2003). Osteoblastic cell stage-specific targeted knockout of ERβ is not yet available.

One way to investigate ERβ's roles in loading-induced bone formation may be to modulate its activity with SERMs prior to loading. One study using this methodology treated mice with tamoxifen, a SERM used as an ERα antagonist in humans with breast cancer, but which also activates nongenomic signaling through ERβ and has mixed agonist/antagonist effects on mouse ERα. *In vitro*, tamoxifen treatment reduces *Sost* expression, in a similar manner to ERβ activation (Galea et al. 2013a). *In vivo*, treating mice with tamoxifen dramatically increases both cortical and trabecular bone mass (Sugiyama et al. 2010). In cortical bone, this osteogenic effect is synergistically enhanced by loading, indicating that tamoxifen and loading interact to produce significantly greater bone formation. In trabecular bone, tamoxifen on its own predominantly increases trabecular number, whereas loading predominantly increases trabecular thickness in young-adult female mice. When combined, the effect of loading predominates, such that the increase in trabecular number caused by a high dose of tamoxifen is significantly reduced by loading but the increase in trabecular thickness is synergistically enhanced relative to loading alone. Thus, pretreatment with tamoxifen, a SERM already in clinical use in humans, is able to enhance the mechanostat in mice.

7.4 Skeletal Consequences of Human Estrogen Receptor Polymorphisms: Human Genetic and Exercise-Intervention Studies

Extrapolation of findings from cells or mice to the clinical setting in humans must be done with caution. To our knowledge, no clinical trials have yet directly investigated the ability of estrogen receptor modulators to enhance the mechanostat in humans. Numerous genetic studies have clearly demonstrated the positive roles played by ERα and ERβ in the human skeleton. The most extreme example is that of the "ERα knockout man": a male patient with mutations in ERα who developed low bone mass and gigantism due to failure of growth plate closure (Smith et al. 1994). More subtle genetic polymorphisms in ERα have repeatedly been associated with differences in bone mineral density (BMD) and fracture risk in humans. Furthermore, a small number of studies have investigated whether naturally occurring ERα polymorphisms influence the mechanostat in humans by assessing the response to exercise. Most commonly, these studies have assessed the impact of the PvuII polymorphism in the first intron of ERα on the increase in BMD associated with exercise. One study of prepubertal Finnish girls found that high levels of physical activity were associated with higher BMD and cortical thickness in girls heterozygous for this polymorphism only (Pp genotype) (Suuriniemi et al. 2004). In middle-aged Finnish men enrolled in a 4-year randomized controlled exercise intervention trial, exercise significantly increased BMD in the lumbar spine of individuals with PP or Pp genotypes, but not the recessive pp genotype (Remes et al. 2003). These small, limited human studies suggest that naturally occurring ERα polymorphisms in humans influence the mechanostat, corroborating the findings in ERα knockout mouse loading studies and *in vitro* mechanistic studies. If confirmed in larger cohorts in different populations, studies such as these could help identify patients likely to benefit from exercise interventions for the treatment of osteoporosis based on their ERα status.

No studies have yet been reported investigating the effects of ERβ polymorphisms on the response to exercise in humans. Nonetheless, ERβ polymorphisms have been associated with differences in BMD in various populations, suggesting that ERβ also influences bone mass in people (Shearman et al. 2004; Rivadeneira et al. 2006; Honma et al. 2013). However, in postmenopausal women, just as in mice, ERβ interacts with ERα and IGF1, as the effect of ERβ polymorphisms on BMD and fracture risk is genetically modulated by polymorphisms in ERα and IGF1 (Rivadeneira et al. 2006). It is hoped that further clarification of these interactions may lead to early identification of individuals at increased likelihood of developing osteoporotic fractures, individuals most likely to benefit from exercise in the prevention of osteoporosis, and, ultimately, pharmacological strategies for rescuing the mechanostat through the modulation of estrogen receptor signaling later in life. In support of this, a recent meta-analysis of clinical trials confirmed that treatment with estrogens through HRT significantly enhances the osteogenic effects of exercise in postmenopausal women (Zhao et al. 2015). Though the studies included in this meta-analysis involved various different exercise regimes and HRT treatments, the combination of HRT and exercise was associated with greater overall gains in BMD at both the hip and the lumbar spine relative to exercise alone.

7.5 Conclusion

In this chapter, we have attempted to describe the historical findings that first suggested estrogen receptors are components of the mechanostat; summarize the wealth of *in vitro* data implicating the ligand-independent functions of estrogen receptors in the facilitation of various strain-responsive pathways in osteoblastic cells, particularly those promoting proliferation; correlate these *in vitro* findings with *in vivo* findings of loading studies in transgenic mice; and, finally, explain the potential relevance of these studies to the clinical situation in humans. Throughout, it is clear that the roles of ERα have been clarified to a far greater extent than those of its interaction partner, ERβ. Understanding the roles of ERβ in the mechanostat is likely to require further *in vitro* mechanistic studies, as well as *in vivo* studies in models including osteoblast lineage-targeted knockouts and pharmacological modulation with SERMs. The osteogenic effects of SERMs such as tamoxifen and tamoxifen's ability to augment bone gain induced by loading in mice (Sugiyama et al. 2010), together with the finding that modulating estrogen receptor signaling with HRT enhances the effects of exercise on BMD in postmenopausal women (Zhao et al. 2015), serve as proof of principle that modulating estrogen receptor function can augment the mechanostat. It remains to be determined whether SERMs can be developed which safely and effectively augment the mechanostat to maintain functionally appropriate levels of bone mass in aging humans.

References

Aguirre, I. J., L. I. Plotkin, A. R. Gortazar, M. M. Millan, C. A. O'Brien, S. C. Manolagas, and T. Bellido. 2007. "A novel ligand-independent function of the estrogen receptor is essential for osteocyte and osteoblast mechanotransduction." *Journal of Biological Chemistry* **282**(35): 25 501–8. PMID:17609204.

Armstrong, V. J., M. Muzylak, A. Sunters, G. Zaman, L. K. Saxon, J. S. Price, and L. E. Lanyon. 2007. "Wnt/beta-catenin signaling is a component of osteoblastic bone cell early responses to load-bearing and requires estrogen receptor alpha." *Journal of Biological Chemistry* **282**(28): 20 715–27. PMID:17491024.

Arnal, J.-F., C. Fontaine, A. Abot, M.-C. Valera, H. Laurell, P. Gourdy, and F. Lenfant. 2013. "Lessons from the dissection of the activation functions (AF-1 and AF-2) of the estrogen receptor alpha in vivo." *Steroids* **78**(6): 576–82. doi:10.1016/j.steroids.2012.11.011.

Beral, V. 2003. "Breast cancer and hormone-replacement therapy in the million women study." *Lancet* **362**(9382): 419–27. PMID:12927427.

Beral, V., D. Bull, J. Green, and G. Reeves. 2007. "Ovarian cancer and hormone replacement therapy in the million women study." *Lancet* **369**(9574): 1703–10. PMID:17512855.

Börjesson, A. E., S. H. Windahl, M. K. Lagerquist, C. Engdahl, B. Frenkel, S. Movérare-Skrtic, et al. 2011. "Roles of transactivating functions 1 and 2 of estrogen receptor-alpha in bone." *Proceedings of the National Academy of Sciences of the United States of America* **108**(15): 6288–93. doi:10.1073/pnas.1100454108.

Castillo, A. B., J. W. Triplett, F. M. Pavalko, and C. H. Turner. 2014. "Estrogen receptor-B regulates mechanical signaling in primary osteoblasts." *American Journal of Physiology, Endocrinology and Metabolism* **306**(8): E937–44. doi:10.1152/ajpendo.00458.2013.

Damien, E., J. S. Price, and L. E. Lanyon. 1998. "The estrogen receptor's involvement in osteoblasts' adaptive response to mechanical strain." *Journal of Bone and Mineral Research* **13**(8): 1275–82. PMID:9718196.

De Souza, R. L., M. Matsuura, F. Eckstein, S. C. F. Rawlinson, L. E. Lanyon, and A. A. Pitsillides. 2005. "Non-invasive axial loading of mouse tibiae increases cortical bone formation and modifies trabecular organization: a new model to study cortical and cancellous compartments in a single loaded element." *Bone* **37**(6): 810–18. PMID:16198164.

Dey, P., R. P. A. Barros, M. Warner, A. Ström, and J.-Å. Gustafsson. 2013. "Insight into the mechanisms of action of estrogen receptor B in the breast, prostate, colon, and CNS." *Journal of Molecular Endocrinology* **51**(3): T61–74. doi:10.1530/JME-13-0150.

Frost, H M. 1987. "Bone 'mass' and the 'mechanostat': a proposal." *Anatomical Record* **219**(1): 1–9. PMID:3688455.

Galea, G. L. and J. S. Price. 2015. "Four-point bending protocols to study the effects of dynamic strain in osteoblastic cells in vitro." *Methods in Molecular Biology* **1226**: 117–30. doi:10.1007/978-1-4939-1619-1_10.

Galea, G. L., L. B. Meakin, T. Sugiyama, N. Zebda, A. Sunters, H. Taipaleenmaki, et al. 2013a. "Estrogen receptor A mediates proliferation of osteoblastic cells stimulated by estrogen and mechanical strain, but their acute down-regulation of the Wnt antagonist Sost is mediated by estrogen receptor B." *Journal of Biological Chemistry* **288**(13): 9035–48. doi:10.1074/jbc.M112.405456.

Galea, G. L., J. S. Price, and L. E. Lanyon. 2013b. "Estrogen receptors' roles in the control of mechanically adaptive bone (re)modeling." *BoneKEy Reports* **2**: 413. doi:10.1038/bonekey.2013.147.

Honma, N., S. Mori, H. Zhou, S. Ikeda, M. N. Mieno, N. Tanaka, et al. 2013. "Association between estrogen receptor-B dinucleotide repeat polymorphism and incidence of femoral fracture." *Journal of Bone and Mineral Metabolism* **31**(1): 96–101. doi:10.1007/s00774-012-0383-z.

Jessop, H. L., C. P. Wheeler-Jones, L. E. Lanyon, M. Sjöberg, G. Zaman, and M. Z. Cheng. 2001. "Mechanical strain and estrogen activate estrogen receptor alpha in bone cells." *Journal of Bone and Mineral Research* **16**(6): 1045–55. PMID:11393781.

Jessop, H. L., R. F. L. Suswillo, S. C. F. Rawlinson, K. Lee, V. Das-Gupta, A. A. Pitsillides, and L. E. Lanyon. 2004. "Osteoblast-like cells from estrogen receptor alpha knockout mice have deficient responses to mechanical strain." *Journal of Bone and Mineral Research* **19**(6): 938–46. PMID:15190886.

Lee, K., H. Jessop, R. Suswillo, G. Zaman, and L. Lanyon. 2003. "Endocrinology: bone adaptation requires oestrogen receptor-alpha." *Nature* **424**(6947): 389. PMID:12879058.

Lee, K. C. L., L. E. Lanyon, H. Jessop, R. Suswillo, and G. Zaman. 2004. "The adaptive response of bone to mechanical loading in female transgenic mice is deficient in the absence of oestrogen receptor-alpha and -beta." *Journal of Endocrinology* **182**(2): 193–201. PMID:15283680.

Lindberg, M. K., Z. Weihua, N. Andersson, S. Movérare, H. Gao, O. Vidal, et al. 2002. "Estrogen receptor specificity for the effects of estrogen in ovariectomized mice." *Journal of Endocrinology* **174**(2): 167–78. PMID:12176656.

Lindberg, M. K., S. Movérare, S. Skrtic, H. Gao, K. Dahlman-Wright, J.-A. Gustafsson, and C. Ohlsson. 2003. "Estrogen receptor (ER)-beta reduces ERalpha-regulated gene transcription, supporting a 'ying yang' relationship between ERalpha and ERbeta in mice." *Molecular Endocrinology* **17**(2): 203–8. PMID:12554748.

Meakin, L. B., G. L. Galea, T. Sugiyama, L. E. Lanyon, and J. S. Price. 2014a. "Age-related impairment of bones' adaptive response to loading in mice is associated with sex-related deficiencies in osteoblasts but no change in osteocytes." *Journal of Bone and Mineral Research* **29**(8): 1859–71. doi:10.1002/jbmr.2222.

Meakin, L. B., J. S. Price, and L. E. Lanyon. 2014b. "The contribution of experimental in vivo models to understanding the mechanisms of adaptation to mechanical loading in bone." *Frontiers in Endocrinology* **5**: 154. doi:10.3389/fendo.2014.00154.

Ohlsson, C. and L. Vandenput. 2009. "Estrogens as regulators of bone health in men." *Nature Reviews. Endocrinology* **5**(8): 437–43. doi:10.1038/nrendo.2009.112.

Remes, T., S. B. Väisänen, A. Mahonen, J. Huuskonen, H. Kröger, J. S. Jurvelin, et al. 2003. "Aerobic exercise and bone mineral density in middle-aged Finnish men: a controlled randomized trial with reference to androgen receptor, aromatase, and estrogen receptor alpha gene polymorphisms." *Bone* **32**(4): 412–20. PMID:12689685.

Rivadeneira, F., J. B. J. van Meurs, J. Kant, M. C. Zillikens, L. Stolk, T. J. Beck, et al. 2006. "Estrogen receptor beta (ESR2) polymorphisms in interaction with estrogen receptor alpha (ESR1) and insulin-like growth factor I (IGF1) variants influence the risk of fracture in postmenopausal women." *Journal of Bone and Mineral Research* **21**(9): 1443–56. PMID:16939403.

Saxon, L. K., A. G. Robling, A. B. Castillo, S. Mohan, and C. H. Turner. 2007. "The skeletal responsiveness to mechanical loading is enhanced in mice with a null mutation in estrogen receptor-beta." *American Journal of Physiology, Endocrinology and Metabolism* **293**(2): E484–91. PMID:17535856.

Saxon, L. K., G. Galea, L. Meakin, J. Price, and L. E. Lanyon. 2012. "Estrogen receptors α and β have different gender-dependent effects on the adaptive responses to load bearing in cancellous and cortical bone." *Endocrinology* **153**(5): 2254–66. doi:10.1210/en.2011-1977.

Shearman, A. M., D. Karasik, K. M. Gruenthal, S. Demissie, L. A. Cupples, D. E. Housman, and D. P. Kiel. 2004. "Estrogen receptor beta polymorphisms are associated with bone mass in women and men: the Framingham Study." *Journal of Bone and Mineral Research* **19**(5): 773–81. PMID:15068501.

Skerry, T. M. 2006. "One mechanostat or many? Modifications of the site-specific response of bone to mechanical loading by nature and nurture." *Journal of Musculoskeletal & Neuronal Interactions* **6**(2): 122–7. PMID:16849820.

Skerry, T. and L. Lanyon. 2001. "Postmenopausal osteoporosis as a failure of bone's adaptation to functional loading: a hypothesis." *Journal of Bone and Mineral Research* **16**(11): 1937–47. PMID:11697789.

Smith, E. P., J. Boyd, G. R. Frank, H. Takahashi, R. M. Cohen, B. Specker, et al. 1994. "Estrogen resistance caused by a mutation in the estrogen-receptor gene in a man." *New England Journal of Medicine* **331**(16): 1056–61. PMID:8090165.

Sugiyama, T., L. K. Saxon, G. Zaman, A. Moustafa, A. Sunters, J. S. Price, and L. E. Lanyon. 2008. "Mechanical loading enhances the anabolic effects of intermittent parathyroid hormone (1-34) on trabecular and cortical bone in mice." *Bone* **43**(2): 238–48. doi:10.1016/j.bone.2008.04.012.

Sugiyama, T., G. L. Galea, L. E. Lanyon, and J. S. Price. 2010. "Mechanical loading-related bone gain is enhanced by tamoxifen but unaffected by fulvestrant in female mice." *Endocrinology* **151**(12): 5582–90. doi:10.1210/en.2010-0645.

Sugiyama, T., L. B. Meakin, G. L. Galea, B. F. Jackson, L. E. Lanyon, F. H. Ebetino, et al. 2011. "Risedronate does not reduce mechanical loading-related increases in cortical and trabecular bone mass in mice." *Bone* **49**(1): 133–9. doi:10.1016/j.bone.2011.03.775.

Sunters, A., V. J. Armstrong, G. Zaman, R. M. Kypta, Y. Kawano, L. E. Lanyon, and J. S. Price. 2009. "Mechano-transduction in osteoblastic cells involves strain-regulated estrogen receptor alpha-mediated control of insulin-like growth factor (IGF) I receptor sensitivity to Ambient IGF, leading to phosphatidylinositol 3-kinase/AKT-dependent Wnt/LRP5 receptor-independent activation of beta-catenin signaling." *Journal of Biological Chemistry* **285**(12): 8743–58. doi:10.1074/jbc.M109.027086.

Suuriniemi, M., A. Mahonen, V. Kovanen, M. Alén, A. Lyytikäinen, Q. Wang, et al. 2004. "Association between exercise and pubertal BMD is modulated by estrogen receptor alpha genotype." *Journal of Bone and Mineral Research* **19**(11): 1758–65. PMID:15476574.

Van Staa, T. P., E. M. Dennison, H. G. Leufkens, and C. Cooper. 2001. "Epidemiology of fractures in England and Wales." *Bone* **29**(6): 517–22. PMID:11728921.

Windahl, S. H., K. Hollberg, O. Vidal, J. A. Gustafsson, C. Ohlsson, and G. Andersson. 2001. "Female estrogen receptor beta-/- mice are partially protected against age-related trabecular bone loss." *Journal of Bone and Mineral Research* **16**(8): 1388–98. PMID:11499861.

Windahl, S. H., L. Saxon, A. E. Börjesson, M. K. Lagerquist, B. Frenkel, P. Henning, et al. 2013a. "Estrogen receptor-A is required for the osteogenic response to mechanical loading in a ligand-independent manner involving its activation function 1 but not 2." *Journal of Bone and Mineral Research* **28**(2): 291–301. doi:10.1002/jbmr.1754.

Windahl, S. H., A. E. Börjesson, H. H. Farman, C. Engdahl, S. Movérare-Skrtic, K. Sjögren, et al. 2013b. "Estrogen receptor-A in osteocytes is important for trabecular bone formation in male mice." *Proceedings of the National Academy of Sciences of the United States of America* **110**(6): 2294–9. doi:10.1073/pnas.1220811110.

Zaman, G., M. Z. Cheng, H. L. Jessop, R. White, and L. E. Lanyon. 2000. "Mechanical strain activates estrogen response elements in bone cells." *Bone* **27**(2): 233–9. PMID:10913916.

Zhao, R., Z. Xu, and M. Zhao. 2015. "Effects of oestrogen treatment on skeletal response to exercise in the hips and spine in postmenopausal women: a meta-analysis." *Sports Medicine* **45**(8): 1163–73.

8

Mechanical Responsiveness of Distinct Skeletal Elements

Possible Exploitation of Low Weight-Bearing Bone

Simon C. F. Rawlinson

Centre for Oral Growth and Development, Institute of Dentistry Barts and The London School of Medicine and Dentistry, London, UK

8.1 Introduction

The average person will recognize the muscular system as the bodily system that is most adaptive to mechanical demands. This is because weight-training can be seen to have dramatic effects on the physique, while loss of muscle mass can be observed following removal of a plaster cast. The skeleton is also dependent on weight-bearing physical activity for maintenance of the structural mass and strength required to carry out the normal activities of daily living. Prolonged bed rest (Smith et al. 2014b) or periods of exposure to low gravitational forces, as experienced by astronauts, lead to disuse bone loss (Bikle et al. 1997; Smith et al. 2014a) – this latter condition presents one of the many problems for long-term space travel. Conversely, weightlifters have proportionally greater bone mass, and tennis players have more bone in their playing arm than in their non-playing arm, demonstrating not only adaptation to mechanical demands, but also local regulation of bone mass (Ducher et al. 2009).

What drives this mechanical adaptation is the subject of much research. While this research has been productive in illustrating metabolites and pathways from various bones and animals (Thompson et al. 2012), there are some inherent differences between distinct skeletal sites that must be investigated further.

Bones develop utilizing one of two methods of primary ossification: endochondral or intramembranous. The former substitutes a cartilaginous template with bone, while the latter produces bone directly within soft tissues. The long bones, vertebrae, and ribs form by endochondral ossification. Skull bones and parts of the jaw and clavicle form by intramembranous ossification. This significant difference in the primary ossification process could superficially provide an explanation for the susceptibility of weight-bearing limb bones to osteoporosis. However, based on the primary ossification and susceptibility to bone loss in the lateral aspect of the clavicle (formed by intramembranous ossification) compared with the medial aspect (formed by endochondral ossification), the preservation of skull bone is not a "simple case" of intramembranous bone being more resistant to bone loss.

Mechanobiology: Exploitation for Medical Benefit, First Edition. Edited by Simon C. F. Rawlinson.

It is obvious that there are anatomical and functional differences in the skeleton: the skull is vastly different to postcranial bones. Importantly, the mechanical integrity of the skull appears to be independent of mechanical input. Whether the skull never had the ability to respond to the mechanical environment or has lost it (and the limbs developed with such a capability) is under-studied. The weight-bearing limb and spinal bone compartments can lose bone due to mechanical disuse, age, and certain drug regimens. While there are numerous investigations aimed at finding mechanisms that could protect the weight-bearing skeleton against bone loss or promote preservation, little attention has focused on the fact that the skull is better able to preserve mass and integrity through life – despite low levels of mechanical loading. It is the aim of this chapter to discuss potential loading-related phenomena, demonstrate differences between bones of the skull and limb, and suggest novel strategies for exploiting the apparent lack of mechanobiological responses in the skull in order to maintain bone mass at weight-bearing sites that are susceptible to osteopenia/osteoporosis.

8.2 Anatomy and Loading-Related Stimuli

The importance of physical activity for the maintenance of a structurally competent weight-bearing skeleton with adequate ability to resist fracture is widely recognized. Reduction in physical activity leads to adaptive remodeling and diminished load-bearing strength. Despite the benefits of habitual physical exercise in achieving and maintaining structural competence, the incidence of osteoporotic fracture, even in active individuals, continues to rise. The exact nature of this increase is undetermined, but that bone loss can occur with continued habitual loading suggests that the mechanism by which bone cells perceive mechanical inputs is lost, or that the ability to promote the "correct" response has failed. It is recognized that a dynamic, intermittent loading stimulus is crucial, as static loads are not osteogenic (Lanyon and Rubin 1984; Forwood and Turner 1995). The loading-generated signal(s) responsible for altering resident bone cell behavior in order to adaptively remodel bone and produce a mechanically competent structure have not been elucidated, but many have been proposed.

The application of a load to a curved bone induces both compressive and tensile mechanical strains on opposite cortices, as well as shifts in the fluid residing within the canaliculi of the matrix. These physical changes induce a number of potential signaling parameters to which resident bone cells might respond.

8.2.1 Mechanochemical Control

It has been suggested that mechanical loads might influence the solubility of bone matrix hydroxyapatite crystals (Justus and Luft 1970; Carter 1984). Increased tension leads to an increase in the solubility of hydroxyapatite crystals and thus the Ca^{2+} concentration in the bathing solution. In compression, solubility decreases, leading to a reduction in Ca^{2+} levels (Justus and Luft 1970). It has been hypothesized that changes in the local mechanical loading environment might alter local Ca^{2+} concentrations and thus influence local bone cell behavior. The calcium-sensing receptors present on bone cell surfaces (Quarles et al. 1994) would then perceive changes in the local extracellular Ca^{2+} concentration and influence cell behavior accordingly.

8.2.2 Microdamage

Repetitive mechanical loading has been shown to produce regionalized damage in bone tissue (Burr et al. 1985). It has been proposed that this damage initiates an osteonal repair response to replace bone that has reached the limit of its fatigue life, constituting an adaptive response. Microdamage need not originate from excessive loading levels: it can result from strains in the physiological range. However, the extensiveness of microdamage is correlated to strain magnitude and strain rate. An argument against microdamage being the main driving force of the physiological adaptive remodeling response is that, whereas 10 000 cycles of repetitive loading were required to initiate visible microdamage (Burr et al. 1985), only 36 cycles were needed to produce an adaptive response (Rubin and Lanyon 1984).

8.2.3 Intermittent Compressive Force

Isolated cells and bone explants have been exposed to intermittent compressive forces, a form of hydrostatic mechanical perturbation that is equal in all directions. Regional differences in bone tissue composition and rigidity might allow for structural deformation and subsequent signal generation in response to such pressure. The biochemical responses studied have shown that intermittent compressive forces increase pro-osteogenic response in fetal mouse calvariae (Klein-Nulend et al. 1987), produce soluble mediators that act to inhibit the growth and differentiation of osteoclasts (Klein-Nulend et al. 1993), increase radiolabeled sulfate release from prelabeled mouse metatarsal bone rudiments, and increase radiolabeled sulfate incorporation in similar cultures (Bagi and Burger 1989). In a developmental model, hydrostatic force induced increased expression of type II collagen, osteogenic markers, and mineralization levels in an embryonic chick femur (Henstock et al. 2013).

8.2.4 Piezoelectricity

In 1957, it was hypothesized that piezoelectricity was responsible for controlling cellular modeling and remodeling activities (Fukada and Yasuda 1957). Though there is evidence that electric fields can modulate cell behavior (Bassett et al. 1982), it was considered that the loading-engendered field strengths were of insufficient magnitude to produce such effects, and that streaming potentials (see Section 8.2.9) would dominate over piezoelectricity in wet bone (Gross and Williams 1982). A recent review interrogates the use of piezoelectric materials for tissue regeneration and provides a discussion of the generation of amplified streaming potentials in bone (Rajabi et al. 2015).

8.2.5 Proteoglycan Reorientation

Mechanical loading of the bone matrix induces a change in matrix proteoglycan reorientation both *in vitro* and *in vivo* (Skerry et al. 1990). It has been suggested that the orientation of proteoglycans to collagen fibers reflects the local loading history, and thus provide a site-specific strain memory. Proteoglycan core proteins attach to cell membrane receptors and link to the cytoskeleton directly (Woods et al. 1984; Rapraeger et al. 1986). Thus, by direct linkage to intracellular components, it can be envisaged that proteoglycan reorientation could influence cellular behavior.

8.2.6 Strain Energy Density

Energy is dissipated throughout bone in response to loading. It has been proposed that this "strain energy density" might be utilized by resident cells (Carter 1984). Rubin's group has correlated the bone growth responses in the loadable, functionally isolated avian ulna model to strain energy density (Gross et al. 1992). How this is manifested as a signal, and how cells can respond to such a "signal," is not known.

8.2.7 Mechanical Strain

The result of load-bearing is a deformation in bone tissue, which produces direct mechanical strain throughout the matrix. Such strain can be measured directly by strain gauges attached to the bone surface. Isolated osteoblasts and osteoblast-like cell lines respond to direct strain *in vitro*, yet the levels of strain applied are generally in a range considered to be supraphysiological, and it has been suggested that resident osteocytes *in situ* are actually only subjected to very low mechanical strains. However, Cowin has proposed a mechanism by which the low mechanical strain at osteocyte surfaces could be amplified by associated collagen fibers (Cowin and Weinbaum 1998). Alternatively, osteocytes are responsive to a consequence of the mechanical deformation, and signal amplification is considered to be achieved by loading-related interstitial fluid flow (Weinbaum et al. 1994; Zeng et al. 1994; Riddle and Donahue 2009).

8.2.8 Fluid Shear Stress

Fluid in bone resides in the lacuna-canalicular (LC) network, surrounding osteocyte–osteocyte and osteocyte–osteoblast processes. Load-related deformation that produces bending will induce fluid flow from regions of high pressure to regions of low pressure. This produces two possible stimuli, postulated as stimulators of resident cells: shear stress and streaming potentials. Fluid shear stresses are generated when a fluid flows over a surface. They have been shown to engender responses in endothelial cells (Busse and Fleming 1998), chondrocytes (Das et al. 1997), and bone cells (Reich et al. 1990; Reich and Frangos 1993) in culture. The responses of endothelial cells to shear stresses include very early (msec) K^+ ion channel activation, which is considered to be a controlling factor of vasorelaxation (Davies 1995). Fluid flow has been shown to elevate production of the second messenger, nitric oxide (NO), from Ca^{2+}-dependent NO synthases in endothelial (Busse and Fleming 1998), osteoblast, and osteocyte cells (Klein-Nulend et al. 1995). Indeed, fluid flow is more potent at increasing NO release than Ca^{2+} ionophores, which has been attributed to the increase in eNOS phosphorylation with flow (but not with exposure to ionophore) (Corson et al. 1996). Exposure of calvarial-derived osteoblasts to fluid shear stresses also leads to prostaglandin-dependent cyclic adenosine monophosphate (cAMP) and inositol triphosphate (IP3) production in these cells (Reich and Frangos 1991).

Calvarial osteoblasts appear to be particularly sensitive to fluid shear: 5 dynes/cm^2 of pulsatile fluid shear stimulated a reduction of alkaline phosphatase mRNA levels within 1 hour, dropping to 30% of non-flow controls after a further 2 hours. Steady shear flow of 4 dynes/cm^2 resulted in a 68% reduction in alkaline phosphatase mRNA expression after 8 hours of flow (Hillsley and Frangos 1997). Osteocytes, with the extended network of processes *in situ*, provide a massive surface area on which shear stresses can act.

8.2.9 Streaming Potentials

These represent an electrochemical phenomenon, first reviewed in bone by Eriksson (1974). Physiological fluids, containing amino acids and proteins, will display a net molecular charge (due to COO^- and NH_3^+ ions) that depends upon the pH. When a physiological fluid comes in contact with a charged solid, there is an electric charge separation and a generation of potential difference. At a solid–liquid interface, the surface charge of the solid attracts counterions in the fluid. Immediately outside the layer of electrostatically bound ions, beyond the hydrodynamic slip plane, an area of diffuse and weakly bound ions extends into the fluid. This region contains an unequal number of positive and negative ions, resulting in an electrostatically charged layer. When induced to flow, the diffuse, charged liquid layer constitutes an electric current, and the voltage produced is termed a "streaming potential," which can measured on the bone surface as "stress-generated potentials" (SGPs). In bone, SGP levels are greater than loading-induced piezoelectricity levels (Gross and Williams 1982). *In vivo* streaming potentials have been measured in the canine tibia: ligation of the femoral artery blocked the streaming-potential oscillations (Otter et al. 1990); when the bone was subjected to applied bending, the stress-generated potentials could be measured, and modulation of the circulatory proteins by injection of protamine sulfate affected SGP levels (Otter et al. 1993). *Ex vivo*, using a four-point mechanical loading system of rat tibia, SGPs monotonically increase with increased loading frequency (Turner et al. 1994); that is, the rate of fluid flow is a determinant of the magnitude of the resultant SGPs.

Given that so many possible osteoregulatory signals can be initiated by mechanical loads in bone *in vivo*, it is sensible to study the early responses in a model that allows these anatomically related variables to be generated in combination.

8.3 Preosteogenic Responses *In Vitro*

The early *in vitro* work in Professor Lance Lanyon's laboratory at the Royal Veterinary College sought to establish loading-related responses in bone and employed a model system of perfused and loadable trabecular bone cores (El Haj et al. 1990). This model demonstrated that bone cells remained alive in culture within their natural matrix for 24 hours, responding to parathyroid hormone with a cAMP release and to loading with an increase in glucose-6-phopshate dehydrogenase (G6PD) activity. The timing of prostaglandin release in response to an applied intermittent (1 Hz) physiological load (3000 microstrain) was determined (Rawlinson et al. 1991) and the potential sites for prostaglandin E_2 (PGE_2) and prostacyclin (PGI_2) release from resident bone cells were identified. While PGI_2 was located to osteocytes and osteoblasts, PGE_2 could only be discerned in osteoblasts. Having established that prostanoid release occurred with the onset of loading, the potential mechanism of this preosteogenic mechanical response here and in a cortical ulna bone model was studied. The data indicate that in osteocytes and osteoblasts, arachidonic acid for PGI_2 synthesis is mediated by pertussis toxin (PTX)-insensitive G-protein-dependent secretary phospholipase A_2 ($sPLA_2$) alone, while in osteoblasts, arachidonic acid for PGE_2 synthesis is released by PTX-sensitive, G-protein-dependent, cytoplasmic PLA_2-mediated activity, which also requires upstream $sPLA_2$

and PKC activities (Rawlinson et al. 2000). Such discrepancies between osteoblast and osteocyte activity would not be observed in cell monoculture and indicate the value of co-culture models of bone.

8.4 Site-Specific, Animal-Strain Differences

Rat calvarial bone cells *in situ* and primary osteoblasts in culture have been found not to respond to physiological mechanical strain with increases in G6PD activity or prostanoid release – in contrast to ulnae and ulnar-derived osteoblasts (Rawlinson et al. 1995). Experiments investigating strain levels in the skull and limb found those in the skull to be much lower than those in the tibia: 0.0192% maximum in the parietal bone and 0.2% in the tibia (Hillam et al. 2015). This finding is consistent with previous strain data derived from rats, where the maximal level of mechanical strains recorded in the parietal bone were over 30 times lower than those in the ulna (Rawlinson et al. 1995). Such strain levels would lead to disuse osteoporosis in the limb, but there is no apparent disuse-induced bone loss or reduction of mechanical integrity in the skull. Evidently, either the skull is extremely sensitive to strain or, as we have postulated, the mechanical integrity of skull bone is modulated by other local mechanoindependent mechanisms.

There are significant differences in the transcriptome of parietal and ulnar bones in the rat (Rawlinson et al. 2009a), and it is the consequence of this local gene expression profile that we propose must confer the ability of the skull to resist low loading-disuse bone loss. Further evidence for local genetics regulating bone mass comes from work using the tibiotarsus of chicks bred for different purposes. These birds demonstrate inherently altered growth rates and mechanical responsiveness. Birds bred for meat production grow quickly and do not register applied mechanical loading with the expected responses. Birds bred for egg-laying and wild types do respond to mechanical loading (Rawlinson et al. 2009b).

A single bone may also demonstrate regional differences in responses to mechanical loading. Tooth loss leads to resorption of the bony mandibular alveolar ridge that once supported the tooth. However, the basal bone of the mandible is preserved for much longer in this disuse state (Reich et al. 2011). Comparison of transcriptomes of the mandible, ulna, and parietal bone hints that the mandible might contain pathways that permit disuse osteoporosis and maintain bone mass despite low mechanical load levels (Kingsmill et al. 2013). These findings are consistent with the recently proposed view that mechanical adaption in bones, at least, might try to achieve a local set point (Hillam et al. 2015), but this would ultimately depend on local transcriptome for that site.

Genetically related control and regulation of mechanoresponsive bone mass may also explain regional responses to specific drugs. An example would be the different efficacies of bisphosphonates in reducing fracture risk in distinct skeletal compartments (Boonen 2007). Radionuclide studies using teriparatide, a recombinant form of parathyroid hormone, indicate that the effect of teriparatide differs at different sites in the skeleton (Blake et al. 2011).

The composition of the bone tissue appears distinct at different regional sites. Recently, preliminary experiments using solid-state nuclear magnetic resonance (NMR) spectroscopy to examine powdered bone matrix from ulna and parietal bones suggested a difference in the chemical speciation of the bones, particularly in the amounts of alanine and glycine (the latter constituting a third of collagen). Furthermore, calvarial bone gene array indicates a sevenfold higher expression of Ccl9, a glycine transmembrane transporter protein. Insulin growth factor binding protein 5 (IGFBP5) expression is also greater in calvarial bone compared with limb. The activities of IGFBP5 include alanine, glycine, and proline transportation and are all consistent with these NMR findings (Niazi et al. 2011; Shaikh et al. 2013). More interesting was the determination of water levels, which were highest in parietal and lowest in rib, with ulna levels in between (Niazi et al. 2011). These studies require support from further experimentation before any significant conclusions can be drawn. However, it appears that water is important in ordering the nanoscale apatite mineral in bone (Wang et al. 2013).

Preliminary analysis of parietal and ulna bone using X-ray diffraction hints that the mineral sizes and organizations in these bones are distinct. In calvaria, mineral is organized in a more parallel arrangement, whereas in the ulna it is less organized (Baber et al. 2013) and is consistent with the NMR findings of water levels in these skeletal elements. Whether this mineral organization impacts on susceptibility to osteoporosis has not been investigated, though one report suggests osteoclasis is related to mineral density (Jones et al. 1995). Nonetheless, regional composition/construction of the bone collagen/noncollagenous matrix and consequent mineral size and order may be based on the local transcriptome, modified by mechanical loading.

8.5 Exploitation of Regional Information

Based on the fact that the skull is resistant to osteoporosis attributed to the local expression of positional identity and transcription factors, it is proposed that by activating specific regional/transcriptional factors of skull bone in other regions of the skeleton, osteoporosis might be prevented in these regions.

In addition, host positional identity markers may also be important in resolving recipient autologous transplant failures. Leucht et al. (2008) have suggested that successful transplantation of skeletal stem cells is based on matching host and recipient Hox expression profiles. Hoxa11-negative skeletal stem cells from the mandible were able to integrate appropriately in Hoxa11-positive tibial sites, whereas Hoxa11-positive skeletal stem cells did not integrate into the mandible. The former transplantation resulted in osteoblasts, and the latter produced chondrocytes that were inappropriate for the site. Thus, this local information has great importance for successful regenerative and tissue-engineering endeavors. However, transplantation of bone means that the cells still reside in the host site, despite being transferred to a recipient site. With time, what happens to those host cells? Do they maintain their positional identity within the host bone, or are they replaced with recipient-site bone cells, while a new regionally appropriate bone matrix is produced?

Finally, the expression of positional identity markers differs in distinct weight-bearing bones, and the bone-marrow stromal cells closely match those of the host bone (Prajaneh et al. 2009). Based on the Hoxa findings of Leucht et al. (2008), if osteoclasts derived from a particular source were to maintain positional identity "status," would they function appropriately, or less efficiently, in non-host environments?

8.6 Conclusion

Bones of the weight-bearing skeleton vary in shape, size, and composition. This is likely the result of inherent local transcriptome and mechanical responsiveness. The anatomy of each bone provides the opportunity for a number of potential stimuli to be generated by mechanical loading. Some skeletal sites are subject to inappropriate bone loss or osteopenia/osteoporosis despite continued mechanical loading. Bone loss at these sites is not usually recognized until fragility fractures occur. However, the low weight-bearing skull bone mass and integrity appear to be independent of mechanical loading and are not subject to bone loss. The question is whether the local regulation that maintains skull bone mass could ever be "transplanted" to sites subject to bone loss in order to preserve bone mass there.

References

Baber, H., S. C. F. Rawlinson, and M. Al-Jawad. 2013. "Variations in mineral crystallites of bone at distinct skeletal sites, and its relationship with diseases like osteoporosis." MSc., Dental Physics Sciences Unit, Queen Mary University of London.

Bagi, C. and E. H. Burger. 1989. "Mechanical stimulation by intermittent compression stimulates sulfate incorporation and matrix mineralization in fetal mouse long-bone rudiments under serum-free conditions." *Calcified Tissue International* **45**(6): 342–7. PMID:2509024.

Bassett, C. A., S. N. Mitchell, and M. M. Schink. 1982. "Treatment of therapeutically resistant non-unions with bone grafts and pulsing electromagnetic fields." *Journal of Bone and Joint Surgery. American Volume* **64**(8): 1214–20. PMID:6752151.

Bikle, D. D., B. P. Halloran, and E. Morey-Holton. 1997. "Spaceflight and the skeleton: lessons for the earthbound." *Gravitational and Space Biology Bulletin* **10**(2): 119–35. PMID:11540113.

Blake, G. M., M. L. Frost, A. E. Moore, M. Siddique, and I. Fogelman. 2011. "The assessment of regional skeletal metabolism: studies of osteoporosis treatments using quantitative radionuclide imaging." *Journal of Clinical Densitometry* **14**(3): 263–71. doi: 10.1016/j.jocd.2011.04.003.

Boonen, S. 2007. "Bisphosphonate efficacy and clinical trials for postmenopausal osteoporosis: Similarities and differences." *Bone* **40**(5 Suppl. 2): S26–31. doi: http://dx.doi.org/10.1016/j.bone.2007.03.003.

Burr, D. B., R. B. Martin, M. B. Schaffler, and E. L. Radin. 1985. "Bone remodeling in response to in vivo fatigue microdamage." *Journal of Biomechanics* **18**(3): 189–200. PMID:3997903.

Busse, R. and I. Fleming. 1998. "Pulsatile stretch and shear stress: physical stimuli determining the production of endothelium-derived relaxing factors." *Journal of Vascular Research* **35**(2): 73–84. PMID:9588870.

Carter, D. R. 1984. "Mechanical loading histories and cortical bone remodeling." *Calcified Tissue International* **36**(Suppl. 1): S19–24.

Corson, M. A., N. L. James, S. E. Latta, R. M. Nerem, B. C. Berk, and D. G. Harrison. 1996. "Phosphorylation of endothelial nitric oxide synthase in response to fluid shear stress." *Circulation Research* **79**(5): 984–91. PMID:8888690.

Cowin, S. C. and S. Weinbaum. 1998. "Strain amplification in the bone mechanosensory system." *American Journal of Medical Science* **316**(3): 184–8. PMID:9749560.
Das, P., D. J. Schurman, and R. L. Smith. 1997. "Nitric oxide and G proteins mediate the response of bovine articular chondrocytes to fluid-induced shear." *Journal of Orthopaedic Research* **15**(1): 87–93. doi: 10.1002/jor.1100150113.
Davies, P. F. 1995. "Flow-mediated endothelial mechanotransduction." *Physiological Reviews* **75**(3): 519–60. PMID:7624393.
Ducher, G., R. M. Daly, and S. L. Bass. 2009. "Effects of repetitive loading on bone mass and geometry in young male tennis players: a quantitative study using MRI." *Journal of Bone Mineral Research* **24**(10): 1686–92. doi: 10.1359/jbmr.090415.
El Haj, A. J., S. L. Minter, S. C. F. Rawlinson, R. Suswillo, and L. E. Lanyon. 1990. "Cellular responses to mechanical loading in vitro." *Journal of Bone Mineral Research* **5**(9): 923–32. doi: 10.1002/jbmr.5650050905.
Eriksson, C. 1974. "Streaming potentials and other water-dependent effects in mineralized tissues." *Annals of the New York Academy of Sciences* **238**: 321–38. PMID:4531266.
Forwood, M. R. and C. H. Turner. 1995. "Skeletal adaptations to mechanical usage: results from tibial loading studies in rats." *Bone* **17**(4 Suppl.): 197S–205S. PMID:8579917.
Fukada, E. and I. Yasuda. 1957. "On the piezoelectric effect of bone." *Journal of the Physical Society of Japan* **12**: 1158–62. doi: http://dx.doi.org/10.1143/JPSJ.12.1158.
Gross, D. and W. S. Williams. 1982. "Streaming potential and the electromechanical response of physiologically-moist bone." *Journal of Biomechanics* **15**(4): 277–95. PMID:7096383.
Gross, T. S., K. J. McLeod, and C. T. Rubin. 1992. "Characterizing bone strain distributions in vivo using three triple rosette strain gages." *Journal of Biomechanics* **25**(9): 1081–7. PMID:1517269.
Henstock, J. R., M. Rotherham, J. B. Rose, and A. J. El Haj. 2013. "Cyclic hydrostatic pressure stimulates enhanced bone development in the foetal chick femur in vitro." *Bone* **53**(2): 468–77. doi: 10.1016/j.bone.2013.01.010.
Hillam, R. A., A. E. Goodship, and T. M. Skerry. 2015. "Peak strain magnitudes and rates in the tibia exceed greatly those in the skull: an in vivo study in a human subject." *Journal of Biomechanics* **48**(12): 3292–8. doi:10.1016/j.jbiomech.2015.06.021.
Hillsley, M. V. and J. A. Frangos. 1997. "Alkaline phosphatase in osteoblasts is down-regulated by pulsatile fluid flow." *Calcified Tissue International* **60**(1): 48–53. PMID:9030480.
Jones, S. J., M. Arora, and A. Boyde. 1995. "The rate of osteoclastic destruction of calcified tissues is inversely proportional to mineral density." *Calcified Tissue International* **56**(6): 554–8. PMID:7648486.
Justus, R. and J. H. Luft. 1970. "A mechanochemical hypothesis for bone remodeling induced by mechanical stress." *Calcified Tissue Research* **5**(3): 222–35. PMID:5433626.
Kingsmill, V. J., I. J. McKay, P. Ryan, M. R. Ogden, and S. C. Rawlinson. 2013. "Gene expression profiles of mandible reveal features of both calvarial and ulnar bones in the adult rat." *Journal of Dentistry* **41**(3): 258–64. doi: 10.1016/j.jdent.2012.11.010.
Klein-Nulend, J., J. P. Veldhuijzen, M. de Jong, and E. H. Burger. 1987. "Increased bone formation and decreased bone resorption in fetal mouse calvaria as a result of intermittent compressive force in vitro." *Bone and Mineral* **2**(6): 441–8. PMID:3505768.

Klein-Nulend, J., C. M. Semeins, J. P. Veldhuijzen, and E. H. Burger. 1993. "Effect of mechanical stimulation on the production of soluble bone factors in cultured fetal mouse calvariae." *Cell and Tissue Research* **271**(3): 513–17. PMID:8472308.

Klein-Nulend, J., C. M. Semeins, N. E. Ajubi, P. J. Nijweide, and E. H. Burger. 1995. "Pulsating fluid flow increases nitric oxide (NO) synthesis by osteocytes but not periosteal fibroblasts – correlation with prostaglandin upregulation." *Biochemical and Biophysical Research Communications* **217**(2): 640–8. PMID:7503746.

Lanyon, L. E. and C. T. Rubin. 1984. "Static vs dynamic loads as an influence on bone remodelling." *Journal of Biomechanics* **17**(12): 897–905. PMID:6520138.

Leucht, P., J. B. Kim, R. Amasha, A. W. James, S. Girod, and J. A. Helms. 2008. "Embryonic origin and Hox status determine progenitor cell fate during adult bone regeneration." *Development* **135**(17): 2845–54. doi: 10.1242/dev.023788.

Niazi, M., S. C. F. Rawlinson, and N. Karpukhina. 2011. "Study on the compositional variations at distinct sites in bones using solid state nuclear magnetic resonance spectroscopy." MSc., Dental Physics Sciences Unit, Queen Mary University of London.

Otter, M. W., V. R. Palmieri, and G. V. Cochran. 1990. "Transcortical streaming potentials are generated by circulatory pressure gradients in living canine tibia." *Journal of Orthopaedic Research* **8**(1): 119–26. doi: 10.1002/jor.1100080115.

Otter, M. W., D. D. Wu, W. A. Bieber, and G. V. Cochran. 1993. "Intraarterial protamine sulfate reduces the magnitude of streaming potentials in living canine tibia." *Calcified Tissue International* **53**(6): 411–15. PMID:8293355.

Prajaneh, S., S. C. F. Rawlinson, M. Ghuman, F. J. Hughes, and I. J. McKay. 2009. "Gene expression in BMSCs isolated from different bones." Conference: British Society for Dental Research, Glasagow.

Quarles, L. D., J. E. Hartle, J. P. Middleton, J. Zhang, J. M. Arthur, and J. R. Raymond. 1994. "Aluminum-induced DNA synthesis in osteoblasts: mediation by a G-protein coupled cation sensing mechanism." *Journal of Cellular Biochemistry* **56**(1): 106–17. doi: 10.1002/jcb.240560115.

Rajabi, A. H., M. Jaffe, and T. L. Arinzeh. 2015. "Piezoelectric materials for tissue regeneration: a review." *Acta Biomaterialia* **24**: 12–23. doi: 10.1016/j.actbio.2015.07.010.

Rapraeger, A., M. Jalkanen, and M. Bernfield. 1986. "Cell surface proteoglycan associates with the cytoskeleton at the basolateral cell surface of mouse mammary epithelial cells." *Journal of Cell Biology* **103**(6 Pt. 2): 2683–96. PMID:3025223.

Rawlinson, S. C. F., A. J. el-Haj, S. L. Minter, I. A. Tavares, A. Bennett, and L. E. Lanyon. 1991. "Loading-related increases in prostaglandin production in cores of adult canine cancellous bone in vitro: a role for prostacyclin in adaptive bone remodeling?" *Journal of Bone Mineral Research* **6**(12): 1345–51. doi: 10.1002/jbmr.5650061212.

Rawlinson, S. C. F., J. R. Mosley, R. F. Suswillo, A. A. Pitsillides, and L. E. Lanyon. 1995. "Calvarial and limb bone cells in organ and monolayer culture do not show the same early responses to dynamic mechanical strain." *Journal of Bone Mineral Research* **10**(8): 1225–32. doi:10.1002/jbmr.5650100813.

Rawlinson, S. C. F., C. P. Wheeler-Jones, and L. E. Lanyon. 2000. "Arachidonic acid for loading induced prostacyclin and prostaglandin E(2) release from osteoblasts and osteocytes is derived from the activities of different forms of phospholipase A(2)." *Bone* **27**(2): 241–7. PMID:10913917.

Rawlinson, S. C. F., I. J. McKay, M. Ghuman, C. Wellmann, P. Ryan, S. Prajaneh, et al. 2009a. "Adult rat bones maintain distinct regionalized expression of markers associated with their development." *PLoS One* **4**(12): e8358. doi:10.1371/journal.pone.0008358.

Rawlinson, S. C. F., D. H. Murray, J. R. Mosley, C. D. Wright, J. C. Bredl, L. K. Saxon, et al. 2009b. "Genetic selection for fast growth generates bone architecture characterised by enhanced periosteal expansion and limited consolidation of the cortices but a diminution in the early responses to mechanical loading." *Bone* **45**(2): 357–66. doi: 10.1016/j.bone.2009.04.243.

Reich, K. M. and J. A. Frangos. 1991. "Effect of flow on prostaglandin E2 and inositol trisphosphate levels in osteoblasts." *American Journal of Physiology* **261**(3 Pt. 1): C428–32. PMID:1887871.

Reich, K. M. and J. A. Frangos. 1993. "Protein kinase C mediates flow-induced prostaglandin E2 production in osteoblasts." *Calcified Tissue International* **52**(1): 62–6. PMID:8453507.

Reich, K. M., C. V. Gay, and J. A. Frangos. 1990. "Fluid shear stress as a mediator of osteoblast cyclic adenosine monophosphate production." *Journal of Cellular Physiology* **143**(1): 100–4. doi:10.1002/jcp.1041430113.

Reich, K. M., C. D. Huber, W. R. Lippnig, C. Ulm, G. Watzek, and S. Tangl. 2011. "Atrophy of the residual alveolar ridge following tooth loss in an historical population." *Oral Diseases* **17**(1): 33–44. doi: 10.1111/j.1601-0825.2010.01699.x.

Riddle, R. C. and H. J. Donahue. 2009. "From streaming-potentials to shear stress: 25 years of bone cell mechanotransduction." *Journal of Orthopaedic Research* **27**(2): 143–9. doi: 10.1002/jor.20723.

Rubin, C. T. and L. E. Lanyon. 1984. "Regulation of bone formation by applied dynamic loads." *Journal of Bone and Joint Surgery. American Volume* **66**(3): 397–402. PMID:6699056.

Shaikh, M. S., S. C. F. Rawlinson, and N. Karpukhina. 2013. "A 13C NMR spectroscopy study on the amino acids composition of the mandible: comparison with the ulna and calvaria bones." MSc., Dental Physics Sciences Unit, Queen Mary University of London.

Skerry, T. M., R. Suswillo, A. J. el Haj, N. N. Ali, R. A. Dodds, and L. E. Lanyon. 1990. "Load-induced proteoglycan orientation in bone tissue in vivo and in vitro." *Calcified Tissue International* **46**(5): 318–26. PMID:2110854.

Smith, S. M., S. A. Abrams, J. E. Davis-Street, M. Heer, K. O. O'Brien, M. E. Wastney, and S. R. Zwart. 2014a. "Fifty years of human space travel: implications for bone and calcium research." *Annual Review of Nutrition* **34**: 377–400. doi: 10.1146/annurev-nutr-071813-105440.

Smith, S. M., C. Castaneda-Sceppa, K. O. O'Brien, S. A. Abrams, P. Gillman, N. E. Brooks, et al. 2014b. "Calcium kinetics during bed rest with artificial gravity and exercise countermeasures." *Osteoporosis International* **25**(9): 2237–44. doi: 10.1007/s00198-014-2754-x.

Thompson, W. R., C. T. Rubin, and J. Rubin. 2012. "Mechanical regulation of signaling pathways in bone." *Gene* **503**(2): 179–93. doi: 10.1016/j.gene.2012.04.076.

Turner, C. H., M. R. Forwood, and M. W. Otter. 1994. "Mechanotransduction in bone: do bone cells act as sensors of fluid flow?" *FASEB Journal* **8**(11): 875–8.

Wang, Y., S. Von Euw, F. M. Fernandes, S. Cassaignon, M. Selmane, G. Laurent, et al. 2013. "Water-mediated structuring of bone apatite." *Nature Materials* **12**(12): 1144–53. doi: 10.1038/nmat3787.

Weinbaum, S., S. C. Cowin, and Y. Zeng. 1994. "A model for the excitation of osteocytes by mechanical loading-induced bone fluid shear stresses." *Journal of Biomechanics* **27**(3): 339–60. PMID:8051194.

Woods, A., M. Höök, L. Kjellén, C. G. Smith, and D. A. Rees. 1984. "Relationship of heparan sulfate proteoglycans to the cytoskeleton and extracellular matrix of cultured fibroblasts." *Journal of Cell Biology* **99**(5): 1743–53. PMID:6238037.

Zeng, Y., S. C. Cowin, and S. Weinbaum. 1994. "A fiber matrix model for fluid flow and streaming potentials in the canaliculi of an osteon." *Annals of Biomedical Engineering* **22**(3): 280–92. PMID:7978549.

9

Pulmonary Vascular Mechanics in Pulmonary Hypertension

Zhijie Wang[1]*, Lian Tian*[2]*, and Naomi C. Chesler*[1]

[1] *Department of Biomedical Engineering, University of Wisconsin–Madison, Madison, WI, USA*
[2] *Department of Medicine, Queen's University, Kingston, ON, Canada*

9.1 Introduction

Biomechanics is the study of the structure and function of biological systems by means of the principles and methods of mechanics. It often involves the use of traditional engineering sciences to analyze biological systems, such as measuring the mechanical properties of cells, tissues, and entire organs. Understanding of the biomechanical changes between healthy and diseased states reveals the mechanical mechanisms of the disease and offers tools for diagnosis and therapy.

Mechanobiology is the study of the biological effects of mechanical forces on cells, tissues, and organs. It is critical to understanding disease processes, since altered biological structural mechanics with disease are not only metrics of altered function but also directly affect biological processes such as signaling and remodeling of the structures themselves. Furthermore, better understanding of mechanobiology can lead to novel therapies aimed at interfering with the biological response to pathological mechanical environments.

In this chapter, we will give a brief overview of the application of biomechanics and mechanobiology in the pulmonary vasculature, with a focus on the disease, pulmonary hypertension (PH).

9.2 Pulmonary Vascular Mechanics

9.2.1 Anatomical and Structural Character of Pulmonary Vessels

Blood ejected from the right ventricle (RV) enters the main pulmonary artery (PA), which splits into left and right PAs. After entering the left and right lungs, respectively, these left and right main PAs branch repeatedly into smaller and smaller PAs, pulmonary arterioles, and alveolar capillaries. After leaving the alveolar capillaries, blood enters pulmonary venules and pulmonary veins, which have a branching structure similar to that of the PA tree, and finally returns to the left atrium (LA) (Singhal et al. 1973; Huang et al. 1996).

Mechanobiology: Exploitation for Medical Benefit, First Edition. Edited by Simon C. F. Rawlinson.

While all these types of pulmonary vessel are important to pulmonary vascular function, PAs, and specifically PA mechanics and mechanobiology, are the subject of interest in this review. In general, PAs consist of three layers: tunica intima, tunica media, and tunica adventitia. The intima is the innermost layer, consisting of endothelial cells resting on a thin basal membrane. This layer is very thin in healthy states and contributes insignificantly to the mechanical properties of the arterial wall (Holzapfel et al. 2000). The middle or media layer is relatively thick and consists of smooth-muscle cells (SMCs), elastin, and collagen fibrils. Due to the dominant elastin and collagen components, this layer is resilient and has high strength, and it carries most of the mechanical load in a healthy artery (Holzapfel et al. 2000). The adventitia is the outermost layer, and consists of collagen fibrils, fibroblasts, fibrocytes, nerves, and vasa vasorum (Humphrey 1995, 2002; Holzapfel et al. 2000). This layer is thicker than the intima layer but thinner than the media layer in healthy arteries; it is much less stiff than the media at low pressures, but it protects the artery from overstretch and rupture at high pressures (Humphrey 1995, 2002; Holzapfel et al. 2000).

9.2.2 Mechanical Properties of PAs

PAs in general are much more compliant than systemic arteries, since they experience lower blood pressure (normal systolic PA pressure ≈ 15–25 mmHg) than the systemic arteries. In physiological conditions, the PA in the radial direction is under compressive stress. The stress–strain behavior in the radial direction mainly depends on elastin, since collagen cannot carry a load in compression. The mechanical behavior in this direction is less well studied than the circumferential or longitudinal directions. In both circumferential and longitudinal directions, the typical experimental J-shaped stress–strain curve of the PA displays highly nonlinear behavior, with an approximately linear region at low strain, a transition region, and another approximately linear region at high strain, with higher modulus before fracture (Figure 9.1a). This mechanical behavior is mainly due to elastin and collagen fibers. As in systemic arteries, the first linear region is dominated by elastin, which has a relatively low modulus. Collagen fibers begin to carry load in the transition region, due to the recruitment and straightening of wavy collagen fibers – a process known as "collagen recruitment." As more collagen fibers are recruited, the second linear region is dominated by the recruited collagen fibers, which have a relatively high modulus.

Besides the extracellular matrix (ECM) proteins collagen and elastin, the vascular SMCs can also contribute significantly to the PA's mechanical properties. In an active state, SMCs contract and the vessel decreases in diameter and shortens in the longitudinal direction, if allowed, leading to a shift to the left of the stress–strain curve (Tabima and Chesler 2010). When the SMCs relax, the vessel increases in diameter and lengthens in the longitudinal direction, if allowed, leading to a shift to the right of the curve (Tabima and Chesler 2010). In contrast to most systemic arteries, the SMCs of the large conduit PAs are not contracted in a basal tone state; therefore, the dilated and basal tone states show no significant differences in stress–strain curves (Tabima and Chesler 2010).

Like other biological tissues, the PA is viscoelastic, as is evident from the stress–strain hysteresis loop under dynamic mechanical test conditions (Figure 9.1b). This viscoelastic behavior results from both SMCs and ECM components (Cox 1982, 1984; Silver et al. 2001; Santana et al. 2005). The elastic component of the PA can be measured by

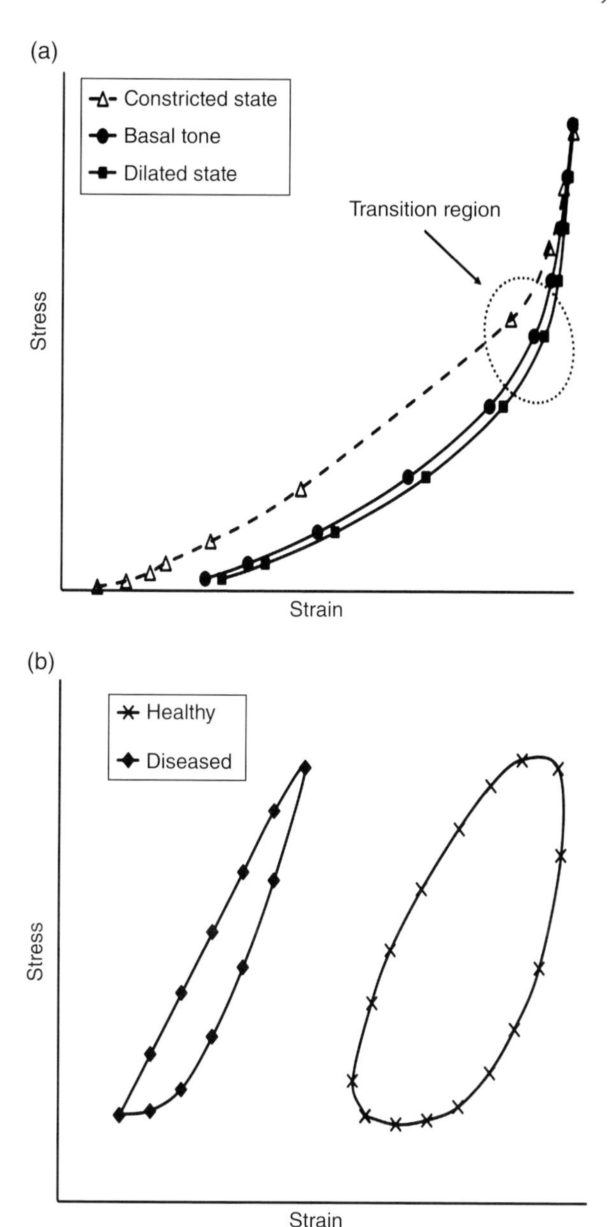

Figure 9.1 (a) Stress–strain curves of a large conduit PA, with SMCs at dilated state, basal tone, and constricted state. Note that the transition region is marked. (b) Stress–strain loops of a large conduit PA under dynamic loading in healthy and disease conditions at SMC dilated state.

the slope of the hysteresis loop, whereas the viscous component can be measured by its area or area ratio. The damping capacity, for example, measures the dissipated energy over an entire cycle, and generally increases as the frequency of dynamic loading increases (Wang et al. 2013b; Tian et al. 2013). The viscoelastic properties of the PA are critical to conducting and buffering the pulsatile blood flow from proximal to the distal

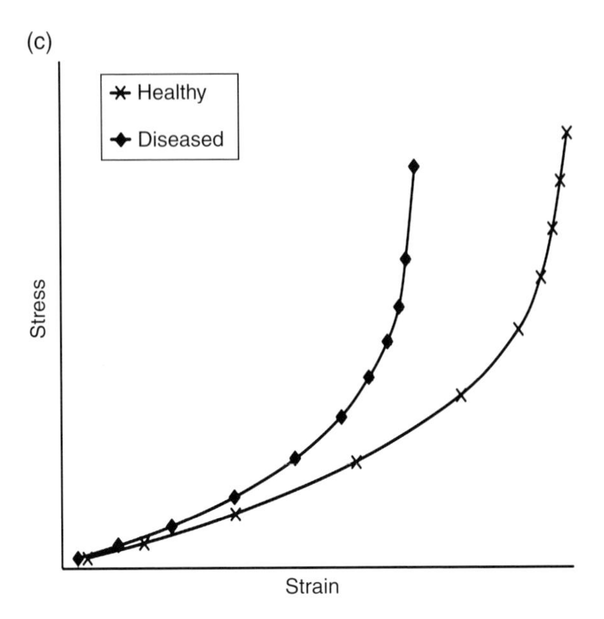

Figure 9.1 (Continued) (c) Stress–strain curves of a large conduit PA in healthy and disease conditions at SMC dilated state.

arteries. These properties have been found to change in the diseased state, but the impact on cardiopulmonary function is not fully understood.

9.2.3 How Diseases Alter the Biomechanics of Pulmonary Arteries

PH is a complex disorder that manifests as abnormally high blood pressure in the vasculature of the lungs. Based on its causes, the World Health Organization (WHO) has classified PH into five categories: group I, pulmonary arterial hypertension (PAH), resulting from increased pulmonary vascular resistance; group II, PH associated with left heart disease; group III, PH associated with lung diseases and/or hypoxemia; group IV, PH due to chronic thrombotic and/or embolic disease; and group V, other miscellaneous causes of PH (McLaughlin et al. 2009). PAH is a rare and deadly disease with high mortality (Humbert et al. 2010). It is characterized by sustained increases in resting mean pulmonary artery pressure (mPAP) of >25 mmHg, with a pulmonary capillary wedge pressure (PCWP) <15 mmHg (Simonneau et al. 2013). In PAH, the elevated mPAP is mostly due to arterial or arteriolar obstruction and constriction, evident from an increase in pulmonary vascular resistance (PVR) and a decrease in pulmonary vascular compliance or capacitance (Mahapatra et al. 2006a; Gan et al. 2007; Hunter et al. 2008; Swift et al. 2012). During the progression of PAH, dramatic remodeling occurs in both large proximal and small distal PAs in all layers of the vessel wall: medial thickening is attributed to SMC hypertrophy and proliferation, as well as accumulation of ECM components such as elastin and collagen; adventitial thickening is attributed to fibroblast proliferation and the accumulation of ECM components such as collagen; and intimal changes occur as a result of endothelial cell dysfunction (Humbert et al. 2004; Stenmark et al. 2006; Ryan et al. 2013).

As a result of this arterial remodeling, PAs become stiffer and the stress–strain curve shifts to the left (Figure 9.1c) when tested under static loading conditions. If tested under dynamic loading conditions, the PAs' viscoelastic behavior also changes; this is evident in their increased stiffness (data not shown) and decreased damping capacity (Wang et al. 2013b) (Figure 9.1b).

In general, there is no change in elastin content that leads to no change in the elastic modulus in the first linear region (Ooi et al. 2010). Collagen fibers are engaged earlier (at a smaller strain), possibly due to increased collagen fiber crosslinking, as seen in a left shift of the transition region (Wang and Chesler 2012; Wang et al. 2013a). Also, the modulus at the second linear region is increased due to increased collagen content and/or collagen crosslinking (Ooi et al. 2010; Wang and Chesler 2012; Wang et al. 2013a). In persistent PH of the newborn (the similarities and differences between this and adult PAH are unknown), evidence from a neonatal calf PH model suggests elastin could become stiffer, resulting in an increase in the modulus in the first linear region but no significant changes in collagen fibers (Lammers et al. 2008).

9.3 Measurements of Pulmonary Arterial Mechanics

9.3.1 Measurement at the Single Artery Level

The common mechanical tests and parameters used to quantify the arterial segment's mechanical properties have been reviewed recently by our group (Wang and Chesler 2011; Tian and Chesler 2012). Here, we will focus on the measurement of PA diameter as a function of pressure, from which the pressure–diameter (PD) curve or stress–strain curve can be derived, because this approach can be applied both *in vivo* and *in vitro*.

In humans and large animals, the measurement of *in vivo* pulmonary arterial pressure (PAP) is typically achieved by right heart catheterization (RHC) (Kim et al. 2000; Gust and Schuster 2001; Hunter et al. 2010; Rain et al. 2013), which provides accurate measurements of systolic, diastolic, and mean PAP but is invasive, or by tricuspid regurgitant (TR) jet velocity via continuous-wave Doppler (Dyer et al. 2006; Friedberg et al. 2006), which is noninvasive but less accurate and has additional limitations. Recently, our group described a novel correlation between stroke volume (SV) and relative area change (RAC) that allows us to calculate PA pulse pressure in dogs using only noninvasive imaging techniques (Bellofiore et al. 2013). *In vivo*, the diameter or cross-sectional area is measured over the cardiac cycle as a function of pressure, and wall thickness is usually assumed (Hunter et al. 2010) or ignored. In humans, only noninvasive imaging methods are used to determine diameter, including cineangiography, computed tomography (CT), phase-contrast magnetic resonance imaging (MRI), standard transthoracic echocardiogram (TTE), intravascular ultrasound (IVUS), and color Motion-Mode (CMM) Doppler tissue imaging (DTI).

Similar *in vivo* approaches have been applied in small animals, and tremendous advances have been made in using the RHC approach to obtain RV pressure or PAP with closed and open chest preparations (Champion et al. 2000; Tabima et al. 2010; Wang and Chesler 2012). However, with the current techniques, the difficulties and challenges lie in the relatively poor spatial precision in the measurement of PA diameter, due to the small size of rodent blood vessels.

In vitro, inner and outer diameter are typically measured as a function of pressure, and factors that influence the mechanical behavior of the artery are examined as well (Hudetz 1979; Faury et al. 1999; Schulze-Bauer and Holzapfel 2003; Yuan et al. 2011). In large animals, both large (proximal) and small (distal) PAs can be tested (Shimoda, Norins, and Madden 1997); in small animals, only large PAs can be (Kobs et al. 2005; Ooi et al. 2010; Wang and Chesler 2012; Wang et al. 2013a). In the isolated vessel mechanical test, the PA is harvested and mounted in a vessel testing chamber (Shimoda et al. 1997; Kobs et al. 2005; Herrera et al. 2007; Ooi et al. 2010; Wang and Chesler 2012; Wang et al. 2013a) and the vessel diameters are measured over a range of physiological and pathological pressures. The *in vitro* measurement allows fine control of the test conditions, such that the effect of a single parameter (e.g., SMC tone, drug treatment, etc.) on the mechanical properties of the PA can be determined. Biaxial (in two directions) tests of arterial sections and uniaxial (in one direction) tests of tissue, either in strips or in rings, can also be performed (Lally et al. 2004; Lammers et al. 2008). Most biaxial and strip test methods do not allow the effects of SMC tone to be investigated, but isolated vessel and ring tests do (Griffith et al. 1994; Packer et al. 1998; Boutouyrie et al. 1998; Ooi et al. 2010; Tabima and Chesler 2010).

9.3.2 Measurement of the Mechanical Properties of the Whole Pulmonary Vasculature

Because the pulmonary vascular bed consists of a network of PAs, veins, and capillaries, a systemic measurement of pulmonary vascular mechanics should not be limited to a single vessel segment but should include changes in proximal PAs and distal PAs and the interactions between the two, such as pulse wave reflections. Therefore, in order to assess the complete mechanical function of the pulmonary circulation, the pressure–flow (P-Q) relationship, known as pulmonary vascular impedance (PVZ), is obtained *in vivo* (Pace 1971; Dujardin et al. 1982; Ewalenko et al. 1993, 1997; Maggiorini et al. 1998; Zhao et al. 2001; Tabima et al. 2012; Schreier et al. 2014) or *ex vivo* in isolated whole lungs (Zhao et al. 1993, 2001; Nossaman et al. 1994; Berkenbosch et al. 2000; Fagan et al. 2004; Tuchscherer et al. 2007; Vanderpool et al. 2011a).

The advantage of the *ex vivo* isolated, perfused, and ventilated lung preparation is that P-Q relationships are not affected by anesthesia (Ewalenko et al. 1993), volume status (Dujardin et al. 1982), or level of sympathetic nervous system activation (Pace 1971). In addition, the effect of drugs on the pulmonary vasculature can be investigated independent of their effect on the systemic vasculature (Vanderpool et al. 2011b). However, in most isolated, ventilated, perfused lung preparations, only steady P-Q relationships are obtained, from which only distal arterial caliber and stiffness can be derived in a global (spatially averaged) way. Our group has used pulsatile flow waveforms to obtain pulsatile P-Q relationships, which allow estimates of proximal artery stiffness to be made, but these are still difficult to compare to *in vivo* measurements obtained with physiological flow waveforms (Tuchscherer et al. 2007; Vanderpool and Chesler 2011; Vanderpool et al. 2011a,b).

From the synchronized pressure and flow measurements, which are typically obtained *in vivo* with an RHC and either ultrasound (Nakayama et al. 1997; Huez et al. 2004) or a catheter-based flow sensor (Laskey et al. 1993; Syyed et al. 2008) or *ex vivo* by direct recording of pressure and flow, impedance (PVZ) can be derived as the ratio of the pulsatile blood pressure to flow. The calculation of PVZ requires a spectral analysis of the PAP and flow waveforms and a mathematical elaboration (Fourier analysis) to derive a PVZ spectrum, which is expressed as the ratio of P to Q moduli and a phase angle (θ), both as a function of frequency (O'Rourke 1982):

$$PVZ(\omega) = \frac{|P(\omega)|}{|Q(\omega)|} \tag{9.1}$$

$$\theta(\omega) = \Phi(\omega) - \varphi(\omega) \tag{9.2}$$

where ω is the frequency, Φ is the pressure phase, and φ is the flow phase. This approach has been adopted in PH animal models, using *in vivo* and *ex vivo* techniques, and in patients (Naeije et al. 1990; Huez et al. 2004; Wauthy et al. 2004; Tuchscherer et al. 2007; Hunter et al. 2008, 2010). Details of how to calculate impedance for the pulmonary circulation are reviewed elsewhere (Champion et al. 2009; Chesler et al. 2009; Yuan et al. 2011).

The impedance spectra in systemic and pulmonary circulations share a similar, classic pattern of a high 0 Hz value (Z_0) followed by a local minimum and oscillations at high frequencies (Figure 9.2). Z_0, the input impedance in the absence of flow oscillations, is obtained as the magnitude of PVZ at 0 Hz, while the characteristic impedance Z_C is obtained by averaging PVZ from the 4th to the 10th harmonic (Reddy et al. 2003).

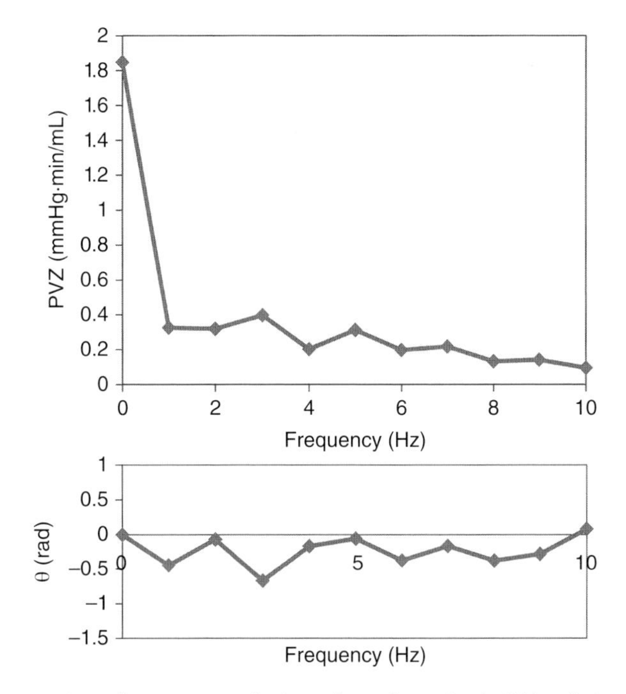

Figure 9.2 Representative pulmonary vascular impedance (magnitude PVZ and phase θ) spectra obtained from a healthy mouse.

PVZ represents the opposition of a vascular bed to pulsatile flow, which is generated by friction in small-diameter vessels, stiff vessels that do not accommodate pulsations, branching and tapering vessels that generate wave reflections, and other phenomena. The steady component of impedance, which forms the opposition to steady flow (i.e., the pulmonary vascular resistance), is represented by Z_0, while the impedance that provides opposition to pulsatile flow (i.e., the characteristic impedance) is represented by Z_C. Based on Z_C, the PAP waveform is separated into forward (P_f) and backward (P_b) traveling components using the linear wave separation method (Westerhof et al. 1972). The degree of global pulse wave reflection can then be calculated as the ratio of the amplitude of P_b to P_f (P_b/P_f) or as pulse wave velocity (PWV) (Nichols and O'Rourke 2005).

9.4 Mechanobiology in Pulmonary Hypertension

The biological response of an artery to increased blood pressure, which not only increases circumferential and axial wall stress but also increases luminal shear stress in the axial direction, has been critically reviewed by Humphrey (2008). The biological responses that lead to changes in arterial geometry, structure, and function include altered cell proliferation, migration, differentiation, apoptosis, vasoactivity (including contractile state, reactivity, and the ability to relax or dilate), synthesis and degradation of ECM, and crosslinking of ECM and integrin bindings. Similar responses have been found in PAs with PH development (Berkenbosch et al. 2000; Stenmark et al. 2006; Rabinovitch 2008). For instance, there is increased accumulation of collagen and elastin in large PAs in animals with PH (Kobs et al. 2005; Stenmark et al. 2006; Lammers et al. 2008). Using a transgenic mouse model, our group has found that mechanical changes (e.g., stiffening) of large PAs in PAH are associated with increased collagen deposition (Ooi et al. 2010), and particularly with increased collagen crosslinking (Wang et al. 2013a).

The mechanical changes in PAs with PAH mainly involve vascular stiffening (reduced compliance) and narrowing (increased resistance) (Wang and Chesler 2011). Furthermore, it has been found recently that there are interactions between the proximal and distal PAs in the pulmonary vascular bed. For instance, large proximal PA stiffening leads to distal arterial cyclic strain damage (Li et al. 2009), which promotes SMC proliferation and narrowing of distal PAs. At the same time, increased flow pulsatility in distal arteries induces inflammatory gene expression, leukocyte adhesion, and cell proliferation in endothelial cells (Li et al. 2009) and vascular fibrosis through endothelial–mesenchymal transdifferentiation (Elliott et al. 2015), which may alter remodeling in proximal PAs. Regardless of the mechanisms, distal arterial narrowing increases mPAP, which dilates the proximal arteries, and this arterial dilation increases circumferential stress and promotes SMC-mediated wall thickening (Stenmark et al. 2006; Humphrey 2008), leading to increased arterial stiffness (Kobs et al. 2005; Kobs and Chesler 2006; Ooi et al. 2010; Tabima and Chesler 2010).

The luminal shear stress and/or strain is also important for vascular cell signaling. There are some pilot studies measuring the shear stress in the PA *in vivo* using MRI (Truong et al. 2013; Barker et al. 2014) and contrast-enhanced CT imaging (Kheyfets et al. 2015). In particular, a recent computational fluid dynamics (CFD) study in PH patients showed that wall shear stress is correlated with PVR, arterial compliance, and wave reflection index, suggesting a new prognostic parameter in the management of PH (Kheyfets et al. 2015). However, how the shear stress affects arterial remodeling in PAH and affects disease progression is not clear.

9.5 Computational Modeling in Pulmonary Circulation

Pulmonary vascular diseases (PVDs) involve biomechanical changes at the cellular level (e.g., cell metabolism), tissue level (e.g., single artery), and organ level (e.g., heart and lungs). To understand the biomechanical mechanisms of these diseases, computational modeling serves as a useful tool for integrating information on molecular and cellular mechanisms with understanding at a larger scale (e.g., mechanical properties of a tissue).

Constitutive models of a single artery are mathematical models of the artery's mechanical properties. Various constitutive models, either phenomenological (i.e., capturing the observed mechanical function, or in this case stress–strain behavior) or structural (i.e., incorporating the known mechanical function of biological components such as elastin, collagen, and SMCs), have been proposed to study the physiological and pathological mechanical properties associated with hypertension, aging, and other conditions. The strain–energy function (SEF), which describes the strain energy per unit volume stored in a material, is a useful structural model of the artery wall, because it contains terms representing elastin and collagen and can be expressed as a function of strain in the radial, circumferential, and longitudinal directions of an artery. It is well known that the synthesis and degradation of ECM components are critical contributors to the arterial mechanical properties. Since our experimental data suggest an important role for collagen crosslinking in PA stiffening during PAH development (Wang and Chesler 2012; Wang et al. 2013a), we have developed a constitutive model that includes material parameters related to collagen crosslinking, validated with experimental data. Previously, an eight-chain orthotropic-element model that captured the hyperplastic behavior of macromolecules (Arruda and Boyce 1993; Bischoff et al. 2002) was applied to predict ECM crosslinking in large PAs using a rat PH model (Zhang et al. 2005). However, this model did not distinguish between elastin and collagen content or crosslinking. We recently revised the model (Figure 9.3) by adding a neo-Hookean form to represent elastin fibers, and then investigated the revised model's ability to distinguish the effects of collagen content from collagen crosslinking on the elastic modulus of mouse PAs. Our results show that the material properties (e.g., collagen content and crosslinking) predicted by the revised model are consistent with the experimental measurements (unpublished data).

In contrast to the modeling of a single segment of artery, multiscale computational modeling allows the simulation of the whole cardiopulmonary system at different physiological scales through integration of several models with overlapping scales (genes to molecules, molecules to cells, cells to tissues, and tissues to organs) (Beard et al. 2012). For instance, using a TriSeg model in which the circulation is simulated by an adapted simple lumped parameter model (Lumens et al. 2009), Tewari et al. (2013) analyzed the cardiovascular dynamics in healthy and PAH mice. The model invokes a total of 26 adjustable parameters, which are estimated based on least-squares fitting of the RV pressure and volume (PV) experimental data. While the

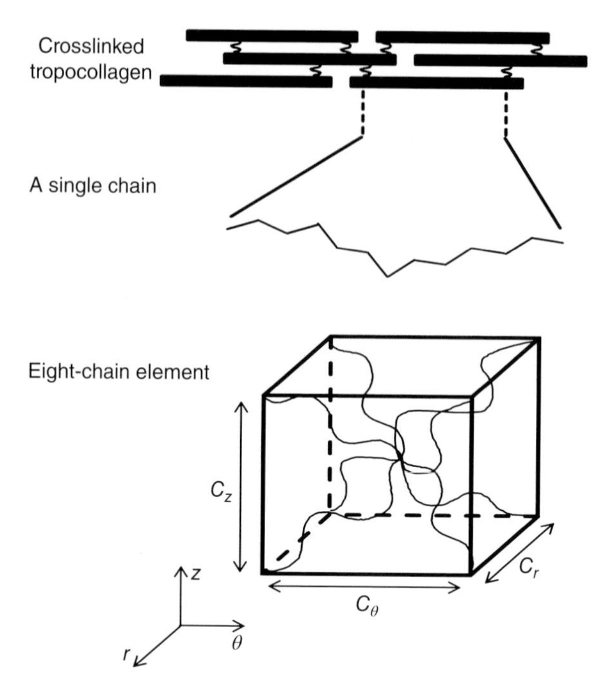

Figure 9.3 Eight-chain model of crosslinked collagen in an artery wall. A single chain represents the tropocollagen molecules between neighboring crosslinks. This single chain is made up of multiple subunits with fixed length, which represent the repeating amino acid motif of a tropocollagen molecule. The length of each chain in the eight-chain element depends on the number of subunits, which in turn depends on the density of crosslinks. The element has normalized dimensions $Cr \times C\theta \times Cz$ along the material axes r, θ, are z, respectively.

in vivo PV data only provide a measure of the RV wall mechanics at the organ level, the model predicts ventricular contractile properties at both organ and sarcomere levels, as well as vascular mechanical properties such as pulmonary vascular resistance and elastance.

9.6 Impact of Pulmonary Arterial Biomechanics on the Right Heart

A rising field of research in PAH is the investigation of the impact of PA biomechanics on RV function. RV afterload is a critical metric of PAH progression, because the most common cause of death in PAH patients is right heart failure due to RV overload.

RV afterload is closely related to the dynamic interplay between vascular resistance, compliance, and wave reflections. It is measured by hydraulic load or hydraulic power, which is work per unit time generated by the heart to sustain forward blood flow (Nichols and O'Rourke 2005). The power provided by RV consists of two components: the steady power required to produce net forward flow and the oscillatory power required to produce zero-mean oscillations in flow (Milnor et al. 1966). Historically, the increase in RV afterload in PAH has been attributed to increased PVR, which only reflects the steady component of total RV power. However, over a third of the RV workload increase in PAH is caused by large PA stiffening (Stenmark et al. 2006), which mostly influences the oscillatory RV power. In clinical studies, an increase in proximal PA stiffness is found to be an excellent predictor of mortality in patients with PAH (Mahapatra et al. 2006a,b; Gan et al. 2007), which suggests an important role for proximal PA stiffening in RV failure. Therefore, a comprehensive investigation of the impact of PA mechanics on RV function requires the measurement of both steady and oscillatory components of the RV power. An open area of research is how altered PA mechanics precede and promote RV failure due to chronic pressure overload.

9.7 Conclusion

There has been substantial advancement in the understanding of pulmonary vascular biomechanics and mechanobiology over the past few decades. It is increasingly recognized that the investigation of the structure–function relationship of pulmonary vascular system components should be placed in the context of RV function, which predicts clinical outcomes and ultimately determines mortality in PAH. This requires an integrated approach to examining overall cardiopulmonary function via multiple experiments or multiscale computational modeling. It is well known that the pulmonary circulation is unique, and is different from the left-sided systemic circulation due to its low blood pressure and high arterial compliance. In addition, the RV is embryologically (Zaffran et al. 2004), structurally (Roche and Redington 2013; Walker and Buttrick 2013), and functionally (Redington et al. 1988; Walker and Buttrick 2013) distinct from the left ventricle. The entire field of right heart failure research is very young (Voelkel et al. 2006; Vandenheuvel et al. 2013), and the biomechanical mechanism of RV failure due to altered pulmonary vasculature biomechanics and mechanobiology is an open and important area.

References

Arruda, E. M. and M. C. Boyce. 1993. “A 3-dimensional constitutive model for the large stretch behavior of rubber elastic-materials.” *Journal of the Mechanics and Physics of Solids* **41**(2): 389–412. doi:10.1016/0022-5096(93)90013-6.

Barker, A. J., A. Roldan-Alzate, P. Entezari, S. J. Shah, N. C. Chesler, O. Wieben, et al. 2014. “Four-dimensional flow assessment of pulmonary artery flow and wall shear stress in adult pulmonary arterial hypertension: results from two institutions.” *Magnetic Resonance in Medicine* **73**(5): 1904–13. doi:10.1002/mrm.25326.

Beard, D. A., M. L. Neal, N. Tabesh-Saleki, C. T. Thompson, J. B. Bassingthwaighte, M. Shimoyama, and B. E. Carlson. 2012. “Multiscale modeling and data integration in the virtual physiological rat project.” *Annals of Biomedical Engineering* **40**(11): 2365–78. doi:10.1007/s10439-012-0611-7.

Bellofiore, A., A. Roldan-Alzate, M. Besse, H. B. Kellihan, D. W. Consigny, C. J. Francois, and N. C. Chesler. 2013. “Impact of acute pulmonary embolization on arterial stiffening and right ventricular function in dogs.” *Annals of Biomedical Engineering* **41**(1): 195–204. doi:10.1007/s10439-012-0635-z.

Berkenbosch, J. W., J. Baribeau, and T. Perreault. 2000. "Decreased synthesis and vasodilation to nitric oxide in piglets with hypoxia-induced pulmonary hypertension." *American Journal of Physiology. Lung Cellular and Molecular Physiology* **278**(2): L276–83. PMID:10666111.

Bischoff, J. E., E. M. Arruda, and K. Grosh. 2002. "A microstructurally based orthotropic hyperelastic constitutive law." *Journal of Applied Mechanics* **69**(5): 570–9. doi:10.1115/1.1485754.

Boutouyrie, P., S. Boumaza, P. Challande, P. Lacolley, and S. Laurent. 1998. "Smooth muscle tone and arterial wall viscosity: an in vivo/in vitro study." *Hypertension* **32**(2): 360–4. PMID:9719068.

Champion, H. C., D. J. Villnave, A. Tower, P. J. Kadowitz, and A. L. Hyman. 2000. "A novel right-heart catheterization technique for in vivo measurement of vascular responses in lungs of intact mice." *American Journal of Physiology. Heart and Circulatory Physiology* **278**(1): H8–15. PMID:10644578.

Champion, H. C., E. D. Michelakis, and P. M. Hassoun. 2009. "Comprehensive invasive and noninvasive approach to the right ventricle-pulmonary circulation unit: state of the art and clinical and research implications." *Circulation* **120**(11): 992–1007. doi:10.1161/CIRCULATIONAHA.106.674028.

Chesler, N. C., A. Roldan, R. R. Vanderpool, and R. Naeije. 2009. "How to measure pulmonary vascular and right ventricular function." *Conference Proceedings: Annual International Conference of the IEEE Engineering in Medicine and Biology Society* **2009**: 177–80. doi:10.1109/IEMBS.2009.5333835.

Cox, R. H. 1982. "Comparison of mechanical and chemical properties of extra- and intralobar canine pulmonary arteries." *American Journal of Physiology* **242**(2): H245–53. PMID:7065159.

Cox, R. H. 1984. "Viscoelastic properties of canine pulmonary arteries." *American Journal of Physiology* **246**(1 Pt. 2): H90–6. PMID:6696094.

Dujardin, J. P., D. N. Stone, C. D. Forcino, L. T. Paul, and H. P. Pieper. 1982. "Effects of blood volume changes on characteristic impedance of the pulmonary artery." *American Journal of Physiology* **242**(2): H197–202. PMID:7065152.

Dyer, K., C. Lanning, B. Das, P. F. Lee, D. D. Ivy, L. Valdes-Cruz, and R. Shandas. 2006. "Noninvasive Doppler tissue measurement of pulmonary artery compliance in children with pulmonary hypertension." *Journal of the American Society of Echocardiography* **19**(4): 403–12. doi:10.1016/j.echo.2005.11.012.

Elliott, W. H., Y. Tan, M. Li, and W. Tan. 2015. "High pulsatility flow promotes vascular fibrosis by triggering endothelial EndMT and fibroblast activation." *Cellular and Molecular Bioengineering* **8**(2): 285–95. doi:10.1007/s12195-015-0386-7.

Ewalenko, P., C. Stefanidis, A. Holoye, S. Brimioulle, and R. Naeije. 1993. "Pulmonary vascular impedance vs. resistance in hypoxic and hyperoxic dogs: effects of propofol and isoflurane." *Journal of Applied Physiology* **74**(5): 2188–93. PMID:8335547.

Ewalenko, P., S. Brimioulle, M. Delcroix, P. Lejeune, and R. Naeije. 1997. "Comparison of the effects of isoflurane with those of propofol on pulmonary vascular impedance in experimental embolic pulmonary hypertension." *British Journal of Anaesthesia* **79**(5): 625–30. PMID:9422903.

Fagan, K. A., M. Oka, N. R. Bauer, S. A. Gebb, D. D. Ivy, K. G. Morris, and I. F. McMurtry. 2004. "Attenuation of acute hypoxic pulmonary vasoconstriction and hypoxic pulmonary hypertension in mice by inhibition of Rho-kinase." *American Journal of Physiology. Lung Cellular and Molecular Physiology* **287**(4): L656–64. doi:10.1152/ajplung.00090.2003.

Faury, G., G. M. Maher, D. Y. Li, M. T. Keating, R. P. Mecham, and W. A. Boyle. 1999. "Relation between outer and luminal diameter in cannulated arteries." *American Journal of Physiology* **277**(5 Pt. 2): H1745–53. PMID:10564127.

Friedberg, M. K., J. A. Feinstein, and D. N. Rosenthal. 2006. "A novel echocardiographic Doppler method for estimation of pulmonary arterial pressures." *Journal of the American Society of Echocardiography* **19**(5): 559–62. doi:10.1016/j.echo.2005.12.020.

Gan, C. T., J. W. Lankhaar, N. Westerhof, J. T. Marcus, A. Becker, J. W. Twisk, et al. 2007. "Noninvasively assessed pulmonary artery stiffness predicts mortality in pulmonary arterial hypertension." *Chest* **132**(6): 1906–12. doi:10.1378/chest.07-1246.

Griffith, S. L., R. A. Rhoades, and C. S. Packer. 1994. "Pulmonary arterial smooth muscle contractility in hypoxia-induced pulmonary hypertension." *Journal of Applied Physiology* **77**(1): 406–14. PMID:7961262.

Gust, R. and D. P. Schuster. 2001. "Vascular remodeling in experimentally induced subacute canine pulmonary hypertension." *Experimental Lung Research* **27**(1): 1–12. PMID:11202060.

Herrera, E. A., V. M. Pulgar, R. A. Riquelme, E. M. Sanhueza, R. V. Reyes, G. Ebensperger, et al. 2007. "High-altitude chronic hypoxia during gestation and after birth modifies cardiovascular responses in newborn sheep." *American Journal of Physiology. Regulatory, Integrative and Comparative Physiology* **292**(6): R2234–40. doi:10.1152/ajpregu.00909.2006.

Holzapfel, G. A., T. C. Gasser, and R. W. Ogden. 2000. "A new constitutive framework for arterial wall mechanics and a comparative study of material models." *Journal of Elasticity* **61**(1–3): 1–48. doi:10.1023/A:1010835316564.

Huang, W., R. T. Yen, M. McLaurine, and G. Bledsoe. 1996. "Morphometry of the human pulmonary vasculature." *Journal of Applied Physiology* **81**(5): 2123–33. PMID:8941537.

Hudetz, A. G. 1979. "Incremental elastic modulus for orthotropic incompressible arteries." *Journal of Biomechanics* **12**(9): 651–5. doi:0021-9290(79)90015-0.

Huez, S., S. Brimioulle, R. Naeije, and J. L. Vachiery. 2004. "Feasibility of routine pulmonary arterial impedance measurements in pulmonary hypertension." *Chest* **125**(6): 2121–8. PMID:15189931.

Humbert, M., N. W. Morrell, S. L. Archer, K. R. Stenmark, M. R. MacLean, I. M. Lang, et al. 2004. "Cellular and molecular pathobiology of pulmonary arterial hypertension." *Journal of the American College of Cardiology* **43**(12 Suppl.): 13S–24S. doi:10.1016/j.jacc.2004.02.029 S0735109704004383.

Humbert, M., O. Sitbon, A. Chaouat, M. Bertocchi, G. Habib, V. Gressin, et al. 2010. "Survival in patients with idiopathic, familial, and anorexigen-associated pulmonary arterial hypertension in the modern management era." *Circulation* **122**(2): 156–63. doi:0.1161/CIRCULATIONAHA.109.911818.

Humphrey, J. D. 1995. "Mechanics of the arterial wall: review and directions." *Critical Reviews in Biomedical Engineering* **23**(1–2): 1–162. PMID:8665806.

Humphrey, JD. 2002. *Cardiovascular Solid Mechanics: Cells, Tissues and Organs.* New York: Springer-Verlag.

Humphrey, J. D. 2008. "Mechanisms of arterial remodeling in hypertension: coupled roles of wall shear and intramural stress." *Hypertension* **52**(2): 195–200. doi:10.1161/HYPERTENSIONAHA.107.103440.

Hunter, K. S., P. F. Lee, C. J. Lanning, D. D. Ivy, K. S. Kirby, L. R. Claussen, et al. 2008. "Pulmonary vascular input impedance is a combined measure of pulmonary vascular resistance and stiffness and predicts clinical outcomes better than pulmonary vascular resistance alone in pediatric patients with pulmonary hypertension." *American Heart Journal* **155**(1): 166–74. doi:10.1016/j.ahj.2007.08.014.

Hunter, K. S., J. A. Albietz, P. F. Lee, C. J. Lanning, S. R. Lammers, S. H. Hofmeister, et al. 2010. "In vivo measurement of proximal pulmonary artery elastic modulus in the neonatal calf model of pulmonary hypertension: development and ex vivo validation." *Journal of Applied Physiology* **108**(4): 968–75. doi:10.1152/japplphysiol.01173.2009.

Kheyfets, V., M. Thirugnanasambandam, L. Rios, D. Evans, T. Smith, T. Schroeder, et al. 2015. "The role of wall shear stress in the assessment of right ventricle hydraulic workload." *Pulmonary Circulation* **5**(1): 90–100. doi:10.1086/679703.

Kim, H., G. L. Yung, J. J. Marsh, R. G. Konopka, C. A. Pedersen, P. G. Chiles, et al. 2000. "Endothelin mediates pulmonary vascular remodelling in a canine model of chronic embolic pulmonary hypertension." *European Respiratory Journal* **15**(4): 640–8. PMID:10780753.

Kobs, R. W. and N. C. Chesler. 2006. "The mechanobiology of pulmonary vascular remodeling in the congenital absence of eNOS." *Biomechanics and Modeling in Mechanobiology* **5**(4): 217–25. doi:10.1007/s10237-006-0018-1.

Kobs, R. W., N. E. Muvarak, J. C. Eickhoff, and N. C. Chesler. 2005. "Linked mechanical and biological aspects of remodeling in mouse pulmonary arteries with hypoxia-induced hypertension." *American Journal of Physiology. Heart and Circulatory Physiology* **288**(3): H1209–17. doi:10.1152/ajpheart.01129.2003.

Lally, C., A. J. Reid, and P. J. Prendergast. 2004. "Elastic behavior of porcine coronary artery tissue under uniaxial and equibiaxial tension." *Annals of Biomedical Engineering* **32**(10): 1355–64. PMID:15535054.

Lammers, S. R., P. H. Kao, H. J. Qi, K. Hunter, C. Lanning, J. Albietz, et al. 2008. "Changes in the structure-function relationship of elastin and its impact on the proximal pulmonary arterial mechanics of hypertensive calves." *American Journal of Physiology. Heart and Circulatory Physiology* **295**(4): H1451–9. doi:10.1152/ajpheart.00127.2008.

Laskey, W. K., V. A. Ferrari, H. I. Palevsky, and W. G. Kussmaul. 1993. "Pulmonary artery hemodynamics in primary pulmonary hypertension." *Journal of the American College of Cardiology* **21**(2): 406–12. doi:0735-1097(93)90682-Q.

Li, M., D. E. Scott, R. Shandas, K. R. Stenmark, and W. Tan. 2009. "High pulsatility flow induces adhesion molecule and cytokine mRNA expression in distal pulmonary artery endothelial cells." *Annals of Biomedical Engineering* **37**(6): 1082–92. doi:10.1007/s10439-009-9684-3.

Lumens, J., T. Delhaas, B. Kirn, and T. Arts. 2009. "Three-wall segment (TriSeg) model describing mechanics and hemodynamics of ventricular interaction." *Annals of Biomedical Engineering* **37**(11): 2234–55. doi:10.1007/s10439-009-9774-2.

Maggiorini, M., S. Brimioulle, D. De Canniere, M. Delcroix, and R. Naeije. 1998. "Effects of pulmonary embolism on pulmonary vascular impedance in dogs and minipigs." *Journal of Applied Physiology* **84**(3): 815–21. PMID:9480938.

Mahapatra, S., R. A. Nishimura, J. K. Oh, and M. D. McGoon. 2006a. "The prognostic value of pulmonary vascular capacitance determined by Doppler echocardiography in patients with pulmonary arterial hypertension." *Journal of the American Society of Echocardiography* **19**(8): 1045–50. doi:10.1016/j.echo.2006.03.008.

Mahapatra, S., R. A. Nishimura, P. Sorajja, S. Cha, and M. D. McGoon. 2006b. "Relationship of pulmonary arterial capacitance and mortality in idiopathic pulmonary arterial hypertension." *Journal of the American College of Cardiology* **47**(4): 799–803. doi:10.1016/j.jacc.2005.09.054.

McLaughlin, V. V., S. L. Archer, D. B. Badesch, R. J. Barst, H. W. Farber, J. R. Lindner, et al. 2009. "ACCF/AHA 2009 expert consensus document on pulmonary hypertension: a report of the American College of Cardiology Foundation Task Force on Expert Consensus Documents and the American Heart Association: developed in collaboration with the American College of Chest Physicians, American Thoracic Society, Inc., and the Pulmonary Hypertension Association." *Circulation* **119**(16): 2250–94. doi:10.1161/CIRCULATIONAHA.109.192230.

Milnor, W. R., D. H. Bergel, and J. D. Bargainer. 1966. "Hydraulic power associated with pulmonary blood flow and its relation to heart rate." *Circulation Research* **19**(3): 467–80. PMID:5925148.

Naeije, R., J. M. Maarek, and H. K. Chang. 1990. "Pulmonary vascular impedance in microembolic pulmonary hypertension: effects of synchronous high-frequency jet ventilation." *Respiratory Physiology* **79**(3): 205–17. PMID:2356361.

Nakayama, Y., N. Nakanishi, M. Sugimachi, H. Takaki, S. Kyotani, T. Satoh, et al. 1997. "Characteristics of pulmonary artery pressure waveform for differential diagnosis of chronic pulmonary thromboembolism and primary pulmonary hypertension." *Journal of the American College of Cardiology* **29**(6): 1311–16. doi:S0735109797000545.

Nichols, W.W. and M.F. O'Rourke. 2005. *McDonald's Blood Flow in Arteries: Theoretical, Experimental and Clinical Principles*, 5th edn. London: Hodder Arnold.

Nossaman, B. D., C. J. Feng, and P. J. Kadowitz. 1994. "Analysis of responses to bradykinin and influence of HOE 140 in the isolated perfused rat lung." *American Journal of Physiology* **266**(6 Pt. 2): H2452–61.

O'Rourke, M. F. 1982. "Vascular impedance in studies of arterial and cardiac function." *Physiological Reviews* **62**(2): 570–623. PMID:6461866.

Ooi, C. Y., Z. Wang, D. M. Tabima, J. C. Eickhoff, and N. C. Chesler. 2010. "The role of collagen in extralobar pulmonary artery stiffening in response to hypoxia-induced pulmonary hypertension." *American Journal of Physiology. Heart and Circulatory Physiology* **299**(6): H1823–31. doi:10.1152/ajpheart.00493.2009.

Pace, J. B. 1971. "Sympathetic control of pulmonary vascular impedance in anesthetized dogs." *Circulation Research* **29**(5): 555–68. PMID:5120617.

Packer, C. S., J. E. Roepke, N. H. Oberlies, and R. A. Rhoades. 1998. "Myosin isoform shifts and decreased reactivity in hypoxia-induced hypertensive pulmonary arterial muscle." *American Journal of Physiology* **274**(5 Pt. 1): L775–85. PMID:9612293.

Rabinovitch, M. 2008. "Molecular pathogenesis of pulmonary arterial hypertension." *Journal of Clinical Investigation* **118**(7):2372-9. doi:10.1172/JCI33452.

Rain, S., M. L. Handoko, P. Trip, C. T. Gan, N. Westerhof, G. J. Stienen, et al. 2013. "Right ventricular diastolic impairment in patients with pulmonary arterial hypertension." *Circulation* **128**(18): 2016–25, 1–10. doi:10.1161/CIRCULATIONAHA.113.001873.

Reddy, A. K., Y. H. Li, T. T. Pham, L. N. Ochoa, M. T. Trevino, C. J. Hartley, et al. 2003. "Measurement of aortic input impedance in mice: effects of age on aortic stiffness." *American Journal of Physiology. Heart and Circulatory Physiology* **285**(4): H1464–70. doi:10.1152/ajpheart.00004.2003.

Redington, A. N., H. H. Gray, M. E. Hodson, M. L. Rigby, and P. J. Oldershaw. 1988. "Characterisation of the normal right ventricular pressure-volume relation by biplane angiography and simultaneous micromanometer pressure measurements." *British Heart Journal* **59**(1): 23–30. PMID:3342146.

Roche, S. L. and A. N. Redington. 2013. "The failing right ventricle in congenital heart disease." *Canadian Journal of Cardiology* **29**(7): 768–78. doi:10.1016/j.cjca.2013.04.018.

Ryan, J. J., G. Marsboom, and S. L. Archer. 2013. "Rodent models of group 1 pulmonary hypertension." *Pharmacotherapy of Pulmonary Hypertension* **218**: 105–49. doi:10.1007/978-3-642-38664-0_5.

Santana, D. B., J. G. Barra, J. C. Grignola, F. F. Gines, and R. L. Armentano. 2005. "Pulmonary artery smooth muscle activation attenuates arterial dysfunction during acute pulmonary hypertension." *Journal of Applied Physiology* **98**(2): 605–13. doi:10.1152/japplphysiol.00361.2004.

Schreier, D. A., T. A. Hacker, K. Hunter, J. Eickoff, A. Liu, G. Song, and N. Chesler. 2014. "Impact of increased hematocrit on right ventricular afterload in response to chronic hypoxia." *Journal of Applied Physiology* **117**(8): 833–9. doi:10.1152/japplphysiol.00059.2014.

Schulze-Bauer, C. A. and G. A. Holzapfel. 2003. "Determination of constitutive equations for human arteries from clinical data." *Journal of Biomechanics* **36**(2):165–9. doi:S0021929002003676.

Shimoda, L. A., N. A. Norins, and J. A. Madden. 1997. "Flow-induced responses in cat isolated pulmonary arteries." *Journal of Applied Physiology* **83**(5): 1617–22. PMID:9375329.

Silver, F. H., I. Horvath, and D. J. Foran. 2001. "Viscoelasticity of the vessel wall: the role of collagen and elastic fibers." *Critical Reviews in Biomedical Engineering* **29**(3): 279–301. PMID:11730097.

Simonneau, G., M. A. Gatzoulis, I. Adatia, D. Celermajer, C. Denton, A. Ghofrani, et al. 2013. "Updated clinical classification of pulmonary hypertension." *Journal of the American College of Cardiology* **62**(25 Suppl.): D34–41. doi:10.1016/j.jacc.2013.10.029.

Singhal, S., R. Henderson, K. Horsfield, K. Harding, and G. Cumming. 1973. "Morphometry of the human pulmonary arterial tree." *Circulation Research* **33**(2): 190–7. PMID:4727370.

Stenmark, K. R., K. A. Fagan, and M. G. Frid. 2006. "Hypoxia-induced pulmonary vascular remodeling: cellular and molecular mechanisms." *Circulation Research* **99**(7): 675–91. doi:10.1161/01.RES.0000243584.45145.3f.

Swift, A. J., S. Rajaram, R. Condliffe, D. Capener, J. Hurdman, C. Elliot, et al. 2012. "Pulmonary artery relative area change detects mild elevations in pulmonary vascular resistance and predicts adverse outcome in pulmonary hypertension." *Investigative Radiology* **47**(10): 571–7. doi:10.1097/RLI.0b013e31826c4341.

Syyed, R., J. T. Reeves, D. Welsh, D. Raeside, M. K. Johnson, and A. J. Peacock. 2008. "The relationship between the components of pulmonary artery pressure remains constant under all conditions in both health and disease." *Chest* **133**(3): 633–9. doi:10.1378/chest.07-1367.

Tabima, D. M. and N. C. Chesler. 2010. "The effects of vasoactivity and hypoxic pulmonary hypertension on extralobar pulmonary artery biomechanics." *Journal of Biomechanics* **43**(10): 1864–9. doi:10.1016/j.jbiomech.2010.03.033.

Tabima, D. M., T. A. Hacker, and N. C. Chesler. 2010. "Measuring right ventricular function in the normal and hypertensive mouse hearts using admittance-derived pressure-volume loops." *American Journal of Physiology. Heart and Circulatory Physiology* **299**(6): H2069–75. doi:10.1152/ajpheart.00805.2010.

Tabima, D. M., A. Roldan-Alzate, Z. J. Wang, T. A. Hacker, R. C. Molthen, and N. C. Chesler. 2012. "Persistent vascular collagen accumulation alters hemodynamic recovery from chronic hypoxia." *Journal of Biomechanics* **45**(5): 799–804. doi:10.1016/j.jbiomech.2011.11.020.

Tewari, S. G., S. M. Bugenhagen, Z. Wang, D. A. Schreier, B. E. Carlson, N. C. Chesler, and D. A. Beard. 2013. "Analysis of cardiovascular dynamics in pulmonary hypertensive C57BL6/J mice." *Frontiers in Physiology* **4**: 355. doi:10.3389/fphys.2013.00355.

Tian, L. and N. C. Chesler. 2012. "In vivo and in vitro measurements of pulmonary arterial stiffness: a brief review." *Pulmonary Circulation* **2**(4): 505–17. doi:10.4103/2045-8932.105040.

Tian, L., Z. Wang, R. S. Lakes, and N. C. Chesler. 2013. "Comparison of approaches to quantify arterial damping capacity from pressurization tests on mouse conduit arteries." *Journal of Biomechanical Engineering* **135**(5): 54504. doi:10.1115/1.4024135.

Truong, U., B. Fonseca, J. Dunning, S. Burgett, C. Lanning, D. D. Ivy, et al. 2013. "Wall shear stress measured by phase contrast cardiovascular magnetic resonance in children and adolescents with pulmonary arterial hypertension." *Journal of Cardiovascular Magnetic Resonance* **15**: 81. doi:10.1186/1532-429X-15-81.

Tuchscherer, H. A., R. R. Vanderpool, and N. C. Chesler. 2007. "Pulmonary vascular remodeling in isolated mouse lungs: effects on pulsatile pressure-flow relationships." *Journal of Biomechanics* **40**(5): 993–1001. PMID:16756983.

Vandenheuvel, M. A., S. Bouchez, P. F. Wouters, and S. G. De Hert. 2013. "A pathophysiological approach towards right ventricular function and failure." *European Journal of Anaesthesiology* **30**(7): 386–94. doi:10.1097/EJA.0b013e3283607a2d.

Vanderpool, R. R. and N. C. Chesler. 2011. "Characterization of the isolated, ventilated, and instrumented mouse lung perfused with pulsatile flow." *Journal of Visualized Experiments* **50**. doi:10.3791/2690.

Vanderpool, R. R., A. R. Kim, R. Molthen, and N. C. Chesler. 2011a. "Effects of acute Rho kinase inhibition on chronic hypoxia-induced changes in proximal and distal pulmonary arterial structure and function." *Journal of Applied Physiology* **110**(1):188–98. doi:10.1152/japplphysiol.00533.2010.

Vanderpool, R. R., R. Naeije, and N. C. Chesler. 2011b. "Impedance in isolated mouse lungs for the determination of site of action of vasoactive agents and disease." *Annals of Biomedical Engineering* **38** (5): 1854–61. doi:10.1007/s10439-010-9960-2.

Voelkel, N. F., R. A. Quaife, L. A. Leinwand, R. J. Barst, M. D. McGoon, D. R. Meldrum, et al. 2006. "Right ventricular function and failure: report of a National Heart, Lung, and Blood Institute working group on cellular and molecular mechanisms of right heart failure." *Circulation* **114**(17): 1883–91. doi:10.1161/CIRCULATIONAHA.106.632208.

Walker, L. A. and P. M. Buttrick. 2013. "The right ventricle: biologic insights and response to disease: updated." *Current Cardiology Reviews* **9**(1): 73–81. PMID:23092273.

Wang, Z. and N.C. Chesler. 2011. "Pulmonary vascular wall stiffness: an important contributor to the increased right ventricular afterload with pulmonary hypertension." *Pulmonary Circulation* **1**: 212–23. doi:10.4103/2045-8932.83453.

Wang, Z. and N. C. Chesler. 2012. “Role of collagen content and cross-linking in large pulmonary arterial stiffening after chronic hypoxia.” *Biomechanics and Modeling in Mechanobiology* **11** (1-2):279-89. doi:10.1007/s10237-011-0309-z.

Wang, Z., R. S. Lakes, J. C. Eickhoff, and N. C. Chesler. 2013a. “Effects of collagen deposition on passive and active mechanical properties of large pulmonary arteries in hypoxic pulmonary hypertension.” *Biomechanics and Modeling in Mechanobiology* **12**(6): 1115–25. doi:10.1007/s10237-012-0467-7.

Wang, Z., R. S. Lakes, M. Golob, J. C. Eickhoff, and N. C. Chesler. 2013b. “Changes in large pulmonary arterial viscoelasticity in chronic pulmonary hypertension.” *PLoS One* **8**(11): e78569. doi:10.1371/journal.pone.0078569.

Wauthy, P., A. Pagnamenta, F. Vassalli, R. Naeije, and S. Brimioulle. 2004. “Right ventricular adaptation to pulmonary hypertension: an interspecies comparison.” *American Journal of Physiology. Heart and Circulatory Physiology* **286**(4): H1441–7. doi:10.1152/ajpheart.00640.2003.

Westerhof, N., P. Sipkema, G. C. van den Bos, and G. Elzinga. 1972. “Forward and backward waves in the arterial system.” *Cardiovascular Research* **6**(6): 648–56. PMID:4656472.

Yuan, J. X.-J., J. G. N. Garcia, C. A. Hales, S. Rich, S. L. Archer, and J. B. West (eds.). 2011. *Textbook of Pulmonary Vascular Disease*. New York: Springer.

Zaffran, S., R. G. Kelly, S. M. Meilhac, M. E. Buckingham, and N. A. Brown. 2004. “Right ventricular myocardium derives from the anterior heart field.” *Circulation Research* **95**(3): 261–8. doi:10.1161/01.RES.0000136815.73623.BE.

Zhang, Y., M. L. Dunn, E. S. Drexler, C. N. McCowan, A. J. Slifka, D. D. Ivy, and R. Shandas. 2005. “A microstructural hyperelastic model of pulmonary arteries under normo- and hypertensive conditions.” *Annals of Biomedical Engineering* **33**(8): 1042–52. doi:10.1007/s10439-005-5771-2.

Zhao, L., D. E. Crawley, J. M. Hughes, T. W. Evans, and R. J. Winter. 1993. “Endothelium-derived relaxing factor activity in rat lung during hypoxic pulmonary vascular remodeling.” *Journal of Applied Physiology* **74**(3): 1061–5. PMID:8482643.

Zhao, L., N. A. Mason, N. W. Morrell, B. Kojonazarov, A. Sadykov, A. Maripov, et al. 2001. “Sildenafil inhibits hypoxia-induced pulmonary hypertension.” *Circulation* **104**(4): 424–8. PMID:11468204.

10

Mechanobiology and the Kidney Glomerulus

Franziska Lausecker[1,2], Christoph Ballestrem[1], and Rachel Lennon[1,2,3]

[1] *Wellcome Trust Centre for Cell-Matrix Research, Faculty of Life Sciences University of Manchester, Manchester, UK*
[2] *Institute of Human Development, Faculty of Human Sciences, University of Manchester, Manchester, UK*
[3] *Department of Paediatric Nephrology, Central Manchester University Hospitals NHS Foundation Trust (CMFT), Manchester Academic Health Science Centre (MAHSC), Manchester, UK*

10.1 Introduction

The kidney is responsible for filtering the blood to enable the excretion of metabolic waste products and to maintain homeostasis of fluids and electrolytes. There are approximately 1 million nephrons in every kidney (Keller et al. 2003), and a single nephron constitutes the functional unit of the kidney. Each nephron broadly consists of two distinct compartments: the glomerulus and the tubular system. Filtration of the blood leads to the production of the primary filtrate, and this takes place in the glomerulus (Figure 10.1a); reabsorption of small molecules and water from the primary filtrate back into the blood is controlled by the tubular system, which follows immediately downstream from the glomerulus. Efficient filtration in the glomerulus depends on two factors: first, the integrity of the glomerular filtration barrier, which consists of specialized epithelial cells, endothelial cells, and an intervening basement membrane (Jarad and Miner 2009); and second, a filtration force, directed from the bloodstream to the outer surface of the capillary wall. The magnitude of this force is dependent upon the opposing action of hydrostatic pressure from within the capillary lumen forcing filtration against oncotic pressure, which maintains water in the capillaries. In health, the glomeruli filter an extraordinary volume of water and small molecules; in a 70 kg adult, this is typically 180 L in 24 hours. The tubular system reabsorbs the vast majority of the filtrate, leaving a final urine volume of 1–2 L for excretion. Given the task of filtration, cells in the glomerulus need to sense and respond to force, and this chapter describes our current understanding of the machinery required for this regulation.

10.2 Glomerular Filtration Barrier

The integrity of all three components of the glomerular filtration barrier is required for normal filtration. From the luminal aspect of the capillary, the first component of the barrier is the glomerular endothelium, where glomerular endothelial cells (GEnCs) are

Mechanobiology: Exploitation for Medical Benefit, First Edition. Edited by Simon C. F. Rawlinson.

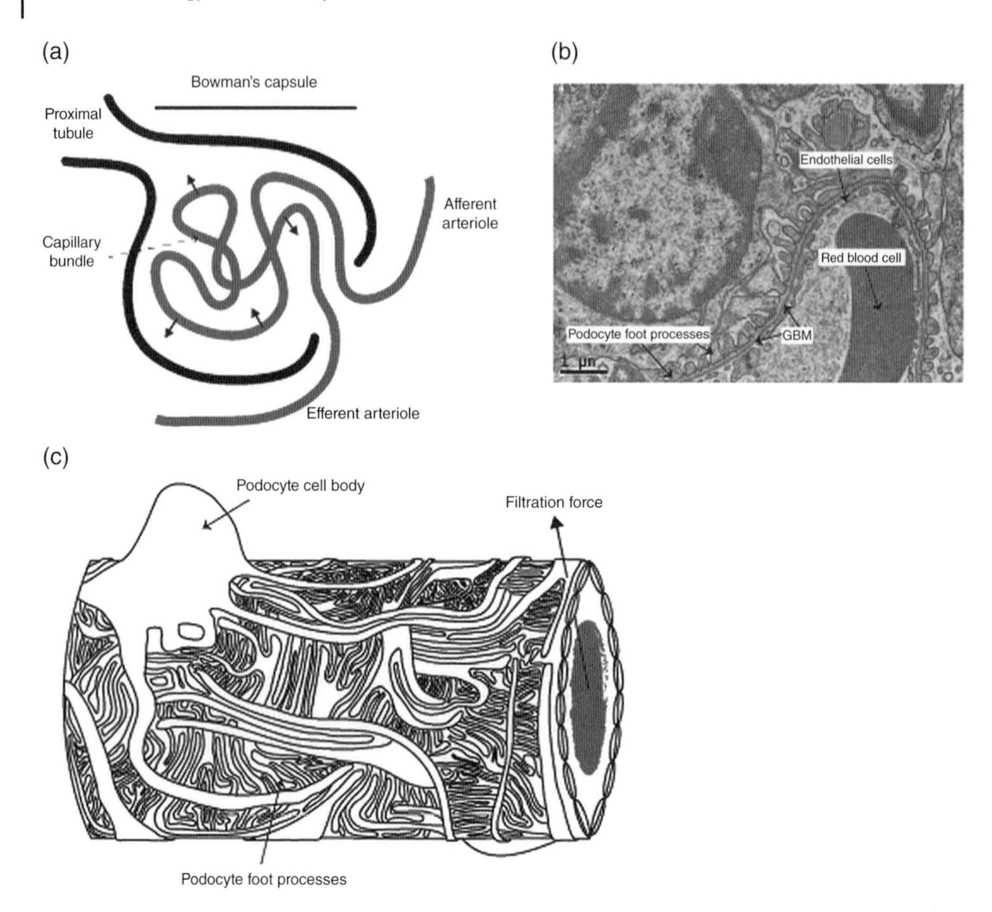

Figure 10.1 Glomerular filtration barrier. (a) The kidney glomerulus consists of a bundle of capillaries enclosed by Bowman's capsule. Blood from the systemic circulation enters capillaries via afferent arterioles. Filtration occurs across specialized capillary walls, and the primary filtrate flows into the proximal tubule. Filtered blood returns to the circulation via efferent arterioles. (b) With electron microscopy, it is possible to visualize the ultrastructure of the filtration barrier, which comprises fenestrated endothelial cells (GEnCs), glomerular basement membrane (GBM), and specialized epithelial cells known as podocytes. (c) Podocytes cover the outer surface of glomerular capillaries with interdigitating foot processes extending from the cell's body. The capillary wall experiences stretch associated with shear force within the lumen, in addition to the force associated with filtration across the wall.

specialized to allow filtration. GEnCs adhere to the second component, the glomerular basement membrane (GBM), a condensed meshwork of extracellular matrix (ECM) that is rich in collagen α3,4,5 (IV) and laminin-521. This separates the GEnCs from the final component: the specialized epithelial cells known as podocytes.

The GBM consists of proteins secreted by both GEnCs and podocytes (Suh and Miner 2013). With its very distinct protein composition, the GBM prevents macromolecules from diffusing into the primary filtrate. It also functions as an anchoring scaffold for both podocytes and GEnCs.

Glomerular-specific, highly differentiated podocytes are a vital component of the filtration barrier. Podocytes extend primary processes that branch ultimately into long

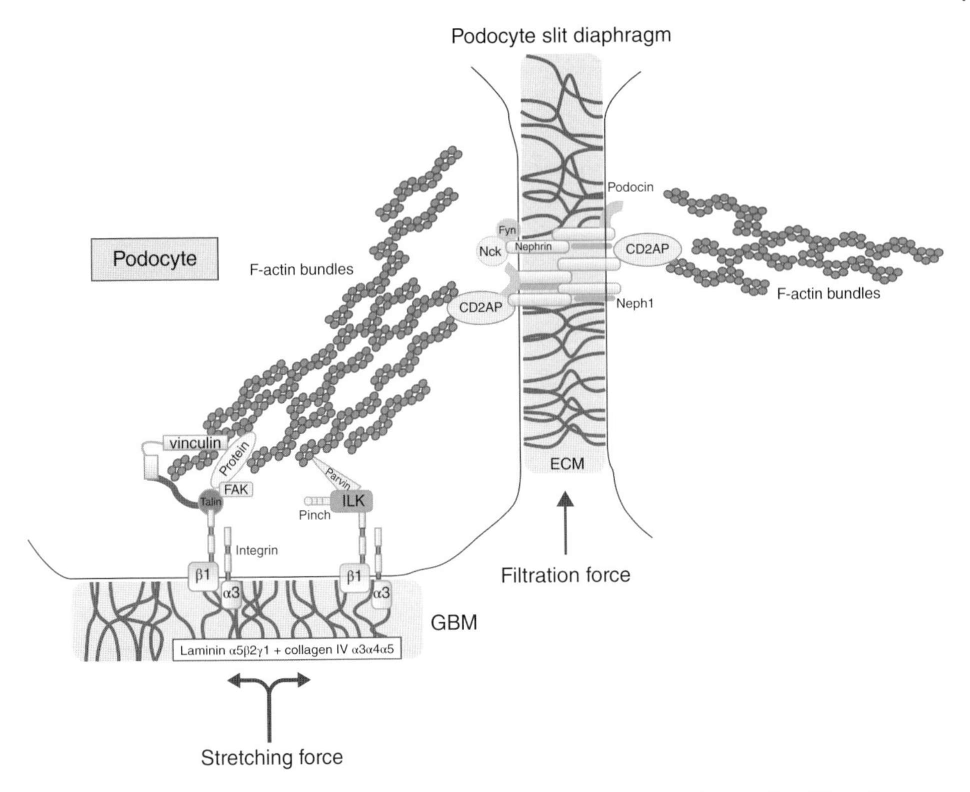

Figure 10.2 Podocyte adhesion. Podocyte foot processes connect via a unique cell–cell junction known as the slit diaphragm. This protein complex includes nephrin, its homologue neph-1, podocin, CD2AP, Nck, and Fyn. Adhesion to the GBM is via focal adhesion (FA) complexes that link to the actin cytoskeleton.

foot processes, which generate a complete meshlike cover over the outer surface of the capillaries (Figure 10.1b,c).

A very specific cell–cell junction, the slit diaphragm (Figure 10.2), mediates the adhesion between adjacent podocytes (Grahammer et al. 2013). The organization of podocyte foot processes on the outer capillary allows these cells to have a major role in the regulation of intracapillary pressure, and therefore glomerular filtration. Environmental changes such as increased blood pressure require adaption and reorganization of the filtration barrier (Kriz and Lemley 2015). Therefore, the cells need to sense changes in forces and translate them into adaptive responses.

Current understanding of mechanotransduction suggests that the sites of both cell–cell and cell–ECM adhesion in the glomerulus are key to the relay of adaptive responses. To date, these responses have been mostly studied in podocytes.

10.3 Podocyte Adhesion

Podocytes form both cell–ECM adhesions and cell–cell junctions, both of which are critical for intact barrier function. The slit diaphragm is a unique cell–cell junction that is only present between adjacent podocyte foot processes. Though the slit diaphragm contains components of tight junctions, gap junctions, and adherens junctions, the most prominent components are podocyte (and neuron)-specific proteins. These include nephrin and the homologues neph-1, neph-2, and neph-3, among others (Lennon et al. 2014).

The slit diaphragm facilitates glomerular filtration in two ways. First, it forms a physical barrier to plasma proteins: the transmembrane proteins nephrin and neph-1 protrude into the extracellular space, building zipper-like junctions (Figure 10.2) (Grahammer et al. 2013). These transmembrane proteins are intracellularly linked through structural and regulatory components to the actin cytosketleton. The resulting protein complex builds a sievelike structure that prevents protein diffusion into the primary filtrate. Second, the slit diaphragm is a signaling hub and facilitates signal transduction to modulate cellular function. For example, nephrin binds to the stomatin domain-containing protein podocin, which enables the cell to receive and translate signals from the external environment (Boute et al. 2000; Huber et al. 2003). In addition, adaptor proteins couple slit diaphragm components to the actin cytoskeleton. Therefore, proteins such as Nck (Jones et al. 2006) and Crk1/2 (George et al. 2012), together with Src family kinases (Yes and Fyn), can initiate reorganization of the actin cytoskeleton (Grahammer et al. 2013), depending on signals received from the slit diaphragm. Subsequent changes in cell shape alter the relative contact between the podocyte and its microenvironment, and therefore influence filtration.

Many components of the slit diaphragm are indispensible for intact barrier function. Nephrin was the first essential component to be described, in 1998, when genetic mutations in *NPHS1* were found to cause severe disruption of the glomerular filtration barrier (Kestila et al. 1998). Since then, there has been regular reporting of key protein components of the slit diaphragm complex, in addition to proteins that link to the cytoskeleton. For example, podocyte-specific deletion of the adaptor protein Nck in mice led to defective foot processes and early-onset severe barrier dysfunction (Jones et al. 2006). This observation stresses the importance of facilitation of the link between the adhesion sites and the actin cytoskeleton by adaptor proteins in the slit diaphragm of podocytes in maintaining normal barrier function. A number of slit diaphragm components are thought to be key for regulating filtration in response to changes in applied force, but the dominant mechanosensor in these cells remains unclear (Endlich and Endlich 2012).

In addition to forming junctions with neighboring cells, podocytes also connect to the GBM via focal adhesions (FAs). These connections are also likely to contribute to the adaptive response to force in the glomerulus. FAs are sites where activated transmembrane receptors, such as integrins, are coupled to the actin cytoskeleton via adaptor proteins. These integrin-mediated adhesions are of key importance for the maintenance of normal filtration. In podocytes, integrin $\alpha3\beta1$ is the predominant mediator of adhesion to laminin and collagen proteins in the GBM (Sterk et al. 1998). Evidence for a major role of FAs in the glomerulus was demonstrated by the deletion of the integrin $\beta1$ chain, specifically in the podocytes (Pozzi et al. 2008). In this study, severe early-onset glomerular barrier dysfunction rapidly progressed to end-stage kidney disease. The glomeruli had reduced podocyte numbers and the foot processes were effaced. Effaced or flattened foot processes are a general feature of barrier dysfunction and indicate defective podocyte signaling to the cytoskeleton. The reduced podocyte numbers suggest cell detachment, highlighting the importance of podocyte–ECM adhesions in maintaining glomerular function.

Activated integrin leads to the recruitment and activation of several FA proteins, which facilitate bridging to the actin cytoskeleton (Figure 10.2). Vinculin and talin are ubiquitous, key adaptor proteins that bridge integrins to the actin cytoskeleton. These proteins have been shown to have mechanosensitive functions and to enable the cell to react to environmental changes (Humphries et al. 2007; Carisey et al. 2013). This is of major interest in the context of the glomerular filtration barrier, as the glomerulus is exposed to variation in blood pressures, and therefore these proteins may play a major function in the maintenance of normal filtration.

In addition to linking the ECM to the actin cytoskeleton, integrin-mediated adhesions also have a signaling role. Following the engagement of integrin with an ECM ligand, there is a coordinated assembly of kinases, scaffolding, and signaling proteins, which ultimately regulate the cytoskeleton (Case et al. 2015). For example, integrin-linked kinase (ILK) is the nidus for the ILK–PINCH–Parvin complex, which localizes to FAs. This complex has a significant role in the regulation of glomerular filtration, and its formation has been shown to be required for normal podocyte adhesion (Yang et al. 2005). Podocyte-specific knockout of ILK is associated with thickened, abnormal basement membranes, podocyte effacement, and barrier dysfunction (El-Aouni et al. 2006). Furthermore, knockout of this complex in cultured podocytes reduced cell adhesion and altered cytoskeletal organization (Zha et al. 2013).

Podocyte adhesion at the slit diaphragm and at the interface with ECM is thus key to maintaining the integrity of the glomerular filtration barrier. It is likely that these adhesion complexes sense changes in force, and it is also plausible that there is molecular crosstalk between the cell–cell and cell–ECM complexes, which coordinates cellular responses to physical cues from the external environment.

10.4 Glomerular Disease

Across the spectrum of glomerular disease, there is disruption of the filtration barrier. Different diseases affect distinct components of the barrier, but the resulting clinical phenotype is often similar. There is a persistently high level of protein in the urine, which is termed "proteinuria." This is usually quantified by screening the urine for the presence of albumin. Albumin, with a molecular weight of 66 kDa, is a small protein. It is retained by the glomerular filtration barrier under physiological conditions, but under pathological conditions it leaks across the barrier and can be detected in the urine, making it suitable as a disease biomarker (Singh and Satchell 2011). Diseases affecting glomerular filtration are typically asymptomatic in the early stages, and patients only present with chronic kidney disease (CKD) when the process is already established. When kidneys eventually fail, patients require renal replacement therapy (RRT) with either dialysis or kidney transplantation. Kidney failure occurs when the glomerular filtration rate (GFR), normalized to the body surface area, is $<15\,\text{mL/min}/1.73\,\text{m}^2$. A healthy GFR is considered $>90\,\text{mL/min}/1.73\,\text{m}^2$. While RRT is lifesaving, there are associated complications and high costs, and lifelong immunosuppressive therapy is required after transplant. Due to the high cost, there is a huge disparity in access across the world.

Glomerular disease can occur as a result of genetic mutations affecting the filtration barrier. Table 10.1 provides examples of mutations in genes encoding for proteins

Table 10.1 Many kidney diseases originate from mutations in genes encoding for adhesion proteins in the glomerular filtration barrier; these are some examples. Gene names in capitals indicate where human mutations have been described; those in lower case indicate where a phenotype has been observed in mice. FSGS, focal segmental glomerulosclerosis.

Gene	Protein	Clinical phenotype
Itga3/ITGA3	Integrin-α3	Congenital nephrotic syndrome
Itgb1	Integrin-β1	Early onset nephrotic syndrome
Ilk	Integrin-linked kinase	FSGS
ACTN4	Actinin α-4	FSGS
MYH9	Non-muscle myosin IIA	Basement membrane defects, FSGS
NPHS1	Nephrin	Congenital nephrotic syndrome
NPHS2	Podocin	Early-onset nephrotic syndrome

involved in podocyte adhesion and their associated phenotype. Systemic diseases can also lead to glomerular dysfunction; these include immunological defects and inflammatory diseases such as systemic lupus erythematous (SLE). However, the commonest worldwide cause of glomerular disease is diabetes mellitus (DM): up to 40% of patients with type 1 or type 2 DM will develop CKD (Ayodele et al. 2004).

While the presentation of glomerular disease is often silent, it can also be dramatic if there is massive protein loss across the filtration barrier. This clinical presentation is called "nephrotic syndrome" and is characterized by massive proteinuria, low circulating protein, and consequent body swelling or edema. When this presents in the first few weeks of life, it is called "congenital nephrotic syndrome." This phenotype is associated with mutations in a number of genes encoding adhesion proteins. Slit-diaphragm mutations in *NPHS1*, which encodes nephrin, lead to congenital nephrotic syndrome (Kestila et al. 1998), while mutations in *NPHS2*, which encodes podocin, cause early-onset nephrotic syndrome (Boute et al. 2000). At the cell–ECM interface, mutations in integrin-α3 cause congenital nephrotic syndrome in association with a severe skin and lung phenotype (Has et al. 2012). Mutations in the ECM component *LAMB2* also lead to congenital nephrotic syndrome and early-onset kidney failure (Zenker et al. 2004). In experimental mouse models, many more mutations of genes encoding proteins that are of importance in podocyte adhesion are described; these include the FA proteins ILK, integrin-β1, talin 1, and α-actinin 4, as well as the slit-diaphragm components Neph-1 and FAT atypical cadherin-1 (Lennon et al. 2014). These human and mouse phenotypes highlight the importance of normal podocyte adhesion for a functioning kidney.

10.5 Forces in the Glomerulus

Variations in blood pressure affect kidney function, and sustained high blood pressure can lead to more rapid progression of kidney disease (Jafar et al. 2003). Blood enters and exits the glomerulus via arterioles. Normally, renal autoregulation prevents systemic increases in blood pressure from perturbing glomerular filtration. This autoregulation is intrinsic to the kidney, causing vasoconstriction and vasodilatation of the arterioles in order to maintain a stable GFR. With hypertension, where the systemic blood pressure (SBP) is increased, the glomeruli must adapt to abnormal conditions. However, autoregulation has its limitations, and hypertension remains a major cause of glomerular damage and progression of CKD.

It is important, therefore, that the glomerular filtration barrier be able to react to changes in forces in order to regulate filtration. In particular, podocytes may have a significant role in regulating capillary hydrostatic pressure, and therefore filtration.

With an increase in blood pressure, podocytes must sense two main forces: a force associated with filtration in the glomerulus (requiring the podocytes to adapt their adhesions) and a stretching force across the whole glomerular filtration barrier (requiring the podocytes to cover an increased GBM surface area, which involves reorganization of their FAs) Under physiological conditions, podocytes are capable of adapting to these changes by lengthening their foot processes and slit diaphragms (Kriz and Lemley 2015), but the underlying regulatory mechanisms that drive this reorganization remain unclear. Though the specific glomerular cell type responsible for adapting to changes in forces is unknown, there is significant evidence that the organelles responsible are the cell–cell or cell–ECM adhesions. Variations in FA morphology were observed when plating cells on ECMs with differing rigidities. Such variations also led to changes in traction forces generated by cells, revealing a direct relationship between FAs and traction forces (Plotnikov and Waterman 2013). Furthermore, FAs were larger in cells seeded on a stiffer substrate than in those on a very soft substrate (Han et al. 2012), demonstrating a cell's ability to sense the stiffness of its environment, transduce this information, and translate it into a specific response. Similar results were observed when investigating the influence of forces on cell–cell junctions. Cells responded to pulling forces with increased growth of adherens junctions (Liu et al. 2010), which can be interpreted as the way in which they resist and counteract applied forces. Upon changes in glomerular haemodynamics such as hyperfiltration, it is proposed that podocytes substitute parts of the slit diaphragm with occludens junctions to avoid a pathological loss of proteins and fluids (Kriz et al. 2014; Kriz and Lemley 2015). Nevertheless, the regulatory mechanisms for such reorganization at the slit diaphragm need to be clarified.

10.6 Mechanosensitive Components and Prospects for Therapy

Across different tissue systems, a variety of mechanosensitive proteins have been described at cell adhesion complexes (Schwartz and DeSimone 2008). Their localization, bridging adhesions to the actin cytoskeleton, and their ability to translate extracellular signals make them interesting targets for study in the context of glomerular filtration.

Both cell–cell junctions and FAs contain two major mechanosensitive proteins: talin and vinculin. Via its N-terminal, FERM-domain talin binds to the cytoplasmic tail of integrins; it also contains C-terminal binding sites for both actin and vinculin (Lee et al. 2007; del Rio et al. 2009). In addition, there is increasing evidence that stretching of talin reveals cryptic binding sites, allowing the transduction of mechanical signals (Yan et al. 2015). The relevance of talin to podocyte function was demonstrated in a recent study

(Tian et al. 2014), which showed that podocyte-specific talin-1 knockout mice had severe glomerular abnormalities (resulting in proteinuria, kidney failure, and lethality within the first 8 weeks in the majority of animals), effacement of the podocyte foot process (with thickening and splitting of the GBM), and altered organization of the actin cytoskeleton. These observations indicate that a known mechanosensitive adaptor protein plays a crucial role in maintaining glomerular filtration and is not dispensable for the relay of adhesion signals to the cytoskeleton. Dysregulation of the actin cytoskeleton in podocytes may well be a common feature of glomerular disease processes, and a recent study demonstrated that targeting of actin regulation could be exploited for therapy in glomerular disease (Schiffer et al. 2015).

In comparison to talin, vinculin provides a more indirect link between adhesions and the actin cytoskeleton, and it may therefore have a role in fine-tuning the response to force. A crucial role for vinculin as a mechanosensor has been observed in the heart, an organ that constantly manages variations in blood pressure (Zemljic-Harpf et al. 2007). During episodes of hypertension, cells must react rapidly to stretch in blood vessel walls and so avoid disruption of the cellular barrier; they effect this response by increasing the amount and strength of adhesions (Collins et al. 2014). Targeted knockout of vinculin in cardiac myocytes resulted in embryonic lethality in most mice in one study. However, those that survived showed abnormal adherens junctions in the cardiac myocytes (Zemljic-Harpf et al. 2007). This highlights the important mechanotransductive role of vinculin *in vivo* in a tissue constantly exposed to differing forces. Vinculin is also known to accumulate in cadherin-containing cell junctions in response to force, and phosphorylation of vinculin has been found to regulate its presence at cell junctions versus FAs (Bays et al. 2014). The consequences of vinculin deletion in the podocyte have not yet been examined, but it will be necessary to determine the role of this adhesion component in maintaining both slit-diaphragm and ECM contacts in the glomerulus.

At the cell–ECM interface, the tetraspannin CD151 has an important role to play in the response of podocytes to force. This protein has a close association with integrin-α3, and deletion of CD151 in mice resulted in a severe glomerular phenotype and early kidney failure (Sachs et al. 2006). Furthermore the effect of CD151 deletion was exacerbated by hypertension (Sachs et al. 2012). αV-integrins have also been proposed as mediators of the mechanoprotective effects of the ECM ligand osteopontin in podocytes (Schordan et al. 2011).

At the podocyte slit diaphragm, a number of proteins may contribute to mechanosensation. The cation channel TRPC6 is expressed at the slit diaphragm, and mutations here cause glomerular dysfunction with adult-onset nephrotic syndrome (Winn et al. 2005). Interacting with podocin, TRPC6 has been proposed to contribute to a mechanosensing complex at the slit diaphragm (Moller et al. 2009). More generally, TRP channels are proposed to be mechanosenensitive, undergoing conformational change with applied force, which leads to channel-gating and signal transduction (Liu and Montell 2015). However, it is not yet known whether TRPC6-associated mechanosensation in podocytes also requires interaction with other mechanosensors, such as talin and vinculin.

Additional slit-diaphragm adaptor proteins may also be involved in mechanotransduction. CD2AP is a scaffolding protein that localizes to the slit diaphragm; it has been shown to play a role in the regulation of the actin cytoskeleton in podocytes (Zhao et al. 2013). Though CD2AP is expressed in almost every tissue of the body, ubiquitous knockout of CD2AP leads to a severe glomerular phenotype in mice (Shih et al. 1999) and in humans (Akilesh et al. 2007). Furthermore, reduced expression of CD2AP in kidney cells decreases cell–cell adhesion following the application of mechanical force (Tang and Brieher 2013). Overall, there are a number of candidate mechanosensors acting at cell–ECM and cell–cell adhesions in podocytes, and further investigation is required to determine how they interact to provide a coordinated response to force.

Existing therapies for CKD focus on delaying the progression of disease, and currently there are no curative treatments. The most effective agents for preserving kidney function are drugs that block the renin–angiotensin–aldosterone system (RAAS). Both the rate of disease progression and proteinuria can be attenuated with RAAS inhibition (Taguma et al. 1985; Bjorck et al. 1986; Heeg et al. 1987; Casas et al. 2005); one putative indirect mechanism for this is the reduction of intracapillary hydrostatic pressure (Brown et al. 1993). Since podocytes express RAAS components (Durvasula et al. 2004), direct effects of RAAS inhibitors on glomerular cells have also been proposed (Wennmann et al. 2012). In cultured podocytes exposed to mechanical stimuli, angiotensin II treatment altered podocyte stiffness, suggesting that a local RAAS pathway may participate in force regulation (Eekhoff et al. 2011). Further consideration of the effects of current efficacious therapies on force regulation in the glomerulus may lead to the identification of novel therapeutic strategies.

10.7 Conclusion

The kidney is constantly exposed to mechanical forces, which facilitate filtration of the blood. These mechanical forces are directly applied to the glomerular filtration barrier, which must manage alterations in the applied force. Glomerular podocytes, with their unique foot processes, have a key role in the regulation of glomerular filtration, and filtration efficiency is determined by podocyte adhesion to the basement membrane and the sizes of the sieving pores, which are formed by podocyte cell–cell adhesions. Podocyte adhesions must respond to forces at the capillary wall, and key mechanosensors at these adhesions sites have been proposed. Elucidating the detailed mechanisms of mechanotransduction in the glomerulus and regulating the proteins involved may ultimately reveal new targets for the treatment of patients with kidney disease.

References

Akilesh, S., A. Koziell, and A. S. Shaw. 2007. "Basic science meets clinical medicine: identification of a CD2AP-deficient patient." *Kidney International* **72**(10): 1181–3. doi:10.1038/sj.ki.5002575.

Ayodele, O. E., C. O. Alebiosu, and B. L. Salako. 2004. "Diabetic nephropathy – a review of the natural history, burden, risk factors and treatment." *Journal of the National Medical Association* **96**(11): 1445–54. PMID:15586648.

Bays, J. L., X. Peng, C. E. Tolbert, C. Guilluy, A. E. Angell, Y. Pan, et al. 2014. "Vinculin phosphorylation differentially regulates mechanotransduction at cell-cell and cell-matrix adhesions." *Journal of Cell Biology* **205**(2): 251–63. doi:10.1083/jcb.201309092.

Bjorck, S., G. Nyberg, H. Mulec, G. Granerus, H. Herlitz, and M. Aurell. 1986. "Beneficial effects of angiotensin converting enzyme inhibition on renal function in patients with diabetic nephropathy." *British Medical Journal (Clinical Research Edition)* **293**(6545): 471–4. PMID:3017501.

Boute, N., O. Gribouval, S. Roselli, F. Benessy, H. Lee, A. Fuchshuber, et al. 2000. "NPHS2, encoding the glomerular protein podocin, is mutated in autosomal recessive steroid-resistant nephrotic syndrome." *Nature Genetics* **24**(4): 349–54. doi:10.1038/74166.

Brown, S. A., C. L. Walton, P. Crawford, and G. L. Bakris. 1993. "Long-term effects of antihypertensive regimens on renal hemodynamics and proteinuria." *Kidney International* **43**(6): 1210–18. PMID:8391095.

Carisey, A., R. Tsang, A. M. Greiner, N. Nijenhuis, N. Heath, A. Nazgiewicz, et al. 2013. "Vinculin regulates the recruitment and release of core focal adhesion proteins in a force-dependent manner." *Current Biology* **23**(4): 271–81. doi:10.1016/j.cub.2013.01.009.

Casas, J. P., W. Chua, S. Loukogeorgakis, P. Vallance, L. Smeeth, A. D. Hingorani, and R. J. MacAllister. 2005. "Effect of inhibitors of the renin-angiotensin system and other antihypertensive drugs on renal outcomes: systematic review and meta-analysis." *Lancet* **366**(9502): 2026–33. doi:10.1016/S0140-6736(05)67814-2.

Case, L. B., M. A. Baird, G. Shtengel, S. L. Campbell, H. F. Hess, M. W. Davidson, and C. M. Waterman. 2015. "Molecular mechanism of vinculin activation and nanoscale spatial organization in focal adhesions." *Nature Cell Biology* **17**(7): 880–92. doi:10.1038/ncb3180.

Collins, C., L. D. Osborne, C. Guilluy, Z. Chen, E. T. O'Brien, J. S. Reader, et al. 2014. "Haemodynamic and extracellular matrix cues regulate the mechanical phenotype and stiffness of aortic endothelial cells." *Nature Communications* **5**: 3984. doi:10.1038/ncomms4984.

del Rio, A., R. Perez-Jimenez, R. Liu, P. Roca-Cusachs, J. M. Fernandez, and M. P. Sheetz. 2009. "Stretching single talin rod molecules activates vinculin binding." *Science* **323**(5914): 638–41. doi:10.1126/science.1162912.

Durvasula, R. V., A. T. Petermann, K. Hiromura, M. Blonski, J. Pippin, P. Mundel, et al. 2004. "Activation of a local tissue angiotensin system in podocytes by mechanical strain." *Kidney International* **65**(1): 30–9. doi:10.1111/j.1523-1755.2004.00362.x.

Eekhoff, A., N. Bonakdar, J. L. Alonso, B. Hoffmann, and W. H. Goldmann. 2011. "Glomerular podocytes: a study of mechanical properties and mechano-chemical signaling." *Biochemical and Biophysical Research Communications* **406**(2): 229–33. doi:10.1016/j.bbrc.2011.02.022.

El-Aouni, C., N. Herbach, S. M. Blattner, A. Henger, M. P. Rastaldi, G. Jarad, et al. 2006. "Podocyte-specific deletion of integrin-linked kinase results in severe glomerular basement membrane alterations and progressive glomerulosclerosis." *Journal of the American Society of Nephrology* **17**(5): 1334–44. doi:10.1681/ASN.2005090921.

Endlich, N. and K. Endlich. 2012. "The challenge and response of podocytes to glomerular hypertension." *Seminars in Nephrology* **32**(4): 327–41. doi:10.1016/j.semnephrol.2012.06.004.

George, B., R. Verma, A. A. Soofi, P. Garg, J. Zhang, T. J. Park, et al. 2012. "Crk1/2-dependent signaling is necessary for podocyte foot process spreading in mouse models of glomerular disease." *Journal of Clinical Investigation* **122**(2): 674–92. doi:10.1172/JCI60070.

Grahammer, F., C. Schell, and T. B. Huber. 2013. "The podocyte slit diaphragm – from a thin grey line to a complex signalling hub." *Nature Reviews. Nephrology* **9**(10): 587–98. doi:10.1038/nrneph.2013.169.

Han, S. J., K. S. Bielawski, L. H. Ting, M. L. Rodriguez, and N. J. Sniadecki. 2012. "Decoupling substrate stiffness, spread area, and micropost density: a close spatial relationship between traction forces and focal adhesions." *Biophysical Journal* **103**(4): 640–8. doi:10.1016/j.bpj.2012.07.023.

Has, C., G. Sparta, D. Kiritsi, L. Weibel, A. Moeller, V. Vega-Warner, et al. 2012. "Integrin alpha3 mutations with kidney, lung, and skin disease." *New England Journal of Medicine* **366**(16): 1508–14. doi:10.1056/NEJMoa1110813.

Heeg, J. E., P. E. de Jong, G. K. van der Hem, and D. de Zeeuw. 1987. "Reduction of proteinuria by angiotensin converting enzyme inhibition." *Kidney International* **32**(1): 78–83. PMID:3041097.

Huber, T. B., M. Simons, B. Hartleben, L. Sernetz, M. Schmidts, E. Gundlach, et al. 2003. "Molecular basis of the functional podocin-nephrin complex: mutations in the NPHS2 gene disrupt nephrin targeting to lipid raft microdomains." *Human Molecular Genetics* **12**(24): 3397–405. doi:10.1093/hmg/ddg360.

Humphries, J. D., P. Wang, C. Streuli, B. Geiger, M. J. Humphries, and C. Ballestrem. 2007. "Vinculin controls focal adhesion formation by direct interactions with talin and actin." *Journal of Cell Biology* **179**(5): 1043–57. doi:10.1083/jcb.200703036.

Jafar, T. H., P. C. Stark, C. H. Schmid, M. Landa, G. Maschio, P. E. de Jong, et al. 2003. "Progression of chronic kidney disease: the role of blood pressure control, proteinuria, and angiotensin-converting enzyme inhibition: a patient-level meta-analysis." *Annals of Internal Medicine* **139**(4): 244–52. PMID:12965979.

Jarad, G. and J. H. Miner. 2009. "Update on the glomerular filtration barrier." *Current Opinion in Nephrology and Hypertension* **18**(3): 226–32. PMID:19374010.

Jones, N., I. M. Blasutig, V. Eremina, J. M. Ruston, F. Bladt, H. Li, et al. 2006. "Nck adaptor proteins link nephrin to the actin cytoskeleton of kidney podocytes." *Nature* **440**(7085): 818–23. doi:10.1038/nature04662.

Keller, G., G. Zimmer, G. Mall, E. Ritz, and K. Amann. 2003. "Nephron number in patients with primary hypertension." *New England Journal of Medicine* **348**(2): 101–8. doi:10.1056/NEJMoa020549.

Kestila, M., U. Lenkkeri, M. Mannikko, J. Lamerdin, P. McCready, H. Putaala, et al. 1998. "Positionally cloned gene for a novel glomerular protein – nephrin – is mutated in congenital nephrotic syndrome." *Molecular Cell* **1**(4): 575–82. PMID:9660941.

Kriz, W. and K. V. Lemley. 2015. "A potential role for mechanical forces in the detachment of podocytes and the progression of CKD." *Journal of the American Society of Nephrology* **26**(2): 258–69. doi:10.1681/ASN.2014030278.

Kriz, W., B. Hähnel, H. Hosser, S. Rösener, and R. Waldherr. 2014. "Structural analysis of how podocytes detach from the glomerular basement membrane under hypertrophic stress." *Frontiers in Endocrinology* **5**: 207. doi:10.3389/fendo.2014.00207.

Lee, S. E., R. D. Kamm, and M. R. Mofrad. 2007. "Force-induced activation of talin and its possible role in focal adhesion mechanotransduction." *Journal of Biomechanics* **40**(9): 2096–106. doi:10.1016/j.jbiomech.2007.04.006.

Lennon, R., M. J. Randles, and M. J. Humphries. 2014. "The importance of podocyte adhesion for a healthy glomerulus." *Frontiers in Endocrinology* **5**: 160. doi:10.3389/fendo.2014.00160.

Liu, C. and C. Montell. 2015. “Forcing open TRP channels: mechanical gating as a unifying activation mechanism.” *Biochemical and Biophysical Research Communications* **460**(1): 22–5. doi:10.1016/j.bbrc.2015.02.067.

Liu, Z., J. L. Tan, D. M. Cohen, M. T. Yang, N. J. Sniadecki, S. A. Ruiz, et al. 2010. “Mechanical tugging force regulates the size of cell-cell junctions.” *Proceedings of the National Academy of Sciences United States of America* **107**(22): 9944–9. doi:10.1073/pnas.0914547107.

Moller, C. C., J. Flesche, and J. Reiser. 2009. “Sensitizing the slit diaphragm with TRPC6 ion channels.” *Journal of the American Society of Nephrology* **20**(5): 950–3. doi:10.1681/ASN.2008030329.

Plotnikov, S. V. and C. M. Waterman. 2013. “Guiding cell migration by tugging.” *Current Opinion in Cell Biology* **25**(5): 619–26. doi:10.1016/j.ceb.2013.06.003.

Pozzi, A., G. Jarad, G. W. Moeckel, S. Coffa, X. Zhang, L. Gewin, et al. 2008. “Beta1 integrin expression by podocytes is required to maintain glomerular structural integrity.” *Developmental Biology* **316**(2): 288–301. doi:10.1016/j.ydbio.2008.01.022.

Sachs, N., M. Kreft, M. A. van den Bergh Weerman, A. J. Beynon, T. A. Peters, J. J. Weening, and A. Sonnenberg. 2006. “Kidney failure in mice lacking the tetraspanin CD151.” *Journal of Cell Biology* **175**(1): 33–9. doi:10.1083/jcb.200603073.

Sachs, N., N. Claessen, J. Aten, M. Kreft, G. J. Teske, A. Koeman, et al. 2012. “Blood pressure influences end-stage renal disease of Cd151 knockout mice.” *Journal of Clinical Investigation* **122**(1): 348–58. doi:10.1172/JCI58878.

Schiffer, M., B. Teng, C. Gu, V. A. Shchedrina, M. Kasaikina, V. A. Pham, et al. 2015. “Pharmacological targeting of actin-dependent dynamin oligomerization ameliorates chronic kidney disease in diverse animal models.” *Nature Medicine* **21**(6): 601–9. doi:10.1038/nm.3843.

Schordan, S., E. Schordan, K. Endlich, and N. Endlich. 2011. “AlphaV-integrins mediate the mechanoprotective action of osteopontin in podocytes.” *American Journal of Physiology. Renal Physiology* **300**(1): F119–32. doi:10.1152/ajprenal.00143.2010.

Schwartz, M. A. and D. W. DeSimone. 2008. “Cell adhesion receptors in mechanotransduction.” *Current Opinion in Cell Biology* **20**(5): 551–6. doi:10.1016/j.ceb.2008.05.005.

Shih, N. Y., J. Li, V. Karpitskii, A. Nguyen, M. L. Dustin, O. Kanagawa, et al. 1999. “Congenital nephrotic syndrome in mice lacking CD2-associated protein.” *Science* **286**(5438): 312–15. PMID:10514378.

Singh, A. and S. C. Satchell. 2011. “Microalbuminuria: causes and implications.” *Pediatric Nephrology* **26**(11): 1957–65. doi:10.1007/s00467-011-1777-1.

Sterk, L. M., A. A. de Melker, D. Kramer, I. Kuikman, A. Chand, N. Claessen, et al. 1998. “Glomerular extracellular matrix components and integrins.” *Cell Adhesion Communications* **5**(3): 177–92. PMID:9686316.

Suh, J. H. and J. H. Miner. 2013. “The glomerular basement membrane as a barrier to albumin.” *Nature Reviews. Nephrology* **9**(8): 470–7. doi:10.1038/nrneph.2013.109.

Taguma, Y., Y. Kitamoto, G. Futaki, H. Ueda, H. Monma, M. Ishizaki, et al. 1985. “Effect of captopril on heavy proteinuria in azotemic diabetics.” *New England Journal of Medicine* **313**(26): 1617–20. doi:10.1056/NEJM198512263132601.

Tang, V. W. and W. M. Brieher. 2013. “FSGS3/CD2AP is a barbed-end capping protein that stabilizes actin and strengthens adherens junctions.” *Journal of Cell Biology* **203**(5): 815–33. doi:10.1083/jcb.201304143.

Tian, X., J. J. Kim, S. M. Monkley, N. Gotoh, R. Nandez, K. Soda, et al. 2014. "Podocyte-associated talin1 is critical for glomerular filtration barrier maintenance." *Journal of Clinical Investigation* **124**(3): 1098–113. doi:10.1172/JCI69778.

Wennmann, D. O., H. H. Hsu, and H. Pavenstadt. 2012. "The renin-angiotensin-aldosterone system in podocytes." *Seminars in Nephrology* **32**(4): 377–84. doi:10.1016/j.semnephrol.2012.06.009.

Winn, M. P., P. J. Conlon, K. L. Lynn, M. K. Farrington, T. Creazzo, A. F. Hawkins, et al. 2005. "A mutation in the TRPC6 cation channel causes familial focal segmental glomerulosclerosis." *Science* **308**(5729): 1801–4. doi:10.1126/science.1106215.

Yan, J., M. Yao, B. T. Goult, and M. P. Sheetz. 2015. "Talin dependent mechanosensitivity of cell focal adhesions." *Cellular and Molecular Bioengineering* **8**(1): 151–9. doi:10.1007/s12195-014-0364-5.

Yang, Y., L. Guo, S. M. Blattner, P. Mundel, M. Kretzler, and C. Wu. 2005. "Formation and phosphorylation of the PINCH-1-integrin linked kinase-alpha-parvin complex are important for regulation of renal glomerular podocyte adhesion, architecture, and survival." *Journal of the American Society of Nephrology* **16**(7): 1966–76. doi:10.1681/ASN.2004121112.

Zemljic-Harpf, A. E., J. C. Miller, S. A. Henderson, A. T. Wright, A. M. Manso, L. Elsherif, et al. 2007. "Cardiac-myocyte-specific excision of the vinculin gene disrupts cellular junctions, causing sudden death or dilated cardiomyopathy." *Molecular and Cellular Biology* **27**(21): 7522–37. doi:10.1128/MCB.00728-07.

Zenker, M., T. Aigner, O. Wendler, T. Tralau, H. Muntefering, R. Fenski, et al. 2004. "Human laminin beta2 deficiency causes congenital nephrosis with mesangial sclerosis and distinct eye abnormalities." *Human Molecular Genetics* **13**(21): 2625–32. doi:10.1093/hmg/ddh284.

Zha, D., C. Chen, W. Liang, X. Chen, T. Ma, H. Yang, et al. 2013. "Nephrin phosphorylation regulates podocyte adhesion through the PINCH-1-ILK-alpha-parvin complex." *BMB Reports* **46**(4): 230–5. PMID:23615266.

Zhao, J., S. Bruck, S. Cemerski, L. Zhang, B. Butler, A. Dani, et al. 2013. "CD2AP links cortactin and capping protein at the cell periphery to facilitate formation of lamellipodia." *Molecular and Cellular Biology* **33**(1): 38–47. doi:10.1128/MCB.00734-12.

11

Dynamic Remodeling of the Heart and Blood Vessels

Implications of Health and Disease

Ken Takahashi[1], *Hulin Piao*[1,2], *and Keiji Naruse*[1]

[1] *Department of Cardiovascular Physiology, Graduate School of Medicine, Dentistry and Pharmaceutical Sciences, Okayama University, Okayama, Japan*
[2] *Department of Cardiovascular Surgery, The Second Affiliated Hospital of Jilin University, Changchun, China*

11.1 Introduction

Tissues are not just agglomerates resulting from random cell division. Matched pairs of arms, legs, and eyes and a four-chambered heart all result from a precisely controlled process of cellular proliferation, differentiation, and apoptosis. In this process, the mechanical environment (e.g., stiffness, pressure) plays a crucial role (Cowin 2004). For example, fluid forces in the heart are an essential factor in the appropriate development of cardiac chambers and valves (Hove et al. 2003), while the contractile force of fetal or neonatal cardiomyocytes adapts to the hardness of surroundings (Majkut and Discher 2012). Our lives are impossible without response to mechanical stimuli. Tissue differentiation and proliferation persist after the developmental period. The same is true for cardiomyocytes, which were once considered incapable of cell division. The heart and blood vessels are different from other organs or tissues in one aspect, however: they are constantly exposed to mechanical stimuli derived from blood flow and pressure. Alteration in the mechanical environment (e.g., by a factor such as high blood pressure, if of sufficient magnitude) dramatically changes the structure of the heart and blood vessels.

Remodeling is the process of structural and functional modification of an organ or tissue. For example, cardiac hypertrophy (thickening of the heart muscle) is an adaptive response to increased pressure and/or volume workload. While cardiac remodeling often refers to a process of transition to heart failure in a limited sense, it also includes physiological hypertrophy caused by exercise. Each component of the heart (e.g., cardiomyocyte, cardiac fibroblast, blood vessel, and extracellular matrix (ECM)) undergoes structural and functional remodeling. The consequences of cardiac remodeling range from enhanced cardiac capacity, sufficient to finish a triathlon ironman race, to sudden death by heart failure.

At the end of the 20th century, while it was known that mechanical work causes cardiac hypertrophy and that physiological remodeling can switch to pathological remodeling, the underlying mechanism was still a mystery (Richey and Brown 1998).

Mechanobiology: Exploitation for Medical Benefit, First Edition. Edited by Simon C. F. Rawlinson.

In this chapter, we will explore the dynamic responses of the heart (both healthy and pathological) to external stimuli (mainly mechanical).

11.2 Causes of Remodeling

Remodeling of the heart is caused by a wide variety of specific preceding factors, including aging, pregnancy, and high blood pressure. In the aged heart, the number of small arteries decreases (Anversa et al. 1994) and heart cells become more likely to develop programmed cell death (apoptosis) (Sheydina et al. 2011), so cardiac pumping function is lost. In contrast, cardiac output dramatically increases in the heart of a pregnant woman, due to significant remodeling (Thornburg et al. 2000): there is a fine molecular mechanism that makes the heart function efficient during pregnancy and recovers it to normal after delivery (Eghbali et al. 2006). Pathological conditions such as metabolic syndromes (Asrih et al. 2013) and diabetes (Battiprolu et al. 2013) also cause remodeling of the heart. Furthermore, certain data suggest that cigarette smoking exacerbates pathological remodeling, due to volume overload (Bradley et al. 2012).

11.2.1 Mechanical Stimuli

Transient elevation in blood pressure due to exercise or chronic systemic hypertension is a significant cause of cardiac remodeling (Force et al. 2002; Mann 2004; Hoshijima 2006; McCain and Parker 2011). Excess mechanical stimuli are the direct cause of heart failure, which is the terminal form of pathological cardiac remodeling. Mechanical stimuli to the heart can be excessive in two main cases: volume overload and pressure overload. Valve regurgitation (leakage of blood at the heart valve) is a common cause of volume overload. This is a pathological process that elicits eccentric hypertrophy and ECM remodeling in left ventricular (LV) myocytes, both of which contribute to increases in myocardial wall stress and to disproportionate increases in the ratio of LV end-diastolic diameter (EDD) to wall thickness (wt) (LVEDD/wt ratio), ultimately leading to LV dilatation and congestive heart failure (Hutchinson et al. 2010). Volume overload generally elicits eccentric hypertrophy: enlargement of the heart, with the cardiac wall thickening outwards.

11.2.2 Pressure Overload

Pressure overload is a state of increased cardiac afterload. Hypertension and stenosis are primary causes of pressure overload. To accommodate the increased afterload, the LV undergoes structural and functional changes, such as concentric hypertrophy and increases in wall thickness and mass. This is followed by ECM remodeling and myocardial fibrosis, leading to reduced ventricular compliance and diastolic dysfunction (Yarbrough et al. 2012). Pressure overload generally elicits concentric hypertrophy, with the cardiac wall thickening inwards.

Interestingly, a chronically failed heart can be recovered (reverse remodeling) by unloading pressure and volume using an LV assist device (Matsumiya et al. 2009; Birks 2010). This suggests that mechanical overload, by volume or by pressure, is a direct cause of cardiac remodeling.

11.3 Mechanical Transduction in Cardiac Remodeling

In 1980s, it was not certain whether pure mechanical stimulus was the primary cause of cardiac hypertrophy, as cardiac hypertrophy was also associated with disturbances in neurohumoral factors (Yamazaki et al. 1995). Today, it is an established fact that mechanical stimuli such as elevation in blood pressure or stretch of cardiac tissue can cause cardiac remodeling, because cardiac tissue is capable of sensing mechanical stimuli. Mechanical stimuli, comprising stress and strain, elicit not only passive change in a way that can deform a spring or clay, but also an inherent active response from the cells or tissues. The heart is equipped with a number of molecular mechanisms for detecting mechanical stimuli, including the ECM (MacKenna et al. 2000; Kresh and Chopra 2011), focal adhesion (FA) (Samarel 2005; Romer et al. 2006; Seong et al. 2013), titin (Linke 2008), AT1 receptor (Zou et al. 2004), and mechanosensitive ion channels (Inoue et al. 2006; Takahashi and Naruse 2012; Takahashi et al. 2013).

The ECM is not simply a static structure filling gaps between cells: rather, it dynamically influences cardiac function. Though it was originally considered a simple collagen network, it turns out to be made up of a number of components, including proteoglycans, glycoproteins, elastin, fibronectin, laminin, growth factors, cytokines, chemokines, and proteases (Borg and Baudino 2011). The ECM determines the stiffness of the cardiac tissue and plays an important role in mechanotransduction in the sarcomere during cardiac contraction (Spinale et al. 2013). It also provides scaffolds composed of basement membrane adhesion proteins (including fibronectin, collagen, and laminin) to the cardiomyocytes. The mechanical environment of cardiomyocytes is defined by their adhesion to ECM and other cells. Inhibition of the normal binding between the ECM and cardiomyocytes leads to an abnormal cardiomyocyte structure, causing hypertrophic and dilated cardiac myopathy (Kresh and Chopra 2011). As already mentioned, the ECM is both an effector of remodeling, in that it is part of the mechanotransduction machinery, and a receptor of remodeling, in that it changes its stiffness.

FAs are structural and functional complexes composed of a specific disposition of multiple proteins at the adhesion site between cell and cell or between cell and ECM (Ross 2004). They are one origin of the cellular mechanotransduction pathway. Cardiac hypertrophy due to pressure overload is attenuated without focal adhesion kinase (FAK) (DiMichele et al. 2006); FAK is accumulated when the surrounding cardiomyocytes becomes rigid (McCain et al. 2012). Titin is a huge protein that acts as an elastic spring inside the cardiomyocytes. Mutation of the titin gene causes arrhythmogenic right ventricular cardiomyopathy (ARVC), which replaces right ventricular (RV) wall muscle with fatty tissue and fibrosis (Taylor et al. 2011).

Certain ion channel subtypes for the transient receptor potential (TRP) family are considered mechanosensitive. The TRPC1 (Maroto et al. 2005) and TRPC6 (Spassova et al. 2006) channels, for example, open in response to bilayer tension. Generally, mechanosensitivity plays an important role in tissue development. TRPC1 is implicated in the growth of spinal axon, which is influenced by the stiffness of the surrounding tissue (Kerstein et al. 2013). TRPC3 and TRPC6 are expressed on the cellular membrane of cardiomyocytes, and they are upregulated in pathological remodeling (Inoue et al. 2009).

11.4 The Remodeling Process

Cardiac remodeling is thought to be an adaptive response to increased workload during cardiac contractile activity. The two types of pathway – response to moderate exercise and response to pathological pressure/volume overload – are rather similar, rather than totally distinct. For example, though cardiac troponin is considered a marker of pathological cardiac damage, it is detected in healthy subjects who undergo relatively long endurance exercise, such as running a marathon (Shave et al. 2010; George et al. 2012), while the decreased level of miR-133 seen in physiological hypertrophy on exercise is also observed in pathological remodeling (Care et al. 2007). It is suggested that athletes undergoing intense training may develop arrhythmogenic remodeling (Rowland 2011). Impaired RV function can last 1 week after an intense endurance race, such as a triathlon, and may lead to fibrosis of the septum (La Gerche et al. 2012). In an animal model, long periods of intense training are shown to induce pathological remodeling (Benito et al. 2011).

Remodeling of the heart encompasses a broad range of alterations, including hypertrophy of cardiomyocytes, fibrosis, vascularization, inflammation, mitochondrial energetics, generation of reactive oxygen species (ROS), and cardiogenesis. These processes are not independent, but are mutually related and together make up a whole. In this section, we discuss first pathological remodeling and then physiological remodeling.

11.4.1 Pathological Remodeling

11.4.1.1 Hypertrophy

Pressure or volume overload induces hypertrophy of cardiomyocytes. There are two types of hypertrophy: increase in thickness and increase in length. Generally, volume overload leads to increase in length, while pressure overload leads to increase in thickness.

Animal experiments have revealed the mechanism of cardiomyocyte hypertrophy (Oparil 1985). Cardiac load and RNA polymerase activity increase within 12–24 hours. The ratio of mitochondria to cellular volume then increases within another 48 hours. This is followed by the elongation of the z-band. These changes result in hypertrophy of cardiomyocytes. An *in vitro* experiment using rat cardiomyocytes was carried out to reveal the molecular entity responsible for mechanosensing. However, neither blockade of stretch-activated (SA) channels nor disruption of microtubules and actin filaments inhibited the hypertrophic response of the cardiomyocytes (Sadoshima et al. 1992). This suggests that other mechanisms distinct from SA channels and cytoskeleton are responsible for mechanosensing the mechanical stimulus.

11.4.1.2 Interstitial Fibrosis

Pressure overload to the LV imposes enlargement of the cardiac tissue through the accumulation of ECM, mainly composed of fibrillar collagen. Matrix metalloproteinases (MMPs) play a crucial role in the accumulation of ECM. Collagen deposition occurs as a result of increased generation versus decomposition (Jugdutt 2003; Hou and Kang 2012) and can inhibit cardiac contraction and relaxation. Though MMPs are proteinases, one of the subtypes, MMP-2 (gelatinase A), elicits fibrosis of ECM by increasing the level of expression of collagen I (Hori et al. 2012). Increased MMP-2 expression was observed in a rat neonatal primary culture of cardiomyocytes in response to 20% repetitive stretch (Wang et al. 2004). A relatively intense stretch of 20% is considered to mimic an excessive mechanical stimulus derived by pressure overload, leading to pathological fibrosis. Indeed, the expression and activity of MMP-2 were increased in the cardiomyocytes of patients suffering pressure overload due to aortic stenosis (Polyakova et al. 2004). While expression levels of MMP-2 were unchanged in the LV of terminal dilated cardiomyopathy (DCM), those of MMP-9 (gelatinase B) implicated in inflammation were increased (Thomas et al. 1998). As already mentioned, pressure and volume overload induce distinct types of MMP, resulting in different types of remodeling.

Myocardial inflammation is an early response to pathological stimulation, such as cardiac injury. Under the inflammatory condition, cytokines and chemokines further augment the inflammatory effects, and cell loss (apoptosis and necrosis) triggers compensatory responses. All of these factors can significantly increase the activity of fibroblasts and cause excessive deposition of fibrillar collagens (Hou and Kang 2012). There are two origins for cardiac fibroblasts: resident and nonresident. The resident, existing fibroblasts proliferate rapidly in response to cardiac injury. The nonresident, circulating, bone marrow-derived cells drafted to the cardiac tissue can transform into fibroblasts and myofibroblasts. When mechanical stress is applied, fibroblasts turn into myofibroblasts (Chen and Frangogiannis 2013). Myofibroblasts are more mobile and have a greater synthetic ability to produce ECM proteins (Petrov et al. 2002). They do not exist in the healthy heart and only appear after cardiac injury, contributing to the development of fibrosis (Fan et al. 2012). During pathological ECM remodeling, fibroblasts synthesize procollagen and secrete it into the myocardial interstitium, where it is cleaved by procollagen N- and C-proteinases (PCP) in the end-terminal propeptide sequence, enabling formation of collagen fiber (Fan et al. 2012). In addition, fibrotic areas in the myocardium act as a substrate for lethal ventricular arrhythmias (Wu et al. 1998; Hsia and Marchlinski 2002), causing adverse cardiac remodeling.

11.4.1.3 Apoptosis

Cell death is an important cause of decreasing cardiomyocyte numbers, and nonexistent or insufficient cell regeneration accelerates the phenomenon. Traditionally, two types of cell death are necrosis and apoptosis based on morphologic characteristics. Necrosis is a feature of early heart failure, especially of the ischemic variety, but the cause of cell death in chronic heart failure is mainly apoptotic (Chandrashekhar 2005). Apoptosis of cardiomyocytes is increased in patient hearts exposed to pressure or volume overload (Yamamoto et al. 2000; Gonzalez et al. 2002). Mechanical stimulus imposes cellular apoptosis. In rats, excessive stretching of cardiac papillary muscle elicits apoptosis by increasing superoxide anion formation (Cheng et al. 1995). Pressure overload to the LV imposes apoptosis on both cardiomyocytes and non-cardiomyocytes in experiments using rats and dogs (Gelpi et al. 2011).

The classical view is that the apoptotic cell dies and this contributes to heart failure through a reduced contractile cell mass. There may also be other mechanisms through which the apoptotic cascade mediates heart failure. Because much of the apoptotic machinery can cleave contractile proteins, including actin, myosin, and troponins (Communal et al. 2002), it is possible that activation of apoptotic pathways mediates contractile dysfunction that is to some extent independent of cell death. Recently, it has been postulated that such cells in systolic dysfunction may precede any breakdown of DNA or that irreversible cell death may show contractile dysfunction, producing so-called "zombie myocytes" (Narula et al. 2001). This suggests that apoptosis may play a much broader role in heart failure – it might cause contractile failure in surviving cells, in addition to loss of contractile cell mass.

11.4.1.4 Mitochondria Failure

Mitochondrial dysfunction is the hallmark of pathological cardiac remodeling. A number of studies have focused on the contribution of the mitochondrion in the progression of cardiac remodeling, due to its central role in energy production, metabolism, calcium homeostasis, and oxidative stress, all of which may have an effect on pathogenesis. There are three aspects to mitochondrial function: ROS signaling, Ca^{2+} handling, and mitochondrial dynamics.

11.4.1.5 Oxidative Stress and Cardiac Remodeling

Though the tightly regulated production of relatively low levels of ROS acts as a secondary messenger-amplifying signal that is crucial for normal cell function, oxidative stress has direct effects on cellular structure and function, and may activate integral signaling molecules in myocardial remodeling and failure. First, ROS activate a broad variety of hypertrophy signaling kinases and transcription factors (Sabri et al. 2003), such as tyrosine kinase Src, GTP-binding protein Ras, protein kinase C, mitogen-activated protein kinases (MAPKs), and Jun-nuclear kinase (JNK). Second, ROS induces apoptosis, another important contributor to remodeling and dysfunction, which is induced by ROS-mediated DNA and mitochondrial damage, as well as by activation of pro-apoptotic signaling kinases (Cesselli et al. 2001). Third, ROS cause DNA strand breaks, activating the nuclear enzyme poly (ADP-ribose) polymerase 1 (PARP-1). PARP-1 regulates the expression of a variety of inflammatory mediators, which facilitate the progression of cardiac remodeling. Fourth, ROS can activate MMPs, a family of proteolytic enzymes (Spinale et al. 1998). Fifth, ROS directly influence contractile function by modifying proteins involved in excitation–contraction coupling (Zima and Blatter 2006; Tsutsui et al. 2011).

11.4.1.6 Ca^{2+} Signaling

Rises in cytoplasmic Ca^{2+} activate the Ca^{2+}-dependent proteins CAMKII, calcineurin (CaN), and protein kinase C (PKC), which in turn induce a characteristic genetic program involved in cardiac hypertrophy development. Also, the activation of CAMKII and CaN promotes mitochondrial fission and mitochondrial permeability transition pore (MPTP) opening, while the release of cytochrome c activates cardiomyocyte apoptosis, further contributing to the remodeling process (Verdejo et al. 2012).

11.4.1.7 Mitochondrial Dynamics and Cardiac Remodeling

The participation of mitochondrial dynamics in remodeling seems to depend on the nature of the injury. For example, while a decrease in mitochondrial fusion protein mitofusin-2 (Mfn2) promotes cardiac hypertrophy in the pressure-overload model, it improves the functional recovery of cardiac injury in the myocardial ischemia–reperfusion model (Verdejo et al. 2012). Mitochondrial ATP-sensitive K^+ (K_{ATP}) channels appear mandatory in acute and chronic cardiac adaptation to imposed hemodynamic load, protecting against congestive heart failure and death (Yamada et al. 2006). Knockout of Kir6.2, the pore-forming subunit of the K_{ATP} channel, shows an aberrant prolongation of action potentials with intracellular calcium overload and ATP depletion, whereas wild-type maintains ionic and energetic handling.

11.4.1.8 Gene Expression

One of the key features of pathological cardiac remodeling is altered cardiac gene expression. This has been investigated for many years. The gene-expression pattern in pathological remodeling (e.g., dilated failing human heart) resembles a fetal gene program, including such genes as ANP, BNP, skeletal α-actin, and β-myosin heavy chain. Reactivation of the fetal gene program is initially an adaptive process that increases the contractility, excitability, and plasticity of cardiac myocytes in response to pathological stress, but when these are sustained, they contribute to the progression of maladaptive processes that ultimately lead to cardiac dysfunction (Hou and Kang 2012; Kuwahara et al. 2012).

11.4.2 Physiological Remodeling

In 1899, Henschen reported that the size of the heart increases in response to exercise, based on auscultatory findings (Rowland 2011; Weiner and Baggish 2012). Today, advanced imaging technologies allow us to detect changes in the heart structure before and after training in athletes (De Luca et al. 2011). Different types of exercise result in distinct structural alterations (Ellison et al. 2011). For example, endurance training (also referred to as "isotonic exercise"), such as marathon-running and swimming, imposes volume load (namely, preload) on the heart and generally elicits eccentric hypertrophy. On the other hand, resistance training (also referred to as "isometric exercise" or "strength training"), such as weightlifting and wrestling, imposes afterload on the heart by transiently increasing vascular resistance systemically, thereby eliciting concentric hypertrophy generally. Meanwhile, certain data question the development of cardiac remodeling in response to resistance training (Spence et al. 2011).

The consequence of cardiac remodeling is not just hypertrophy of existing cardiomyocytes. Cardiomyocytes were once considered to be the same age as the individual in whom they are found, since they do not divide and are not newly created after development. Recently, however, the existence of cardiac stem cells (CSCs), which enable the generation of new cardiomyocytes, was discovered (Torella et al. 2007), and the apparent formation of new cardiomyocytes from CSCs in the cardiac muscle of patients suffering aortic stenosis was reported (Urbanek et al. 2003). An unexpected consequence of nuclear bomb testing during the Cold War was the finding that half of all cardiomyocytes are replaced by newly generated ones across the course of a lifetime, as evidenced by the radioactive isotope ^{14}C (Bergmann et al. 2009).

Importantly, exercise does not just decrease the risk of coronary artery disease (CAD) by enhancing cardiac function, but also attenuates the symptoms of already-developed cardiovascular diseases (CVDs) (Thompson et al. 2003). Though exercise elicits broad systemic changes, such as protection against diabetes mellitus (DM) and breast cancer, we focus in this section on its effect on cardiac remodeling.

11.4.2.1 Apoptosis

Cardiomyocytes undergo apoptosis with age and decrease in number without specific cardiac disease (Sheydina et al. 2011). We discussed apoptosis due to pressure or volume overload in pathological remodeling in the previous section. In pathological remodeling, it has been reported that moderate exercise does not elicit apoptosis in the cardiac tissue, and may even inhibit it. In experiments using rats, 13-week treadmill training regimens caused augmented cardiac function without any trace of apoptosis (Jin et al. 2000). Other researchers reported that exercise increased the expression of Hsp70, which belongs to the apoptosis inhibition system, and attenuated apoptosis (Siu et al. 2004).

11.4.2.2 Arrhythmias

Pathological remodeling due to pressure overload is considered to elicit heterogeneity of action potential firing and to cause arrhythmia. In a rat model of hypertension, it was suggested that exercise improved heterogeneity (Roman-Campos et al. 2012).

11.4.2.3 Mitochondrial Function

Proliferation of mitochondria during the process of cardiac hypertrophy delays the development of heart failure in response to pressure overload (Rosca, Tandler, and Hoppel 2013). ROS generated by mitochondria augment cardiac contractility via the β-adrenergic pathway.

11.4.2.4 Ca^{2+} Signaling

Cardiac contractility is an important factor that defines cardiac function. Calcium dynamics in cardiomyocytes directly influences cardiac contractility (Kranias and Hajjar 2012). In animal models, it is suggested that the increase in cardiac contractility is due to the modulation of calcium ion concentration in the cardiomyocytes, namely quick change in calcium concentration and low concentration during diastole (Kemi and Wisloff 2010). The molecular mechanism is considered to be elevation of sarcoplasmic reticulum Ca^{2+}-ATPase (SERCA) 2a activity and phospholamban (PLB) phosphorylation. Phosphorylation of PLB decreases inhibition of SERCA2a.

11.4.2.5 Vascular Remodeling

Pressure overload in the ventricle induces vascularization in both pathological and physiological remodeling. Pressure overload promotes the release of angiogenic factors according to hypoxia-induced factor Hif-1α and causes angiogenesis. However, long-lasting pressure overload accumulates the tumor-suppressor gene p53 and inhibits Hif-1α, resulting in decreased angiogenesis (Sano et al. 2007). This facilitates the development of heart failure. Interestingly, cyclic stretch of vascular smooth-muscle cells (SMCs) increases the expression of p53 and induces apoptosis *in vitro* (Sedding et al. 2008). Though a precise biophysical description of the pathway from detection of mechanical stimuli to increased p53 expression remains to be established, it has been suggested by experiments on protein interactions. There is a region called the N2A-mechanosensory complex within the huge protein titin, expressed in the sarcomeres of cardiomyocytes (Linke 2008). Ankrd1/CARP, one of the muscle ankyrin repeat proteins (MARP), binds to titin N2A and translocates to the nucleus in response to mechanical load. It works as a transcriptional coactivator and upregulates p53 (Kojic et al. 2010). A similar protein to titin is expressed in SMCs (Kim and Keller 2002).

Mechanical stress is imposed on the coronary artery during exercise. Filling of the vessel with blood during diastole stretches the vessels attached to the surrounding tissue along the long axis (Brown 2003). Exercise causes remodeling of the coronary artery. In one study, endurance training for 5 months in healthy male subjects significantly increased the cross-sectional areas of the right branch, the left main trunk, and the left circumflex coronary artery (Windecker et al. 2002).

11.5 Conclusion

We have discussed how moderate exercise augments cardiac function in both healthy subjects and those with CVD, while long-lasting pressure or volume overload imposes pathological remodeling of the heart. The vast amount of knowledge we have regarding cardiac remodeling should contribute to the prevention of pathological remodeling and heart failure. Considering the powerful preventive and therapeutic effects evidenced by detailed molecular mechanisms, exercise should be applied in a much more coordinated manner – it is a measure that can be taken by anyone, without special license. Further understanding of cardiac mechanotransduction will help rectify pathology and improve health.

References

Anversa, P., P. Li, E. H. Sonnenblick, and G. Olivetti. 1994. "Effects of aging on quantitative structural properties of coronary vasculature and microvasculature in rats." *American Journal of Physiology* **267**(3 Pt. 2): H1062–73. PMID:8092271.

Asrih, M., F. Mach, A. Nencioni, F. Dallegri, A. Quercioli, and F. Montecucco. 2013. "Role of mitogen-activated protein kinase pathways in multifactorial adverse cardiac remodeling associated with metabolic syndrome." *Mediators of Inflammation* **2013**: 367245. doi:10.1155/2013/367245.

Battiprolu, P. K., C. Lopez-Crisosto, Z. V. Wang, A. Nemchenko, S. Lavandero, and J. A. Hill. 2013. "Diabetic cardiomyopathy and metabolic remodeling of the heart." *Life Sciences* **92**(11): 609–15. doi:S0024-3205(12)00614-5.

Benito, B., G. Gay-Jordi, A. Serrano-Mollar, E. Guasch, Y. Shi, J. C. Tardif, et al. 2011. "Cardiac arrhythmogenic remodeling in a rat model of long-term intensive exercise training." *Circulation* **123**(1): 13–22. doi:CIRCULATIONAHA.110.938282.

Bergmann, O., R. D. Bhardwaj, S. Bernard, S. Zdunek, F. Barnabe-Heider, S. Walsh, et al. 2009. "Evidence for cardiomyocyte renewal in humans." *Science* **324**(5923): 98–102. doi:324/5923/98 [pii].

Birks, E. J. 2010. "Myocardial recovery in patients with chronic heart failure: is it real?" *Journal of Cardiac Surgery* **25**(4): 472–7. doi:JCS1051.

Borg, T. K. and T. A. Baudino. 2011. "Dynamic interactions between the cellular components of the heart and the extracellular matrix." *Pflugers Archiv* **462**(1): 69–74. doi:10.1007/s00424-011-0940-7.

Bradley, J. M., J. B. Nguyen, A. C. Fournett, and J. D. Gardner. 2012. "Cigarette smoke exacerbates ventricular remodeling and dysfunction in the volume overloaded heart." *Microscopy and Microanalysis* **18**(1): 91–8. doi:10.1017/S1431927611012207.

Brown, M. D. 2003. "Exercise and coronary vascular remodelling in the healthy heart." *Experimental Physiology* **88**(5): 645–58. doi:EPH_2618.

Care, A., D. Catalucci, F. Felicetti, D. Bonci, A. Addario, P. Gallo, et al. 2007. "MicroRNA-133 controls cardiac hypertrophy." *Nature Medicine* **13**(5): 613–18. doi:nm1582.

Cesselli, D., I. Jakoniuk, L. Barlucchi, A. P. Beltrami, T. H. Hintze, B. Nadal-Ginard, et al. 2001. "Oxidative stress-mediated cardiac cell death is a major determinant of ventricular dysfunction and failure in dog dilated cardiomyopathy." *Circulation Research* **89**(3): 279–86. PMID:11485979.

Chandrashekhar, Y. 2005. "Role of apoptosis in ventricular remodeling." *Current Heart Failure Reports* **2**(1): 18–22. doi:10.1007/s11897-005-0003-5.

Chen, W. and N. G. Frangogiannis. 2013. "Fibroblasts in post-infarction inflammation and cardiac repair." *Biochimica Biophysica Acta* **1833**(4): 945–53. doi:S0167-4889(12)00257-1.

Cheng, W., B. Li, J. Kajstura, P. Li, M. S. Wolin, E. H. Sonnenblick, et al. 1995. "Stretch-induced programmed myocyte cell death." *Journal of Clinical Investigation* **96**(5): 2247–59. doi:10.1172/JCI118280.

Communal, C., M. Sumandea, P. de Tombe, J. Narula, R. J. Solaro, and R. J. Hajjar. 2002. "Functional consequences of caspase activation in cardiac myocytes." *Proceedings of the National Academy of Sciences United States of America* **99**(9): 6252–6. doi:10.1073/pnas.092022999.

Cowin, S. C. 2004. "Tissue growth and remodeling." *Annual Reviews in Biomedical Engineering* **6**: 77–107. doi:10.1146/annurev.bioeng.6.040803.140250.

De Luca, A., L. Stefani, G. Pedrizzetti, S. Pedri, and G. Galanti. 2011. "The effect of exercise training on left ventricular function in young elite athletes." *Cardiovascular Ultrasound* **9**: 27. doi:1476-7120-9-27.

DiMichele, L. A., J. T. Doherty, M. Rojas, H. E. Beggs, L. F. Reichardt, C. P. Mack, and J. M. Taylor. 2006. "Myocyte-restricted focal adhesion kinase deletion attenuates pressure overload-induced hypertrophy." *Circulation Research* **99**(6): 636–45. doi:01 RES.0000240498.44752.d6.

Eghbali, M., Y. Wang, L. Toro, and E. Stefani. 2006. "Heart hypertrophy during pregnancy: a better functioning heart?" *Trends Cardiovascular Medicine* **16**(8): 285–91. doi:S1050-1738(06)00113-7.

Ellison, G. M., C. D. Waring, C. Vicinanza, and D. Torella. 2011. "Physiological cardiac remodelling in response to endurance exercise training: cellular and molecular mechanisms." *Heart* **98**(1): 5–10. doi:heartjnl-2011-300639.

Fan, D., A. Takawale, J. Lee, and Z. Kassiri. 2012. "Cardiac fibroblasts, fibrosis and extracellular matrix remodeling in heart disease." *Fibrogenesis Tissue Repair* **5**(1): 15. doi:10.1186/1755-1536-5-15.

Force, T., A. Michael, H. Kilter, and S. Haq. 2002. "Stretch-activated pathways and left ventricular remodeling." *Journal of Cardiovascular Failure* **8**(6 Suppl.): S351–8. doi:10.1054/jcaf.2002.129272.

Gelpi, R. J., M. Park, S. Gao, S. Dhar, D. E. Vatner, and S. F. Vatner. 2011. "Apoptosis in severe, compensated pressure overload predominates in nonmyocytes and is related to the hypertrophy but not function." *American Journal of Physiology. Heart and Circulatory Physiology* **300**(3): H1062–8. doi:ajpheart.00998.2010.

George, K., G. P. Whyte, D. J. Green, D. Oxborough, R. E. Shave, D. Gaze, and J. Somauroo. 2012. "The endurance athletes heart: acute stress and chronic adaptation." *British Journal of Sports Medicine* **46**(Suppl. 1): i29–36. doi:46/Suppl_1/i29.

Gonzalez, A., B. Lopez, S. Ravassa, R. Querejeta, M. Larman, J. Diez, and M. A. Fortuno. 2002. "Stimulation of cardiac apoptosis in essential hypertension: potential role of angiotensin II." *Hypertension* **39**(1): 75–80. PMID:11799082.

Hori, Y., T. Kashimoto, T. Yonezawa, N. Sano, R. Saitoh, S. Igarashi, et al. 2012. "Matrix metalloproteinase-2 stimulates collagen-I expression through phosphorylation of focal adhesion kinase in rat cardiac fibroblasts." *American Journal of Physiology. Cell Physiology* **303**(9): C947–53. doi:ajpcell.00401.2011.

Hoshijima, M. 2006. "Mechanical stress-strain sensors embedded in cardiac cytoskeleton: Z disk, titin, and associated structures." *American Journal of Physiology. Heart and Circulatory Physiology* **290**(4): H1313–25. doi:290/4/H1313.

Hou, J. and Y. J. Kang. 2012. "Regression of pathological cardiac hypertrophy: signaling pathways and therapeutic targets." *Pharmacology & Therapeutics* **135**(3): 337–54. doi:10.1016/j.pharmthera.2012.06.006.

Hove, J. R., R. W. Koster, A. S. Forouhar, G. Acevedo-Bolton, S. E. Fraser, and M. Gharib. 2003. "Intracardiac fluid forces are an essential epigenetic factor for embryonic cardiogenesis." *Nature* **421**(6919): 172–7. doi:10.1038/nature01282.

Hsia, H. H. and F. E. Marchlinski. 2002. "Characterization of the electroanatomic substrate for monomorphic ventricular tachycardia in patients with nonischemic cardiomyopathy." *Pacing and Clinical Electrophysiology* **25**(7): 1114–27. PMID:12164454.

Hutchinson, K. R., J. A. Stewart Jr., and P. A. Lucchesi. 2010. "Extracellular matrix remodeling during the progression of volume overload-induced heart failure." *Journal of Molecular and Cellular Cardiology* **48**(3): 564–9. doi:10.1016/j.yjmcc.2009.06.001.

Inoue, R., L. J. Jensen, J. Shi, H. Morita, M. Nishida, A. Honda, and Y. Ito. 2006. "Transient receptor potential channels in cardiovascular function and disease." *Circulation Research* **99**(2): 119–31. doi:99/2/119.

Inoue, R., Z. Jian, and Y. Kawarabayashi. 2009. "Mechanosensitive TRP channels in cardiovascular pathophysiology." *Pharmacology & Therapeutics* **123**(3): 371–85. doi:S0163-7258(09)00113-2.

Jin, H., R. Yang, W. Li, H. Lu, A. M. Ryan, A. K. Ogasawara, et al. 2000. "Effects of exercise training on cardiac function, gene expression, and apoptosis in rats." *American Journal of Physiology. Heart and Circulatory Physiology* **279**(6): H2994–3002. PMID:11087257.

Jugdutt, B. I. 2003. "Remodeling of the myocardium and potential targets in the collagen degradation and synthesis pathways." *Current Drug Targets. Cardiovascular & Haematological Disorders* **3**(1): 1–30. PMID:12769643.

Kemi, O. J. and U. Wisloff. 2010. "Mechanisms of exercise-induced improvements in the contractile apparatus of the mammalian myocardium." *Acta Physiologica* **199**(4): 425–39. doi:10.1111/j.1748-1716.2010.02132.x.

Kerstein, P. C., B. T. Jacques-Fricke, J. Rengifo, B. J. Mogen, J. C. Williams, P. A. Gottlieb, et al. 2013. "Mechanosensitive TRPC1 channels promote calpain proteolysis of talin to regulate spinal axon outgrowth." *Journal of Neuroscience* **33**(1): 273–85. doi:10.1523/JNEUROSCI.2142-12.2013.

Kim, K. and T. C. Keller 3rd. 2002. "Smitin, a novel smooth muscle titin-like protein, interacts with myosin filaments in vivo and in vitro." *Journal of Cell Biology* **156**(1): 101–11. doi: 10.1083/jcb.200107037.

Kojic, S., A. Nestorovic, L. Rakicevic, A. Belgrano, M. Stankovic, A. Divac, and G. Faulkner. 2010. "A novel role for cardiac ankyrin repeat protein Ankrd1/CARP as a co-activator of the p53 tumor suppressor protein." *Archives of Biochemistry and Biophysics* **502**(1): 60–7. doi:10.1016/j.abb.2010.06.029.

Kranias, E. G. and R. J. Hajjar. 2012. "Modulation of cardiac contractility by the phospholamban/SERCA2a regulatome." *Circulation Research* **110**(12): 1646–60. doi:110/12/1646.

Kresh, J. Y. and A. Chopra. 2011. "Intercellular and extracellular mechanotransduction in cardiac myocytes." *Pflugers Archiv* **462**(1): 75–87. doi:10.1007/s00424-011-0954-1.

Kuwahara, K., T. Nishikimi, and K. Nakao. 2012. "Transcriptional regulation of the fetal cardiac gene program." *Journal of Pharmacological Sciences* **119**(3): 198–203. PMID:22786561.

La Gerche, A., A. T. Burns, D. J. Mooney, W. J. Inder, A. J. Taylor, J. Bogaert, et al. 2012. "Exercise-induced right ventricular dysfunction and structural remodelling in endurance athletes." *European Heart Journal* **33**(8): 998–1006. doi:10.1093/eurheartj/ehr397.

Linke, W. A. 2008. "Sense and stretchability: the role of titin and titin-associated proteins in myocardial stress-sensing and mechanical dysfunction." *Cardiovascular Research* **77**(4): 637–48. doi:10.1016/j.cardiores.2007.03.029.

MacKenna, D., S. R. Summerour, and F. J. Villarreal. 2000. "Role of mechanical factors in modulating cardiac fibroblast function and extracellular matrix synthesis." *Cardiovascular Research* **46**(2): 257–63. doi:S0008-6363(00)00030-4.

Majkut, S. F. and D. E. Discher. 2012. "Cardiomyocytes from late embryos and neonates do optimal work and striate best on substrates with tissue-level elasticity: metrics and mathematics." *Biomechanics and Modeling in Mechanobiology* **11**(8): 1219–25. doi:10.1007/s10237-012-0413-8.

Mann, D. L. 2004. "Basic mechanisms of left ventricular remodeling: the contribution of wall stress." *Journal of Cardiovascular Failure* **10**(6 Suppl.): S202–6. PMID:15803551.

Maroto, R., A. Raso, T. G. Wood, A. Kurosky, B. Martinac, and O. P. Hamill. 2005. "TRPC1 forms the stretch-activated cation channel in vertebrate cells." *Nature Cell Biology* **7**(2): 179–85. doi:10.1038/ncb1218.

Matsumiya, G., S. Saitoh, Y. Sakata, and Y. Sawa. 2009. "Myocardial recovery by mechanical unloading with left ventricular assist system." *Circulation Journal* **73**(8): 1386–92. doi:JST.JSTAGE/circj/CJ-09-0396.

McCain, M. L. and K. K. Parker. 2011. "Mechanotransduction: the role of mechanical stress, myocyte shape, and cytoskeletal architecture on cardiac function." *Pflugers Archiv* **462**(1): 89–104. doi:10.1007/s00424-011-0951-4.

McCain, M. L., H. Lee, Y. Aratyn-Schaus, A. G. Kleber, and K. K. Parker. 2012. "Cooperative coupling of cell-matrix and cell-cell adhesions in cardiac muscle." *Proceedings of the National Academy of Sciences United States of America* **109**(25): 9881–6. doi:1203007109.

Narula. J., E. Arbustini, Y. Chandrashekhar, and M. Schwaiger. 2001. "Apoptosis and the systolic dysfunction in congestive heart failure. Story of apoptosis interruptus and zombie myocytes." *Cardiol Clin* **19**(1): 113–26. PMID:11787805.

Oparil, S. 1985. "Pathogenesis of ventricular hypertrophy." *Journal of the American College of Cardiology* **5**(6 Suppl.): 57B–65B. PMID:3158693.

Petrov, V. V., R. H. Fagard, and P. J. Lijnen. 2002. "Stimulation of collagen production by transforming growth factor-beta1 during differentiation of cardiac fibroblasts to myofibroblasts." *Hypertension* **39**(2): 258–63. PMID:11847194.

Polyakova, V., S. Hein, S. Kostin, T. Ziegelhoeffer, and J. Schaper. 2004. "Matrix metalloproteinases and their tissue inhibitors in pressure-overloaded human myocardium during heart failure progression." *Journal of the American College of Cardiology* **44**(8): 1609–18. doi:S0735-1097(04)01509-8.

Richey, P. A. and S. P. Brown. 1998. "Pathological versus physiological left ventricular hypertrophy: a review." *Journal of Sports Science* **16**(2): 129–41. doi:10.1080/026404198366849.

Roman-Campos, D., M. A. Carneiro-Junior, T. N. Primola-Gomes, K. A. Silva, J. F. Quintao-Junior, A. N. Gondim, et al. 2012. "Chronic exercise partially restores the transmural heterogeneity of action potential duration in left ventricular myocytes of spontaneous hypertensive rats." *Clinical and Experimental Pharmacology & Physiology* **39**(2): 155–7. doi:10.1111/j.1440-1681.2011.05669.x.

Romer, L. H., K. G. Birukov, and J. G. Garcia. 2006. "Focal adhesions: paradigm for a signaling nexus." *Circulation Research* **98**(5): 606–16. doi:10.1161/01.RES.0000207408.31270.db.

Rosca, M. G., B. Tandler, and C. L. Hoppel. 2013. "Mitochondria in cardiac hypertrophy and heart failure." *Journal of Molecular and Cellular Cardiology* **55**: 31–41. doi:S0022-2828(12)00335-5.

Ross, R. S. 2004. "Molecular and mechanical synergy: cross-talk between integrins and growth factor receptors." *Cardiovascular Research* **63**(3): 381–90. doi:10.1016/j.cardiores.2004.04.027.

Rowland, T. 2011. "Is the 'athlete's heart' arrhythmogenic? Implications for sudden cardiac death." *Sports Medicine* **41**(5): 401–11. doi:10.2165/11583940-000000000-00000.

Sabri, A., H. H. Hughie, and P. A. Lucchesi. 2003. "Regulation of hypertrophic and apoptotic signaling pathways by reactive oxygen species in cardiac myocytes." *Antioxidants & Redox Signaling* **5**(6): 731–40. doi:10.1089/152308603770380034.

Sadoshima, J., T. Takahashi, L. Jahn, and S. Izumo. 1992. "Roles of mechano-sensitive ion channels, cytoskeleton, and contractile activity in stretch-induced immediate-early gene expression and hypertrophy of cardiac myocytes." *Proceedings of the National Academy of Sciences United States of America* **89**(20): 9905–9. PMID:1384064.

Samarel, A. M. 2005. "Costameres, focal adhesions, and cardiomyocyte mechanotransduction." *American Journal of Physiology. Heart and Circulatory Physiology* **289**(6): H2291–301. doi:10.1152/ajpheart.00749.2005.

Sano, M., T. Minamino, H. Toko, H. Miyauchi, M. Orimo, Y. J. Qin, et al. 2007. "p53-induced inhibition of Hif-1 causes cardiac dysfunction during pressure overload." *Nature* **446**(7134): 444–8. doi:10.1038/nature05602.

Sedding, D. G., M. Homann, U. Seay, H. Tillmanns, K. T. Preissner, and R. C. Braun-Dullaeus. 2008. "Calpain counteracts mechanosensitive apoptosis of vascular smooth muscle cells in vitro and in vivo." *FASEB Journal* **22**(2): 579–89. doi:10.1096/fj.07-8853com.

Seong, J., N. Wang, and Y. Wang. 2013. "Mechanotransduction at focal adhesions: from physiology to cancer development." *Journal of Cellular and Molecular Medicine* **17**(5): 597–604. doi:10.1111/jcmm.12045.

Shave, R., A. Baggish, K. George, M. Wood, J. Scharhag, G. Whyte, et al. 2010. "Exercise-induced cardiac troponin elevation: evidence, mechanisms, and implications." *Journal of the American College of Cardiology* **56**(3): 169–76. doi:S0735-1097(10)01705-5.

Sheydina, A., D. R. Riordon, and K. R. Boheler. 2011. "Molecular mechanisms of cardiomyocyte aging." *Clinical Science* **121**(8): 315–29. doi:10.1042/CS20110115.

Siu, P. M., R. W. Bryner, J. K. Martyn, and S. E. Alway. 2004. "Apoptotic adaptations from exercise training in skeletal and cardiac muscles." *FASEB Journal* **18**(10): 1150–2. doi:10.1096/fj.03-1291fje.

Spassova, M. A., T. Hewavitharana, W. Xu, J. Soboloff, and D. L. Gill. 2006. "A common mechanism underlies stretch activation and receptor activation of TRPC6 channels." *Proceedings of the National Academy of Sciences United States of America* **103**(44): 16 586–91. doi:10.1073/pnas.0606894103.

Spence, A. L., L. H. Naylor, H. H. Carter, C. L. Buck, L. Dembo, C. P. Murray, et al. 2011. "A prospective randomised longitudinal MRI study of left ventricular adaptation to endurance and resistance exercise training in humans." *Journal of Physiology* **589**(Pt. 22): 5443–52. doi:jphysiol.2011.217125.

Spinale, F. G., M. L. Coker, C. V. Thomas, J. D. Walker, R. Mukherjee, and L. Hebbar. 1998. "Time-dependent changes in matrix metalloproteinase activity and expression during the progression of congestive heart failure: relation to ventricular and myocyte function." *Circulation Research* **82**(4): 482–95. PMID:9506709.

Spinale, F. G., J. S. Janicki, and M. R. Zile. 2013. "Membrane-associated matrix proteolysis and heart failure." *Circulation Research* **112**(1): 195–208. doi:10.1161/CIRCRESAHA.112.266882.

Takahashi, K. and K. Naruse. 2012. "Stretch-activated BK channel and heart function." *Progress in Biophysics and Molecular Biology* **110**(2–3): 239–44. PMID:23281538.

Takahashi, K., Y. Kakimoto, K. Toda, and K. Naruse. 2013. "Mechanobiology in cardiac physiology and diseases." *Journal of Cellular and Molecular Medicine* **17**(2): 225–32. doi:10.1111/jcmm.12027.

Taylor, M., S. Graw, G. Sinagra, C. Barnes, D. Slavov, F. Brun, et al. 2011. "Genetic variation in titin in arrhythmogenic right ventricular cardiomyopathy-overlap syndromes." *Circulation* **124**(8): 876–85. doi:CIRCULATIONAHA.110.005405.

Thomas, C. V., M. L. Coker, J. L. Zellner, J. R. Handy, A. J. Crumbley, and F. G. Spinale. 1998. "Increased matrix metalloproteinase activity and selective upregulation in LV myocardium from patients with end-stage dilated cardiomyopathy." *Circulation* **97**(17): 1708–15. PMID:9591765.

Thompson, P. D., D. Buchner, I. L. Pina, G. J. Balady, M. A. Williams, B. H. Marcus, et al. 2003. "Exercise and physical activity in the prevention and treatment of atherosclerotic cardiovascular disease: a statement from the Council on Clinical Cardiology (Subcommittee on Exercise, Rehabilitation, and Prevention) and the Council on Nutrition, Physical Activity, and Metabolism (Subcommittee on Physical Activity)." *Circulation* **107**(24): 3109–16. doi:10.1161/01.CIR.0000075572.40158.77.

Thornburg, K. L., S. L. Jacobson, G. D. Giraud, and M. J. Morton. 2000. "Hemodynamic changes in pregnancy." *Seminars in Perinatology* **24**(1): 11–14. PMID:10709851.

Torella, D., G. M. Ellison, I. Karakikes, and B. Nadal-Ginard. 2007. "Resident cardiac stem cells." *Cellular and Molecular Life Sciences* **64**(6): 661–73. doi:10.1007/s00018-007-6519-y.

Tsutsui, H., S. Kinugawa, and S. Matsushima. 2011. "Oxidative stress and heart failure." *American Journal of Physiology. Heart and Circulatory Physiology* **301**(6): H2181–90. doi:10.1152/ajpheart.00554.2011.

Urbanek, K., F. Quaini, G. Tasca, D. Torella, C. Castaldo, B. Nadal-Ginard, et al. 2003. "Intense myocyte formation from cardiac stem cells in human cardiac hypertrophy." *Proceedings of the National Academy of Sciences United States of America* **100**(18): 10 440–5. doi:10.1073/pnas.1832855100.

Verdejo, H. E., A. del Campo, R. Troncoso, T. Gutierrez, B. Toro, C. Quiroga, et al. 2012. "Mitochondria, myocardial remodeling, and cardiovascular disease." *Current Hypertension Reports* **14**(6): 532–9. doi:10.1007/s11906-012-0305-4.

Wang, T. L., Y. H. Yang, H. Chang, and C. R. Hung. 2004. "Angiotensin II signals mechanical stretch-induced cardiac matrix metalloproteinase expression via JAK-STAT pathway." *Journal of Molecular and Cellular Cardiology* **37**(3): 785–94. doi:10.1016/j.yjmcc.2004.06.016.

Weiner, R. B. and A. L. Baggish. 2012. "Exercise-induced cardiac remodeling." *Progress in Cardiovascular Disease* **54**(5): 380–6. doi:S0033-0620(12)00007-2.

Windecker, S., Y. Allemann, M. Billinger, T. Pohl, D. Hutter, T. Orsucci, et al. 2002. "Effect of endurance training on coronary artery size and function in healthy men: an invasive followup study." *American Journal of Physiology. Heart and Circulatory Physiology* **282**(6): H2216–23. doi:10.1152/ajpheart.00977.2001.

Wu, T. J., J. J. Ong, C. Hwang, J. J. Lee, M. C. Fishbein, L. Czer, et al. 1998. "Characteristics of wave fronts during ventricular fibrillation in human hearts with dilated cardiomyopathy: role of increased fibrosis in the generation of reentry." *Journal of the American College of Cardiology* **32**(1): 187–96. PMID:9669269.

Yamada, S., G. C. Kane, A. Behfar, X. K. Liu, R. B. Dyer, R. S. Faustino, et al. 2006. "Protection conferred by myocardial ATP-sensitive K+ channels in pressure overload-induced congestive heart failure revealed in KCNJ11 Kir6.2-null mutant." *Journal of Physiology* **577**(Pt. 3): 1053–65. doi:jphysiol.2006.119511.

Yamamoto, S., K. Sawada, H. Shimomura, K. Kawamura, and T. N. James. 2000. "On the nature of cell death during remodeling of hypertrophied human myocardium." *Journal of Molecular and Cellular Cardiology* **32**(1): 161–75. doi:10.1006/jmcc.1999.1064.

Yamazaki, T., I. Komuro, and Y. Yazaki. 1995. "Molecular mechanism of cardiac cellular hypertrophy by mechanical-stress." *Journal of Molecular and Cellular Cardiology* **27**(1): 133–40. PMID:7760338.

Yarbrough, W. M., R. Mukherjee, J. S. Ikonomidis, M. R. Zile, and F. G. Spinale. 2012. "Myocardial remodeling with aortic stenosis and after aortic valve replacement: mechanisms and future prognostic implications." *Journal of Thoracic and Cardiovascular Surgery* **143**(3): 656–64. doi:10.1016/j.jtcvs.2011.04.044.

Zima, A. V. and L. A. Blatter. 2006. "Redox regulation of cardiac calcium channels and transporters." *Cardiovascular Research* **71**(2): 310–21. doi:10.1016/j.cardiores.2006.02.019.

Zou, Y., H. Akazawa, Y. Qin, M. Sano, H. Takano, T. Minamino, et al. 2004. "Mechanical stress activates angiotensin II type 1 receptor without the involvement of angiotensin II." *Nature Cell Biology* **6**(6): 499–506. doi:10.1038/ncb1137.

12

Aortic Valve Mechanobiology

From Organ to Cells

K. Jane Grande-Allen[1], Daniel Puperi[1], Prashanth Ravishankar[2], and Kartik Balachandran[2]

[1] *Department of Bioengineering, Rice University, Houston, TX, USA*
[2] *Department of Biomedical Engineering, University of Arkansas, Fayetteville, AR, USA*

12.1 Introduction

The aortic valve is an elegant structure that opens and closes more than 3 billion times during the normal human lifetime. Its primary role is to control unidirectional flow between the left ventricle (LV) and the aortic outflow during systole (contraction of the ventricle) and to prevent retrograde flow during diastole (filling of the ventricle). Comprising three semilunar-shaped cusps with a multilayered tissue architecture of matrix interspersed with interstitial cells and covered with endothelial cells, the valve was originally thought to be a mere passive structure that opened and closed in response to the pressures exerted by the flow of blood. It is now well accepted that proper, healthy function of the valve, as well as progression toward disease, is controlled by the intricate interaction between the valve tissue, cells, and the surrounding mechanical environment. Indeed, understanding this interaction, termed "mechanobiology," is critical for a complete description of healthy heart function and disease progression and for the development of effective treatment strategies.

Experimental and computational methods have aided immensely in the characterization of the mechanical environment of the aortic valve (Figure 12.1), which includes transvalvular pressure gradients, pulsatile and oscillatory shear stresses, and cyclic tensile and bending stresses (Yap et al. 2010, 2012; Weiler et al. 2011). It was noted very early on that the locations of aortic valve pathologies correlated with spatiotemporally distinct mechanical environments. Disease of the aortic valve is primarily classified as either stenotic disease, where forward flow is impeded due to incomplete opening, or regurgitation, where incomplete closure during diastole results in backward flow. Calcific aortic stenosis, characterized by the presence of subendothelial calcific nodules or lesions and fibrotic matrix organization, has become the most common indication of valvular disease in the United States (Rajamannan 2009). Furthermore, it was reported that these calcific lesions occurred preferentially on the outflow side of the valve, which experiences low-magnitude oscillatory shear stresses (Freeman and Otto 2005; Otto 2008). These lesions were also often observed to aggregate in the regions of highest tensile and

Mechanobiology: Exploitation for Medical Benefit, First Edition. Edited by Simon C. F. Rawlinson.

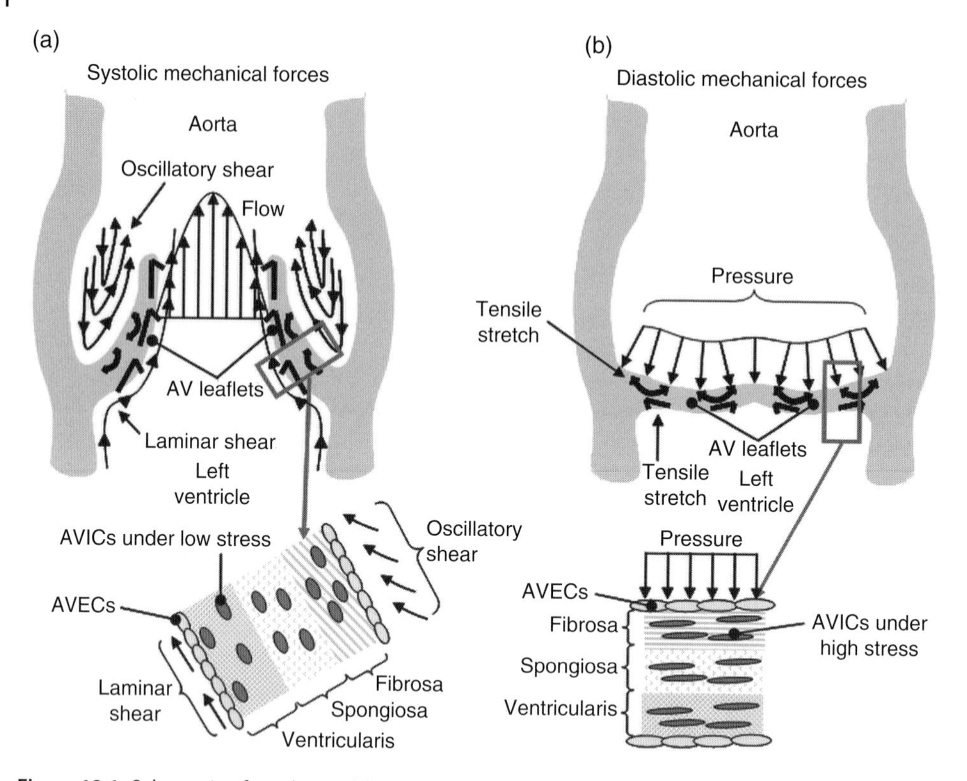

Figure 12.1 Schematic of mechanical forces experienced by the aortic valve during (a) systole and (b) diastole. Insets show the effect of these forces on the valve endothelial and interstitial cells. *Source:* Figure from Balachandran et al. (2011b). Reprinted with permission from Hindawi.

bending stress (Thubrikar et al. 1986; Dal-Bianco et al. 2009). Additionally, bicuspid valves, which have significantly different mechanics than a normal tricuspid valve, are prone to accelerated calcific disease progression (Otto 2002). The current treatment standard for aortic valve disease is surgical replacement, as there are no effective treatments for its prevention or regression (Rajamannan et al. 2011).

All of this evidence clearly points to a pressing need to understand valve mechanobiology as part of an overall strategy for developing an effective treatment regime for aortic valve disease. In addition, the existence of organ, tissue, and cell-level architecture within the aortic valve dictates a multiscale approach in the study of valve mechanobiology. This chapter will address the multiscale mechanobiology of the aortic valve, focusing on the existing state of knowledge and future areas for research, and how these can potentially be exploited in the design and development of treatments for valvulopathy.

12.2 Mechanobiology at the Organ Level

Aortic valve cusps have a trilayered morphology comprising a collagen-rich fibrosa, a middle spongiosa layer that is predominantly glycosaminoglycans, and an elastic ventricularis (Figure 12.2). Valve endothelial cells (VECs) line the surface of the cusps,

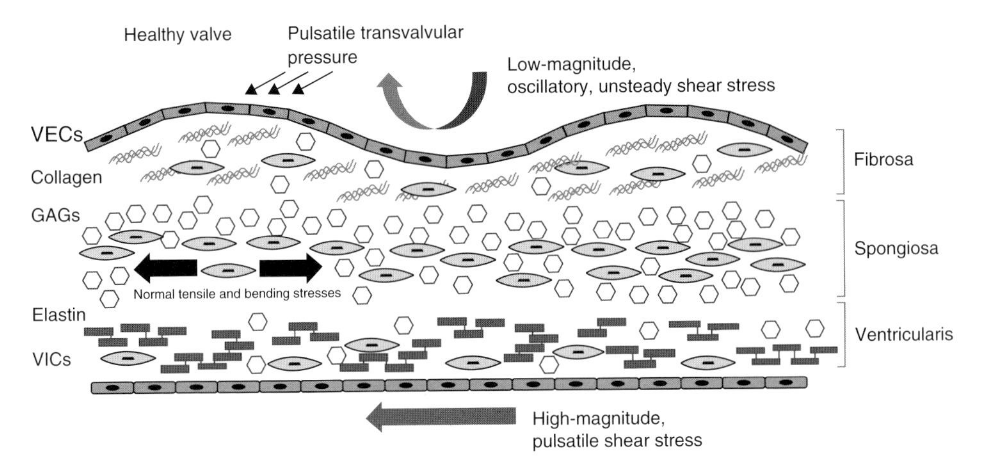

Figure 12.2 Schematic depicting the complex cross-sectional structure of the aortic valve cusp and the various mechanical forces acting at this scale. The fibrosa layer serves for the load-bearing function and is made up of type I and III collagen. The spongiosa layer located in between the fibrosa and ventricularis provides lubrication to the supporting layers during cardiac cycle and is made up of glycosaminoglycans. The layer closest to the LV, the ventricularis, serves to reduce the radial strain of the valve during a cardiac cycle and is made up of elastin. *Source:* Figure adapted from Lam and Balachandran (2015).

are nonthrombogenic, and regulate the inflammatory response in the valve. They are also the primary mechanosensors of fluid shear stress (Butcher et al. 2008). Valve interstitial cells (VICs) are interspersed throughout the cusp within the three layers. They are a heterogeneous population of cells, and can exhibit quiescent, smooth muscle-like, activated, or osteoblast-like phenotypes (Chester and Taylor 2007; Liu et al. 2007). In this section, whole-organ and tissue mechanobiological studies encompassing all these components are discussed.

To study the relationship between valve cells, native extracellular matrix (ECM), and the structural arrangement of both, many researchers use organ culture systems, which allow for the culture of whole heart valves or of isolated heart valve leaflets *in vitro*. Organ cultures offer a variety of advantages over other systems, including the less complex two-dimensional (2D) cell culture. Through organ culture, it is possible to study valve cells in their native environment, which is especially important given the heterogeneous nature of heart valves and their ECM. Additionally, since the hearts used for such studies are generally purchased from abattoirs serving the commercial meat industry, organ culture studies cost a fraction of the price of animal studies.

Early organ-cultured valve studies involved growing isolated mitral valve leaflets freely floating in a tissue culture dish for less than a week (Lester et al. 1992). It was later demonstrated that with valve leaflets cultured in this manner, without any mechanical stimulation, the cells within remained viable for several weeks (Allison et al. 2004) but tended to migrate from the interior toward the outer layers (Barzilla et al. 2010). In addition, over several weeks in this freely floating culture model, substantial ECM remodeling within the cultured valve tissues was observed (Barzilla et al. 2010). Though this simple organ culture system is still in use for many short-term studies of aortic valve calcification (Rodriguez et al. 2011), many investigators have

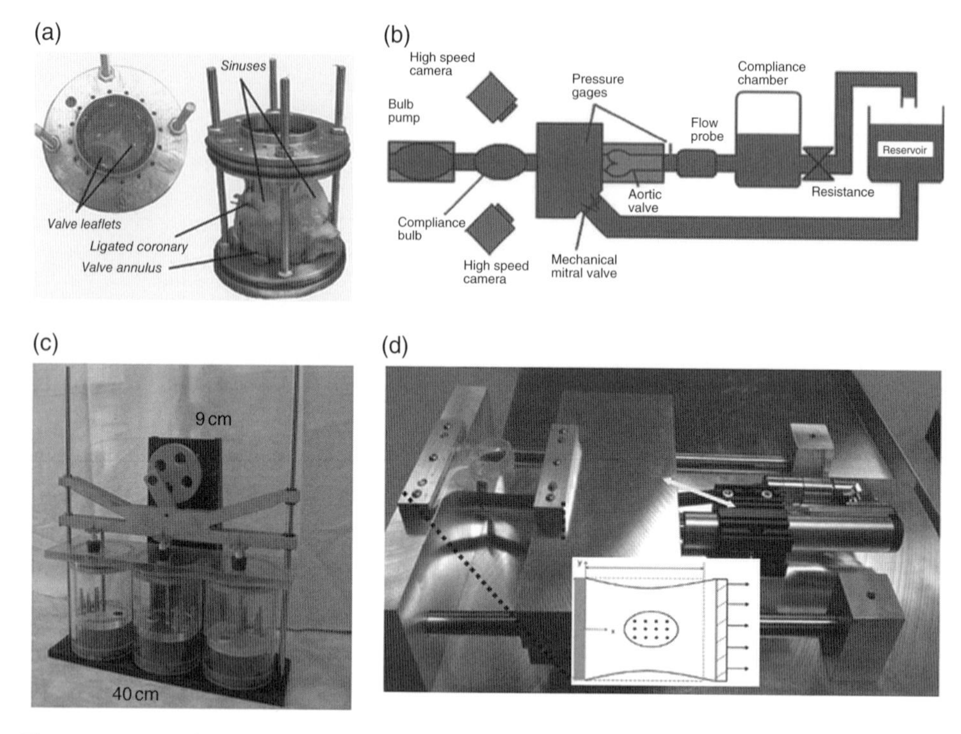

Figure 12.3 Examples of bioreactor systems currently in use. (a) Aortic valve anchoring module. *Source:* Weiler et al. (2011). Reproduced with permission of Elsevier. (b) Flow-loop schematic of a pulsatile left heart simulator. *Source:* Weiler et al. (2011). Reproduced with permission of Elsevier. (c) Synchronous multivalve aortic valve culture system. *Source:* Durst and Grande-Allen (2010). Reproduced with permission of Springer. (d) Uniaxial cyclic biostretcher, used to stretch cell monolayers grown on an elastic silicone rubber substrate.

worked to develop organ culture bioreactor systems (Figure 12.3) that provide dynamic mechanical stimulation to the leaflet tissues.

12.2.1 *In Vitro* Studies at the Whole Organ Level

A few organ culture systems have been designed to culture whole, intact aortic valves, meaning that the three valve leaflets are still located within the aortic root (Figure 12.3a). These bioreactors are generally complex in design and function, and often have the ability to control pressure and/or flow rates using a computer interface that regulates pumping and resistance components through feedback mechanisms (Figure 12.3b). These systems can therefore culture the valve within an environment that mimics several aspects of cardiac and cardiovascular physiology, and allow the valve to undergo opening and closing motions. Some of these systems resemble the circulatory system (Figure 12.3b), moving culture medium through the valve in a loop that tightly mimics physiologic pressure and flow waveforms (Hildebrand et al. 2004).

An alternative bioreactor system design placed greater importance upon replicating the opening and closing motions of the intact aortic valve, by moving a piston structure containing a mount for the intact valve/root back and forth through a cylindrical chamber – a simple system that could be replicated to culture multiple valves in parallel

(Figure 12.3c) (Durst and Grande-Allen 2010). Though this scalable system was operated using very simple drive and control mechanisms, and was able to culture multiple aortic valves/roots in parallel for up to 3 weeks under sterile conditions, it lacked robust flow-control and feedback systems and subjected all of the parallel-cultured valves to the same stroke volume and flow rate.

While few experimental studies using these organ culture bioreactor systems have been published, they have made important contributions to the field. A flow loop-based bioreactor that simulated the LV physiological function was used to demonstrate that aortic valves cultured under dynamic, mechanically stimulated conditions for 48 hours maintained proportions of collagen, elastin, and glycosaminoglycan that were not statistically different from those of fresh control valves (Konduri et al. 2005), while valves that were statically cultured without mechanical stimulation showed lower-than-normal concentrations of glycosaminoglycans and elastin. The statically cultured valves also contained a significantly greater concentration of apoptotic cells than did the dynamically cultured valves in the bioreactors, which is consistent with the dogma that cells lacking stimulation will undergo apoptosis.

Unique and promising variations on this approach have moved beyond the intact aortic root and valve to culturing and applying flow to nascent valve cushions isolated from the developing hearts of chick embryos (Goodwin et al. 2005; Biechler et al. 2014), as well as the development of bioreactors that can monitor the opening and closing motion of the valve endoscopically (Konig et al. 2012). Though this chapter does not focus on tissue engineering of heart valves, many investigators studying that subject are driving advances in bioreactor development that could also be applied toward the organ culture of intact aortic valves.

12.2.2 *In Vitro* Studies at the Tissue Level

Studies at the tissue level adopt an approach wherein tissue sections are excised from the valve cusps and mechanically stimulated using bioreactor systems. These studies have the advantage of preserving the various components of the valve in their native three-dimensional (3D) structural arrangements, while applying various controlled modes of mechanical stimulation. Elevated cyclic stretch has been reported to have a role in regulating aortic valve remodeling (Balachandran et al. 2009) and calcification (Balachandran et al. 2010) via a bone morphogenetic protein (BMP)-mediated pathway. Additionally, low and oscillatory shear stresses potentiate valve inflammation and transforming growth factor beta (TGF-β) signaling in an endothelium-dependent manner (Sucosky et al. 2009). In yet another study, elevated cyclic stretch was implicated in activating the TGF-β pathway (Merryman et al. 2007). While these studies made significant insights into the importance of the TGF-β pathway as a common theme, *in vitro* and *ex vivo* tissue studies are limited in the maintenance of tissue viability and integrity for long-term culture. Chronic studies on the mechanobiology of the aortic valve have to be undertaken using *in vivo* methods.

12.2.3 *In Vivo* Studies of Aortic Valve Mechanobiology

Though the use of bioreactors has opened the door to very controllable, mechanistic studies of aortic valve mechanobiology, there is still an important role for *in vivo* investigations.

There can be substantial expense associated with animal studies, but the healthy function of the heart valves is inextricably linked with the function of the heart, and indeed that of the entire cardiovascular system. Because the renal system regulates fluid balance (volume and pressure) within the body, including in the blood, it also shares a relationship with the health of the heart valves. To understand the *in vivo* development of aortic valve disease, and in turn the physiological impact of this process, numerous animal models have been developed. These models require careful analysis in order to understand the organism-level relationships with heart valves, but they provide highly relevant context for investigations of aortic valve mechanobiology.

The majority of *in vivo* valve mechanobiology studies involve rodents (usually mice, rats, or rabbits) that have been manipulated surgically, genetically, metabolically, or pharmacologically to produce a mechanically altered or otherwise dysfunctional valve. For example, hypertension, the elevation of systolic blood pressure above 140 mmHg, can be produced by interfering with renal function in several ways (Johns et al. 1996), such as controlling the regulation of the renin–angiotensin system (Fujisaka et al. 2013). Aortic constriction, in which a band or suture is tightened around the ascending aorta, will increase the pressure load on the aortic valve, as well as the heart overall, leading to heart failure (Wang et al. 2014). One of the most common methods of generating aortic valve disease in mice is to alter their genetic susceptibility to this condition by "knocking out" a gene for a critical aspect of lipid metabolism and then feeding the animals a so-called "Western diet" rich in cholesterol. Apolipoproteins, particularly E and B, and the low-density lipoprotein (LDL) receptor are among the most common genes to knock out in efforts to develop thickened, diseased aortic valves in mice (Fujisaka et al. 2013; Wang et al. 2014).

These models can be manipulated further to produce various combinations of genetic, pharmacological, and mechanical stimuli and thus tease out their relative contributions to disease development, which must be understood in order to develop appropriately targeted medical treatments for valve disease. Such models have contributed significantly to our understanding of calcific aortic stenosis. For example, one investigation of apolipoprotein E-knockout mice showed that activation of the renin–angiotensin system promoted aortic valve calcification regardless of whether or not the mouse had hypertension (Fujisaka et al. 2013). Another model was used to show that the noncanonical TGF-β1 signaling pathway was activated within the valve cells as a result of increased shear stress on the valve leaflets (Wang et al. 2014).

Though rodents are the most common animal model for these *in vivo* investigations, the miniscule size of their heart valves has presented a challenge for analysis. For this reason, one research group has examined the suitability of pigs fed a hypercholesterolemic diet (Sider et al. 2014). In this model, after 5 months the aortic valve leaflets demonstrated an increased presence of proteoglycans within the fibrosa layer, consistent with the current model for the early stages of calcific aortic valve disease.

Investigations of aortic valve mechanobiology at the tissue level can be complicated to perform, but they offer the unique opportunity to study the valve cells in the context of their native surrounding ECM, and often under conditions mimicking their native mechanics. These contextual factors add to the levels of analysis required to understand these models, but are an essential complement to cell-based studies in providing insight into the dynamic relationships between valve biology and organism-level health.

12.3 Mechanobiology at the Cellular Level

In intact heart valve tissues, VECs and VICs interact with each other and the surrounding tissue matrix to respond to any perturbations in their mechanical environment. While elucidating the overall mechanobiology between the cells and the matrix is important, it is also crucial to isolate each cellular component in order to understand its individual role in the healthy and pathological response to mechanical stimuli. The main questions for such a "divide and conquer" approach are: (i) What is the role of the mechanical environment in inducing VECs to undergo a phenomenon known as endothelial–mesenchymal transformation (EndoMT), wherein they transform into an activated VIC-like cell that is more prone to activation and osteoblastic differentiation and disease progression? (Kaden et al. 2005), (ii) What causes VICs to transition between quiescent, activated and osteoblast-like phenotypes, eventually leading to disease progression?, and (iii) How do VECs and VICs interact to maintain valve health?

12.3.1 2D *In Vitro* Models

Early work on the mechanobiology of VECs and VICs was undertaken using extensions of standard static 2D monolayer culture models. Commercially available or custom-designed bioreactor systems were used to impose the various modes of mechanical stimulation (Figure 12.1) on the cells in a controlled environment (Figure 12.3d). These models are discussed in this section in the context of the important questions just raised.

12.3.1.1 Mechanosensitivity of EndoMT

One of the more active areas of research today is in understanding the mechanobiology of VECs in light of their ability to undergo EndoMT. During this process, VECs delaminate from the endothelium and transform into an activated, motile VIC-like cell, in a process akin to that observed during fetal heart valve development (Armstrong and Bischoff 2004). These transformed VECs are more prone to activation, osteoblastic differentiation, and disease progression (Kaden et al. 2005). Interest in EndoMT also stems from the hypothesis that this process is an early indicator of disease initiation in the aortic valve. Low oscillatory shear stresses (2 dyn/cm^2) have been noted to upregulate markers of EndoMT (smooth-muscle alpha-actin (α-SMA), snail, TGF-β) and inflammation (intracellular adhesion molecule (ICAM), nuclear factor κB (NF-κB)) in a microfluidic culture model (Mahler et al. 2014). Furthermore, EndoMT was reported to be modulated via the TGF-β or the Wnt/β-catenin pathways, depending on the magnitude of applied cyclic stretch (Balachandran et al. 2011a); the upregulation of these pathways happened acutely 24–48 hours after mechanical stimulation. Interestingly, these pathways are also key regulators of subsequent osteogenic differentiation and calcification within the valve, suggesting the importance of EndoMT as an early regulator of aortic valve disease, and as a potential future therapeutic target for the treatment of aortic valve disease progression.

12.3.1.2 VECs in Remodeling and Chronic Inflammation

Apart from their role in EndoMT, VECs have also been identified as one of the key regulators in aortic valve disease progression via the recruitment of immune cells, dysregulation of protective nitric oxide signaling, and expression of procalcific proteins

(Hjortnaes et al. 2015). Several studies have reported on the role of low and oscillatory fluid shear stresses in the potentiation of VEC inflammatory responses, as well as in the inhibition of matrix remodeling markers. These studies were performed in laminar-flow parallel-plate chambers (Butcher et al. 2004; Platt et al. 2006a,b) and in cone-and-plate viscometer bioreactors (Sorescu et al. 2004; Holliday et al. 2011). One exciting recent report was the discovery of shear- and side-specific microribonucleic acids (microRNAs) as unique regulators of the inflammatory and calcification process (Holliday et al. 2011). MicroRNAs, which regulate gene expression by directing their target mRNAs for degradation or translational inhibition, are thought to have immense potential in molecular-based therapies for valve disease, and represent a new area of research in valve mechanobiology.

12.3.1.3 Activation and Osteogenic Differentiation of VICs

VICs do not directly experience fluid shear stress, but undergo elongation and bending as a result of cyclic stretching and pulsatile pressures (Sacks et al. 2009). These modes of mechanical stimulation are primarily experienced by cells during diastole, as the valve tissue extends to form a coaptive seal, and typically increase with incidence of hypertension or developing valve sclerosis (Yap et al. 2010). The majority of cyclic stretch investigations have been conducted with VICs seeded on equibiaxially stretched flexible membranes, producing observations similar to those seen in the organ-level studies. Under elevated cyclic stretch, VICs have been reported to become activated and to increase their collagen synthesis (Ku et al. 2006). Elevated stretch has also been observed to induce osteogenic differentiation of VICs (Ferdous et al. 2011). In more detailed mechanistic studies, it was reported that Notch1, serotonin, and TGF-β1 were key regulators of elevated stretch-mediated disease progression in the valve (Hutcheson et al. 2012; Chen et al. 2015).

12.3.2 The Push toward Co-Culture and 3D Models

While the cell-level studies discussed so far were all monoculture studies that yielded important insights into the mechanobiology of valve disease, the interaction between VECs and VICs is crucial in modulating the overall response. Recent studies demonstrated that VICs can suppress the EndoMT and activation of VECs even when the cells were cultured in an osteogenic medium (Hjortnaes et al. 2015). In another study, in which VICs were cultured in monoculture and in co-culture with VECs, the authors reported increased VIC activation in the monoculture but not in co-culture (Butcher and Nerem 2006). Taken as a whole, these results point to the need for benchtop models of valve mechanobiology, in order to more accurately mimic the structure, architecture, and cell types observed in the native aortic valve, such as the 3D and co-culture models addressed in this section.

12.3.2.1 3D Culture and Co-Culture

Valve cells grown in 3D within natural or synthetic hydrogel scaffolds can be mechanically stimulated in two ways: exogenously, through the application of external mechanical loading, and endogenously, through the variation in hydrogel stiffness. For the application of exogenous loading to homogeneous populations of VICs, one of the most common strategies has been to encapsulate cells within nascent collagen gels.

The VICs will bind to and reorganize the collagen, creating a neotissue that can be readily constrained and subjected to various loading conditions. One of the most advanced applications of this approach has been to cast the collagen gels within spring coils of varying shapes, so that the gels can be stretched under a range of biaxial loading conditions (Gould et al. 2012). Using this technique, researchers determined that stretching the valve cells under more anisotropic conditions increased their turnover (as demonstrated by increased proliferation and apoptosis) and led to activation toward a myofibroblast phenotype (as demonstrated by greater staining for α-SMA).

Collagen gels have also served as the basis for co-culture mechanobiology studies: VICs were encapsulated within a collagen gel scaffold, VECs were seeded atop the scaffold surface, and the endothelial surface was subjected to 20 dyn/cm^2 of steady shear stress (Butcher and Nerem 2006). Compared to VICs cultured alone under the same flow conditions, the VECs promoted a more quiescent, nonactivated, less proliferative state. A comparable effect of VECs on VICs was shown within a structured co-culture hydrogel model prepared from a synthetic polymer: biofunctionalized poly(ethylene glycol) diacrylate (PEGDA) (Puperi et al. 2015). Though this system lacked exogenous mechanical stimulation, it was structured in a similar manner, as the VICs were encapsulated within the hydrogel interior and the VECs were seeded on the surface (Figure 12.4).

12.3.2.2 The Role of Substrate Stiffness

Several studies have shown that valve cells respond to the stiffness of their environment in both 2D and 3D culture. VICs cultured on soft substrates are known to exhibit a phenotype that is more similar to their native phenotype than that shown when they are cultured on very stiff tissue-culture polystyrene (Wang et al. 2013). Research has shown that substrate stiffness modulates valve cell response to other stimuli, as well. For example, in calcifying media, VICs cultured on softer substrates formed calcific lesions with osteoblast-like gene expression, whereas VICs cultured on stiffer substrates took on a myofibroblastic phenotype and were more responsive to TGF-β treatment (Yip et al. 2009). VICs cultured on softer gels also demonstrated a greater response to stretch than those cultured on stiff gels (Throm Quinlan et al. 2011).

The effect of substrate stiffness has also been studied by culturing VICs on polyethylene glycol (PEG)-based hydrogels and then changing the substrate modulus by either stiffening or degrading the gel using ultraviolet (UV) light. Significantly more activated myofibroblast phenotype cells were observed among VICs cultured in 2D on stiff gels (32 kPa) than in those cultured on either soft gels (7 kPa) or stiff gels (32 kPa) that were weakened to 7 kPa after 5 days in culture (Wang et al. 2012). Conversely, another study showed that when VICs were cultured in 3D, those in very soft gels (0.24 kPa) expressed significantly more α-SMA and demonstrated a more spread morphology than those in stiffer gels (4 and 13 kPa) (Mabry et al. 2015). When the softest gels were later stiffened by further crosslinking, the VICs retained their spread morphology, but lost expression of α-SMA, suggesting that the cells reverted to a quiescent phenotype. The effect of substrate stiffness is not as well studied for VECs, but cells on softer substrates exhibited significantly higher expression of hemostatic genes, von Willebrand factor (VWF), and a disintegrin and metalloproteinase with thrombospondin motifs (ADAMTS)-13, while VECs on stiffer gels formed more confluent monolayers and were not as responsive to stimulation by histamine (Balaoing et al. 2015).

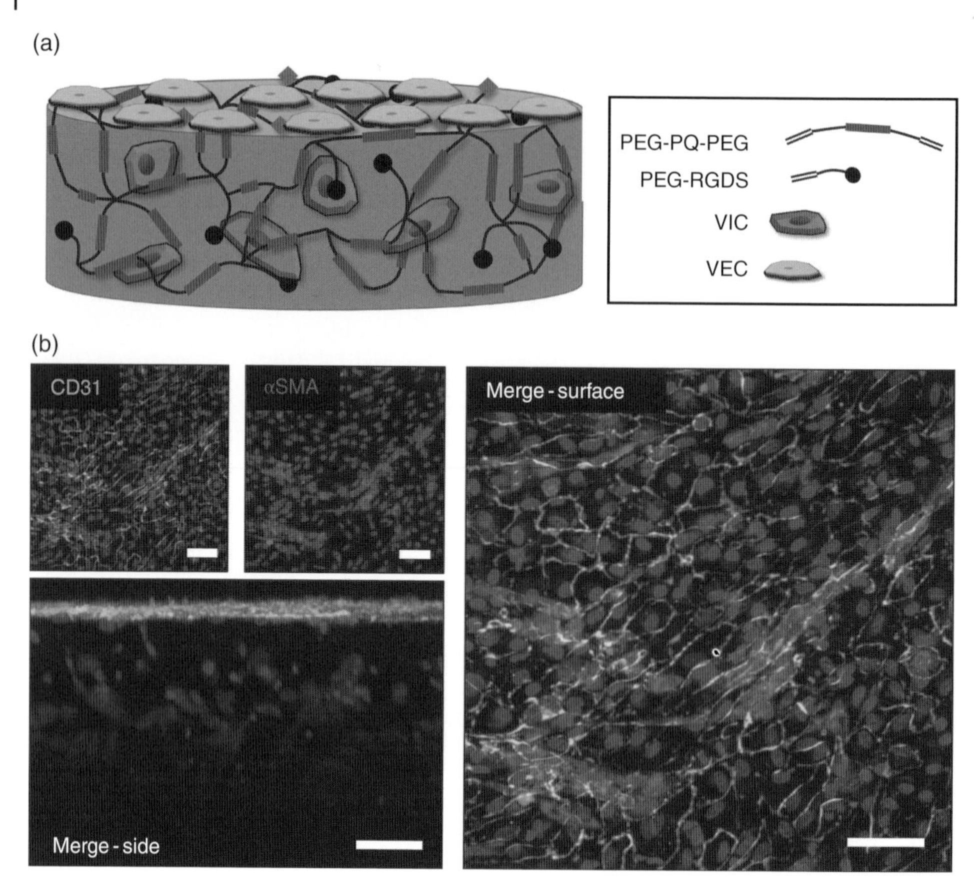

Figure 12.4 (a) Schematic depicting a valve cell co-culture model, with VECs seeded atop VIC-encapsulated hydrogel. (b) Co-culture scaffold demonstrating zonally organized cell populations after 7 days in culture. CD31-expressing VECs form a confluent monolayer on top of the gel, while encapsulated VICs express low levels of α-SMA (scale bars = 50 μm). *Source:* Puperi et al. (2015). Reproduced with permission of Elsevier. (*See insert for color representation of the figure.*)

In the native valve, the stiffness varies across its regions and throughout its thickness due to ECM composition changes. A more complex model of the substrate stiffness that native valve cells sense *in vivo* is needed if we are to understand how the stiffness plays a role in valve disease. In one study, VICs responded to a composite scaffold constructed from soft PEGDA with embedded stiff polycaprolactone (PCL) electrospun fibers by sensing the stiffness of the encapsulated fibers and orienting themselves along their direction (Tseng et al. 2014). The changing stiffness of the valve throughout its thickness was modeled by laminating hydrogels with different moduli. A softer PEGDA layer was laminated between stiff PEGDA layers to roughly approximate the layered nature of the valve ECM, with a softer spongiosa layer between the stiffer fibrosa and ventricularis layers (Tseng et al. 2013). Further studies are needed to demonstrate how the heterogeneous mechanical composition of the valve affects valve cells and contributes to valve disease.

All in all, the approaches used to investigate heart valve cell mechanobiology have grown more sophisticated and complex over the last decade, with a move toward novel synthetic material platforms with fine-tuning of the elastic modulus, as well as increasing emphasis on co-culturing VICs and VECs together. Though this diverse range of techniques is gradually mimicking the tissue structure and mechanical setting of the valve cells, there are still challenges in integrating these approaches. Ideally, it will be possible in the future to develop a system that regulates both the exogenous and the endogenous mechanical environment of the VECs and VICs. Such a system would be invaluable in dissecting details concerning mechanically driven cell signaling pathways.

12.4 Conclusion

The last 2 decades have witnessed tremendous growth in and excitement about the subject of mechanobiology of heart valves. This growth has largely been driven by the development of tissue-engineered heart valves, but there is also considerable interest in understanding the roles of mechanical factors in the initiation and progression of valve disease. Though there have been significant advances in developing experimental approaches to the investigation of valve mechanobiology, this subject is still very much in its infancy. There is ample room for innovation in overcoming current limitations, such as the inability to conduct organ culture studies that are truly long-term (i.e., on the order of months). Within the scope of the research discussed in this chapter, there are topics that are just beginning to be investigated, such as the mechanobiology of VECs and the use of altered mechanical environments for the aortic valve *in vivo*. With time, this research will illuminate the myriad ways in which the biology and function of heart valves can be mechanically regulated, and this knowledge will be used to design novel treatments and predict when a particular patient is predisposed to early onset of valve disease.

Acknowledgments

The authors appreciate the editorial assistance of Dr. Jennifer Connell.

References

Allison, D. D., J. A. Drazba, I. Vesely, K. N. Kader, and K. J. Grande-Allen. 2004. "Cell viability mapping within long-term heart valve organ cultures." *Journal of Heart Valve Disease* **13**(2): 290–6. PMID:15086269.

Armstrong, E. J. and J. Bischoff. 2004. "Heart valve development: endothelial cell signaling and differentiation." *Circulation Research* **95**(5): 459–70. doi:10.1161/01.RES.0000141146.95728.da.

Balachandran, K., P. Sucosky, H. Jo, and A. P. Yoganathan. 2009. "Elevated cyclic stretch alters matrix remodeling in aortic valve cusps: implications for degenerative aortic valve disease." *American Journal of Physiology. Heart and Circulatory Physiology* **296**(3): H756–64. doi:10.1152/ajpheart.00900.2008.

Balachandran, K., P. Sucosky, H. Jo, and A. P. Yoganathan. 2010. "Elevated cyclic stretch induces aortic valve calcification in a bone morphogenic protein-dependent manner." *American Journal of Pathology* **177**(1): 49–57. doi:10.2353/ajpath.2010.090631.

Balachandran, K., P. W. Alford, J. Wylie-Sears, J. A. Goss, A. Grosberg, J. Bischoff, et al. 2011a. "Cyclic strain induces dual-mode endothelial-mesenchymal transformation of the cardiac valve." *Proceedings of the National Academy of Sciences of the United States of America* **108**(50): 19 943–8. doi:10.1073/pnas.1106954108.

Balachandran, K., P. Sucosky, and A. P. Yoganathan. 2011b. "Hemodynamics and mechanobiology of aortic valve inflammation and calcification." *International Journal of Inflammation* **2011**: 263870. doi:10.4061/2011/263870.

Balaoing, L. R., A. D. Post, A. Y. Lin, H. Tseng, J. L. Moake, and K. J. Grande-Allen. 2015. "Laminin peptide-immobilized hydrogels modulate valve endothelial cell hemostatic regulation." *PLoS One* **10**(6): e0130749. doi:10.1371/journal.pone.0130749.

Barzilla, J. E., A. S. McKenney, A. E. Cowan, C. A. Durst, and K. J. Grande-Allen. 2010. "Design and validation of a novel splashing bioreactor system for use in mitral valve organ culture." *Annals of Biomedical Engineering* **38**(11): 3280–94. doi:10.1007/s10439-010-0129-9.

Biechler, S. V., L. Junor, A. N. Evans, J. F. Eberth, R. L. Price, J. D. Potts, et al. 2014. "The impact of flow-induced forces on the morphogenesis of the outflow tract." *Frontiers in Physiology* **5**: 225. doi:10.3389/fphys.2014.00225.

Butcher, J. T. and R. M. Nerem. 2006. "Valvular endothelial cells regulate the phenotype of interstitial cells in co-culture: effects of steady shear stress." *Tissue Engineering* **12**(4): 905–15. doi:10.1089/ten.2006.12.905.

Butcher, J. T., A. M. Penrod, A. J. Garcia, and R. M. Nerem. 2004. "Unique morphology and focal adhesion development of valvular endothelial cells in static and fluid flow environments." *Arteriosclerosis, Thrombosis, and Vascular Biology* **24**(8): 1429–34. doi:10.1161/01.ATV.0000130462.50769.5a.

Butcher, J. T., C. A. Simmons, and J. N. Warnock. 2008. "Mechanobiology of the aortic heart valve." *Journal of Heart Valve Disease* **17**(1): 62–73. PMID:18365571.

Chen, J., L. M. Ryzhova, M. K. Sewell-Loftin, C. B. Brown, S. S. Huppert, H. S. Baldwin, and W. D. Merryman. 2015. "Notch1 mutation leads to valvular calcification through enhanced myofibroblast mechanotransduction." *Arteriosclerosis, Thrombosis, and Vascular Biology* **35**(7): 1597–605. doi:10.1161/ATVBAHA.114.305095.

Chester, A. H. and P. M. Taylor. 2007. "Molecular and functional characteristics of heart-valve interstitial cells." *Philosophical Transactions of the Royal Society of London. Series B, Biological Sciences* **362**(1484): 1437–43. doi:10.1098/rstb.2007.2126.

Dal-Bianco, J. P., E. Aikawa, J. Bischoff, J. L. Guerrero, M. D. Handschumacher, S. Sullivan, et al. 2009. "Active adaptation of the tethered mitral valve: insights into a compensatory mechanism for functional mitral regurgitation." *Circulation* **120**(4): 334–42. doi:10.1161/CIRCULATIONAHA.108.846782.

Durst, C. A. and K. J. Grande-Allen. 2010. "Design and physical characterization of a synchronous multivalve aortic valve culture system." *Annals of Biomedical Engineering* **38**(2): 319–25. doi:10.1007/s10439-009-9846-3.

Ferdous, Z., H. Jo, and R. M. Nerem. 2011. "Differences in valvular and vascular cell responses to strain in osteogenic media." *Biomaterials* **32**(11): 2885–93. doi:10.1016/j.biomaterials.2011.01.030.

Freeman, R. V. and C. M. Otto. 2005. "Spectrum of calcific aortic valve disease: pathogenesis, disease progression, and treatment strategies." *Circulation* **111**(24): 3316–26. doi:10.1161/CIRCULATIONAHA.104.486738.

Fujisaka, T., M. Hoshiga, J. Hotchi, Y. Takeda, D. Jin, S. Takai, et al. 2013. "Angiotensin II promotes aortic valve thickening independent of elevated blood pressure in apolipoprotein-E deficient mice." *Atherosclerosis* **226**(1): 82–7. doi:10.1016/j.atherosclerosis.2012.10.055.

Goodwin, R. L., T. Nesbitt, R. L. Price, J. C. Wells, M. J. Yost, and J. D. Potts. 2005. "Three-dimensional model system of valvulogenesis." *Developmental Dynamics* **233**(1): 122–9. doi:10.1002/dvdy.20326.

Gould, R. A., K. Chin, T. P. Santisakultarm, A. Dropkin, J. M. Richards, C. B. Schaffer, and J. T. Butcher. 2012. "Cyclic strain anisotropy regulates valvular interstitial cell phenotype and tissue remodeling in three-dimensional culture." *Acta Biomaterialia* **8**(5): 1710–19. doi:10.1016/j.actbio.2012.01.006.

Hildebrand, D. K., Z. J. Wu, J. E. Mayer Jr., and M. S. Sacks. 2004. "Design and hydrodynamic evaluation of a novel pulsatile bioreactor for biologically active heart valves." *Annals of Biomedical Engineering* **32**(8): 1039–49. PMID:15446500.

Hjortnaes, J., K. Shapero, C. Goettsch, J. D. Hutcheson, J. Keegan, J. Kluin, et al. 2015. "Valvular interstitial cells suppress calcification of valvular endothelial cells." *Atherosclerosis* **242**(1): 251–60. doi:10.1016/j.atherosclerosis.2015.07.008.

Holliday, C. J., R. F. Ankeny, H. Jo, and R. M. Nerem. 2011. "Discovery of shear- and side-specific mRNAs and miRNAs in human aortic valvular endothelial cells." *American Journal of Physiology. Heart and Circulatory Physiology* **301**(3): H856–67. doi:10.1152/ajpheart.00117.2011.

Hutcheson, J. D., R. Venkataraman, F. J. Baudenbacher, and W. D. Merryman. 2012. "Intracellular Ca(2+) accumulation is strain-dependent and correlates with apoptosis in aortic valve fibroblasts." *Journal of Biomechanics* **45**(5): 888–94. doi:10.1016/j.jbiomech.2011.11.031.

Johns, C., I. Gavras, D. E. Handy, A. Salomao, and H. Gavras. 1996. "Models of experimental hypertension in mice." *Hypertension* **28**(6): 1064–9. PMID:8952597.

Kaden, J. J., C. E. Dempfle, R. Grobholz, C. S. Fischer, D. C. Vocke, R. Kilic, et al. 2005. "Inflammatory regulation of extracellular matrix remodeling in calcific aortic valve stenosis." *Cardiovascular Pathology* **14**(2): 80–7. doi:10.1016/j.carpath.2005.01.002.

Konduri, S., Y. Xing, J. N. Warnock, Z. He, and A. P. Yoganathan. 2005. "Normal physiological conditions maintain the biological characteristics of porcine aortic heart valves: an ex vivo organ culture study." *Annals of Biomedical Engineering* **33**(9): 1158–66. doi:10.1007/s10439-005-5506-4.

Konig, F., T. Hollweck, S. Pfeifer, B. Reichart, E. Wintermantel, C. Hagl, and B. Akra. 2012. "A pulsatile bioreactor for conditioning of tissue-engineered cardiovascular constructs under endoscopic visualization." *Journal of Functional Biomaterials* **3**(3): 480–96. doi:10.3390/jfb3030480.

Ku, C. H., P. H. Johnson, P. Batten, P. Sarathchandra, R. C. Chambers, P. M. Taylor, et al. 2006. "Collagen synthesis by mesenchymal stem cells and aortic valve interstitial cells in response to mechanical stretch." *Cardiovascular Research* **71**(3): 548–56. doi:10.1016/j.cardiores.2006.03.022.

Lam, N. T. and K. Balachandran. 2015. "The mechanobiology of drug-induced cardiac valve disease." *Journal of Long-Term Effects of Medical Implants* **25**(1–2): 27–40. PMID:25955005.

Lester, W. M., A. A. Damji, M. Tanaka, and I. Gedeon. 1992. "Bovine mitral valve organ culture: role of interstitial cells in repair of valvular injury." *Journal of Molecular and Cellular Cardiology* **24**(1): 43–53. doi:0022-2828(92)91158-2.

Liu, A. C., V. R. Joag, and A. I. Gotlieb. 2007. "The emerging role of valve interstitial cell phenotypes in regulating heart valve pathobiology." *American Journal of Pathology* **171**(5): 1407–18. doi:10.2353/ajpath.2007.070251.

Mabry, K. M., R. L. Lawrence, and K. S. Anseth. 2015. "Dynamic stiffening of poly(ethylene glycol)-based hydrogels to direct valvular interstitial cell phenotype in a three-dimensional environment." *Biomaterials* **49**: 47–56. doi:10.1016/j.biomaterials.2015.01.047.

Mahler, G. J., C. M. Frendl, Q. Cao, and J. T. Butcher. 2014. "Effects of shear stress pattern and magnitude on mesenchymal transformation and invasion of aortic valve endothelial cells." *Biotechnology and Bioengineering* **111**(11): 2326–37. doi:10.1002/bit.25291.

Merryman, W. D., H. D. Lukoff, R. A. Long, G. C. Engelmayr Jr., R. A. Hopkins, and M. S. Sacks. 2007. "Synergistic effects of cyclic tension and transforming growth factor-beta1 on the aortic valve myofibroblast." *Cardiovascular Pathology* **16**(5): 268–76. doi:10.1016/j.carpath.2007.03.006.

Otto, C. M. 2002. "Calcification of bicuspid aortic valves." *Heart* **88**(4): 321–2. PMCID:PMC1767390.

Otto, C. M. 2008. "Calcific aortic stenosis – time to look more closely at the valve." *New England Journal of Medicine* **359**(13): 1395–8. doi:10.1056/NEJMe0807001.

Platt, M. O., R. F. Ankeny, and H. Jo. 2006a. "Laminar shear stress inhibits cathepsin L activity in endothelial cells." *Arteriosclerosis, Thrombosis, and Vascular Biology* **26**(8): 1784–90. doi:10.1161/01.ATV.0000227470.72109.2b.

Platt, M. O., R. F. Ankeny, G. P. Shi, D. Weiss, W. R. Taylor, J. D. Vega, and H. Jo. 2006b. "Expression of cathepsin K is regulated by shear stress in cultured endothelial cells and is increased in endothelium in human atherosclerosis." *American Journal of Physiology. Heart and Circulatory Physiology* **292**(3): H1479–86. doi:10.1152/ajpheart.00954.2006.

Puperi, D. S., L. R. Balaoing, R. W. O'Connell, J. L. West, and K. J. Grande-Allen. 2015. "3-dimensional spatially organized PEG-based hydrogels for an aortic valve co-culture model." *Biomaterials* **67**: 354–64. doi:10.1016/j.biomaterials.2015.07.039.

Rajamannan, N. M. 2009. "Calcific aortic stenosis: lessons learned from experimental and clinical studies." *Arteriosclerosis, Thrombosis, and Vascular Biology* **29**(2): 162–8. doi:10.1161/ATVBAHA.107.156752.

Rajamannan, N. M., F. J. Evans, E. Aikawa, K. J. Grande-Allen, L. L. Demer, D. D. Heistad, et al. 2011. "Calcific aortic valve disease: not simply a degenerative process: A review and agenda for research from the National Heart and Lung and Blood Institute Aortic Stenosis Working Group. Executive summary: calcific aortic valve disease-2011 update." *Circulation* **124**(16): 1783–91. doi:10.1161/CIRCULATIONAHA.110.006767.

Rodriguez, K. J., L. M. Piechura, and K. S. Masters. 2011. "Regulation of valvular interstitial cell phenotype and function by hyaluronic acid in 2-D and 3-D culture environments." *Matrix Biology* **30**(1): 70–82. doi:10.1016/j.matbio.2010.09.001.

Sacks, M. S., W. David Merryman, and D. E. Schmidt. 2009. "On the biomechanics of heart valve function." *Journal of Biomechanics* **42**(12): 1804–24. doi:10.1016/j.jbiomech.2009.05.015.

Sider, K. L., C. Zhu, A. V. Kwong, Z. Mirzaei, C. F. de Lange, and C. A. Simmons. 2014. "Evaluation of a porcine model of early aortic valve sclerosis." *Cardiovascular Pathology* **23**(5): 289–97. doi:10.1016/j.carpath.2014.05.004.

Sorescu, G. P., H. Song, S. L. Tressel, J. Hwang, S. Dikalov, D. A. Smith, et al. 2004. "Bone morphogenic protein 4 produced in endothelial cells by oscillatory shear stress induces monocyte adhesion by stimulating reactive oxygen species production from a nox1-based NADPH oxidase." *Circulation Research* **95**(8): 773–9. doi:10.1161/01.RES.0000145728.22878.45.

Sucosky, P., K. Balachandran, A. Elhammali, H. Jo, and A. P. Yoganathan. 2009. "Altered shear stress stimulates upregulation of endothelial VCAM-1 and ICAM-1 in a BMP-4- and TGF-beta1-dependent pathway." *Arteriosclerosis, Thrombosis, and Vascular Biology* **29**(2): 254–60. doi:10.1161/ATVBAHA.108.176347.

Throm Quinlan, A. M., L. N. Sierad, A. K. Capulli, L. E. Firstenberg, and K. L. Billiar. 2011. "Combining dynamic stretch and tunable stiffness to probe cell mechanobiology in vitro." *PLoS One* **6**(8): e23272. doi:10.1371/journal.pone.0023272.

Thubrikar, M. J., J. Aouad, and S. P. Nolan. 1986. "Patterns of calcific deposits in operatively excised stenotic or purely regurgitant aortic valves and their relation to mechanical stress." *American Journal of Cardiology* **58**(3): 304–8.

Tseng, H., M. L. Cuchiara, C. A. Durst, M. P. Cuchiara, C. J. Lin, J. L. West, and K. J. Grande-Allen. 2013. "Fabrication and mechanical evaluation of anatomically-inspired quasilaminate hydrogel structures with layer-specific formulations." *Annals of Biomedical Engineering* **41**(2): 398–407. doi:10.1007/s10439-012-0666-5.

Tseng, H., D. S. Puperi, E. J. Kim, S. Ayoub, J. V. Shah, M. L. Cuchiara, et al. 2014. "Anisotropic poly(ethylene glycol)/polycaprolactone hydrogel-fiber composites for heart valve tissue engineering." *Tissue Engineering. Part A* **20**(19–20): 2634–45. doi:10.1089/ten.TEA.2013.0397.

Wang, H., S. M. Haeger, A. M. Kloxin, L. A. Leinwand, and K. S. Anseth. 2012. "Redirecting valvular myofibroblasts into dormant fibroblasts through light-mediated reduction in substrate modulus." *PLoS One* **7**(7): e39969. doi:10.1371/journal.pone.0039969.

Wang, H., M. W. Tibbitt, S. J. Langer, L. A. Leinwand, and K. S. Anseth. 2013. "Hydrogels preserve native phenotypes of valvular fibroblasts through an elasticity-regulated PI3K/AKT pathway." *Proceedings of the National Academy of Sciences of the United States of America* **110**(48): 19 336–41. doi:10.1073/pnas.1306369110.

Wang, W., S. Vootukuri, A. Meyer, J. Ahamed, and B. S. Coller. 2014. "Association between shear stress and platelet-derived transforming growth factor-beta1 release and activation in animal models of aortic valve stenosis." *Arteriosclerosis, Thrombosis, and Vascular Biology* **34**(9): 1924–32. doi:10.1161/ATVBAHA.114.303852.

Weiler, M., C. H. Yap, K. Balachandran, M. Padala, and A. P. Yoganathan. 2011. "Regional analysis of dynamic deformation characteristics of native aortic valve leaflets." *Journal of Biomechanics* **44**(8): 1459–65. doi:10.1016/j.jbiomech.2011.03.017.

Yap, C. H., H. S. Kim, K. Balachandran, M. Weiler, R. Haj-Ali, and A. P. Yoganathan. 2010. "Dynamic deformation characteristics of porcine aortic valve leaflet under normal and hypertensive conditions." *American Journal of Physiology. Heart and Circulatory Physiology* **298**(2): H395–405. doi:10.1152/ajpheart.00040.2009.

Yap, C. H., N. Saikrishnan, G. Tamilselvan, and A. P. Yoganathan. 2012. "Experimental measurement of dynamic fluid shear stress on the aortic surface of the aortic valve leaflet." *Biomechanics and Modeling in Mechanobiology* **11**(1–2): 231–44. doi:10.1007/s10237-011-0306-2.

Yip, C. Y., J. H. Chen, R. Zhao, and C. A. Simmons. 2009. "Calcification by valve interstitial cells is regulated by the stiffness of the extracellular matrix." *Arteriosclerosis, Thrombosis, and Vascular Biology* **29**(6): 936–42. doi:10.1161/ATVBAHA.108.182394.

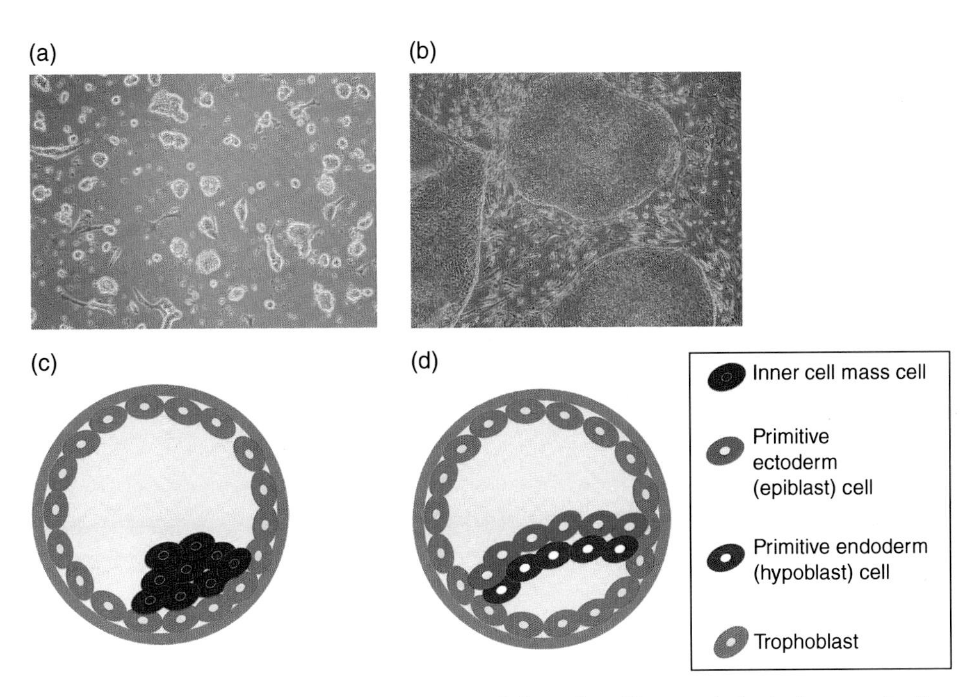

Figure 1.3 ESCs: differences in origin. (a) Murine ESCs form domelike, rounded colonies several cell layers thick, whereas (b) human ESCs form flattened, epithelial colonies. This may reflect differences in their origins. (c) Murine ESCs are thought to be analogous to cells of the inner cell mass of the embryo, which has no obvious polarity. (d) On the other hand, human ESCs (and murine EpiSCs) are likely to be more closely related to cells of the epiblast of the blastocyst. This structure is a polarized epithelium covering a basement membrane on the surface of the primitive endoderm (hypoblast).

Mechanobiology: Exploitation for Medical Benefit, First Edition. Edited by Simon C. F. Rawlinson.

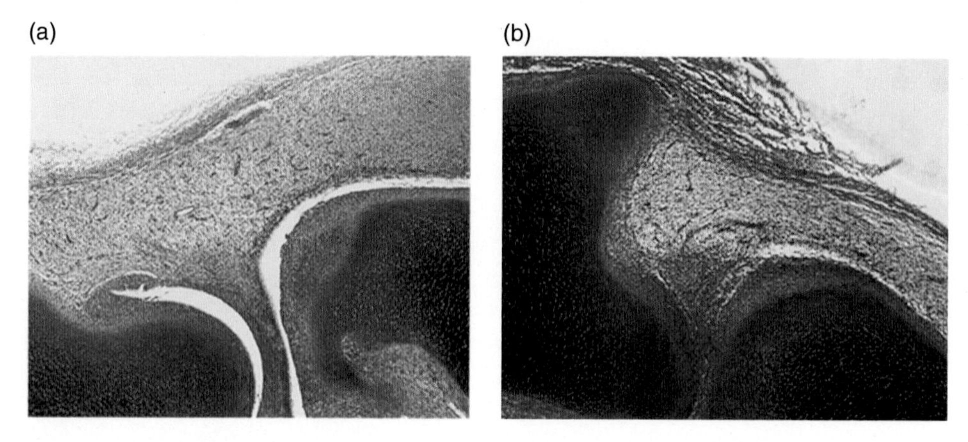

Figure 6.1 Joint cavitation is dependent upon embryo movement. Knee joints of embryonic chickens at 11 days into development. The distal femur and proximal tibia are visible in the sagittal plane. (a) A fully formed joint cavity in a normal embryo. (b) Failure of joint cavitation in response to pharmacological immobilization.

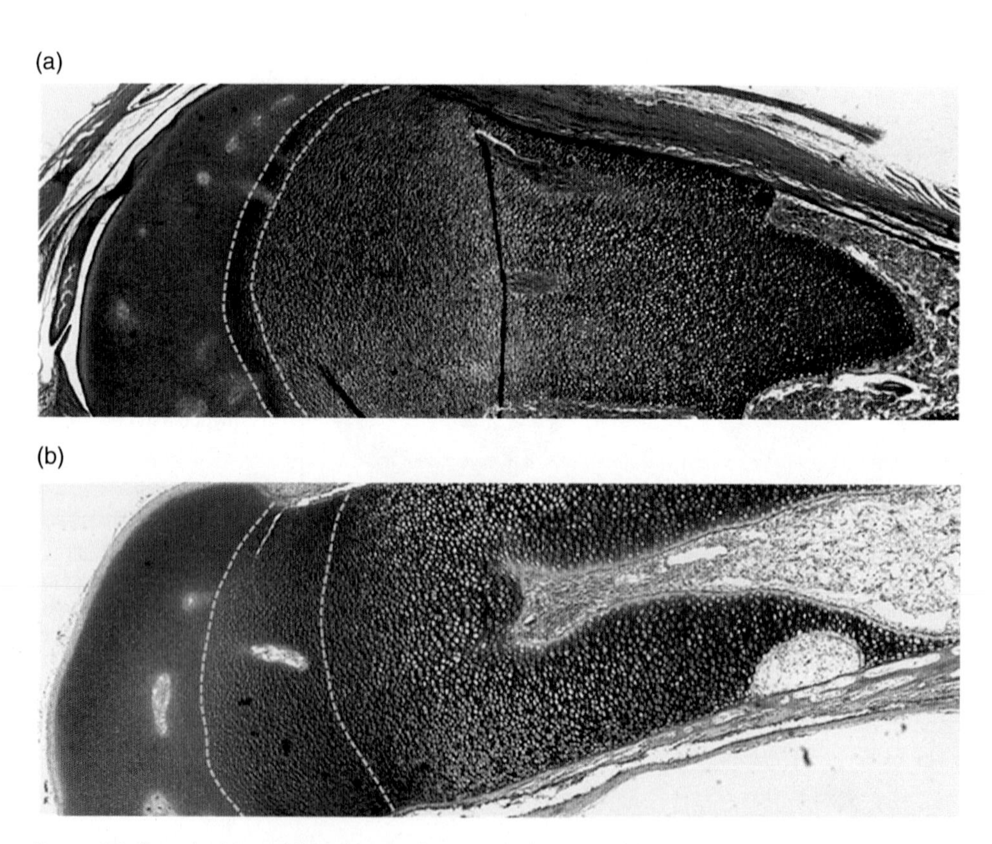

Figure 6.2 Growth plate cartilage from embryonic chickens at 18 days' incubation. The "proliferative zone" where cells express proliferative markers such as PCNA, which is expressed in the S-phase of the cell cycle, is indicated by dotted lines. (a) Growth plate of a normal embryo. (b) Growth plate of a pharmacologically immobilized embryo, demonstrating an expanded proliferative zone, resulting from the failure of cells to complete the cell cycle and progress through the growth plate.

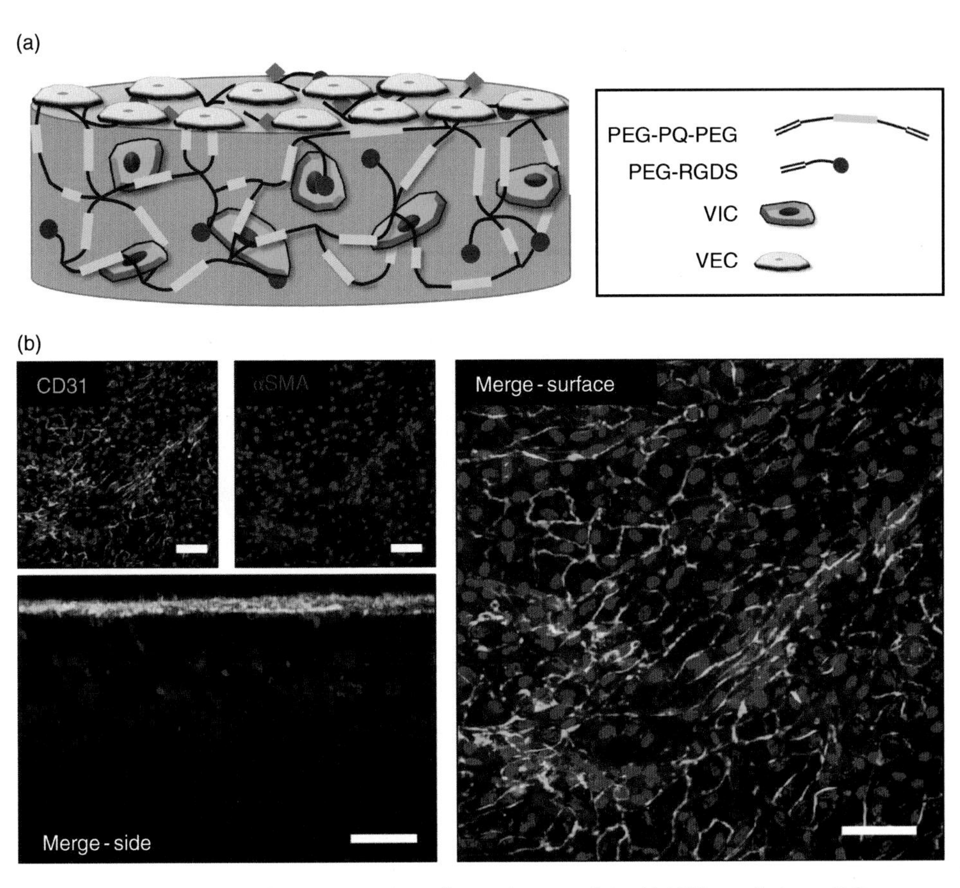

Figure 12.4 (a) Schematic depicting a valve cell co-culture model, with VECs seeded atop VIC-encapsulated hydrogel. (b) Co-culture scaffold demonstrating zonally organized cell populations after 7 days in culture. CD31-expressing VECs form a confluent monolayer on top of the gel, while encapsulated VICs express low levels of α-SMA (scale bars = 50 μm). *Source:* Puperi et al. (2015). Reproduced with permission of Elsevier.

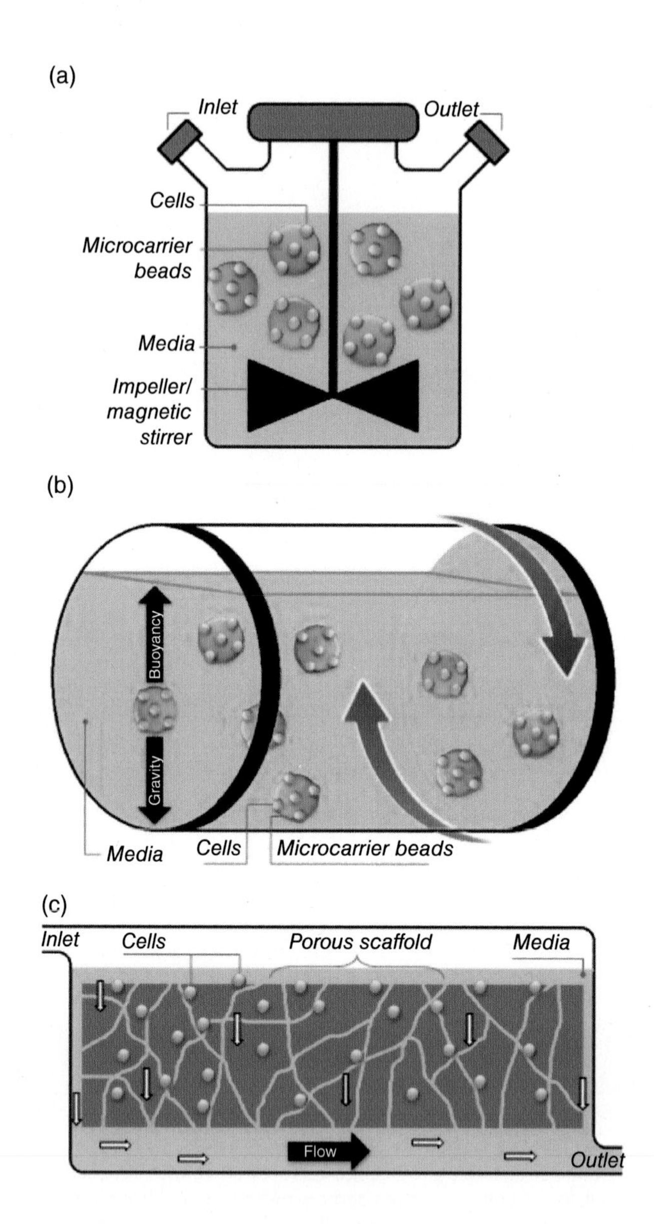

Figure 18.2 Bioreactors can deliver shear forces in many different ways. (a) Stirred spinner flask. (b) Rotating-wall bioreactor. These were initially developed for the industrial expansion of microorganisms, but adherent mammalian cells must first be seeded on to microcarriers or biomaterial scaffolds or cultured as tissue explants. (c) A simple method for providing shear is to perfuse the cell-seeded scaffold with fluid, which can be pressurized; the nonuniform effects on flow rate and the subsequent cell responses are a common theme of mathematical modeling. (d) Laminar microfluidic flow system, which utilizes compressed air to drive media flow across an adherent cell monolayer. *Source:* Adapted from Martinez et al. (2013). (e,f) Commercial version of a rotating-wall bioreactor, marketed by Synthecon. *Source:* Courtesy of Dr. Yvonne Reinwald. All of these types of bioreactor deliver shear forces in different ways, and each has unique challenges in delivering consistent, measurable flow rates.

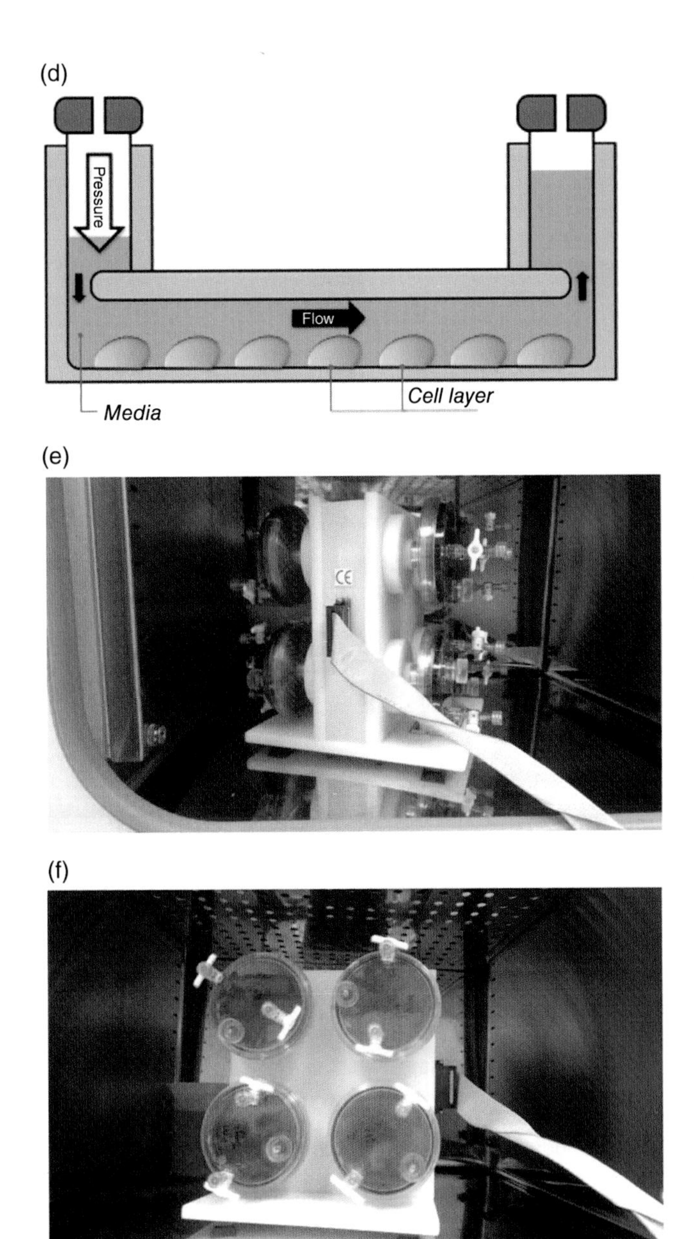

Figure 18.2 (Continued)

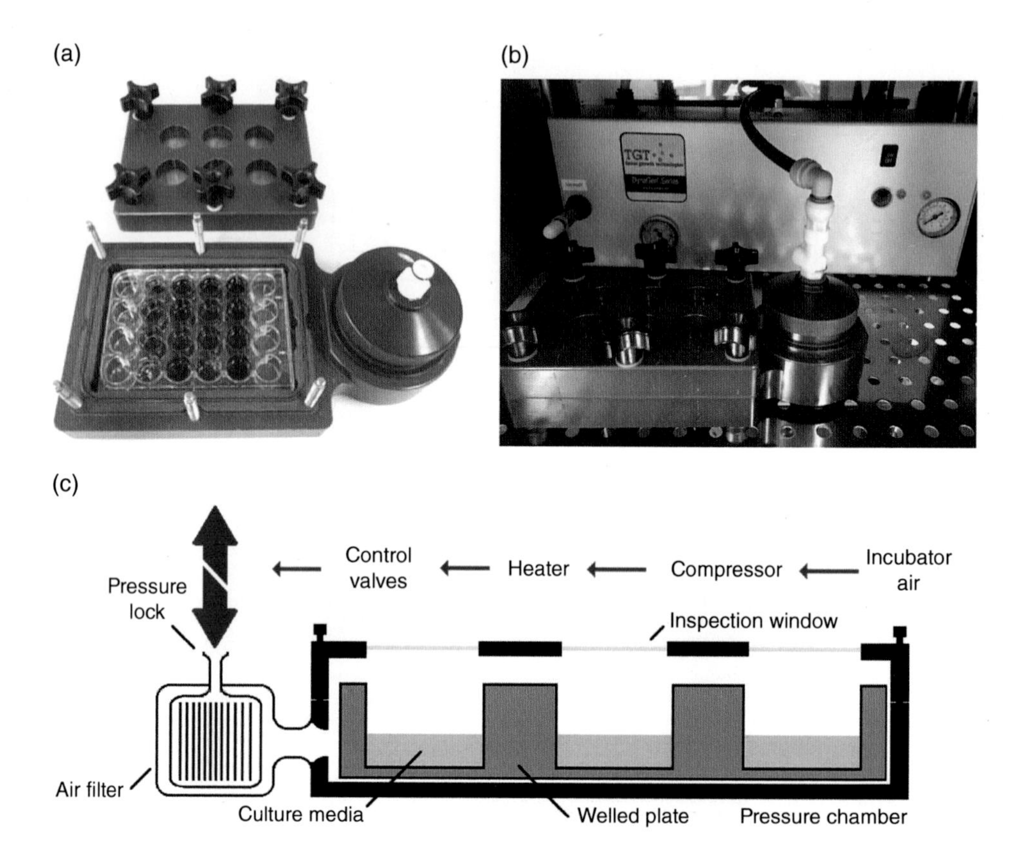

Figure 18.3 Hydrostatic bioreactor designed and created in partnership between Keele University (Professor Alica El-Haj) and TissueGrowthTechnologies (now a part of Instron). The key features of this bioreactor have been optimized to enhance high laboratory throughput and maintain maximum sterility, with the majority of the bioreactor and most of the moving hardware being located outside of the incubator environment, including the compressor and the computer control system. The bioreactor chamber itself exists as a "mini-incubator": a sealed, autoclavable chamber accommodating a standard multiwelled plate, allowing for standard experiments to be run under conventional, static incubator cultures and under dynamic hydrostatic loading. Separation from the outside space is provided by a replaceable and autoclavable filter, and is maintained by wide flanges to the chamber and a gasket, which helps reduce microbial infiltration when the chamber is opened in order to replace culture plates. The addition of inspection windows allows the cultures to be viewed, but reflections from the glass and the height above the culture restrict opportunities to derive measurable data or high-resolution images – this will be a point of optimization of future models in this range, which has now been commercialized as "CartiGen HP" (hydrostatic pressure). (a) Bioreactor chamber, containing cells in a standard culture plate. (b) Chamber and valve-control box – the only parts of the bioreactor placed within the incubator. (c) Schematic of the bioreactor.

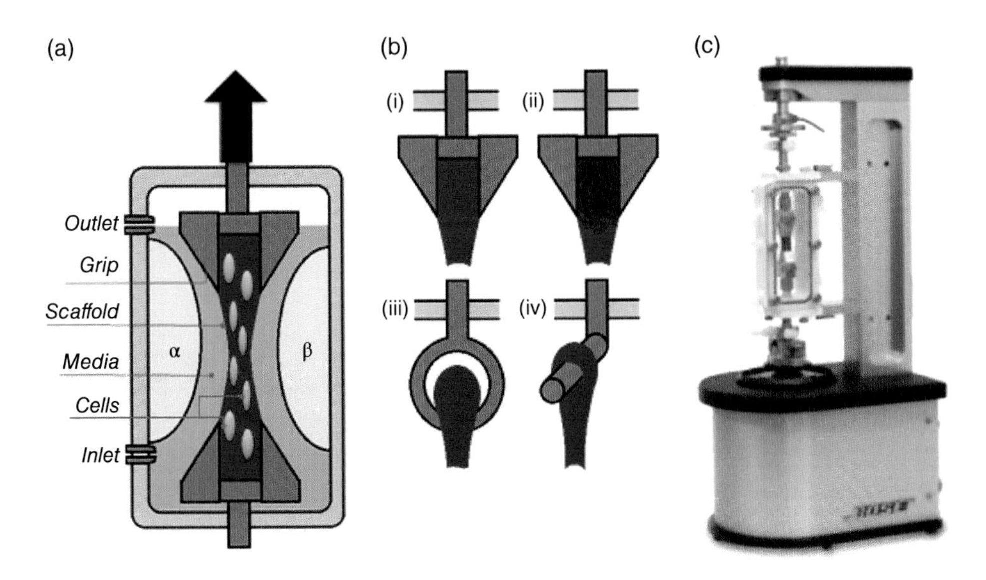

Figure 18.4 Schematic of a typical tension bioreactor, in which cells in a biomaterial scaffold such as collagen are cultured between two grips, which can be pulled apart to generate strain within the cell-seeded construct. (a) A useful adaptation of the second-generation Bose Electroforce series is the addition of space-filling solid baffles (α and β), which require less medium in the bioreactor chamber, and thus provide substantial savings on the expensive growth factors used in culture. (b) A particular engineering challenge is (i) the grip–biomaterial interface, where a great deal of mechanical failure occurs. Researchers have thus developed various ways of reinforcing this region. Options include (ii) the use of composite materials with an integrated solid scaffold or sacrificial zone, which is mechanically more resilient than the region under tensile load. Alternatively, the cell-seeded scaffold can be fabricated as a circular band and (iii) connected to the tension actuator via a loop or (iv) pinned on to the grips. Often, researchers will use a combination of these approaches, depending on the application (see Table 18.1). (c) The Electroforce series (now owned by TA Instruments) has been extensively used by researchers, as it conveniently attaches to existing mechanical testing systems, which include adaptable commercial software.

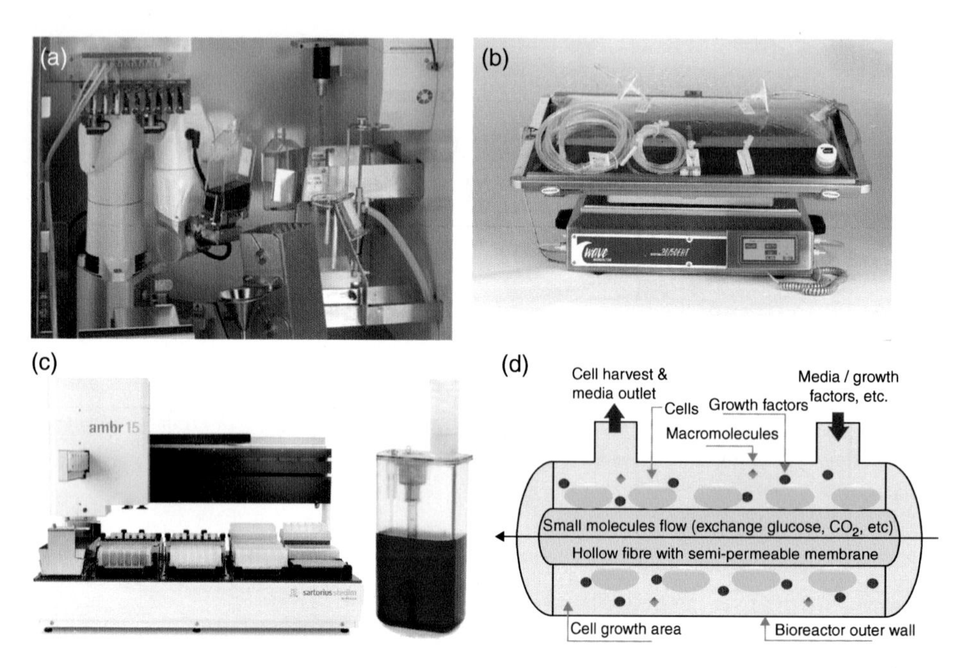

Figure 18.5 Commercial bioreactors are often designed principally to improve bioproduction methods – increasing cell yields and reducing costs through the extensive use of automation, in order to achieve consistent, reliable cell growth *in vitro*. Most of these systems use an element of perfusion to provide nutrients to cells, though the effects of this as a mechanical stimulus are generally a secondary consideration and are often limited to ameliorating the negative effects of excessive shear forces on cell viability. Nevertheless, this is likely to be the direction in which cells become commercialized as manufactured therapeutic products. Cell factories may be relatively easy to re-engineer with more appropriate mechanical characteristics which support and enhance cell growth, ultimately leading to more effective cell products. (a) Robotic T-flask handling (CompacT SelecT TAP Biosystems). (b) GE Healthcare's WAVE Bioreactor 2050. *Source:* Courtesy of GE Healthcare. (c) Ambr 15 microscale (10–15 mL) disposable microbioreactors for cells cultured on microcarriers and stirred by an impeller (TAP Biosystems). (d) Generalized schematic of a hollow-fiber bioreactor that allows cell culture with nutrient perfusion and a degree of shear/flow through the cell growth area while keeping expensive growth factors and important biological macromolecules within the culture.

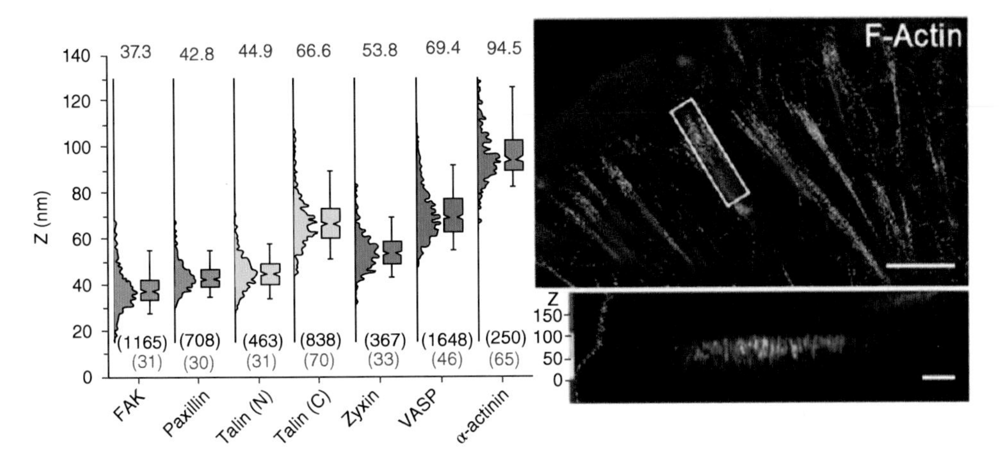

Figure 19.1 Nanoscale architecture of focal adhesions (FAs). Left: average z-position of different FA-associated proteins (Liu et al. 2015). Right: super-resolution image (top: top view; bottom: side view) of the actin network, with color-coded z-spatial information, at cell protrusions and FAs (scale bars top: 2 μm; bottom: 250 nm). *Source:* Reproduced with permission from Liu et al. (2015).

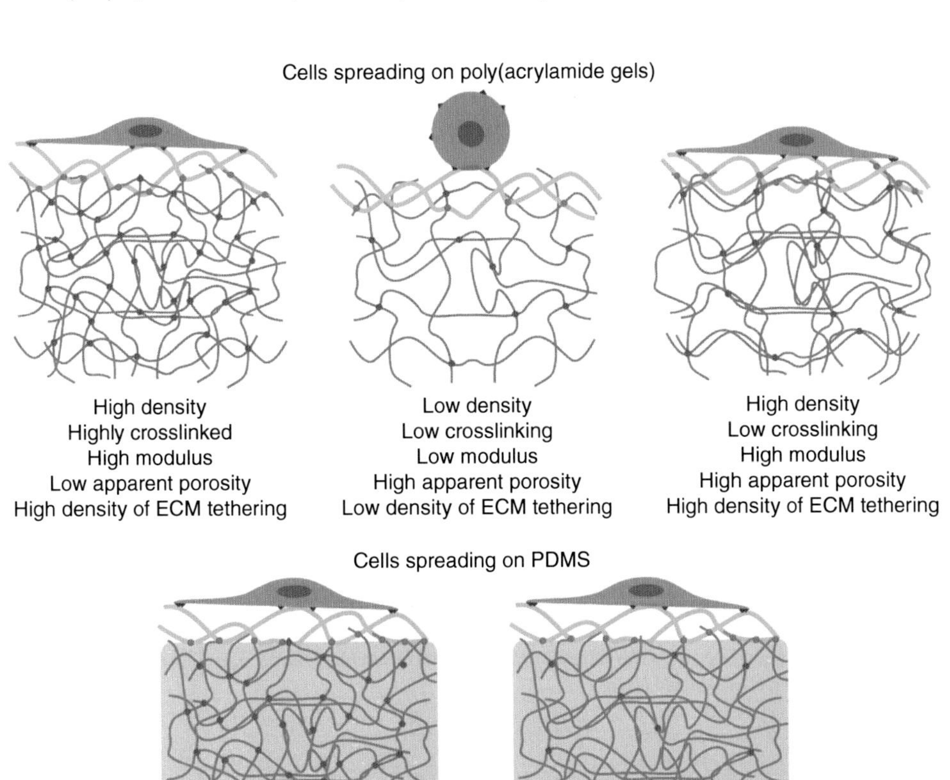

Figure 19.3 Proposed model to account for the distinct cell response to the bulk moduli of ECM protein-functionalized PAAm gels and PDMS substrates. Key parameters influencing the bulk modulus and cell response are summarized.

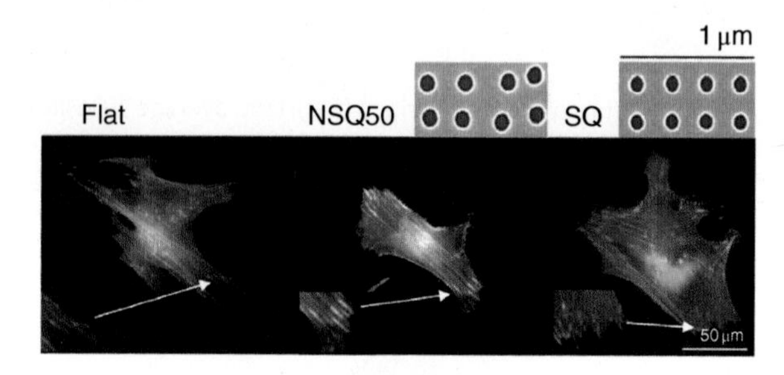

Figure 19.5 Impact of substrate topography on FA formation and cell spreading. *Source:* Reprinted with permission from Tsimbouri et al. (2014). Copyright 2014 John Wiley and Sons.

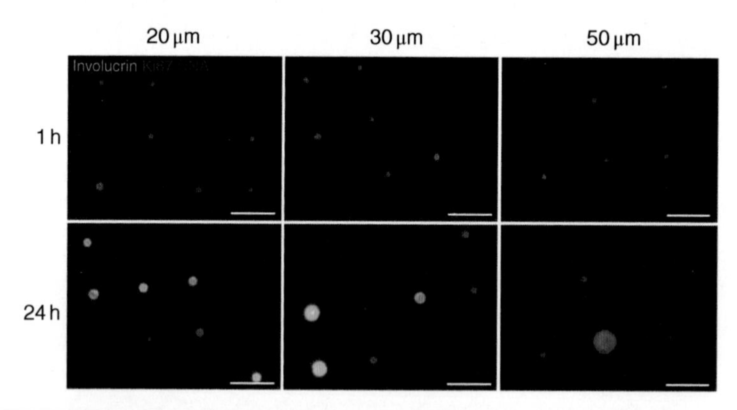

Figure 19.9 Cell shape controls the differentiation of keratinocytes. *Source:* Connelly et al. (2010). Reproduced with permission of *Nature*.

Human micro-epidermis

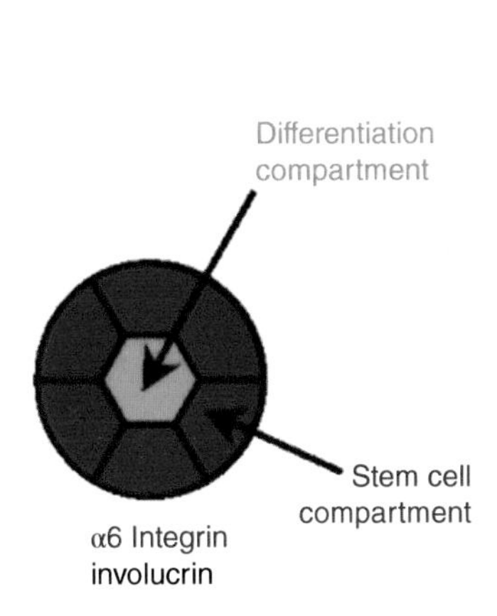

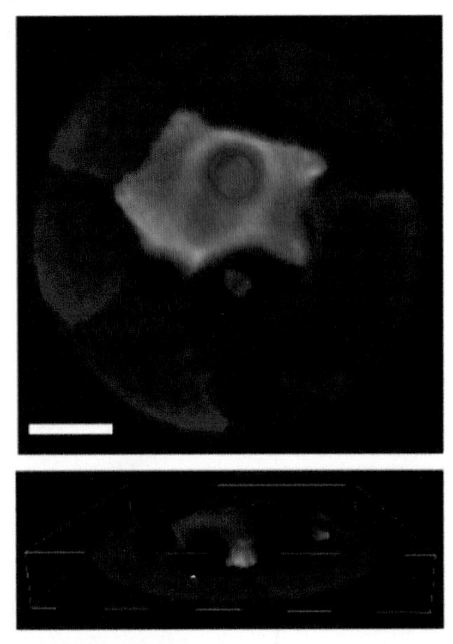

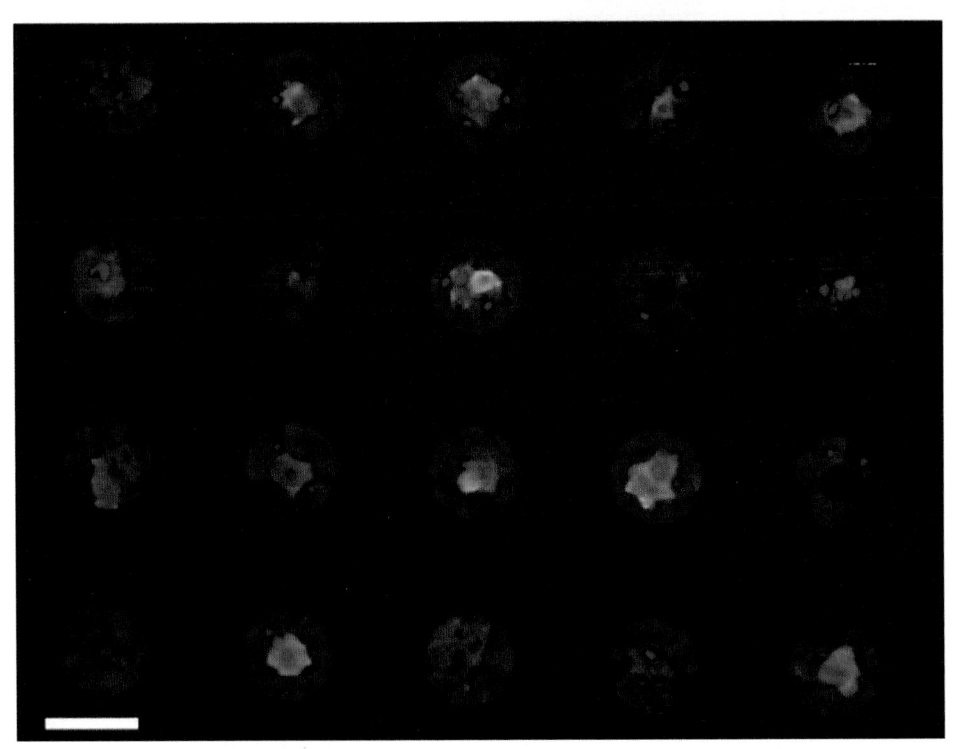

Figure 19.10 Arrays of microepidermis with controlled partitioning of differentiated keratinocytes. *Source:* Gautrot et al. (2012). Reproduced with permission of Elsevier.

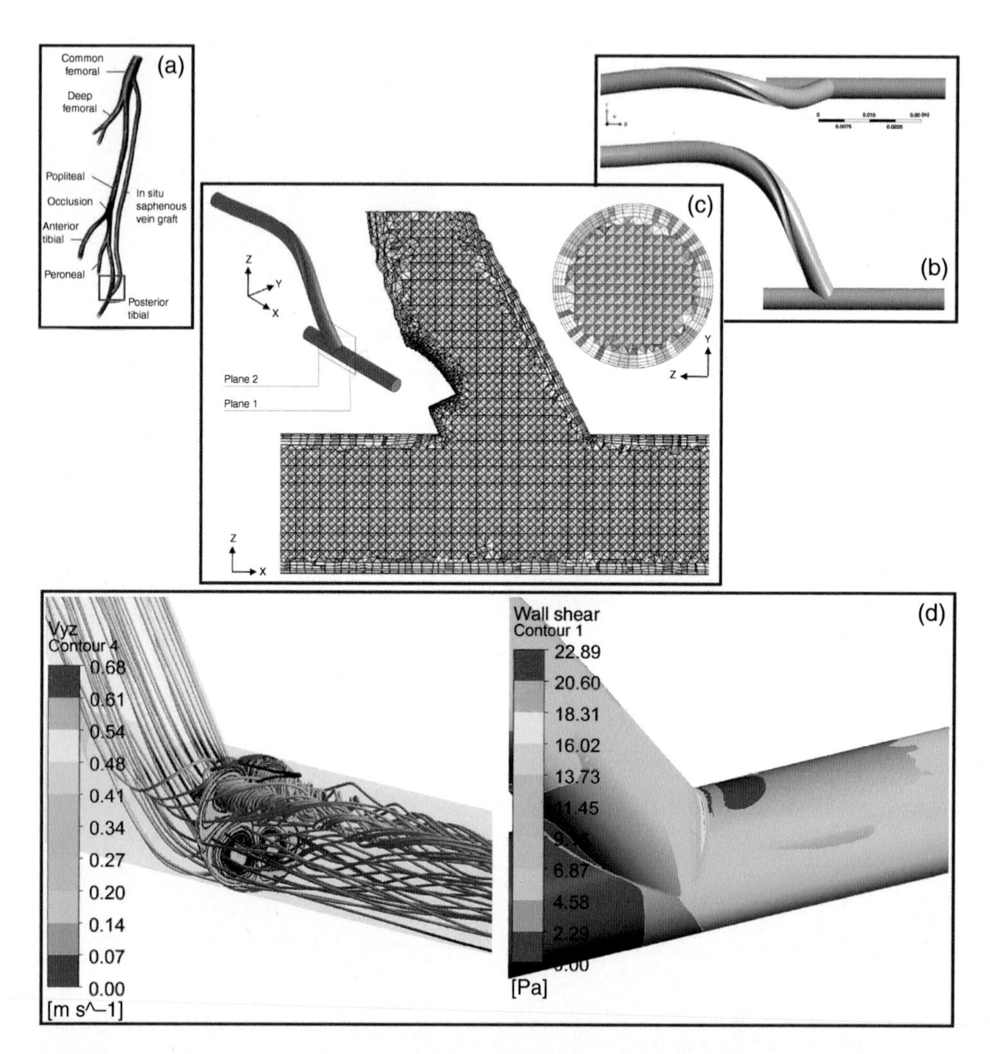

Figure 22.2 Typical CFD process in a biomedical application. (a) Typical peripheral bypass grafting. (b) Solid computer-aided model (CAD) for the computational domain. (c) Hybrid mesh consisting of quadrilateral cells near the wall boundaries and tetrahedral cells further from the wall. (d) Post-processing of the CFD simulation results using streamlines and contours of the hemodynamic parameters.

13

Testing the Perimenopause Ageprint using Skin Visoelasticity under Progressive Suction

Gérald E. Piérard[1,2], Claudine Piérard-Franchimont[3,4], Ulysse Gaspard[5], Philippe Humbert[2,6,7], and Sébastien L. Piérard[8]

[1] *Laboratory of Skin Bioengineering and Imaging (LABIC), Department of Clinical Sciences, University of Liège, Liège, Belgium*
[2] *University of Franche-Comté, Besançon, France*
[3] *Department of Dermatopathology, Unilab Lg, Liège University Hospital, Liège, Belgium*
[4] *Department of Dermatology, Regional Hospital of Huy, Huy, Belgium*
[5] *Department of Gynecology, Liège University Hospital, Liège, Belgium*
[6] *Department of Dermatology, University Hospital Saint-Jacques, Besançon, France*
[7] *Inserm Research Unit U645, IFR133, Besançon, France*
[8] *Telecommunications and Imaging Laboratory INTELSIG, Montefiore Institute, University of Liège, Liège, Belgium*

13.1 Introduction

There is some confusion about so-called "healthy skin aging": a physiological process characterized by a progressive decrease in the homeostatic capacity of the organism, ultimately increasing its vulnerability to certain environmental threats and leading to a series of physiopathological conditions (Pierard et al. 2014d). The physiological effects of aging should be separated from the social and emotional effects. Both physical growth and regular senescence combine cumulative progressions of interlocking biologic events. Normally, they do not proceed inescapably in distinct parts of the life-cycle, but rather occur occasionally in intricate association with one another. Healthy aging in many organs is frequently perceived as a progressive linear reduction in both maximal function and reserve capacity.

The concept of healthy biological aging involves three key components: (i) survival to old age; (ii) delay of the onset of chronic disorders; and (iii) maintenance of optimal functioning for an extended period. Like the presentation of skin, the physical capability of many internal organs is clearly involved in the healthy aging process. The casual aging rate evolves distinctly among different individuals of the same age, and proceeds unevenly even among the organs and constituent tissues, cells, and subcellular structures of a given individual (Mac-Mary et al. 2010). The various facets of healthy skin aging are affected by a range of physiological decrements, resulting in part from acute and chronic environmental insults. Indeed, a whole range of different domains of health, function, and well-being are involved, in complex interrelationships.

In the past, global skin aging was occasionally perceived as a single basic process of physiological decline with age. In recent decades, however, our understanding of healthy skin aging has been considerably refined (Naylor et al. 2011; Pierard et al. 2014d).

Mechanobiology: Exploitation for Medical Benefit, First Edition. Edited by Simon C. F. Rawlinson.

Emphasis was once placed on the distinction between intrinsic chronologic aging and photoaging due to chronic sun exposure (Farage et al. 2008), but this duality of skin aging was challenged as an oversimplification. A more diversified classification of seven types of skin aging was offered in its place (Pierard 1996; Pierard et al. 2014d), including endocrine and other metabolic effects, past and present lifestyles, and various environmental threats (including cumulative ultraviolet (UV) and infrared (IR) radiation) (Frantz et al. 2010), as well as repeat mechanical stress expressed by muscles and external forces, including gravity. In this framework, global skin aging results from the synergistic combination of these various factors. Awareness of the diversity of physiological skin changes with age likely improves the promotion of efficient preventive measures, allows the development of more effective skin care regimens, and helps refine dermatological treatment strategies (Pierard et al. 2014d).

Targeting of early cutaneous signs of wear and tear is common in affluent societies, with a variety of novel corrective treatments promising to reverse the effects of aging. However, since time immemorial, only a few such treatments have fulfilled the majority of their promises. Beyond new health management advances in this field, the forefront of scientific knowledge relies on an increased understanding of the relationships between cell biology, the structural organization of native biomolecules, the overall physiology of the dermal extracellular matrix (ECM), and any perceptible change in the ultimate clinical presentation (Frantz et al. 2010; Graham et al. 2010; Oh et al. 2011; Pierard-Franchimont et al. 2013).

13.2 Gender-Linked Skin Aging

In both genders, three main endocrine functions are modulated with age: (i) the hypothalamic–pituitary–gonadal (HPG) axis, which directly affects gonadal functions; (ii) the adrenals, which produce most of the sex hormone precursor corresponding to dehydroepiandrosterone (DHEA); and (iii) the growth hormone (GH)/insulin-like growth factor I (IGF-I) axis, which affects both GH production and IGF-I release predominating in the liver (Farage et al. 2012). A complex interplay exists between these endocrine changes and other hormonal systems affected by aging, such as production of melatonin and leptin.

Aging in women represents a multifaceted topic for laypeople, the media, cosmetic scientists, and the medical community. Any purported advance in this field is avidly watched by a lot of anti-aging worshippers. Notably, specific alterations in the HPG axis initiate the menopausal period, a seminal transition in the aging of women. This period of life corresponds to the permanent cessation of menstruation following the decline and loss of cyclic ovarian activity. Perimenopause is associated with vasomotor alterations ("hot flashes"), osteoporosis, and cardiovascular and immune system effects. Changes also occur in mood and sleep patterns, and there are impairments in sexual and cognitive function.

The transition from regular ovulatory cycles to the menopausal stage is not an instantaneous event. Rather, a series of progressive hormonal and clinical modifications supervene during the ongoing decline in ovarian activity. The time between the reproductive life period and the postmenopausal phase is referred to as "perimenopause." This includes the last few years preceding menopause, when specific endocrine, clinical, and biological changes develop, and the first year following the installation of amenorrhea (Harlow et al. 2012). The perimenopausal period is a milestone in the aging of women: a universal and global evolution characterized by a series of features.

A woman's appearance is appreciated in large measure through her skin presentation, but her perceived age is difficult to assess by clinical inspection alone (Coma et al. 2014). Supposedly, such perceptions reflect in part her general state of health health. Our current understanding of how functional measures of aging are altered across the lifespan is limited by a lack of validated and standardized scientific assessment procedures relevant to healthy aging. Inevitably, human skin – like any other organ or tissue – is subjected to regressive changes with aging. Accordingly, most women associate menopause with a negative experience, due to a progressive decline in the appearance and physical properties of the skin. Changes in the gonadal, adrenal, and peripheral production of the sex hormones impact overall skin physiology (Zouboulis and Makrantonaki 2011). Estrogens (specifically, estradiol) and androgens (specifically, testosterone and 5α-dihydrotestosterone) mediate their skin effects by activating specific cellular receptors. Peri- and postmenopausal aging particularly affects dermal tensile strength, through tissue atrophy and wrinkling (Pierard et al. 2001b, 2014a,b; Doubal and Klemera 2002; Hermanns-Le et al. 2004; Krueger et al. 2011).

13.3 Dermal Aging, Thinning, and Wrinkling

A decline in dermal thickness accounts for most of the measurable thinning of aging skin. The major ECM components of the dermis (collagen, elastin, and hyaluronic acid) are all affected by age. Collagen fibers become more disorganized; in photoaged skin, numbers of collagen bundles and fibers decrease markedly, primarily in relation to the upregulation of matrix metalloproteinases (MMPs). Consequently, the balance between collagen synthesis and degradation is deranged and the residual collagen fibers break up, disrupting the tensegrity of dermal fibroblasts found in a healthy collagen matrix, causing fibroblasts to collapse (Fisher et al. 2008; Farage et al. 2012).

Sex steroids notoriously affect the skin's structure, thickness, and elasticity. DHEA plays a role in maintaining skin structure by regulating the synthesis and degradation of ECM proteins. It promotes procollagen synthesis and limits collagen degradation by decreasing the MMP synthesis and increasing the production of tissue inhibitor of matrix metalloproteinase (TIMP). Consequently, the substantial decline in DHEA with age results in lower procollagen synthesis and higher collagen degradation.

Estrogens appear to affect skin thickness and elasticity primarily through their impact on constituents of the dermis. Hormone replacement therapy (HRT) maintains or improves skin thickness following menopause, largely by influencing dermal thickness. Wrinkling is related to loss of connective tissue and the resultant altered elasticity. Some studies of wrinkling have focused on the potential benefits of estrogen supplementation in perimenopausal women.

13.4 Skin Viscoelasticity under Progressive Suction

Measurements of a number of physical parameters characterizing human skin have been attempted in recent decades. Among the procedures used were many devices assessing skin viscoelasticity both *in vitro* and *in vivo*, which proved to be useful tools for both scientists and medical practitioners (Pierard 1999; Rodrigues 2001). Skin tensile strength can be assessed by a number of methods, including stretching, elevation, indentation, vibration, torsion, and suction procedures. The latter approach has been extensively used to study the physiological tensile properties of the dermis, while objective analytic methods have been used to collect noninvasive measurements of specific skin functions (Delalleau et al. 2008; Firooz et al. 2012; Pierard et al. 2013c). Some quandaries related to aging appear more complex and puzzling when skin has altered its regular mechanical function.

The dermo–epidermal atrophy related to late postmenopause is known as "transparent skin" or "dermatoporosis" (Kaya and Saurat 2007). Its clinical manifestations encompass morphologic evidence of fragility, senile purpura, so-called "stellate pseudoscars," and prominent skin atrophy. Functional alterations leading to skin wounds may ensue following even minor trauma. They present as skin lacerations, delayed wound-healing, nonhealing atrophic ulcers, and subcutaneous bleeding followed by dissecting hematomas and extended tissue necrosis. The latter clinical signs are associated with prominent morbidity, potentially bringing about surgical repair as a matter of emergency.

Data collected prospectively across all stages of life are valuable in that they enable different markers of health and changes in the aging process to be measured across the lifespan. Over a large part of the body, the overall viscoelastic behavior of the skin primarily depends on the structures of the dermo–hypodermal ECM, with minimal contribution from the epidermis (Silver et al. 2003; Hendriks et al. 2006; Pierard et al. 2013c, 2014a). Each skin viscoelastic parameter is under a time-dependent evolution and is likely related to the thickness of the tissues involved. Globally, data vary with body site, the subject's age and gender, and the duration and repetition of mechanical promptings. They are further influenced by the impact of various specific environmental conditions.

The suction method is widely used in determining the clinically relevant biomechanical characteristics of human skin in health and disease (Ryu et al. 2008; Iivarinen et al. 2013; Ohshima et al. 2013; Pierard et al. 2013a,b). From an engineering viewpoint, the skin and subcutaneous tissues represent a complex integrated and heterogeneous load-transmitting structure. In most controlled *in vitro* biomechanical study designs, the crude information received from an experiment is the relationship linking an applied force to relative deformation over time. In such instances, stress corresponds to the ratio between the suction and the test area of skin in a plane at right angles to the direction of the force. Strain represents the ratio between tissue elongation and its original length; this parameter is dimensionless, as it is measured as millimetres per millimetre. These definitions are different in the *in vivo* Cutometer application: the negative pressure applied to the skin is called "stress" irrespective of the size of the probe aperture, while "strain" corresponds to the vertical elevation of skin (Pierard et al. 2013a).

The progressive suction mode, using a stress–strain graphic recording, is convenient to use here. Following this procedure, a progressive increase in stress suction for a defined period of time is followed by a symmetrical rate of suction release. Skin deformation, defined as the strain, is recorded across the whole process. A purely elastic material is typically characterized by a straight linear relationship between stress and strain, which is independent of time. "Viscosity" refers to changes in skin deformation occurring in time under a sustained constant force. "Plasticity" implies resistance to any deformation for small forces. The combination of such basic properties is expressed by a nonlinearity of the relationship of force (stress) to deformation (strain). Typically, *in vivo* skin viscoelasticity exhibits nonlinear stress–strain properties (Pierard et al. 2013a).

13.5 Skin Tensile Strength during the Perimenopause

It is generally acknowledged that estrogen benefits skin elasticity (Farage et al. 2012). In the first 5 years following menopause, facial skin distensibility increases about 1.1% per year and elasticity decreases by 1.5% per year (Pierard et al. 1995; Henry et al. 1997; Pierard-Franchimont et al. 1999). Women who received HRT during this time period experience no significant changes in skin elasticity. Studies of oral, transdermal, and topical estrogen treatment also show benefits (Creidi et al. 1994; Pierard-Franchimont et al. 1999; Sator et al. 2001; Sumino et al. 2004), though topical estrogen treatment seems to be effective only in sun-protected skin (Rittie et al. 2008). The extent to which the effect is due to improvements in elastin fiber quality is unclear.

We performed a study in 130 healthy Caucasian women aged 29–53 years. Eligible participants had a body mass index (BMI) ranging from 19 to 23, corresponding to a normal range for nonobese women in Western Europe. The women were assigned to two distinct age groups: 65 nonmenopausal women aged 29–47 years (40 ± 4) and 65 perimenopausal women aged 48–53 years (51 ± 1) who were out of HRT (Table 13.1).

The Cutometer MPA 580 (CK electronic, Cologne, Germany) was equipped with a hollow probe centered by a 2 mm-diameter suction aperture. The handheld probe was maintained on the skin surface with a constant pressure guaranteed by a built-in spring. An additional outer concentric 55 mm-diameter steel guard ring was affixed to the skin by a double-sided adhesive film (Pierard et al. 2013a,b). Adhesive silicone tapes were placed in a crosswise pattern between the outer guard ring and the probe. In each woman, duplicate measurements at 1-hour intervals were performed on the volar aspect of both forearms. The respective viscoelastic values (four measurements) were averaged.

The progressive suction modality with a stress–strain recording showed expected shapes. This skin deformation test corresponded to a single cycle comprising a two-step procedure involving the successive application of increasing and decreasing suction forces at a constant pace. Strain always remained more intense when the force was decreasing than in the increasing phase at the same force. At each evaluation time, the maximum vertical skin deformation (MD) representing the skin stiffness was measured after applying a progressive suction force at a 25 mbar/s rate for 20 seconds. This step procedure was followed by a similar linear decrease in suction during a 20–second

Table 13.1 Progressive suction procedure. Median and range values of viscoelastic parameters in nonmenopausal (n = 65) and perimenopausal (n = 65) women.

Parameter	Nonmenopausal women	p	Perimenopausal women
MD (mm)	0.17 (0.14–0.31)	<0.01	0.26 (0.15–0.38)
BE (%)	68.2 (45.8–73.9)	<0.001	43.6 (23.4–96.5)
HY (AU)	93 (66.0–119.0)	<0.01	114 (89.0–130.0)
ed/ei	1.17 (0.91–1.89)	<0.001	3.26 (2.13–12.42)

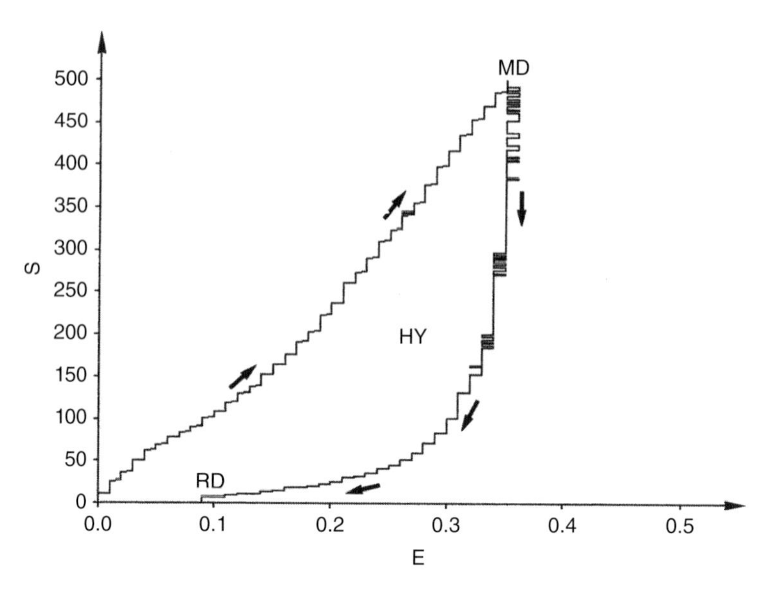

Figure 13.1 Stress–strain curve obtained under progressive suction procedure, showing progressive linear increase in suction (S, mbar) of 25 mbar/s for 20 seconds followed by a relaxation recovery at the same rate. The maximum deformation (MD) and the residual deformation (RD) of the skin extensibility (E, mm) are recorded. Hysteresis (HY) is the area delimited by the suction–relaxation curves.

relaxation phase. During the relaxation period, the strain values did not immediately resume the baseline value, and the intercept of the curve on the strain axis defined the residual deformation (RD). This parameter represented a short-term limitation on elastic recovery. The biologic elasticity (BE) was expressed as a percentage following 10^2(MD-RD)MD^{-1}. The stress–strain curve on suction was not superposed to the relaxation curve. The area delimited by the two curves corresponded to the hysteresis (HY) loop (Figure. 13.1). Planimetry of the area below the upward curve and of the area within the hysteresis loop corresponded to the energy input (ei) and the energy dissipation (ed), respectively (Figure 13.2). The ratio between ei and ed was influenced by age-related ECM changes.

Magnitude, spread, and symmetry of the data were assessed using the Shapiro–Wilks test. Data were expressed as medians and ranges according to their distribution. Statistical comparisons were performed using the nonparametric unpaired Mann–Whitney U test. A p-value lower than 0.05 was considered statistically significant.

In each woman, HY was disclosed under the progressive suction modality. During the upward phase of suction increment, the rate of skin deformation was nearly straight or discretely concave. By contrast, the relaxation curve appeared more bulky and exhibited an aspect different from the stretched portion of the curve. Its initial portion showed plasticity with a near absence or a discrete reduction in strain deformation. By contrast, the rate of strain reduction was boosted during the late phase of recovery. The different patterns in the unloading curve were responsible for the RD, which was always higher than the initial rest position of the upward loading phase.

The median MD value was increased ($p < 0.01$) in perimenopausal women compared with younger, nonmenopausal women. Clearly, the average RD value was significantly

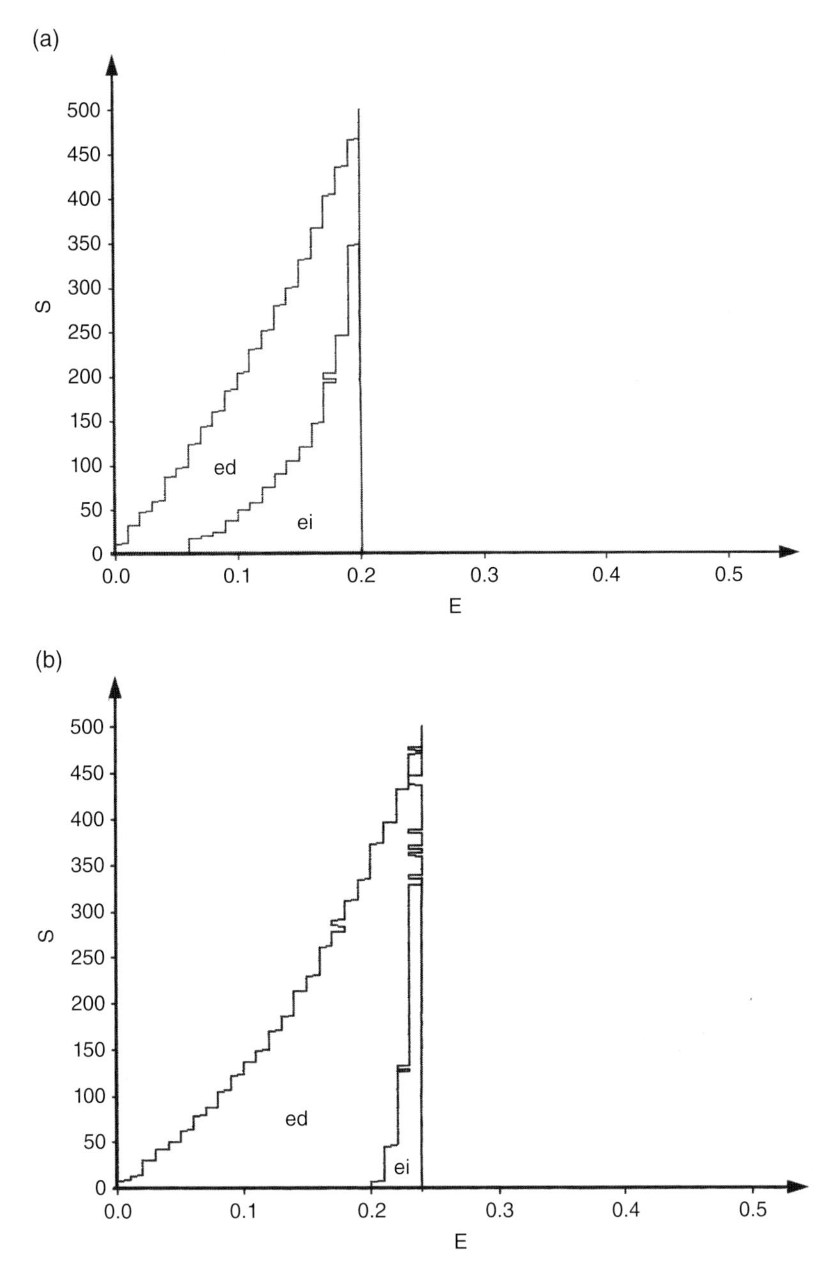

Figure 13.2 Examples of stress–strain relationships showing a variable value of energy dissipation (ed) and energy input (ei): (a) nonmenopausal woman; (b) perimenopausal woman with marked skin slackness; (c) perimenopausal woman with discrete skin atrophy.

($p < 0.01$) higher during perimenopause than before the menopause. As a result, BE was decreased ($p < 0.001$) in menopausal women. HY was increased during perimenopause ($p < 0.01$), particularly when MD was increased. For HY values, the ratio ed/ei was similarly increased ($p < 0.001$).

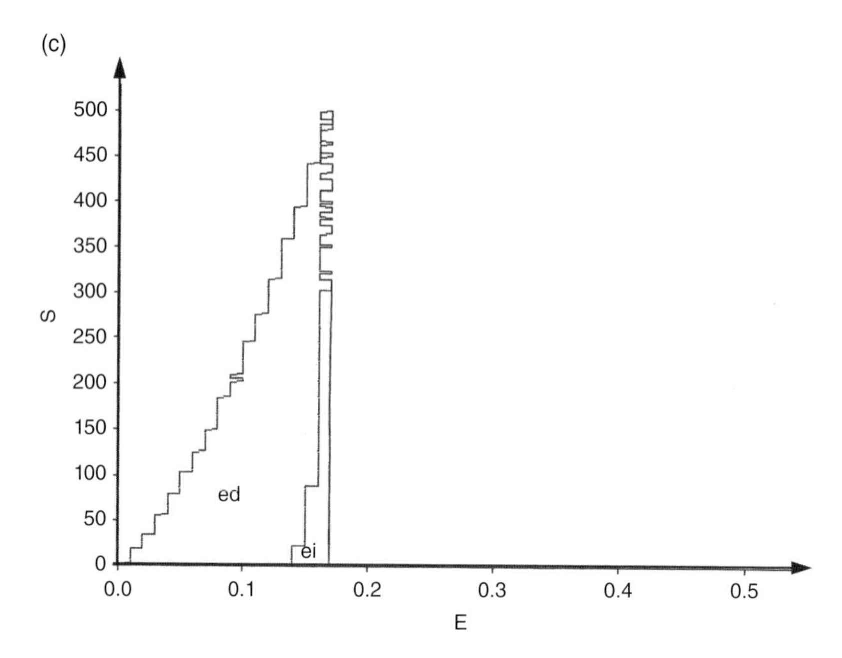

Figure 13.2 (Continued)

13.6 Conclusion

Healthy skin aging is a challenge. Identifying modifiable factors across life that influence aging will hopefully help to change them. By identifying those at risk of poor health much earlier in life, we can design interventions to help them, boosting their chances of having longer, healthier lives.

Clearly, the Cutometer device, in most of its clinical applications is not a diagnostic tool, but rather a functional assessor for ECM disorders (Pierard et al. 2013a; Sandford et al. 2013). For a given skin condition, the interindividual variations expressed by each parameter are quite large. However, the patterns of associated viscoelastic changes are commonly consistent in each of the considered disorders.

In young women, the upward part of the stress–strain curve under constant suction rate ascends in an almost straight progression. This upward phase indicates dependency on Hooke's law. The mechanical parameters are influenced by age, hormones, environmental factors, and various desmotropic drugs (Calleja-Agius and Brincat 2012; Pierard et al. 2013b, 2014c). During perimenopause, BE falls below the values found at younger ages. By contrast, MD and ed/ei are increased. This means that the elastic and viscous properties of skin are altered during perimenopause.

"Perimenopausal aging" refers to the period in a woman's life commonly initiating the atrophic skin withering and slackness responsible for changes in ECM viscoelasticity. Some women benefit from HRT in terms of controlling these unpleasant changes of perimenopause. In particular, HRT has the potential to correct the functional damage caused to dermal tensile strength (Pierard-Franchimont et al. 1999; Pierard et al. 2001b, 2013b, 2014b).

A number of methods can be employed to assess specific characteristics of the skin (Rodrigues 2001). The suction method is currently used by a majority of investigators. Skin distensibility appears to increase during perimenopause regardless of whether the woman is on HRT. In contrast, BE decreases significantly in the absence of HRT. To ensure that the evidence base on which we build policy and practice concerning HRT is as robust as possible, we need to undertake in-depth analyses of the most relevant data and conduct cross-cohort analyses in order to answer specific research questions. To date, intervention studies focused on aging outcomes have been undertaken almost exclusively in populations that were already older at baseline. It is important to identify ways of testing whether interventions earlier in life provide lasting benefits for healthy aging. Further studies are needed to explore the role of dermocosmetic products and cosmeceuticals in improving skin viscoelasticity after menopause.

The effect of photoaging has not been studied to date in menopausal women. The action spectrum of photodamage is quite vast. The spectral dependence of cumulative damages does not parallel the erythema spectrum for acute UV injury on human skin. The cumulative effects of repeat exposures to suberythemal irradiations of human skin by UVA and UVB have been identified by Schroeder et al. (2008). The role of UVB in elastin promoter activation during photoaging is likely operative. Further UVA radiation contributes largely to long-term actinic damages. Near-infrared (IR) radiations likely bring additional deleterious effects to skin aging (Schroeder et al. 2008).

The major viscoelastic properties of skin are governed by the ECM components. Both the dermis and the hypodermis are characterized by their own intimate structures, whose tensile functions are balanced in order to adequately respond to casual mechanical demands. It is acknowledged that a series of physisopathological variables alters the viscoelasticity of the whole skin. Accordingly, the assessment of skin viscoelasticity provides incentives for progress in skin care management.

In hysteresis experiments using a regular suction method, both the elastic and the viscous properties of skin are measured without reference to directional differences (Pierard et al. 2001a). When extension at a constant rate is achieved, the return toward the initial position is controlled with the same velocity. The unloading curve shows a different pattern toward the baseline with a residual RD compared to the curve lifted during the upward phase. In practice, the most valuable parameters altered during the perimenopausal phase correspond to MD, BE, and the ratio between ed (corresponding to HY) and ei. Such analytical information can be used to manage treatments and predict some aspects of the internal consequences of the climacteric period (Pierard et al. 2001a,b, 2014b; Raine-Fenning et al. 2003; Schlangen et al. 2003).

Acknowledgements

The authors have no conflicts of interest that are directly relevant to the content of this review. The authors appreciate the excellent secretarial assistance of Mrs. Ida Leclercq.

References

Calleja-Agius, J. and M. Brincat. 2012. "The effect of menopause on the skin and other connective tissues." *Gynecological Endocrinology* **28**(4): 273–7. doi:10.3109/09513590.2011.613970.

Coma, M., R. Valls, J. M. Mas, A. Pujol, M. A. Herranz, V. Alonso, and J. Naval. 2014. "Methods for diagnosing perceived age on the basis of an ensemble of phenotypic features." *Clinical, Cosmetic and Investigational Dermatology* **7**: 133–7. doi:10.2147/ccid.s52257.

Creidi, P., B. Faivre, P. Agache, E. Richard, V. Haudiquet, and J. P. Sauvanet. 1994. "Effect of a conjugated oestrogen (Premarin) cream on ageing facial skin. A comparative study with a placebo cream." *Maturitas* **19**(3): 211–23. PMID:7799828.

Delalleau, A., G. Josse, J. M. Lagarde, H. Zahouani, and J. M. Bergheau. 2008. "A nonlinear elastic behavior to identify the mechanical parameters of human skin in vivo." *Skin Research and Technology* **14**(2): 152–64. doi:10.1111/j.1600-0846.2007.00269.x.

Doubal, S. and P. Klemera. 2002. "Visco-elastic response of human skin and aging." *Journal of the American Aging Association* **25**(3): 115–17. doi:10.1007/s11357-002-0009-9.

Farage, M. A., K. W. Miller, P. Elsner, and H. I. Maibach. 2008. "Intrinsic and extrinsic factors in skin ageing: a review." *International Journal of Cosmetic Science* **30**(2): 87–95. doi:10.1111/j.1468-2494.2007.00415.x.

Farage, M. A., K. W. Miller, C. C. Zouboulis, G. E. Piérard, and H. I. Maibach. 2012. "Gender differences in skin aging and the changing profile of the sex hormones with age." *Journal of Steroids & Hormonal Science* **3**: 109. doi:10.4172/2157-7536.1000109.

Firooz, A., B. Sadr, S. Babakoohi, M. Sarraf-Yazdy, F. Fanian, A. Kazerouni-Timsar, et al. 2012. "Variation of biophysical parameters of the skin with age, gender, and body region." *Scientific World Journal* **2012**: 386936. doi:10.1100/2012/386936.

Fisher, G. J., J. Varani, and J. J. Voorhees. 2008. "Looking older: fibroblast collapse and therapeutic implications." *Archives of Dermatology* **144**(5): 666–72. doi:10.1001/archderm.144.5.666.

Frantz, C., K. M. Stewart, and V. M. Weaver. 2010. "The extracellular matrix at a glance." *Journal of Cell Science* **123**(Pt. 24): 4195–200. doi:10.1242/jcs.023820.

Graham, H. K., N. W. Hodson, J. A. Hoyland, S. J. Millward-Sadler, D. Garrod, A. Scothern, et al. 2010. "Tissue section AFM: In situ ultrastructural imaging of native biomolecules." *Matrix Biology* **29**(4): 254–60. doi:10.1016/j.matbio.2010.01.008.

Harlow, S. D., M. Gass, J. E. Hall, R. Lobo, P. Maki, R. W. Rebar, et al. 2012. "Executive summary of the Stages of Reproductive Aging Workshop + 10: addressing the unfinished agenda of staging reproductive aging." *Journal of Clinical Endocrinology and Metabolism* **97**(4): 1159–68. doi:10.1210/jc.2011-3362.

Hendriks, F. M., D. Brokken, C. W. Oomens, D. L. Bader, and F. P. Baaijens. 2006. "The relative contributions of different skin layers to the mechanical behavior of human skin in vivo using suction experiments." *Medical Engineering & Physics* **28**(3): 259–66. doi:10.1016/j.medengphy.2005.07.001.

Henry, F., C. Pierard-Franchimont, G. Cauwenbergh, and G. E. Pierard. 1997. "Age-related changes in facial skin contours and rheology." *Journal of the American Geriatric Society* **45**(2): 220–2. PMID:9033524.

Hermanns-Le, T., I. Uhoda, S. Smitz, and G. E. Pierard. 2004. "Skin tensile properties revisited during ageing. Where now, where next?" *Journal of Cosmetic Dermatology* **3**(1): 35–40. doi:10.1111/j.1473-2130.2004.00057.x.

Iivarinen, J. T., R. K. Korhonen, P. Julkunen, and J. S. Jurvelin. 2013. "Experimental and computational analysis of soft tissue mechanical response under negative pressure in forearm." *Skin Research and Technology* **19**(1): e356–65. doi:10.1111/j.1600-0846.2012.00652.x.

Kaya, G. and J. H. Saurat. 2007. "Dermatoporosis: a chronic cutaneous insufficiency/fragility syndrome. Clinicopathological features, mechanisms, prevention and potential treatments." *Dermatology* **215**(4): 284–94. doi:10.1159/000107621.

Krueger, N., S. Luebberding, M. Oltmer, M. Streker, and M. Kerscher. 2011. "Age-related changes in skin mechanical properties: a quantitative evaluation of 120 female subjects." *Skin Research and Technology* **17**(2): 141–8. doi:10.1111/j.1600-0846.2010.00486.x.

Mac-Mary, S., J. M. Sainthillier, A. Jeudy, C. Sladen, C. Williams, M. Bell, and P. Humbert. 2010. "Assessment of cumulative exposure to UVA through the study of asymmetrical facial skin aging." *Clinical Interventions in Aging* **5**: 277–84. PMID:20924436.

Naylor, E. C., R. E. Watson, and M. J. Sherratt. 2011. "Molecular aspects of skin ageing." *Maturitas* **69**(3): 249–56. doi:10.1016/j.maturitas.2011.04.011.

Oh, J. H., Y. K. Kim, J. Y. Jung, J. E. Shin, and J. H. Chung. 2011. "Changes in glycosaminoglycans and related proteoglycans in intrinsically aged human skin in vivo." *Experimental Dermatology* **20**(5): 454–6. doi:10.1111/j.1600-0625.2011.01258.x.

Ohshima, H., S. Kinoshita, M. Oyobikawa, M. Futagawa, H. Takiwaki, A. Ishiko, and H. Kanto. 2013. "Use of Cutometer area parameters in evaluating age-related changes in the skin elasticity of the cheek." *Skin Research and Technology* **19**(1): e238–42. doi:10.1111/j.1600-0846.2012.00634.x.

Pierard, G. E. 1996. "The quandary of climacteric skin ageing." *Dermatology* **193**(4): 273–4. PMID:8993948.

Pierard, G. E. 1999. "EEMCO guidance to the in vivo assessment of tensile functional properties of the skin. Part 1: Relevance to the structures and ageing of the skin and subcutaneous tissues." *Skin Pharmacology and Applied Skin Physiology* **12**(6): 352–62. doi:29897.

Pierard, G. E., C. Letawe, A. Dowlati, and C. Pierard-Franchimont. 1995. "Effect of hormone replacement therapy for menopause on the mechanical properties of skin." *Journal of the American Geriatric Society* **43**(6): 662–5. PMID:7775726.

Pierard, G. E., C. Pierard-Franchimont, S. Vanderplaetsen, N. Franchimont, U. Gaspard, and M. Malaise. 2001a. "Relationship between bone mass density and tensile strength of the skin in women." *European Journal of Clinical Investigation* **31**(8): 731–5. PMID:11473575.

Pierard, G. E., S. Vanderplaetsen, and C. Pierard-Franchimont. 2001b. "Comparative effect of hormone replacement therapy on bone mass density and skin tensile properties." *Maturitas* **40** (3): 221–7. PMID:11731183.

Pierard, G. E., T. Hermanns-Le, and C. Pierard-Franchimont. 2013a. "Scleroderma: skin stiffness assessment using the stress-strain relationship under progressive suction." *Expert Opinions in Medical Diagnosis* 7(2): 119–25. doi:10.1517/17530059.2013.753877.

Pierard, G. E., P. Humbert, E. Berardesca, U. Gaspard, T. Hermanns-Le, and C. Pierard-Franchimont. 2013b. "Revisiting the cutaneous impact of oral hormone replacement therapy." *Biomedical Research International* **2013**: 971760. doi:10.1155/2013/971760.

Pierard, G. E., S. Pierard, P. Delvenne, and C. Pierard-Franchimont. 2013c. "In vivo evaluation of the skin tensile strength by the suction method: pilot study coping with hysteresis and creep extension." *ISRN Dermatology* **2013**: 841217. doi:10.1155/2013/841217.

Pierard, G. E., T. Hermanns-Le, U. Gaspard, and C. Pierard-Franchimont. 2014a. "Asymmetric facial skin viscoelasticity during climacteric aging." *Clinical, Cosmetic and Investigational Dermatology* **7**: 111–18. doi:10.2147/ccid.s60313.

Pierard, G. E., T. Hermanns-Le, P. Paquet, and C. Pierard-Franchimont. 2014b. "Skin viscoelasticity during hormone replacement therapy for climacteric ageing." *International Journal of Cosmetic Science* **36**(1): 88–92. doi:10.1111/ics.12100.

Pierard, G. E., P. Paquet, and C. Pierard-Franchimont. 2014c. "Skin viscoelasticity in incipient gravitational syndrome." *Journal of Cosmetic Dermatology* **13**(1): 52–5. doi:10.1111/jocd.12077.

Pierard, G. E., C. Pierard-Franchimont, and P. Quatresooz. 2014d. "Skin ageprint: the causative factors." In: A. O. Barel and M. Payc (eds.). *Handbook of Cosmetic Science and Technology*, 4th edn. Boca Raton, FL: CRC Press.

Pierard-Franchimont, C., F. Cornil, J. Dehavay, F. Deleixhe-Mauhin, B. Letot, and G. E. Pierard. 1999. "Climacteric skin ageing of the face – a prospective longitudinal comparative trial on the effect of oral hormone replacement therapy." *Maturitas* **32**(2): 87–93. PMID:10465376.

Pierard-Franchimont, C., G. Loussouarn, S. Panhard, D. Saint Leger, M. Mellul, and G. E. Pierard. 2013. "Immunohistochemical patterns in the interfollicular Caucasian scalps: influences of age, gender, and alopecia." *Biomedical Research International* **2013**: 769489. doi:10.1155/2013/769489.

Raine-Fenning, N. J., M. P. Brincat, and Y. Muscat-Baron. 2003. "Skin aging and menopause: implications for treatment." *American Journal of Clinical Dermatology* **4**(6): 371–8. PMID:12762829.

Rittie, L., S. Kang, J. J. Voorhees, and G. J. Fisher. 2008. "Induction of collagen by estradiol: difference between sun-protected and photodamaged human skin in vivo." *Archives of Dermatology* **144**(9): 1129–40. doi:10.1001/archderm.144.9.1129.

Rodrigues, L. 2001. "EEMCO guidance to the in vivo assessment of tensile functional properties of the skin. Part 2: instrumentation and test modes." *Skin Pharmacology and Applied Skin Physiology* **14**(1): 52–67. doi:56334.

Ryu, H. S., Y. H. Joo, S. O. Kim, K. C. Park, and S. W. Youn. 2008. "Influence of age and regional differences on skin elasticity as measured by the Cutometer." *Skin Research and Technology* **14**(3): 354–8. doi:10.1111/j.1600-0846.2008.00302.x.

Sandford, E., Y. Chen, I. Hunter, G. Hillebrand, and L. Jones. 2013. "Capturing skin properties from dynamic mechanical analyses." *Skin Research and Technology* **19**(1): e339–48. doi:10.1111/j.1600-0846.2012.00649.x.

Sator, P. G., J. B. Schmidt, M. O. Sator, J. C. Huber, and H. Honigsmann. 2001. "The influence of hormone replacement therapy on skin ageing: a pilot study." *Maturitas* **39**(1): 43–55. PMID:11451620.

Schlangen, L. J., D. Brokken, and P. M. van Kemenade. 2003. "Correlations between small aperture skin suction parameters: statistical analysis and mechanical model." *Skin Research and Technology* **9**(2): 122–30. PMID:12709130.

Schroeder, P., J. Haendeler, and J. Krutmann. 2008. "The role of near infrared radiation in photoaging of the skin." *Experimental Gerontology* **43**(7): 629–32. doi:10.1016/j.exger.2008.04.010.

Silver, F. H., L. M. Siperko, and G. P. Seehra. 2003. "Mechanobiology of force transduction in dermal tissue." *Skin Research and Technology* **9**(1): 3–23. PMID:12535279.

Sumino, H., S. Ichikawa, M. Abe, Y. Endo, O. Ishikawa, and M. Kurabayashi. 2004. "Effects of aging, menopause, and hormone replacement therapy on forearm skin elasticity in women." *Journal of the American Geriatric Society* **52**(6): 945–9. doi:10.1111/j.1532-5415.2004.52262.x.

Zouboulis, C. C. and E. Makrantonaki. 2011. "Clinical aspects and molecular diagnostics of skin aging." *Clinical Dermatology* **29**(1): 3–14. doi:10.1016/j.clindermatol.2010.07.001.

14

Mechanobiology and Mechanotherapy for Skin Disorders

Chao-Kai Hsu[1,2,3] *and Rei Ogawa*[4]

[1] *Department of Dermatology, National Cheng Kung University Hospital, College of Medicine, National Cheng Kung University, Tainan, Taiwan*
[2] *Institute of Clinical Medicine, College of Medicine, National Cheng Kung University, Tainan, Taiwan*
[3] *International Research Center of Wound Repair and Regeneration, National Cheng Kung University, Tainan, Taiwan*
[4] *Department of Plastic, Reconstructive, and Aesthetic Surgery, Nippon Medical School, Tokyo, Japan*

14.1 Introduction

Human skin is the largest organ of the body, with a total area between 1.6 and 1.8 m^2. It is a complex organ, and provides an effective barrier to microbial invasion and external stimuli such as chemicals, temperature, and gases. It comprises two compartments: the epidermis, where keratinocytes form the skin barrier, and the dermis, which makes up approximately 90% of the thickness of the skin, and provides support and nutrition to the epidermis. The dermis has high levels of collagen and elastic fibers, which are secreted by scattered fibroblasts. This provides the skin with its elastic properties (Figure 14.1). The epidermis and the dermis are separated by a basement membrane. Separating the dermis from the underlying tissues is a layer of fat cells – the subcutaneous layer (hypodermis) – whose accumulation of fat has a cushioning effect. In addition, epidermal appendages (hair follicles, sebaceous glands, and sweat glands) span the epidermis and are embedded in the dermis or hypodermis. These appendages each serve a particular function, including providing lubrication, protection, cooling, and sensation.

Human skin is also a highly specialized mechanoresponsive organ, which constantly senses and adapts to various mechanical stimuli throughout life (Wong et al. 2011). During the growth and development of the human body, the skin responds to the intrinsic forces from the underlying growing skeleton and soft tissues, and expands to cover them. It also responds to the extrinsic forces generated by body movements and external mechanical stimuli. Given these observations, it is unsurprising that mechanical force regulates the differentiation of human epidermal stem cells (Trappmann et al. 2012). Langer was the first to examine the effects of mechanical forces on skin: in 1861, he punctured the skin of a cadaver with numerous holes at short distances from one another. This led him to identify the lines of tension on the skin (Wilhelmi et al. 1999). Nowadays, the direction of Langer's lines is important for surgical operations, particularly cosmetic surgery.

Mechanobiology: Exploitation for Medical Benefit, First Edition. Edited by Simon C. F. Rawlinson.

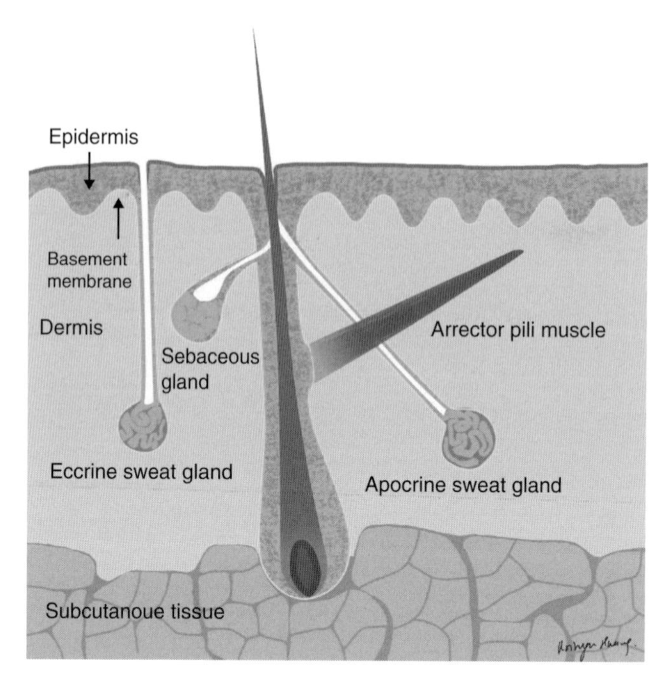

Figure 14.1 Anatomy of skin structure.

In the dermis, fibroblasts secrete collagen and fibronectin, and regulate the volume of the extracellular matrix (ECM) by producing various matrix metalloproteinases (MMPs). The ECM is the substrate for cell adhesion, growth, and differentiation. The binding of a cell to the ECM induces small forces that can deform the cell. If the balance between ECM synthesis and degradation is not carefully maintained, the mechanical properties of the ECM can change. This may induce mechanobiological dysregulation, resulting in skin cell derangements that can cause skin disorders. The extracellular fluid (ECF) also exerts intrinsic mechanical forces, including fluid shear force, hydrostatic pressure, and osmotic pressure. All of these forces are perceived by two types of skin receptor (Ogawa 2008): cellular mechanoreceptors/mechanosensors, such as the cytoskeleton (e.g., actin filaments), cell-adhesion molecules (e.g., integrin), and mechanosensitive ion channels (e.g., calcium channels); and sensory nerve fibers (e.g., mechanosensitive nociceptors), which produce the somatic sensation of mechanical force.

The mechanical forces generated by the ECM and ECF are important in maintaining the structural and functional homeostasis of skin. Thus, mechanobiological dysregulation can result in various skin disorders. For example, autoimmune diseases (e.g., scleroderma and mixed connective-tissue disease) and pathological scarring (keloid and hypertrophic scars) clearly associate with mechanobiological dysfunction of the skin. The skin disorders in Ehlers–Danlos syndrome (EDS), cutis laxa, neurofibromatosis, and diabetes mellitus (DM) are also all caused to some extent by deficiencies in mechanobiological function in the skin. In addition, bullous pemphigus appears to be caused by autoimmune destruction of the mechanobiological environment. Moreover, skin aging can be considered a degenerative process that associates with mechanobiological dysfunction. How mechanobiological dysfunction leads to these skin disorders will be discussed in more detail later.

The association of these diseases with mechanobiology suggests the possibility of new therapeutic strategies, namely, methods that inhibit or accelerate the functions of mechanoreceptors or mechanosensitive nociceptors. These can be implemented by reducing or augmenting relevant mechanical forces or by specific pharmaceutics. Such novel approaches could be used to treat a broad range of cutaneous diseases.

14.2 Skin Disorders Associated with Mechanobiological Dysfunction

14.2.1 Keloid and Hypertrophic Scar

Wound-healing is a dynamic and complex process that consists of three temporal phases: inflammation, tissue formation (cell proliferation), and matrix remodeling (Singer and Clark 1999). During these processes, epithelial cells, fibroblasts, myofibroblasts, granulation tissue, and endothelial cells are subjected to many intrinsic and extrinsic mechanical stimuli. An example of the intrinsic stimuli is the forces that are produced by myofibroblasts, which cause cutaneous wounds to contract. Examples of extrinsic forces on the wound include scratching, compression, and the natural tension of the skin.

Keloids are the classical disorder of abnormal cutaneous wound-healing. Clinically, keloids can be very painful or itchy, and result in disfigurement that is a cosmetic nuisance and places a significant physical and psychological burden on the patient (Bock et al. 2006) (Figure 14.2a). Keloidal scarring is one of the most frustrating unresolved problems in the wound-healing field. Several studies have shown that a genetic predisposition may underlie the development of keloids (Nakashima et al. 2010; Shih and Bayat 2010; Ogawa et al. 2014). However, increasing lines of evidence show that the local microenvironment, particularly mechanical force distribution, also plays a significant role. For example, keloids show a marked preference for particular locations on the body: they usually occur at sites that are constantly or frequently subjected to mechanical forces (such as the anterior chest and scapular regions) and seldom occur in areas where stretching/contraction of the skin is rare (such as the parietal region or anterior lower leg), even in patients with multiple keloids (Ogawa et al. 2012). Keloids also commonly adopt distinct site-specific shapes, namely, butterfly, crab's claw, and dumbbell shapes. These shapes appear to be largely determined by the direction of the mechanical forces on and around the wound site (Akaishi et al. 2008a). The substratum stiffness of the keloid tissue itself also shapes how keloid fibroblasts grow (Figure 14.2b). Taken together, these observations suggest that deficiencies in mechanosignaling pathways explain the formation of keloids (Huang et al. 2012). Since these lesions develop in the dermis, an understanding of dermal mechanobiology will be essential to unraveling their underlying mechanisms.

Hyperreactivity or derangement of the mechanosensitive nociceptors of nerve fibers has also been suggested to cause or contribute to the generation of keloid and hypertrophic scars (Akaishi et al. 2008b). Mechanosensitive nociceptors on unmyelinated axons (C-fibers and Aδ-fibers) can be stimulated while the skin is stretched. In addition, axonal reflexes and the stimulation of antidromic sensory nerves result in the release of vasodilatory factors, including neuropeptides (e.g., substance P and calcitonin

(a)

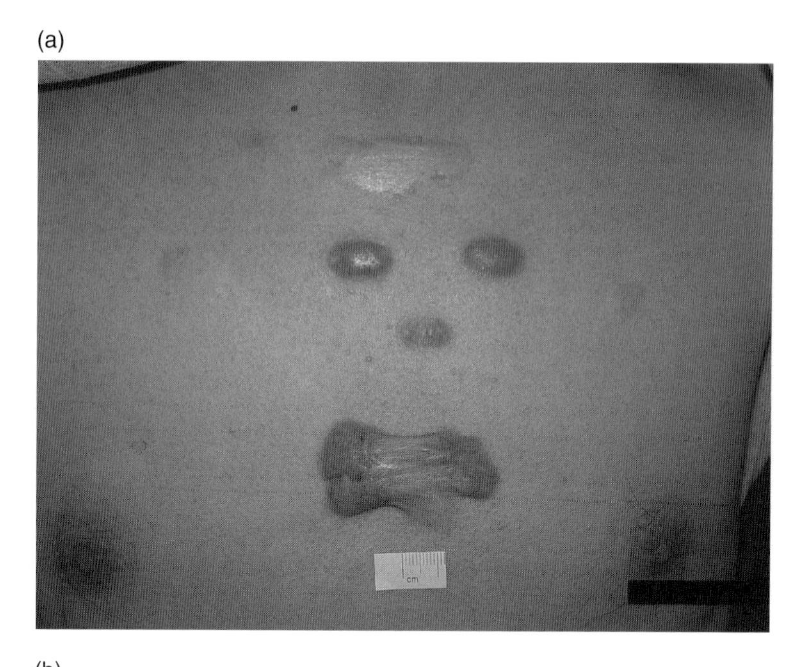

(b)

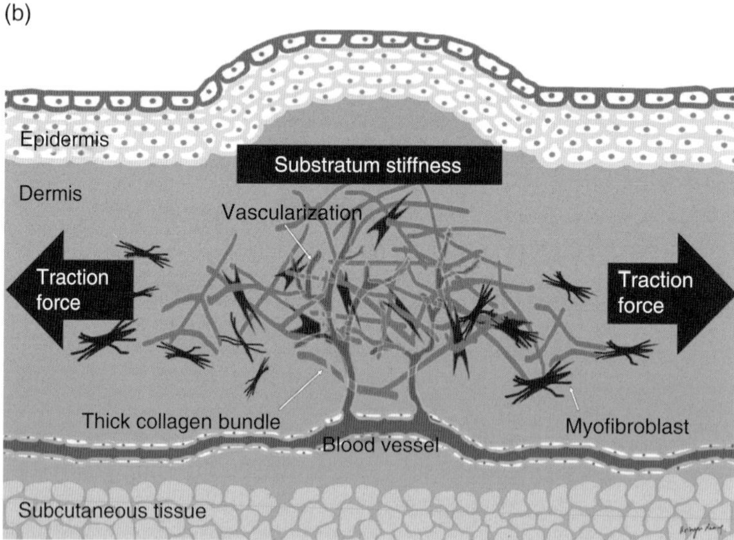

Figure 14.2 (a) Itchy and painful nodules on the anterior chest of a young male with keloids. (b) Keloid fibroblasts encounter both traction force and substratum stiffness.

gene-related peptide), which in turn induce local erythema and may cause fibroblasts and other dermal cells to increase their expression of genes that encode growth factors, such as transforming growth factor beta (TGF-β) and nerve growth factor (NGF). Thus, neurogenic inflammation is a mediator that plausibly connects mechanical forces with the development of abnormal scar progression and/or generation.

Harn et al. (2015) support the notion that perturbation of mechanoreceptor signaling contributes to the pathogenic behavior of keloid fibroblasts, namely, their propensity to migrate beyond the original wound boundaries (Tuan and Nichter 1998). Harn et al. (2015) used atomic force microscopy (AFM) and micropost array detectors to compare keloid fibroblasts with normal fibroblasts in terms of the elasticity of their actin filaments and the amount of force generated by the filaments. They found that when normal fibroblasts migrate, the elasticity and force generation of the actin filaments adopt a characteristic spatial distribution. However, this distribution is disrupted when the cell cannot recognize external forces due to pharmacological inhibition of focal adhesion kinase (FAK), which transmits external forces to the cytoskeleton of the cell. Keloid fibroblasts display the same disrupted distribution. This suggests that keloid fibroblasts are less responsive to mechanical stimuli from the local environment than are normal fibroblasts. This may explain the highly migratory nature of keloid fibroblasts and their ability to migrate outside the limits of the original wound.

Studies by Wong et al. (2011, 2012) have also started to unravel the molecular mechanobiological mechanisms that drive the formation of hypertrophic scars. Though keloids and hypertrophic scars both manifest as red and elevated scars, hypertrophic scars remain within the wound boundaries, and frequently regress spontaneously (Tuan and Nichter 1998). Aarabi et al. (2007) showed that when a full-thickness dorsal incision in mice is subjected to mechanical loading, the scar develops the characteristic morphology and histology of hypertrophic scars. More recently, they showed with this murine model of hypertrophic scar that the mechanical force exerts its fibrosis-inducing effects via the inflammatory FAK–extracellular-related kinase (ERK)–monocyte chemoattractant protein 1 (MCP-1) pathways. Interestingly, they also showed that molecular strategies targeting FAK effectively uncouple mechanical force from pathological scar formation (Wong et al. 2012).

14.2.2 Scleroderma

Scleroderma (systemic sclerosis) is a complex systemic autoimmune disease that primarily manifests in the skin. It is characterized by extensive fibrosis, vascular alterations, and autoantibodies (Gabrielli et al. 2009). It can be divided into two major subgroups: limited and diffuse cutaneous scleroderma. The former was previously termed "CREST syndrome," referring to its five main features: calcinosis, Raynaud's phenomenon, esophageal dysfunction, sclerodactyly, and telangiectasias. The cutaneous manifestations of the limited type mainly affect the hands, arms, and face, and are characterized by swelling, stiffness, or pain (Figure 14.3). In addition, up to one-third of patients develop pulmonary arterial hypertension (PAH), which is the most serious complication of this subtype. Diffuse scleroderma progresses rapidly and can involve a large area of the skin and one or more internal organs (most frequently the kidneys, esophagus, heart, and lungs). This subtype of scleroderma can be quite disabling.

As with the pathogenesis of keloid, dermal fibroblasts are considered to be the principal effector cells in scleroderma because they overproduce ECM components. Reich et al. (2009) investigated whether these functional alterations are accompanied by changes in the mechanical properties and morphology of fibroblasts. AFM of dermal fibroblasts from scleroderma patients or healthy donors showed that dermal fibroblasts derived from sclerodermal lesions were significantly softer than skin fibroblasts from

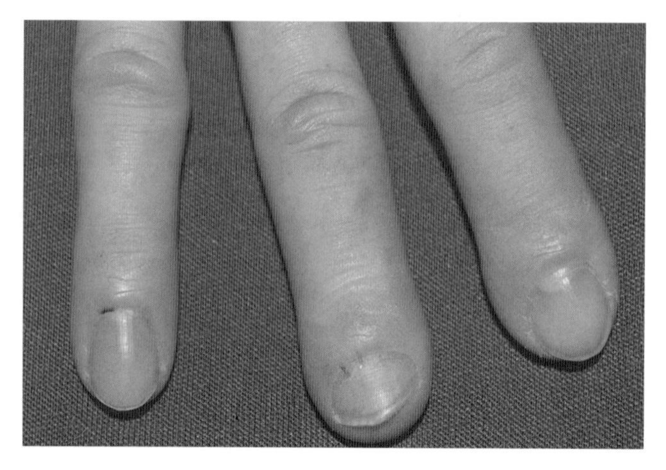

Figure 14.3 Painful swelling fingers in a patient with scleroderma. Periungual telangiectasia is one of the clinical features.

healthy subjects. The authors suggested that the altered stiffness of sclerodermal fibroblasts could lead to an abnormal cellular response to mechanical stimuli, and that this may play an important role in the pathogenesis of this disease.

Another explanation for the overproduction of ECM by sclerodermal lesion fibroblasts is the dysregulation of caveolin-1 (Del Galdo et al. 2008; Tourkina et al. 2008). Caveolin-1 is the most important member of a family of membrane proteins that are the main coating proteins of caveolae. Caveolae are 50 – 100 nm flask-shaped invaginations that form a morphologically identifiable subset of lipid rafts (Parton and Simons 2007). The functions of caveolin-1 include mechanosensing and lipid regulation. These may help caveolae to respond to plasma membrane changes, depending on their specialized lipid-composition and biophysical properties. Del Galdo et al. (2008) found that caveolin-1 expression was markedly decreased in the affected lungs and skin of scleroderma patients and that *Cav1*-knockout mice developed pulmonary and skin fibrosis. Tourkina et al. (2008) showed that restoring caveolin-1 function via treatment with a cell-permeable peptide may be a novel therapeutic approach in scleroderma. These data support the notion that there is a close association between scleroderma and mechanobiology.

14.2.3 Nail Disorders

The nail is part of the skin appendage and is very important in protecting the distal phalanges, enhancing tactile discrimination, and performing fine-manipulation and haptic tasks (Sano and Ogawa 2014). Nails are consistently exposed to physical stimulation during our daily activities, and mechanical force may contribute to normal and disordered nail configurations. For example, pincer nails (defined as the transverse overcurvature of the nail plate) tend to occur in bedridden patients whose feet are not subjected to shoes or weight-bearing (Sano and Ichioka 2012) (Figure 14.4). In addition, the degree of great toenail curvature tends to increase as the duration of the bedridden state rises. Moreover, in subjects with unilateral loading, the great toenail on the unloaded side is more curved than the great toenail on the loaded side. Similarly, in hemiplegia cases, the thumbnail on the palsy side is more curved than the thumbnail on

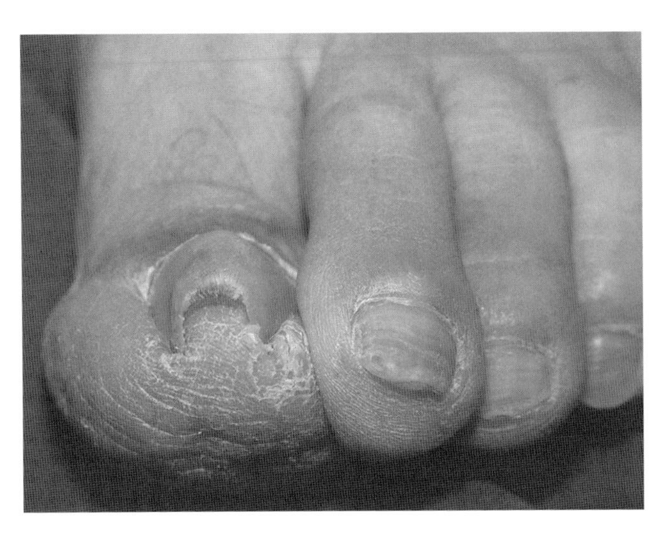

Figure 14.4 Typical appearance of pincer nail. Overcurvature of the nail plate is observed.

the nonpalsy side (Sano and Ogawa 2013). Our comparison of the thumbnails of healthy subjects with office-based versus carpentry professions confirmed that mechanical forces significantly affect nail curvature: the carpenters, who had a higher mean pinch strength than the office workers, also had lower nail curve indices (Sano et al. 2014). Further studies to elucidate the molecular and cellular mechanisms behind nail deformities are warranted.

14.2.4 Bullous Pemphigoid

Bullous pemphigoid is the most common human autoimmune blistering disease. It largely occurs in elderly people (Nishie 2014) and its clinical manifestations include tense blisters, erosions, and crusts, with itchy urticarial plaques or patches over the entire body (Figure 14.5). Bullous pemphigoid associates with a high mortality rate (19–26%).

Histological examination of biopsied lesional skin reveals subepidermal blisters with inflammatory infiltration of the dermis with eosinophils and lymphocytes. Typically, there is a linear deposit of immunoglobulin G (IgG) at the basement membrane zone and there are circulating antibodies to two hemidesmosomal components: transmembrane collagen XVII (BP180 or BPAG2) and plakin-family protein BP230 (BPAG1). Collagen XVII (COL17) is thought to be the main autoantigen (Ujiie et al. 2010; Schmidt and Zillikens 2013). The pathogenic autoantibodies decrease the strength with which keratinocytes adhere to the ECM. This ultimately results in the development of blisters between the keratinocyte layers and the dermis (Iwata et al. 2009).

Hemidesmosomes are very small stud- or rivet-like structures on the inner basal surface of epidermal keratinocytes (Walko et al. 2015). While desmosomes link two cells together, hemidesmosomes bind cells to the ECM. Therefore, hemidesmosomes play an important role in epidermal mechanobiology. The breakdown of the epidermal mechanobiological environment is the main reason for the clinical symptoms of bullous pemphigoid. Interestingly, Zhang et al. (2011) found that the hemidesmosome

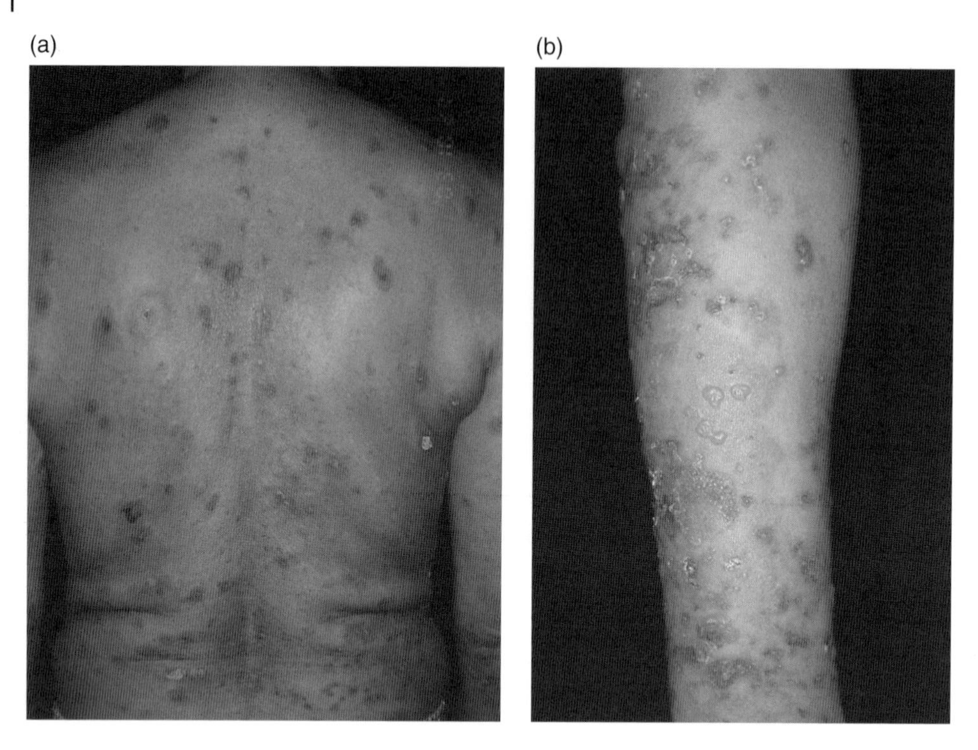

Figure 14.5 Tense blisters, erosions, and crusts with itchy urticarial plaques or patches on (a) the trunk and (b) the forearm of a patient with bullous pemphigoid.

of *Caenorhabditis elegans* is not only an attachment structure, but is also a mechanosensor that responds to tension by triggering signaling processes (Labouesse 2012). Muscle contractions of *C. elegans* stimulate the remodeling of hemidesmosome-like junctions in the epidermis, which may partially explain why bullous pemphigoid tends to occur in the bedridden elderly.

14.2.5 Ehlers–Danlos Syndrome (Cutis Hyperelastica)

EDS is a heterogeneous group of inherited connective-tissue disorders that are caused by genetic defects in the synthesis of type I, III, or V collagen (De Paepe and Malfait 2012; Sobey 2015). There are at least 10 recognized types of EDS. The current classification categorizes them into six subtypes: classic, hypermobility, vascular, kyphoscoliosis, arthrochalasis, and dermatosparaxis (Beighton et al. 1998). The severity of EDS can vary from mild to life-threatening. Treatment of affected individuals is unsatisfactory, and mainly consists of supportive strategies, with close monitoring of the digestive, the excretory, and, particularly, the cardiovascular systems. Therapeutic approaches, including physical therapy and corrective surgery, differ according to the particular disease manifestations.

Abnormal collagen levels lead to increased skin elasticity. A physical examination of the skin in EDS patients can reveal a velvety texture, fragility with easy tearing or bruising, redundant folds, molluscoid pseudotumors (especially on pressure points), subcutaneous spheroids, and fatty growth on the forearms or shins. Moreover, wound-healing and scarring are abnormal: the scars tend to be widened (broad) and atrophic (thin), and have been described as "cigarette-paper scars." The scars also bruise easily and are prone to cutaneous bleeding (Byers et al. 1997; Giunta et al. 1999). These characteristics are the result of deficient dermal ECM-based mechanobiological processes that decrease mechanotransduction input. The low-strength mechanobiological environment of the dermis may also explain why cultured fibroblasts from EDS patients produce lower levels of fibronectin than normal fibroblasts (Cutolo et al. 1986).

14.2.6 Cutis Laxa

Like EDS, cutis laxa (also known as chalazoderma, dermatochalasia, dermatolysis, dermatomegaly, generalized elastolysis, generalized elastorrhexis, and pachydermatocele) is a connective-tissue disorder. However, it is caused by a deficiency of elastic fibers rather than of collagen. Most cutis laxa cases are inherited, but some are acquired. Elastic fibers are bundles of elastin produced by dermal fibroblasts, which form a network in the dermis. They make up 2–5% of the dermis and play a vital role in skin resilience, texture, and quality (Rnjak et al. 2011). The deficiency of cutis laxa results in inelastic skin that hangs loosely in wrinkles and folds. The affected areas of skin may be thickened and dark. In addition, the joints are loose (i.e., hypermobile) due to the lax ligaments and tendons (Berk et al. 2012). Cutis laxa can also affect the internal organs, including inducing several severe impairments that involve the lungs, heart, intestines, and arteries (Berk et al. 2012). Patients have cutaneous wound disruption and poor scarring, though cosmetic surgery procedures (face-lifting) have been shown to be aesthetically and psychologically beneficial (Thomas et al. 1993). The abnormal wound-healing and scarring that occur in cutis laxa can be considered to be the result of dermal deficiencies in ECM-based mechanobiology.

14.2.7 Skin Aging

Aging is a complex multifactorial process that is characterized by an increased susceptibility to disease and tissue dysfunction. The features of skin aging include wrinkling, laxity, and pigmentary irregularities. In addition, the healing capacity of aged skin is markedly diminished, giving a high risk of chronic wounds. The aging process may be induced by environmental factors, such as sun exposure and smoking, or by chronological (intrinsic) factors (Halder and Ara 2003; Perner et al. 2011). The main microscopic feature of aged skin is the degeneration and fragmentation of the dermal collagen matrix, including the organization of the collagen and elastic fibers (solar elastosis); the destruction of the ECM in aged skin means that these cells no longer receive mechanical information, which causes the fibroblasts to collapse in the dermis. Consequently, their collagen production drops and their synthesis of collagen-degrading enzymes (MMPs) increases, leading to further ECM loss (Fligiel et al. 2003; Varani et al. 2001, 2006).

Injection of crosslinked hyaluronic acid dermal filler into photodamaged human skin may stimulate *de novo* collagen production and restore mechanical tension to the dermal matrix (Wang et al. 2007). Moreover, the sites surrounding hyaluronic acid deposition exhibit increased numbers of fibroblasts. Interestingly, these fibroblasts, which show a distinct elongated stretched appearance, express high levels of type I procollagen and of signaling molecules such as TGF-β and connective-tissue growth factor (CTGF)

(Wang et al. 2007; Fisher et al. 2008). Recently, Roper et al. (2015) found that mechanical stimulation of the skin with ultrasound can reverse fibroblast senescence and improve wound-healing rates. In addition, they found that the molecular mechanism underlying this effect was the activation of a calcium/CamKinaseII/Tiam1/Rac1 pathway that replaces fibronectin-dependent signaling and promotes fibroblast migration. Thus, Roper et al. (2015) nicely illustrate how mechanobiology can be manipulated in the clinical setting to yield improved outcomes.

14.2.8 Diabetic Skin Ulcers

Diabetes causes neuropathy, which inhibits nociception and the perception of pain. Consequently, diabetes patients often fail initially to notice small wounds that involve the legs and feet, and may therefore fail to prevent infection or repeated injury. Furthermore, diabetes causes immune compromise, together with characteristic damage to small blood vessels, which prevents adequate tissue oxygenation. This potentiates the development of chronic wounds. Pressure (compression) also plays a role in the formation of diabetic ulcers, as does infection. Regarding the mechanobiological aspects of the disease, fibronectin gene expression has been shown to be enhanced in hypertrophic scars but decreased in diabetic foot ulcers compared with normal skin (Fu et al. 2002). A deficiency in the mechanosensitive nociceptors of nerve fibers has also been postulated as causing or contributing to the generation of diabetic ulcers. Damage to sensory nerve fibers can lead to "neurogenic inflammation," which is a form of mechanical stress-induced inflammation in which there is cutaneous antidromic vasodilatation and plasma extravasation. In neurogenic inflammation, neuropeptides that are released by sensory nerve endings accelerate the production of cytokines by various activated cell types. Under normal conditions, neurogenic inflammation combined with neurovascular control promotes cutaneous wound-healing; this is not the case in diabetic patients (Schaper et al. 2008). Methods that repair or sidestep mechanosensitive nociceptor damage in diabetes may help to improve wound-healing in these patients.

14.2.9 Leprosy

Leprosy, also known as Hansen's disease (HD), is a chronic condition caused by the bacteria *Mycobacterium leprae* and *Mycobacterium lepromatosis* (Sasaki et al. 2001). The bacteria have a preference for cooler temperatures and therefore tend to affect superficial structures (i.e., the skin) rather than the deep visceral organs. Loss of sensation and destruction of the intraepidermal innervation are characteristic findings of leprosy (Facer et al. 2000). An interesting feature of the disease is that extensive scarring is absent. This suggests that mechanosensitive nociceptors involved in initiating the scarring process are denervated in these patients.

14.2.10 Lymphedema

The lymphatic system plays an important role in overall health through its contributions to ECF and protein homeostasis, lipid transport, and immunity (Breslin 2014). The lymphatic system is organized as a vessel tree, with the most distal blind-ended vessels serving as the site where interstitial fluid enters the system to form lymph.

Lymphedema is characterized by localized fluid retention and tissue-swelling, caused by a compromised lymphatic system (Brandon Dixon and Weiler 2015). In lymphedema patients, the malfunction of this system elevates the risk of skin infection.

Lymphatic fluid returns to the blood circulation through the force of osmosis, which acts on the venous capillaries. Prior to its return, the proteins, cellular debris, and bacteria contained in the lymph are filtered through lymph collectors, which are blind-ended and epithelial-lined. Once the lymph enters the valved lymphatic vessels, the primary driving force for its movement is the rhythmic peristaltic-like pumping action of the smooth-muscle cells (SMCs) lining the lymphatic vessel walls. A complex network of innervation regulates this system. In addition, lymphatic endothelial cells have the ability to sense changes in the amount of interstitial fluid. This ability allows them to control lymphatic vessel expansion (Planas-Paz and Lammert 2014). β1 integrin and VEGFR3 signaling have been found to be required for the mechanoinduced proliferation of lymphatic endothelial cells (Planas-Paz et al. 2012).

Lymphedema can be classified as primary or secondary, depending on the underlying mechanism. Primary lymphedema is a rare genetic disease that is caused by mutations in critical lymphangiogenic genes (Brouillard et al. 2014). Most cases belong to secondary lymphedema, which is caused by surgery on the lymph nodes, radiotherapy, or filarial infection. Lymphedema can be considered to be the result of mechanobiological dysfunction that affects the osmosis and fluid pressure within the lymphatic system (Breslin 2014; Brandon Dixon and Weiler 2015). Current palliative treatments for lymphedema are the use of compressive garments and lymphatic massage, which together decrease limb volume by 40–60% (Warren et al. 2007). However, these treatments are not curative. Studies that assess whether lymphedema can be reduced by providing the region that requires lymphatic vessel regeneration with lymphatic massage are needed (Planas-Paz and Lammert 2014).

14.3 Mechanotherapy

The term "mechanotherapy" was first used in the 19th century, and was initially defined as "the employment of mechanical means for the cure of disease." Thanks to the rapid progress in modern molecular biology, biomechanics, and tissue engineering, it can now be extended to denote "therapeutic interventions that reduce and reverse injury to damaged tissues or promote the homeostasis of healthy tissues by mechanical means at the molecular, cellular, or tissue level" (Huang et al. 2013). The future development of novel mechanotherapies will require integration of several disciplines, including:

1) Mechanobiology: The study of the processes by which physical forces are sensed (mechanosensing), transduced (mechanotransduction), and then transformed into intracellular biochemical and gene-expression changes (mechanoresponse).
2) Bioinformatics: The combination of computer science and information technology to aid in the accurate study of the mechanoresponses.
3) Tissue engineering and regenerative medicine: The application and manipulation of mechanobiology to produce mechanically viable and functionally active products capable of preventing or treating diseases.

Several successful mechanotherapies for cutaneous and soft-tissue wounds are currently in use, including microdeformational wound therapy, shockwave therapy, soft-tissue expansion, distraction osteogenesis, and reduction of surgical tension. To date, these methods have not been used to manage the mechanobiological disorders in skin described in the previous section. We believe that such therapies, which control the macroscopic mechanical environment, could be useful for these disorders. Novel future methods that control the microscopic mechanical environment, such as pharmaceutics that block or accelerate mechanosensor function at the molecular level, are also likely to be useful.

14.4 Conclusion

The development of the mechanobiology field has allowed us to view skin disorders in another light. Human skin is a highly specialized mechanoresponsive organ, and several skin disorders have been shown in this chapter to be related to dysregulated mechanosensation, mechanotransduction, and/or mechanoresponsiveness. Clinically, therapeutic strategies wherein mechanical force is provided or reduced, or novel agents inhibit or stimulate mechanoreceptors or mechanosensitive nociceptors, are likely to be successful.

To accelerate skin growth and dermatogenesis, skin-stretching strategies and devices have been developed (Chin et al. 2009b). The optimal amplitude and waveform of skin tension may facilitate skin growth and expansion, but excessive tension can cause heavy scarring (Chin et al. 2009a,b). Static and periodic application of tensile force to rat ears shows vascular remodeling and epidermal proliferation (Pietramaggiori et al. 2007). A gene chip analysis performed on the same rat model suggests tissue-level hypoxia as a possible mechanism for the observed effects (Saxena et al. 2007). In addition, prior *in vitro* studies have shown that mechanotransduction mechanisms can stimulate cell proliferation (Chin et al. 2009b) and angiogenesis (Ingber et al. 1995).

A sophisticated servocontrolled device was used to stretch murine dorsal skin, and stretched samples showed upregulated epidermal proliferation and angiogenesis (Chin et al. 2009b). Real-time polymerase chain reaction (RT-PCR) revealed that epidermal growth factor (EGF), NGF, vascular endothelial growth factor (VEGF), and TGF-β1 were more strongly expressed in cyclically stretched than in statically stretched skin (Chin et al. 2009b). This cyclical stimulation also significantly increased skin neuropeptide accumulation, while the corresponding peptide receptors were downregulated (Chin et al. 2009a). This study demonstrates that neuropeptides are produced in resident skin cells. Though neuropeptide release from the peripheral nerve-fiber terminals was not shown, the study does prove that neuropeptides are associated with the process of skin stretching.

Acknowledgement

We thank Hsin-Yu Huang for his assistance in drawing the figures.

References

Aarabi S., K. A. Bhatt, Y. Shi, J. Paterno, E. I. Chang, S. A. Loh, et al. 2007. "Mechanical load initiates hypertrophic scar formation through decreased cellular apoptosis." *FASEB Journal* **21**(12): 3250–61. doi:10.1096/fj.07-8218com.

Akaishi, S., M. Akimoto, R. Ogawa, and H. Hyakusoku. 2008a. "The relationship between keloid growth pattern and stretching tension: visual analysis using the finite element method." *Annals of Plastic Surgery* **60**(4): 445–51. doi:10.1097/SAP.0b013e3181238dd7.

Akaishi, S., R. Ogawa, and H. Hyakusoku. 2008b. "Keloid and hypertrophic scar: neurogenic inflammation hypotheses." *Medical Hypotheses* **71**(1): 32–8. doi:10.1016/j.mehy.2008.01.032.

Beighton, P., A. De Paepe, B. Steinmann, P. Tsipouras, and R. J. Wenstrup. 1998. "Ehlers-Danlos syndromes: revised nosology, Villefranche, 1997. Ehlers-Danlos National Foundation (USA) and Ehlers-Danlos Support Group (UK)." *American Journal of Medical Genetics* **77**(1): 31–7. PMID:9557891.

Berk, D. R., D. D. Bentley, S. J. Bayliss, A. Lind, and Z. Urban. 2012. "Cutis laxa: a review." *Journal of the American Academy of Dermatology* **66**(5): 842.e1–17. doi:10.1016/j.jaad.2011.01.004.

Bock, O., G. Schmid-Ott, P. Malewski, and U. Mrowietz. 2006. "Quality of life of patients with keloid and hypertrophic scarring." *Archives of Dermatological Research* **297**(10): 433–8. doi:10.1007/s00403-006-0651-7.

Brandon Dixon, J. and M. J. Weiler. 2015. "Bridging the divide between pathogenesis and detection in lymphedema." *Seminars in Cell & Developmental Biology* **38**: 75–82. doi:10.1016/j.semcdb.2014.12.003.

Breslin, J. W. 2014. "Mechanical forces and lymphatic transport." *Microvascular Research* **96**: 46–54. doi:10.1016/j.mvr.2014.07.013.

Brouillard, P., L. Boon, and M. Vikkula. 2014. "Genetics of lymphatic anomalies." *Journal of Clinical Investigation* **124**(3): 898–904. doi:10.1172/JCI71614.

Byers, P. H., M. Duvic, M. Atkinson, M. Robinow, L. T. Smith, S. M. Krane, et al. 1997. "Ehlers-Danlos syndrome type VIIA and VIIB result from splice-junction mutations or genomic deletions that involve exon 6 in the COL1A1 and COL1A2 genes of type I collagen." *American Journal of Medical Genetics* **72**(1): 94–105. PMID:9295084.

Chin M. S., L. Lancerotto, D. L. Helm, P. Dastouri, M. J. Prsa, M. Ottensmeyer, et al. 2009a. "Analysis of neuropeptides in stretched skin." *Plastic and Reconstructive Surgery* **124**(1): 102–13. doi:10.1097/PRS.0b013e3181a81542.

Chin M. S., R. Ogawa, L. Lancerotto, G. Pietramaggiori, K. Schomacker, S. S. Scherer, et al. 2009b. "In vivo acceleration of skin growth using a servo-controlled stretching device." *Tissue Engineering. Part C, Methods* **16**(3): 397–405. doi:10.1089/ten.TEC.2009.0185.

Cutolo, M., P. Castellani, L. Borsi, and L. Zardi. 1986. "Altered fibronectin distribution in cultured fibroblasts from patients with Ehlers-Danlos syndrome." *Clinical and Experimental Rheumatology* **4**(2): 125–8. PMID:3524938.

De Paepe, A. and F. Malfait. 2012. "The Ehlers-Danlos syndrome, a disorder with many faces." *Clinical Genetics* **82**(1): 1–11. doi:10.1111/j.1399-0004.2012.01858.x.

Del Galdo, F., F. Sotgia, C. J. de Almeida, J. F. Jasmin, M. Musick, M. P. Lisanti, and S. A. Jimenez. 2008. "Decreased expression of caveolin 1 in patients with systemic sclerosis: crucial role in the pathogenesis of tissue fibrosis." *Arthritis and Rheumatism* **58**(9): 2854–65. doi:10.1002/art.23791.

Facer, P., D. Mann, R. Mathur, S. Pandya, U. Ladiwala, B. Singhal, et al. 2000. "Do nerve growth factor-related mechanisms contribute to loss of cutaneous nociception in leprosy?" *Pain* **85**(1–2): 231–8. PMID:10692623.

Fisher, G. J., J. Varani, and J. J. Voorhees. 2008. "Looking older: fibroblast collapse and therapeutic implications." *Archives of Dermatology* **144**(5): 666–72. doi:10.1001/archderm.144.5.666.

Fligiel, S. E., J. Varani, S. C. Datta, S. Kang, G. J. Fisher, and J. J. Voorhees. 2003. "Collagen degradation in aged/photodamaged skin in vivo and after exposure to matrix metalloproteinase-1 in vitro." *Journal of Investigative Dermatology* **120**(5): 842–8. doi:10.1046/j.1523-1747.2003.12148.x.

Fu, X., Y. Yang, T. Sun, Y. Wang, and Z. Sheng. 2002. "Comparative study of fibronectin gene expression in tissues from hypertrophic scars and diabetic foot ulcers." *Chinese Medical Sciences Journal* **17**(2): 90–4. PMID:12906161.

Gabrielli, A., E. V. Avvedimento, and T. Krieg. 2009. "Scleroderma." *New England Journal of Medicine* **360**(19): 1989–2003. doi:10.1056/NEJMra0806188.

Giunta, C., A. Superti-Furga, S. Spranger, W. G. Cole, and B. Steinmann. 1999. "Ehlers-Danlos syndrome type VII: clinical features and molecular defects." *Journal of Bone and Joint Surgery. American Volume* **81**(2): 225–38. PMID:10073586.

Halder, R. M. and C. J. Ara. 2003. "Skin cancer and photoaging in ethnic skin." *Dermatologic Clinics* **21**(4): 725–32, x. PMID:14717413.

Harn, H. I., Y. K. Wang, C. K. Hsu, Y. T. Ho, Y. W. Huang, W. T. Chiu, et al. 2015. "Mechanical coupling of cytoskeletal elasticity and force generation is crucial for understanding the migrating nature of keloid fibroblasts." *Experimental Dermatology* **24**(8): 579–84. doi:10.1111/exd.12731.

Huang, C., S. Akaishi, and R. Ogawa. 2012. "Mechanosignaling pathways in cutaneous scarring." *Archives of Dermatological Research* **304**(8): 589–97. doi:10.1007/s00403-012-1278-5.

Huang, C., J. Holfeld, W. Schaden, D. Orgill, and R. Ogawa. 2013. "Mechanotherapy: revisiting physical therapy and recruiting mechanobiology for a new era in medicine." *Trends in Molecular Medicine* **19**(9): 555–64. doi:10.1016/j.molmed.2013.05.005.

Ingber, D. E., D. Prusty, Z. Sun, H. Betensky, N. Wang. 1995 "Cell shape, cytoskeletal mechanics, and cell cycle control in angiogenesis." *Journal of Biomechanics* **28**(12): 1471–84. PMID:8666587.

Iwata, H., N. Kamio, Y. Aoyama, Y. Yamamoto, Y. Hirako, K. Owaribe, and Y. Kitajima. 2009. "IgG from patients with bullous pemphigoid depletes cultured keratinocytes of the 180-kDa bullous pemphigoid antigen (type XVII collagen) and weakens cell attachment." *Journal of Investigative Dermatology* **129**(4): 919–26. doi:10.1038/jid.2008.305.

Labouesse, M. 2012. "Role of the extracellular matrix in epithelial morphogenesis: a view from C. elegans." *Organogenesis* **8**(2): 65–70. doi:10.4161/org.20261.

Nakashima, M., S. Chung, A. Takahashi, N. Kamatani, T. Kawaguchi, T. Tsunoda, et al. 2010. "A genome-wide association study identifies four susceptibility loci for keloid in the Japanese population." *Nature Genetics* **42**(9): 768–71. doi:10.1038/ng.645.

Nishie, W. 2014. "Update on the pathogenesis of bullous pemphigoid: an autoantibody-mediated blistering disease targeting collagen XVII." *Journal of Dermatological Science* **73**(3): 179–86. doi:10.1016/j.jdermsci.2013.12.001.

Ogawa, R. 2008. "Keloid and hypertrophic scarring may result from a mechanoreceptor or mechanosensitive nociceptor disorder." *Medical Hypotheses* **71**(4): 493–500. doi:10.1016/j.mehy.2008.05.020.

Ogawa, R., K. Okai, F. Tokumura, K. Mori, Y. Ohmori, C. Huang, et al. 2012. "The relationship between skin stretching/contraction and pathologic scarring: the important role of mechanical forces in keloid generation." *Wound Repair and Regeneration* **20**(2): 149–57. doi:10.1111/j.1524-475X.2012.00766.x.

Ogawa, R., A. Watanabe, B. Than Naing, M. Sasaki, A. Fujita, S. Akaishi, et al. 2014. "Associations between keloid severity and single-nucleotide polymorphisms: importance of rs8032158 as a biomarker of keloid severity." *Journal of Investigative Dermatology* **134**(7): 2041–3. doi:10.1038/jid.2014.71.

Parton, R. G. and K. Simons. 2007. "The multiple faces of caveolae." *Nature Reviews. Molecular Cell Biology* **8**(3): 185–94. doi:10.1038/nrm2122.

Perner, D., A. Vierkotter, D. Sugiri, M. Matsui, U. Ranft, C. Esser, et al. 2011. "Association between sun-exposure, smoking behaviour and plasma antioxidant levels with the different manifestation of skin ageing signs between Japanese and German women – a pilot study." *Journal of Dermatological Science* **62**(2): 138–40. doi:10.1016/j.jdermsci.2011.02.010.

Pietramaggiori, G., P. Liu, S. S. Scherer, A. Kaipainen, M. J. Prsa, H. Mayer, et al. 2007 "Tensile forces stimulate vascular remodeling and epidermal cell proliferation in living skin." *Annals of Surgery* **246**(5): 896–902. doi:10.1097/SLA.0b013e3180caa47f.

Planas-Paz, L. and E. Lammert. 2014. "Mechanosensing in developing lymphatic vessels." *Advances in Anatomy, Embryology, and Cell Biology* **214**: 23–40. doi:10.1007/978-3-7091-1646-3_3.

Planas-Paz, L., B. Strilic, A. Goedecke, G. Breier, R. Fassler, and E. Lammert. 2012. "Mechanoinduction of lymph vessel expansion." *EMBO Journal* **31**(4): 788–804. doi:10.1038/emboj.2011.456.

Reich, A., M. Meurer, B. Eckes, J. Friedrichs, and D. J. Muller. 2009. "Surface morphology and mechanical properties of fibroblasts from scleroderma patients." *Journal of Cellular and Molecular Medicine* **13**(8B): 1644–52. doi:10.1111/j.1582-4934.2008.00401.x.

Rnjak, J., S. G. Wise, S. M. Mithieux, and A. S. Weiss. 2011. "Severe burn injuries and the role of elastin in the design of dermal substitutes." *Tissue Engineering. Part B, Reviews* **17**(2): 81–91. doi:10.1089/ten.TEB.2010.0452.

Roper, J. A., R. C. Williamson, B. Bally, C. A. Cowell, R. Brooks, P. Stephens, et al. 2015. "Ultrasonic stimulation of mouse skin reverses the healing delays in diabetes and aging by activation of Rac1." *Journal of Investigative Dermatology* **135**(11): 2842–51. doi:10.1038/jid.2015.224.

Sano, H. and S. Ichioka. 2012. "Influence of mechanical forces as a part of nail configuration." *Dermatology* **225**(3): 210–14. doi:10.1159/000343470.

Sano, H. and R. Ogawa. 2013. "Role of mechanical forces in hand nail configuration asymmetry in hemiplegia: an analysis of four hundred thumb nails." *Dermatology* **226**(4): 315–18. doi:10.1159/000350260.

Sano, H. and R. Ogawa. 2014. "Clinical evidence for the relationship between nail configuration and mechanical forces." *Plastic and Reconstructive Surgery. Global Open* **2**(3): e115. doi:10.1097/GOX.0000000000000057.

Sano, H., K. Shionoya, and R. Ogawa. 2014. "Effect of mechanical forces on finger nail curvature: an analysis of the effect of occupation on finger nails." *Dermatologic Surgery* **40**(4): 441–5. doi:10.1111/dsu.12439.

Sasaki, S., F. Takeshita, K. Okuda, and N. Ishii. 2001. "Mycobacterium leprae and leprosy: a compendium." *Microbiology and Immunology* **45**(11): 729–36. PMID:11791665.

Schaper, N. C., M. Huijberts, and K. Pickwell. 2008. "Neurovascular control and neurogenic inflammation in diabetes." *Diabetes/Metabolism Research and Reviews* **24**(Suppl. 1): S40–4. doi:10.1002/dmrr.862.

Schmidt, E. and D. Zillikens. 2013. "Pemphigoid diseases." *Lancet* **381**(9863): 320–32. doi:10.1016/S0140-6736(12)61140-4.

Shih, B. and A. Bayat. 2010. "Genetics of keloid scarring." *Archives of Dermatological Research* **302**(5): 319–39. doi:10.1007/s00403-009-1014-y.

Singer, A. J. and R. A. Clark. 1999. "Cutaneous wound healing." *New England Journal of Medicine* **341**(10): 738–46. doi:10.1056/NEJM199909023411006.

Saxena, V., D. Orgill, and I. Kohane. 2007 "A set of genes previously implicated in the hypoxia response might be an important modulator in the rat ear tissue response to mechanical stretch." *BMC Genomics* **23**(8): 430. doi:10.1186/1471-2164-8-430.

Sobey, G. 2015. "Ehlers-Danlos syndrome: how to diagnose and when to perform genetic tests." *Archives of Disease in Childhood* **100**(1): 57–61. doi:10.1136/archdischild-2013-304822.

Thomas, W. O., M. H. Moses, R. D. Craver, and W. K. Galen. 1993. "Congenital cutis laxa: a case report and review of loose skin syndromes." *Annals of Plastic Surgery* **30**(3): 252–6. PMID:8494307.

Tourkina, E., M. Richard, P. Gooz, M. Bonner, J. Pannu, R. Harley, et al. 2008. "Antifibrotic properties of caveolin-1 scaffolding domain in vitro and in vivo." *American Journal of Physiology. Lung Cellular and Molecular Physiology* **294**(5): L843–61. doi:10.1152/ajplung.00295.2007.

Trappmann, B., J. E. Gautrot, J. T. Connelly, D. G. Strange, Y. Li, M. L. Oyen, et al. 2012. "Extracellular-matrix tethering regulates stem-cell fate." *Nature Materials* **11**(7): 642–9. doi:10.1038/nmat3339.

Tuan, T. L. and L. S. Nichter. 1998. "The molecular basis of keloid and hypertrophic scar formation." *Molecular Medicine Today* **4**(1): 19–24. doi:10.1016/S1357-4310(97)80541-2.

Ujiie, H., A. Shibaki, W. Nishie, and H. Shimizu. 2010. "What's new in bullous pemphigoid." *Journal of Dermatology* **37**(3): 194–204. doi:10.1111/j.1346-8138.2009.00792.x.

Varani, J., D. Spearman, P. Perone, S. E. Fligiel, S. C. Datta, Z. Q. Wang, et al. 2001. "Inhibition of type I procollagen synthesis by damaged collagen in photoaged skin and by collagenase-degraded collagen in vitro." *American Journal of Pathology* **158**(3): 931–42. doi:10.1016/S0002-9440(10)64040-0.

Varani, J., M. K. Dame, L. Rittie, S. E. Fligiel, S. Kang, G. J. Fisher, and J. J. Voorhees. 2006. "Decreased collagen production in chronologically aged skin: roles of age-dependent alteration in fibroblast function and defective mechanical stimulation." *American Journal of Pathology* **168**(6): 1861–8. doi:10.2353/ajpath.2006.051302.

Walko, G., M. J. Castanon, and G. Wiche. 2015. "Molecular architecture and function of the hemidesmosome." *Cell Tissue Research* **360**(2): 363–78. doi:10.1007/s00441-014-2061-z.

Wang, F., L. A. Garza, S. Kang, J. Varani, J. S. Orringer, G. J. Fisher, and J. J. Voorhees. 2007. "In vivo stimulation of de novo collagen production caused by cross-linked hyaluronic acid dermal filler injections in photodamaged human skin." *Archives of Dermatology* **143**(2): 155–63. doi:10.1001/archderm.143.2.155.

Warren, A. G., H.Brorson, L. J. Borud, and S. A. Slavin SA. 2007. “Lymphedema: a comprehensive review.” *Annals of Plastic Surgery* **59**(4): 464–72. doi:10.1097/01.sap.0000257149.42922.7e.

Wilhelmi, B. J., S. J. Blackwell, and L. G. Phillips. 1999. “Langer’s lines: to use or not to use.” *Plastic and Reconstructive Surgery* **104**(1): 208–14. PMID:10597698.

Wong, V. W., S. Akaishi, M. T. Longaker, and G. C. Gurtner. 2011. “Pushing back: wound mechanotransduction in repair and regeneration.” *Journal of Investigative Dermatology* **131**(11): 2186–96. doi:10.1038/jid.2011.212.

Wong, V. W., K. C. Rustad, S. Akaishi, M. Sorkin, J. P. Glotzbach, M. Januszyk, et al. 2012. “Focal adhesion kinase links mechanical force to skin fibrosis via inflammatory signaling.” *Nature Medicine* **18**(1): 148–52. doi:10.1038/nm.2574.

Zhang, H., F. Landmann, H. Zahreddine, D. Rodriguez, M. Koch, and M. Labouesse. 2011. “A tension-induced mechanotransduction pathway promotes epithelial morphogenesis.” *Nature* **471**(7336): 99–103. doi:10.1038/nature09765.

15

Mechanobiology and Mechanotherapy for Cutaneous Wound-Healing

Chenyu Huang[1], Yanan Du[2], and Rei Ogawa[3]

[1] *Department of Plastic, Reconstructive and Aesthetic Surgery, Beijing Tsinghua Changgung Hospital, Tsinghua University, Beijing, China*
[2] *Department of Biomedical Engineering, School of Medicine, Tsinghua University, Beijing, China*
[3] *Department of Plastic, Reconstructive and Aesthetic Surgery, Nippon Medical School, Tokyo, Japan*

15.1 Introduction

Acute or chronic wounds that arise from trauma, surgery, infection, poor circulation, or other causes are common and are a major burden on health care systems. As a result, the healing process has long been investigated by both clinicians and researchers. A "wound" is usually defined as the local disruption of the normal anatomical structure and function (Lazarus et al. 1994). Wounds vary in depth: a skin wound may include only the epidermis, but can also involve the dermis, subcutaneous tissue, and deeper structures, such as muscles and bones, along with their dominant vessels and nerves. It can also extend to internal adjacent organs. Proper wound-healing restores the anatomical continuity and, more importantly, the functions of the tissues that are wounded (Lazarus et al. 1994).

The normal healing cascade is an orderly and efficient process with four phases: hemostasis, inflammation, re-epithelialization/proliferation, and remodeling. The initial inflammatory phase is characterized by infiltration of neutrophils and macrophages, while the proliferation phase involves fibroblast accumulation, angiogenesis, and collagen deposition. Thereafter, collagen is crosslinked in the remodeling phase (Huang et al. 2013c). Chronic wounds, which largely consist of venous stasis ulcers, pressure ulcers in immobile patients, and diabetes mellitus (DM) ulcers wounds (Nwomeh et al. 1998), are the result of a deranged healing cascade that leads to delayed, impaired, incomplete, deficient, or failed healing. Chronic wounds associate particularly with pathologically prolonged inflammation and insufficient matrix deposition. This results in clinical chronicity and a high risk of relapse. Associated complications, such as ischemia and infection, can further impair the healing of these wounds.

Since skin is the largest organ in the human body, it is particularly subject to wounding It is mechanoresponsive and is constantly exposed to both extrinsic and intrinsic mechanical forces. This suggests that cutaneous wound-healing occurs against a backdrop of omnipresent mechanical stimuli that can shape the healing process. Indeed, many lines of evidence show that mechanical stimuli can both promote and derange the

Mechanobiology: Exploitation for Medical Benefit, First Edition. Edited by Simon C. F. Rawlinson.

skin wound-healing process. As a result, there has been a flurry of research in recent years into the mechanobiological mechanisms that influence and promote wound-healing. In particular, researchers in a variety of fields have become interested in mechanotransduction; that is, the mechanism by which external physical forces are converted into internal biochemical signals and integrated into cellular responses (Ingber 2003a; Huang et al. 2004). This research has in turn led to profound interest in the development of mechanotherapies; that is, therapeutic strategies that apply or alter the mechanical forces affecting the skin, with the aim of improving the healing of wounds or promoting the homeostasis of healthy tissues (Huang et al. 2013a).

This chapter will summarize what is currently known about the mechanobiology of cutaneous wound-healing. The mechanotherapies in use today, or which have the potential to be useful clinically, will also be described. A better understanding of basic mechanotransduction and the potential of clinical mechanotherapy will facilitate research into this promising field, and could lead to the development of strategies that restore the mechanical equilibrium of skin and improve wound-healing.

15.2 The Mechanobiology of Cutaneous Wound-Healing

In ideal normal wound-healing, there is a balance between excessive healing, which for example causes hypertrophic scars and keloids, and insufficient healing, which causes diabetic, pressure, and venous ulcers. This reflects an underlying balance at the microscopic level between the individual cells in the wounded epithelium and dermis layers. In part, this balance is shaped by mechanical stimuli that are sensed by the cells and converted into intracellular biochemical signals, which eventually change the shape and function of the cells and their interactions with their neighbors, ultimately leading to changes in the wounded tissues that result in macroscopic healing.

In normal skin, the cells and extracellular matrix (ECM) both collect mechanical information and actively interact with each other. Healthy dermis has an elasticity of 1–5 kPa (Young's modulus), whereas fat has an elasticity of 0.1–1.0 kPa. However, fibroproliferation with concomitant collagen accumulation in the ECM can increase dermis rigidity to 20–100 kPa, which is similar to the elasticity of tendons (Hinz 2009). Thus, the cells shape the mechanical properties of the ECM by secreting its components (e.g., type I collagen) and the ECM in turn markedly influences the behavior of its cellular creators. Multiple studies have shown that the mechanical properties of the ECM influence cell behavior. For example, soft, stiffer, and rigid matrices move the *in vitro* lineage differentiation of human mesenchymal stem cells (MSCs) toward neurogenesis, myogenesis, and osteogenesis, respectively (Engler et al. 2006). Moreover, epidermal keratinocytes migrate faster on harder surfaces than on soft surfaces, and proliferate faster on stiff surfaces (Wang et al. 2012). Similarly, when dermal fibroblasts are placed on flexible polyacrylamide sheets coated with collagen I, they prefer to migrate from the soft side to the stiff side (Lo et al. 2000).

The mechanical forces from the ECM affect cell behavior by triggering cell-surface mechanoreceptors, which in turn alter the architecture and pre-stress (isometric tension) levels of the cytoskeleton (Ingber 2003b). This change in the tensegrity (tense integrity) of the cytoskeleton protects the cell from mechanical stimulus-induced damage by evenly distributing the mechanical load throughout the cell. It also allows small mechanical stimuli to affect a large number of cells (Ingber 1997; Myers et al. 2007). The cell-surface mechanoreceptors that shape cell tensegrity are the integrins and focal adhesions (FAs). Integrins are heterodimeric glycoproteins that span the cell membrane; they are connected at one end to the cytoskeleton and at the other to the ECM (Ingber 2006). FAs act as mechanosensory organelles that connect the ECM to the actin cytoskeleton and link integrins to the ends of contractile microfilament bundles (Geiger and Bershadsky 2002; Ingber 2008).

To achieve the goals of proper wound-healing, both excessive and insufficient wounds must be dealt with. Hypertrophic scars and keloids are fibroproliferative disorders that are characterized by excessive cutaneous wound-healing; that is, the exaggerated accumulation of fibroblasts and their main connective-tissue product, collagen I. Recently, it was shown that the development and progression of these pathological scars correlate closely with the topical physical tension on and movements of the scarred region (Akaishi et al. 2008a; Ogawa et al. 2012). This observation led to the development of surgical techniques that reduce the skin tension on wounds (e.g., the small wave-incision method), which effectively decrease the postsurgical development of pathological scars (Huang et al. 2012b; Huang and Ogawa 2014).

The polar opposite of a pathological scar is a chronic wound; namely, venous, pressure, and diabetic ulcers, which are all caused by insufficient or incomplete wound-healing. A better understanding of the mechanobiology of wound-healing may promote the development of novel therapeutic approaches to chronic ulcers.

Keratinocytes, fibroblasts, and myofibroblasts are the predominant epidermal and dermal cell populations in the skin. They all actively participate in the healing process. They are also sensitive to mechanical stimuli and have numerous mechanosensitive signaling pathways, including the transforming growth factor beta (TGF-β)/Smad, FA/integrin, calcium-ion, mitogen-activated protein kinase (MAPK)/G-protein, Wnt/β-catenin, tumor necrosis factor alpha (TNF-α)/nuclear factor kappa-light-chain-enhancer of activated B cells (NF-κB), and interleukin (IL) pathways (Huang et al. 2012a).

After wounding, the wound edge undergoes re-epithelialization. This process starts with the migration of keratinocytes toward the wound center. Shortly after the migrating epithelial tongue has started to migrate, the keratinocytes proliferate to ensure that there are enough cells to cover the wound (Pastar et al. 2014). The migration of the keratinocytes involves numerous mechanical forces, including the protrusive force of the leading cells, intercellular contractile force, and traction force on the cells that are immediately adjacent to the keratinocytes (du Roure et al. 2005). Keratinocyte migration can be shaped by applying exogenous physical forces. For example, exogenous mechanical stretching accelerates wound closure by activating mechanosensitive hemichannels on the surface of the leading cells around the wound gap; this induces them to release adenosine triphosphate (ATP), which causes waves of Ca^{2+} influx in the cells behind the leading cells. Interestingly, the Ca^{2+} influx in the rear cells is mediated by the TRPC6 ion channel: it has been shown that when the TRPC6 activator hyperforin is applied exogenously, it further accelerates wound closure (Takada et al. 2014). This indicates that pharmaceutics which target the mechanosensitive pathways of keratinocytes can facilitate wound closure. Re-epithelialization can also be promoted by targeting the TGF-β1/Smad3 signaling pathway, since knocking out Smad3 in mice impairs the local inflammatory response in cutaneous wound-healing (Ashcroft et al. 1999) and accelerates wound-healing (Falanga et al. 2004). Similarly, deletion of TGF-β1 gene reduces granulation and increases the epithelialization rate *in vivo* (Koch et al. 2000). This suggests that pharmaceutics which reduce Smad3 or TGF-β1 levels may promote proper wound-healing.

Numerous lines of evidence show that dermal fibroblasts respond sensitively to mechanical stimuli, which suggests that the mechanosensitive pathways in dermal fibroblasts can also be targeted to improve wound-healing. For example, when dermal fibroblasts are subjected to mechanical tension *in vitro*, they express high levels of the proinflammatory mediators IL-1 and IL-6 (Kessler-Becker et al. 2004). Moreover, cyclic strain stimulates fibroblast proliferation and increases the gene expression of TGF-β1 and connective-tissue growth factor (CTGF) (Webb et al. 2006), while cyclic stretch triggers the integrin and Wnt mechanotransduction pathways (Huang et al. 2013b). In addition, exogenous TGF-β1 greatly upregulates the alpha smooth muscle actin (α-SMA) expression of fibroblasts when they are grown on stiff collagen; this effect is minimally seen in fibroblasts grown on soft collagen (Arora et al. 1999). Finally, the incorporation of α-SMA into stress fibers, which capacitates the contractility of cells, only occurs when the cells are subjected to higher mechanical loading (Goffin et al. 2006).

15.3 Mechanotherapy to Improve Cutaneous Wound-Healing

Our understanding of the mechanobiology in normal and aberrant cutaneous wound-healing has led us to identify a number of existing therapeutic methods as tissue-level mechanotherapies. These mechanotherapies are used for chronic cutaneous wounds and include negative-pressure wound therapy (NPWT), shockwave therapy, and ultrasound therapy. Recent research has also identified a number of novel therapeutic strategies that may improve wound-healing. One of these, electrotherapy, is currently being assessed for clinical efficacy.

15.3.1 Negative-Pressure Wound Therapy

NPWT, also called "microdeformational wound therapy," is used to accelerate chronic wound-healing. Its initial development was influenced by the concept of suction drainage and moist wound-healing. Winter (1962) showed with a pig model that keeping the wound moist prevented the formation of the scab and doubled the epithelialization rate. Argenta and Morykwas (1997) and Morykwas et al. (1997) were the first to apply controlled suction on wounds. This force simultaneously creates a moist microenvironment and controls the wound fluids, faciliating wound-healing by preventing dehydration, enhancing angiogenesis, promoting collagen synthesis, and increasing breakdown of dead tissue and fibrin (Junker et al. 2013). All of these benefits are achieved without increasing the risk of infection (Field and Kerstein 1994).

The vacuum-assisted closure system is the most popular NPWT device. It comprises an open-pore foam that fills the wound cavity, a semiocclusive wound dressing, a suction tubing, and a suction device. It causes macrodeformation; that is, the shrinkage of the wound due to the collapse of the foam pore and the centripetal forces on the wound surface. It also induces microdeformation: undulation of the wound surface due to the suction on the porous interface. Moreover, it removes excess fluids and stabilizes the wound environment (Huang et al. 2014). These primary effects are followed by multiple secondary effects that accelerate the healing phases following hemostasis. In the inflammation phase, NPWT removes infiltrating leukocytes; this is shown by the increased cellularity of the wound exudates and the elevated gene expression by the wound of leukocyte chemoattractants (e.g., IL-8) (Nuutila et al. 2013). In the proliferation phase, the NPWT-induced cellular deformation and stretch increase the migration of epithelial cells (Nuutila et al. 2013) and dermal fibroblasts (McNulty et al. 2007) and elevate microvessel density (Greene et al. 2006). They also promote cell proliferation, as shown by the increased levels of the cell proliferation marker KI-67 after short and intermittent NPWT treatment of the cutaneous wounds in diabetic mice (Scherer et al. 2009).

At present, NPWT is widely used to accelerate the healing of various types of wound in a variety of anatomical sites. For example, NPWT of pressure ulcers reduces the wound area (Moues et al. 2004) and depth (Srivastava et al. 2016) and improves granulation (Schwien et al. 2005). It also promotes microbial clearance in chronic wounds on the feet of diabetic patients (Nather et al. 2011). Moreover, it increases the vascularity and reduces the scar height of surgical wounds (Oh et al. 2013) and it effectively immobilizes skin grafts (Azzopardi et al. 2013).

15.3.2 Shockwave Therapy

The history of shockwave therapy dates back to the Second World War, when it was found that depth charges caused lung injuries in submarine crews despite not affecting the structural integrity of the submarines themselves (Coombs 2000). Later, in the 1980s, extracorporeal shockwave started to be used widely as a noninvasive treatment of kidney stones. In the years after 2000, shockwave therapy began being applied to burns, diabetic wounds, and ulcers.

The shockwaves are biphasic high-energy acoustic waves generated by electrohydraulic, electromagnetic, or piezoelectric technologies. Shockwave therapy increases neovascularization (Wang et al. 2003), revascularization (Mittermayr et al. 2011), collagen synthesis (Yang et al. 2011), and cellular proliferation (Kuo et al. 2009a,b). It also reduces apoptosis of ischemic tissue cells (Kuo et al. 2009b). These effects are mediated by mechanotransduction mechanisms (Antonic et al. 2011). The molecular mechanisms involve the upregulation of TGF-β1 in dermal fibroblasts (Berta et al. 2009) and the downregulation of the proinflammatory cytokines IL-1β, IL-6, and TNF-α (Davis et al. 2009).

In terms of treating wounds, shockwave therapy enhances the healing of burns, skin graft donor sites, and chronic wounds. For example, shockwave therapy of a full-thickness burn on a mouse ear accelerated angiogenesis and improves blood perfusion (Goertz et al. 2014), while a randomized phase II clinical trial showed that a single application of defocused shockwave therapy significantly accelerated the re-epithelialization of superficial second-degree burn wounds (Ottomann et al. 2012). Moreover, shockwave therapy also accelerated re-epithelialization of the skin graft donor site immediately after split-thickness harvest (Ottomann et al. 2010). In addition, it produces more pliable scars, with less evidence of color mismatch (Fioramonti et al. 2012). A single-blinded randomized controlled clinical trial showed that shockwave therapy significantly reduced the size of chronic diabetic wounds and their healing time (Omar et al. 2014). Finally, in diabetic rats, low-energy shockwave therapy increased collagen levels and enhanced wound-breaking strength, thereby improving incision wound-healing (Yang et al. 2011).

15.3.3 Ultrasound Therapy

The so-called "ultra" sound is the mechanical vibration whose frequency is above the upper limit of human hearing. Wound-healing-related ultrasound therapies can be classified into four groups: low-power high-frequency ultrasound for mesenchymal disorders, high-power high-frequency ultrasound for cutting, low-power low-frequency ultrasound for cleansing of tissue and biostimulation of cells, and high-power low-frequency ultrasound for operations on the teeth and eyeballs. High-frequency ultrasound acts by heating the tissue, while low-frequency ultrasound has mechanical effects that cause cavitation, which facilitates wound-cleaning, debrides ulcers, and stimulates granulation (Uhlemann et al. 2003).

A number of studies have shown that ultrasound therapy improves wound-healing at both the molecular and the cellular level. At the molecular level, low-intensity pulsed ultrasound (LIPUS) induces mechanotransduction, as shown by the increased intracellular concentrations of calcium (Parvizi et al. 2002), the phosphorylation of FA (Whitney et al. 2012), the activation of the integrin/phosphatidylinositol 3-OH kinase (PI3K)/Akt pathway (Takeuchi et al. 2008), and the improved cell–cell communication via gap junctions, followed by activation of ERK 1/2 and p38 MAPK (Sena et al. 2011; Padilla et al. 2014). At the cellular level, ultrasound therapy increases macrophage responsiveness in terms of releasing fibroblast mitogenic factors (Young and Dyson 1990). Therapeutic ultrasound also induces the proliferation of human fibroblasts and enhances their collagen and noncollagenous protein synthesis and angiogenesis (Doan et al. 1999); moreover, it reduces the colony-forming units (CFUs) of methicillin-resistant *Staphylococcus aureus* (MRSA) *in vitro* (Conner-Kerr et al. 2010), and noncontact nonthermal low-frequency ultrasound has been found to reduce the quantity of bacteria in chronic pressure ulcers (Serena et al. 2009).

Ultrasound therapy also has a number of macroscopic effects. Low-frequency ultrasound as an adjunctive therapy reduces wound area in patients with venous stasis and diabetic foot ulcers (Voigt et al. 2011). Moreover, 1 MHz contact ultrasound decreases hematoma, seroma, and incision-line separation in postoperative flaps (Ennis et al. 2011). Noncontact low-frequency ultrasound treatment improves the closure rates of excisional wounds in diabetic mice (Maan et al. 2014). Noncontact kilohertz ultrasound therapy is particularly effective for recalcitrant diabetic foot ulcers: a randomized double-blind controlled multicenter study showed that this therapy significantly improves the re-epithelialization rate of these chronic wounds (Ennis et al. 2005). Finally, it helps to increase skin graft take rates by controlling the bioburden and increasing angiogenesis (Ennis et al. 2011).

15.3.4 Electrotherapy

Both normal and wounded skin have electrical signals. In normal human skin, the epidermis and dermis are electronegative and -positive, respectively, which leads to a transepithelial potential (TEP) of 20–50 mV (Barker et al. 1982; Nuccitelli 2003).

This TEP is maintained by the epithelium, which resists the passive ion flow down the concentration gradient (Reid and Zhao 2014). On wounding, however, the mechanical disruption of the epidermis causes the TEP at the wound site to drop to zero. This phenomenon was initially described by Burr et al. (1940), who observed that the TEP at abdominal skin wounds is initially positive but then becomes negative after 4 days of healing; thereafter, the TEP remains negative, even after the healing is completed (Weiss et al. 1990). After wounding, the intact epithelium adjacent to the wound site continues to have a high TEP; this, together with the loss of TEP at the wound, generates a lateral electrical field (EF) around the wound and causes large electric currents to flow out of the wound (Reid and Zhao 2014). It is possible that the EF around the wound participates in wound-healing: this is suggested by the fact that diabetic wounds, which have a lower electrical TEP and smaller wound currents than normal wounds, also exhibit impaired wound-healing (Ionescu-Tirgoviste et al. 1985).

A number of studies have assessed the molecular mechanisms by which EF promotes wound-healing. At the cellular level, a physiological EF induces cathodic migration of rat epithelial cells and orients cell division perpendicular to the EF. Moreover, better wound-healing associates with a larger wound EF, and vice versa (Song et al. 2002). In particular, the fibroblasts that align themselves parallel to the EF are more responsive to sinusoidal EF than those that align themselves perpendicular to the EF; the latter cells also only respond to a fivefold more intense EF (Lee et al. 1993). Moreover, human keratinocytes migrate toward a cathode in response to a 100 mV/mm EF (Nuccitelli 2003). In addition, electrical stimulation promotes the inflammation, proliferation, and remodeling phases of wound-healing by inducing the electrotaxis of inflammatory cells (e.g., neutrophils and macrophages) to the wound, increasing the capillary density and blood flow and improving collagen synthesis (Polak et al. 2014). At the molecular level, the response to EFs involves signaling pathways that include epidermal growth factor receptors (EGFRs), integrins (Pullar et al. 2006), ERK1/2 (Zhao et al. 2010), PI3K (Zhao et al. 2006), vascular endothelial growth factor receptors (VEGFRs), and Rho-ROCK (Zhao et al. 2004).

These observations suggest that artificial electrical stimulation could be used to promote proper healing. Devices that generate EF coupled with oscillating magnetic fields have been shown to improve non-union bone fracture healing (Kooistra et al. 2009). However, galvanic current stimulation may be more suited to chronic wounding. In other words, a more suitable device for cutaneous wounds would apply the electrodes to the wound site, thereby directly generating a local EF when the current crosses the wound while flowing between the two electrodes (Lee et al. 1993). Studies on rats have shown that such electrical stimulation enhances the healing of skin incisions (Bach et al. 1991), the survival of full-thickness skin grafts (Politis et al. 1989), and the take of ischemic musculocutaneous flaps (Kjartansson et al. 1988). Moreover, clinical trials have shown that wounds benefit from electrotherapy: it improves the transcutaneous oxygen levels in ischemic wounds (Goldman et al. 2004), enhances the tissue perfusion in diabetic (Peters et al. 1998) and venous (Junger et al. 2008) ulcers, and increases the healing rate of pressure ulcers (Griffin et al. 1991). In addition, electric stimulation reduces the wound bioburden (Kloth 2005), as it has a lethal or sublethal effect on microorganisms (Perni et al. 2007).

15.4 Future Considerations

Though much work has been done to improve our understanding of this subject, there is still a lot that we do not know about the mechanobiological mechanisms that underlie proper and abnormal wound-healing. There are three main lines of research that warrant particular effort. First, studies that compare the molecular changes in excessive and insufficient wound-healing may help to uncover potential mechanisms that could be exploited in therapy. Substance P exemplifies the usefulness of this approach. This neuropeptide, which is known to participate in neurogenic inflammation (Nicoletti et al. 2012), has been shown to be deficient and overexpressed in chronic wounds (Leal et al. 2015) and pathological scars (Akaishi et al. 2008b), respectively. Moreover, mechanical stretching of murine skin increases its substance P levels (Chin et al. 2009). These observations together suggest that substance P also participates in the inflammation that leads to proper wound-healing. Second, identification of pharmaceuticals (mechanopharmaceuticals) and devices that might be used to modulate the mechanosignaling pathways in wound-healing (i.e., mechanomodulations) may help prevent, cure, or reverse aberrant healing. This research may also pave the way for the identification of potential biomarkers that allow poor healing to be diagnosed early. Third, engineering studies may help to identify strategies that improve wound-healing. For example, it is known that during normal wound-healing, the rigidity of the wound rises markedly from 18 to around 40 kPa to ensure wound coverage (Goffin et al. 2006; Evans et al. 2013). A device that induces such rigidity may promote wound-healing. Another example of the potential usefulness of engineering studies is our injectable three-dimensional (3D) microscaffold, whose mechanical properties can be well tuned as an injectable microniche for stem cells (Liu et al. 2014). In mouse models, stem cells delivered in this microscaffold result in better salvage of murine limbs with critical ischemia than do freely injected stem cells (Li et al. 2014). We also recently microengineered a biomimetic *in vitro* cardiac fibrosis model based on substrates with varied stiffness that will allow us to identify pharmaceuticals that prevent cardiac fibrosis (Zhao et al. 2014). Similar engineering concepts and approaches may be highly useful in the field of wound-healing, assisting in meeting the dynamic physical requirements of self-adaptation for the growth of cells and thus leading to a better healing.

References

Akaishi, S., M. Akimoto, R. Ogawa, and H. Hyakusoku. 2008a. "The relationship between keloid growth pattern and stretching tension: visual analysis using the finite element method." *Annals of Plastic Surgery* **60**(4): 445–51. doi:10.1097/SAP.0b013e3181238dd7.

Akaishi, S., R. Ogawa, and H. Hyakusoku. 2008b "Keloid and hypertrophic scar: neurogenic inflammation hypotheses." *Med Hypotheses* **71**(1):32–8. PMID:18406540.

Antonic, V., R. Mittermayr, W. Schaden, and A. Stojadinovic. 2011. "Evidence supporting extracorporeal shock wave therapy for acute and chronic soft tissue wounds." *Wounds* **23**(7): 204–15. PMID:25879174.

Argenta, L. C. and M. J. Morykwas. 1997. "Vacuum-assisted closure: a new method for wound control and treatment: clinical experience." *Annals of Plastic Surgery* **38**(6): 563–76; disc. 577. PMID:9188971.

Arora, P. D., N. Narani, and C. A. McCulloch. 1999. "The compliance of collagen gels regulates transforming growth factor-beta induction of alpha-smooth muscle actin in fibroblasts." *American Journal of Pathology* **154**(3): 871–82. PMID:10079265.

Ashcroft, G. S., X. Yang, A. B. Glick, M. Weinstein, J. L. Letterio, D. E. Mizel, et al. 1999. "Mice lacking Smad3 show accelerated wound healing and an impaired local inflammatory response." *Nature Cell Biology* **1**(5): 260–6. doi:10.1038/12971.

Azzopardi, E. A., D. E. Boyce, W. A. Dickson, E. Azzopardi, J. H. Laing, I. S. Whitaker, and K. Shokrollahi. 2013. "Application of topical negative pressure (vacuum-assisted closure) to split-thickness skin grafts: a structured evidence-based review." *Annals of Plastic Surgery* **70**(1): 23–9. doi:10.1097/SAP.0b013e31826eab9e.

Bach, S., K. Bilgrav, F. Gottrup, and T. E. Jorgensen. 1991. "The effect of electrical current on healing skin incision. An experimental study." *European Journal of Surgery* **157**(3): 171–4. PMID:1678624.

Barker, A. T., L. F. Jaffe, and J. W. Vanable Jr. 1982. "The glabrous epidermis of cavies contains a powerful battery." *American Journal of Physiology* **242**(3): R358–66. PMID:7065232.

Berta, L., A. Fazzari, A. M. Ficco, P. M. Enrica, M. G. Catalano, and R. Frairia. 2009. "Extracorporeal shock waves enhance normal fibroblast proliferation in vitro and activate mRNA expression for TGF-beta1 and for collagen types I and III." *Acta Orthopaedica* **80**(5): 612–17. doi:10.3109/17453670903316793.

Burr, H. S., M. Taffel, and S. C. Harvey. 1940. "An electrometric study of the healing wound in man." *Yale Journal of Biology and Medicine* **12**(5): 483–5. PMID:21433903.

Chin, M. S., L. Lancerotto, D. L. Helm, P. Dastouri, M. J. Prsa, M. Ottensmeyer, et al. 2009. "Analysis of neuropeptides in stretched skin." *Plastic and Reconstructive Surgery* **124**(1): 102–13. doi:10.1097/PRS.0b013e3181a81542.

Conner-Kerr, T., G. Alston, A. Stovall, T. Vernon, D. Winter, J. Meixner, et al. 2010. "The effects of low-frequency ultrasound (35 kHz) on methicillin-resistant Staphylococcus aureus (MRSA) in vitro." *Ostomy/Wound Management* **56**(5): 32–43. PMID:20511683.

Coombs, R. 2000. *Musculoskeletal Shockwave Therapy*. Cambridge: Greenwich Medical Media.

Davis, T. A., A. Stojadinovic, K. Anam, M. Amare, S. Naik, G. E. Peoples, et al. 2009. "Extracorporeal shock wave therapy suppresses the early proinflammatory immune response to a severe cutaneous burn injury." *International Wound Journal* **6**(1): 11–21. doi:10.1111/j.1742-481X.2008.00540.x.

Doan, N., P. Reher, S. Meghji, and M. Harris. 1999. "In vitro effects of therapeutic ultrasound on cell proliferation, protein synthesis, and cytokine production by human fibroblasts, osteoblasts, and monocytes." *Journal of Oral and Maxillofacical Surgery* **57**(4): 409–19; disc. 420. PMID:10199493.

du Roure, O., A. Saez, A. Buguin, R. H. Austin, P. Chavrier, P. Silberzan, and B. Ladoux. 2005. "Force mapping in epithelial cell migration." *Proceedings of the National Academy of Sciences of the United States of America* **102**(7): 2390–5. doi:10.1073/pnas.0408482102.

Engler, A. J., S. Sen, H. L. Sweeney, and D. E. Discher. 2006. "Matrix elasticity directs stem cell lineage specification." *Cell* **126**(4): 677–89. doi:10.1016/j.cell.2006.06.044.

Ennis, W. J., P. Foremann, N. Mozen, J. Massey, T. Conner-Kerr, and P. Meneses. 2005. "Ultrasound therapy for recalcitrant diabetic foot ulcers: results of a randomized, double-blind, controlled, multicenter study." *Ostomy/Wound Management* **51**(8): 24–39. PMID:16234574.

Ennis, W. J., C. Lee, M. Plummer, and P. Meneses. 2011. "Current status of the use of modalities in wound care: electrical stimulation and ultrasound therapy." *Plastic and Reconstructive Surgery* **127**(Suppl. 1): 93S–102S. doi:10.1097/PRS.0b013e3181fbe2fd.

Evans, N. D., R. O. Oreffo, E. Healy, P. J. Thurner, and Y. H. Man. 2013. "Epithelial mechanobiology, skin wound healing, and the stem cell niche." *Journal of the Mechanical Behavior of Biomedical Materials* **28**: 397–409. doi:10.1016/j.jmbbm.2013.04.023.

Falanga, V., D. Schrayer, J. Cha, J. Butmarc, P. Carson, A. B. Roberts, and S. J. Kim. 2004. "Full-thickness wounding of the mouse tail as a model for delayed wound healing: accelerated wound closure in Smad3 knock-out mice." *Wound Repair and Regeneration* **12**(3): 320–6. doi:10.1111/j.1067-1927.2004.012316.x.

Field, F. K. and M. D. Kerstein. 1994. "Overview of wound healing in a moist environment." *American Journal of Surgery* **167**(1A): 2S–6S. PMID:8109679.

Fioramonti, P., E. Cigna, M. G. Onesti, P. Fino, N. Fallico, and N. Scuderi. 2012. "Extracorporeal shock wave therapy for the management of burn scars." *Dermatologic Surgery* **38**(5): 778–82. doi:10.1111/j.1524-4725.2012.02355.x

Geiger, B. and A. Bershadsky. 2002. "Exploring the neighborhood: adhesion-coupled cell mechanosensors." *Cell* **110**(2): 139–42. PMID:12150922.

Goffin, J. M., P. Pittet, G. Csucs, J. W. Lussi, J. J. Meister, and B. Hinz. 2006. "Focal adhesion size controls tension-dependent recruitment of alpha-smooth muscle actin to stress fibers." *Journal of Cell Biology* **172**(2): 259–68. doi:10.1083/jcb.200506179.

Goldman, R., M. Rosen, B. Brewley, and M. Golden. 2004. "Electrotherapy promotes healing and microcirculation of infrapopliteal ischemic wounds: a prospective pilot study." *Advances in Skin and Wound Care* **17**(6): 284–94. PMID:15289715.

Greene, A. K., M. Puder, R. Roy, D. Arsenault, S. Kwei, M. A. Moses, and D. P. Orgill. 2006. "Microdeformational wound therapy: effects on angiogenesis and matrix metalloproteinases in chronic wounds of 3 debilitated patients." *Annals of Plastic Surgery* **56**(4): 418–22. doi:10.1097/01.sap.0000202831.43294.02.

Griffin, J. W., R. E. Tooms, R. A. Mendius, J. K. Clifft, R. Vander Zwaag, and F. el-Zeky. 1991. "Efficacy of high voltage pulsed current for healing of pressure ulcers in patients with spinal cord injury." *Physical Therapy* **71**(6): 433–42; disc. 442–4. PMID:2034707.

Goertz, O., L. von der Lohe, H. Lauer, T. Khosrawipour, A. Ring, A. Daigeler, M. Lehnhardt, J. Kolbenschlag. 2014. "Repetitive extracorporeal shock wave applications are superior in inducing angiogenesis after full thickness burn compared to single application." *Burns.* **40**(7):1365–74.

Hinz, B. 2009. "Tissue stiffness, latent TGF-beta1 activation, and mechanical signal transduction: implications for the pathogenesis and treatment of fibrosis." *Current Rheumatology Reports* **11**(2): 120–6. PMID:19296884.

Huang, C. and R. Ogawa. 2014. "Three-dimensional reconstruction of scar contracture-bearing axilla and digital webs using the square flap method." *Plastic and Reconstructive Surgery. Global Open* **2**(5): e149. doi:10.1097/GOX.0000000000000110.

Huang, H., R. D. Kamm, and R. T. Lee. 2004. "Cell mechanics and mechanotransduction: pathways, probes, and physiology." *American Journal of Physiology. Cell Physiology* **287**(1): C1–11. doi:10.1152/ajpcell.00559.2003.

Huang, C., S. Akaishi, and R. Ogawa. 2012a. "Mechanosignaling pathways in cutaneous scarring." *Archives of Dermatologic Research* **304**(8): 589–97. doi:10.1007/s00403-012-1278-5.
Huang, C., S. Ono, H. Hyakusoku, and R. Ogawa. 2012b. "Small-wave incision method for linear hypertrophic scar reconstruction: a parallel-group randomized controlled study." *Aesthetic Plastic Surgery* **36**(2): 387–95. doi:10.1007/s00266-011-9821-x.
Huang, C., J. Holfeld, W. Schaden, D. Orgill, and R. Ogawa. 2013a. "Mechanotherapy: revisiting physical therapy and recruiting mechanobiology for a new era in medicine." *Trends in Molecular Medicine* **19**(9): 555–64. doi:10.1016/j.molmed.2013.05.005.
Huang, C., K. Miyazaki, S. Akaishi, A. Watanabe, H. Hyakusoku, and R. Ogawa. 2013b. "Biological effects of cellular stretch on human dermal fibroblasts." *Journal of Plastic, Reconstrive & Aesthetic Surgery* **66**(12): e351–61. doi:10.1016/j.bjps.2013.08.002.
Huang, C., G. F. Murphy, S. Akaishi, and R. Ogawa. 2013c. "Keloids and hypertrophic scars: update and future directions." *Plastic and Reconstructive Surgery. Global Open* **1**(4): e25. doi:10.1097/GOX.0b013e31829c4597.
Huang, C., T. Leavitt, L. R. Bayer, and D. P. Orgill. 2014. "Effect of negative pressure wound therapy on wound healing." *Current Problems in Surgery* **51**(7): 301–31. doi:10.1067/j.cpsurg.2014.04.001.
Ingber, D. E. 1997. "Tensegrity: the architectural basis of cellular mechanotransduction." *Annuals Reviews in Physiology* **59**: 575–99. doi:10.1146/annurev.physiol.59.1.575
Ingber, D. E. 2003a. "Mechanobiology and diseases of mechanotransduction." *Annals of Medicine* **35**(8): 564–77. PMID:14708967.
Ingber, D. E. 2003b. "Tensegrity II. How structural networks influence cellular information processing networks." *Journal of Cell Science* **116**(Pt. 8): 1397–408. PMID:12640025.
Ingber, D. E. 2006. "Cellular mechanotransduction: putting all the pieces together again." *FASEB Journal* **20**(7): 811–27. doi:10.1096/fj.05-5424rev.
Ingber, D. E. 2008. "Tensegrity-based mechanosensing from macro to micro." *Progress in Biophysics and Molecular Biology* **97**(2–3): 163–79. doi:10.1016/j.pbiomolbio.2008.02.005.
Ionescu-Tirgoviste, C., O. Bajenaru, I. Zugravescu, E. Dorobantu, D. Hartia, C. Dumitrescu, et al. 1985. "Study of the cutaneous electric potentials and the perception threshold to an electric stimulus in diabetic patients with and without clinical neuropathy." *Médecine Interne* **23**(3): 213–22. PMID:4048802.
Junger, M., A. Arnold, D. Zuder, H. W. Stahl, and S. Heising. 2008. "Local therapy and treatment costs of chronic, venous leg ulcers with electrical stimulation (Dermapulse): a prospective, placebo controlled, double blind trial." *Wound Repair and Regeneration* **16**(4): 480–7. doi:10.1111/j.1524-475X.2008.00393.x.
Junker, J. P., R. A. Kamel, E. J. Caterson, and E. Eriksson. 2013. "Clinical impact upon wound healing and inflammation in moist, wet, and dry environments." *Advances in Wound Care* **2**(7): 348–56. doi:10.1089/wound.2012.0412.
Kessler-Becker, D., T. Krieg, and B. Eckes. 2004. "Expression of pro-inflammatory markers by human dermal fibroblasts in a three-dimensional culture model is mediated by an autocrine interleukin-1 loop." *Biochemical Journal* **379**(Pt. 2): 351–8. doi:10.1042/BJ20031371.
Kjartansson, J., T. Lundeberg, U. E. Samuelson, and C. J. Dalsgaard. 1988. "Transcutaneous electrical nerve stimulation (TENS) increases survival of ischaemic musculocutaneous flaps." *Acta Physiologica Scandinavica* **134**(1): 95–9. doi:10.1111/j.1748-1716.1988.tb08464.x.

Kloth, L. C. 2005. "Electrical stimulation for wound healing: a review of evidence from in vitro studies, animal experiments, and clinical trials." *International Journal of Lower Extremity Wounds* **4**(1): 23–44. doi:10.1177/1534734605275733.

Koch, R. M., N. S. Roche, W. T. Parks, G. S. Ashcroft, J. J. Letterio, and A. B. Roberts. 2000. "Incisional wound healing in transforming growth factor-beta1 null mice." *Wound Repair and Regeneration* **8**(3): 179–91. PMID:10886809.

Kooistra, B. W., A. Jain, and B. P. Hanson. 2009. "Electrical stimulation: nonunions." *Indian Journal of Orthopaedics* **43**(2): 149–55. doi:10.4103/0019-5413.50849.

Kuo, Y. R., C. T. Wang, F. S. Wang, Y. C. Chiang, and C. J. Wang. 2009a. "Extracorporeal shock-wave therapy enhanced wound healing via increasing topical blood perfusion and tissue regeneration in a rat model of STZ-induced diabetes." *Wound Repair and Regeneration* **17**(4): 522–30. doi:10.1111/j.1524-475X.2009.00504.x.

Kuo, Y. R., C. T. Wang, F. S. Wang, K. D. Yang, Y. C. Chiang, and C. J. Wang. 2009b. "Extracorporeal shock wave treatment modulates skin fibroblast recruitment and leukocyte infiltration for enhancing extended skin-flap survival." *Wound Repair and Regeneration* **17**(1): 80–7. doi:10.1111/j.1524-475X.2008.00444.x.

Lazarus, G. S., D. M. Cooper, D. R. Knighton, R. E. Percoraro, G. Rodeheaver, and M. C. Robson. 1994. "Definitions and guidelines for assessment of wounds and evaluation of healing." *Archives of Dermatology* **130**(4): 489–93. PMID:8166487.

Leal, E. C., E. Carvalho, A. Tellechea, A. Kafanas, F. Tecilazich, C. Kearney, et al. 2015. "Substance P promotes wound healing in diabetes by modulating inflammation and macrophage phenotype." *American Journal of Pathology* **185**(6): 1638–48. doi:10.1016/j.ajpath.2015.02.011.

Lee, R. C., D. J. Canaday, and H. Doong. 1993. "A review of the biophysical basis for the clinical application of electric fields in soft-tissue repair." *Journal of Burn Care & Rehabilitation* **14**(3): 319–35. PMID:8360237.

Li, Y., W. Liu, F. Liu, Y. Zeng, S. Zuo, S. Feng, et al. 2014. "Primed 3D injectable microniches enabling low-dosage cell therapy for critical limb ischemia." *Proceedings of the National Academy of Sciences of the United States of America* **111**(37): 13 511–16. doi:10.1073/pnas.1411295111.

Liu, W., Y. Li, Y. Zeng, X. Zhang, J. Wang, L. Xie, et al. 2014. "Microcryogels as injectable 3-D cellular microniches for site-directed and augmented cell delivery." *Acta Biomaterialia* **10**(5): 1864–75. doi:10.1016/j.actbio.2013.12.008.

Lo, C. M., H. B. Wang, M. Dembo, and Y. L. Wang. 2000. "Cell movement is guided by the rigidity of the substrate." *Biophys Journal* **79**(1): 144–52. doi:10.1016/S0006-3495(00)76279-5.

Maan, Z. N., M. Januszyk, R. C. Rennert, D. Duscher, M. Rodrigues, T. Fujiwara, et al. 2014. "Noncontact, low-frequency ultrasound therapy enhances neovascularization and wound healing in diabetic mice." *Plastic and Reconstructive Surgery* **134**(3): 402e–11e. doi:10.1097/PRS.0000000000000467.

McNulty, A. K., M. Schmidt, T. Feeley, and K. Kieswetter. 2007. "Effects of negative pressure wound therapy on fibroblast viability, chemotactic signaling, and proliferation in a provisional wound (fibrin) matrix." *Wound Repair and Regeneration* **15**(6): 838–46. doi:10.1111/j.1524-475X.2007.00287.x.

Mittermayr, R., J. Hartinger, V. Antonic, A. Meinl, S. Pfeifer, A. Stojadinovic, et al. 2011. "Extracorporeal shock wave therapy (ESWT) minimizes ischemic tissue necrosis irrespective of application time and promotes tissue revascularization by stimulating angiogenesis." *Annals of Surgery* **253**(5): 1024–32. doi:10.1097/SLA.0b013e3182121d6e.

Morykwas, M. J., L. C. Argenta, E. I. Shelton-Brown, and W. McGuirt. 1997. "Vacuum-assisted closure: a new method for wound control and treatment: animal studies and basic foundation." *Annals of Plastic Surgery* **38**(6): 553–62. PMID:9188970.

Moues, C. M., M. C. Vos, G. J. van den Bemd, T. Stijnen, and S. E. Hovius. 2004. "Bacterial load in relation to vacuum-assisted closure wound therapy: a prospective randomized trial." *Wound Repair and Regeneration* **12**(1): 11–17. doi:10.1111/j.1067-1927.2004.12105.x.

Myers, K. A., J. B. Rattner, N. G. Shrive, and D. A. Hart. 2007. "Hydrostatic pressure sensation in cells: integration into the tensegrity model." *Biochemistry and Cell Biology* **85**(5): 543–51. doi:10.1139/o07-108.

Nather, A., N. Y. Hong, W. K. Lin, and J. A. Sakharam. 2011. "Effectiveness of bridge V.A.C. dressings in the treatment of diabetic foot ulcers." *Diabetic Foot & Ankle* **2**. doi:10.3402/dfa.v2i0.5893.

Nicoletti, M., G. Neri, G. Maccauro, D. Tripodi, G. Varvara, A. Saggini, et al. 2012. "Impact of neuropeptide substance P an inflammatory compound on arachidonic acid compound generation." *International Journal of Immunopathology and Pharmacology* **25**(4): 849–57. PMID:23298476.

Nuccitelli, R. 2003. "A role for endogenous electric fields in wound healing." *Current Topics in Developmental Biology* **58**: 1–26. PMID:14711011.

Nuutila, K., A. Siltanen, M. Peura, A. Harjula, T. Nieminen, J. Vuola, et al. 2013. "Gene expression profiling of negative-pressure-treated skin graft donor site wounds." *Burns* **39**(4): 687–93. doi:10.1016/j.burns.2012.09.014.

Nwomeh, B. C., D. R. Yager, and I. K. Cohen. 1998. "Physiology of the chronic wound." *Clinics in Plastic Surgery* **25**(3): 341–56. PMID:9696897.

Ogawa, R., K. Okai, F. Tokumura, K. Mori, Y. Ohmori, C. Huang, et al. 2012. "The relationship between skin stretching/contraction and pathologic scarring: the important role of mechanical forces in keloid generation." *Wound Repair and Regeneration* **20**(2): 149–57. doi:10.1111/j.1524-475X.2012.00766.x.

Oh, B. H., S. H. Lee, K. A. Nam, H. B. Lee, and K. Y. Chung. 2013. "Comparison of negative pressure wound therapy and secondary intention healing after excision of acral lentiginous melanoma on the foot." *British Journal of Dermatology* **168**(2): 333–8. doi:10.1111/bjd.12099.

Omar, M. T., A. Alghadir, K. K. Al-Wahhabi, and A. B. Al-Askar. 2014. "Efficacy of shock wave therapy on chronic diabetic foot ulcer: a single-blinded randomized controlled clinical trial." *Diabetes Research and Clinical Practice* **106**(3): 548–54. doi:10.1016/j.diabres.2014.09.024.

Ottomann, C., B. Hartmann, J. Tyler, H. Maier, R. Thiele, W. Schaden, and A. Stojadinovic. 2010. "Prospective randomized trial of accelerated re-epithelization of skin graft donor sites using extracorporeal shock wave therapy." *Journal of the American College of Surgery* **211**(3): 361–7. doi:10.1016/j.jamcollsurg.2010.05.012.

Ottomann, C., A. Stojadinovic, P. T. Lavin, F. H. Gannon, M. H. Heggeness, R. Thiele, et al. 2012. "Prospective randomized phase II trial of accelerated reepithelialization of superficial second-degree burn wounds using extracorporeal shock wave therapy." *Annals of Surgery* **255**(1): 23–9. doi:10.1097/SLA.0b013e318227b3c0.

Padilla, F., R. Puts, L. Vico, and K. Raum. 2014. "Stimulation of bone repair with ultrasound: a review of the possible mechanic effects." *Ultrasonics* **54**(5): 1125–45. doi:10.1016/j.ultras.2014.01.004.

Parvizi, J., V. Parpura, J. F. Greenleaf, and M. E. Bolander. 2002. "Calcium signaling is required for ultrasound-stimulated aggrecan synthesis by rat chondrocytes." *Journal of Orthopaedic Research* **20**(1): 51–7. doi:10.1016/S0736-0266(01)00069-9.

Pastar, I., O. Stojadinovic, N. C. Yin, H. Ramirez, A. G. Nusbaum, A. Sawaya, et al. 2014. "Epithelialization in wound healing: a comprehensive review." *Advances in Wound Care* **3**(7): 445–64. doi:10.1089/wound.2013.0473.

Perni, S., P. R. Chalise, G. Shama, and M. G. Kong. 2007. "Bacterial cells exposed to nanosecond pulsed electric fields show lethal and sublethal effects." *International Journal of Food Microbiology* **120**(3): 311–14. doi:10.1016/j.ijfoodmicro.2007.10.002.

Peters, E. J., D. G. Armstrong, R. P. Wunderlich, J. Bosma, S. Stacpoole-Shea, and L. A. Lavery. 1998. "The benefit of electrical stimulation to enhance perfusion in persons with diabetes mellitus." *Journal of Foot and Ankle Surgery* **37**(5): 396–400; disc. 447–8. PMID:9798171.

Polak, A., A. Franek, and J. Taradaj. 2014. "High-voltage pulsed current electrical stimulation in wound treatment." *Advances in Wound Care* **3**(2): 104–17. doi:10.1089/wound.2013.0445.

Politis, M. J., M. F. Zanakis, and J. E. Miller. 1989. "Enhanced survival of full-thickness skin grafts following the application of DC electrical fields." *Plastic and Reconstructive Surgery* **84**(2): 267–72. PMID:2664831.

Pullar, C. E., B. S. Baier, Y. Kariya, A. J. Russell, B. A. Horst, M. P. Marinkovich, and R. R. Isseroff. 2006. "beta4 integrin and epidermal growth factor coordinately regulate electric field-mediated directional migration via Rac1." *Molecular Biology of the Cell* **17**(11): 4925–35. doi:10.1091/mbc.E06-05-0433.

Reid, B. and M. Zhao. 2014. "The electrical response to injury: molecular mechanisms and wound healing." *Advances in Wound Care* **3**(2): 184–201. doi:10.1089/wound.2013.0442.

Scherer, S. S., G. Pietramaggiori, J. C. Mathews, and D. P. Orgill. 2009. "Short periodic applications of the vacuum-assisted closure device cause an extended tissue response in the diabetic mouse model." *Plastic and Reconstructive Surgery* **124**(5): 1458–65. doi:10.1097/PRS.0b013e3181bbc829.

Schwien, T., J. Gilbert, and C. Lang. 2005. "Pressure ulcer prevalence and the role of negative pressure wound therapy in home health quality outcomes." *Ostomy/Wound Management* **51**(9): 47–60. PMID:16230764.

Sena, K., S. R. Angle, A. Kanaji, C. Aher, D. G. Karwo, D. R. Sumner, and A. S. Virdi. 2011. "Low-intensity pulsed ultrasound (LIPUS) and cell-to-cell communication in bone marrow stromal cells." *Ultrasonics* **51**(5): 639–44. doi:10.1016/j.ultras.2011.01.007.

Serena, T., S. K. Lee, K. Lam, P. Attar, P. Meneses, and W. Ennis. 2009. "The impact of noncontact, nonthermal, low-frequency ultrasound on bacterial counts in experimental and chronic wounds." *Ostomy/Wound Management* **55**(1): 22–30. PMID:19174586.

Song, B., M. Zhao, J. V. Forrester, and C. D. McCaig. 2002. "Electrical cues regulate the orientation and frequency of cell division and the rate of wound healing in vivo." *Proceedings of the National Academy of Sciences of the United States of America* **99**(21): 13 577–82. doi:10.1073/pnas.202235299.

Srivastava, R. N., M. K. Dwivedi, A. K. Bhagat, S. Raj, R. Agarwal, and A. Chandra. 2016. "A non-randomised, controlled clinical trial of an innovative device for negative pressure wound therapy of pressure ulcers in traumatic paraplegia patients." *International Wound Journal* **13**(3): 343–8. doi:10.1111/iwj.12309.

Takada, H., K. Furuya, and M. Sokabe. 2014. "Mechanosensitive ATP release from hemichannels and Ca(2)(+) influx through TRPC6 accelerate wound closure in keratinocytes." *Journal of Cell Science* **127**(Pt. 19): 4159–71. doi:10.1242/jcs.147314.

Takeuchi, R., A. Ryo, N. Komitsu, Y. Mikuni-Takagaki, A. Fukui, Y. Takagi, et al. 2008. "Low-intensity pulsed ultrasound activates the phosphatidylinositol 3 kinase/Akt pathway and stimulates the growth of chondrocytes in three-dimensional cultures: a basic science study." *Arthritis Research & Therapy* **10**(4): R77. doi:10.1186/ar2451.

Uhlemann, C., B. Heinig, and U. Wollina. 2003. "Therapeutic ultrasound in lower extremity wound management." *International Journal of Lower Extremity Wounds* **2**(3): 152–7. doi:10.1177/1534734603257988.

Voigt, J., M. Wendelken, V. Driver, and O. M. Alvarez. 2011. "Low-frequency ultrasound (20-40 kHz) as an adjunctive therapy for chronic wound healing: a systematic review of the literature and meta-analysis of eight randomized controlled trials." *International Journal of Lower Extremity Wounds* **10**(4): 190–9. doi:10.1177/1534734611424648.

Wang, C. J., F. S. Wang, K. D. Yang, L. H. Weng, C. C. Hsu, C. S. Huang, and L. C. Yang. 2003. "Shock wave therapy induces neovascularization at the tendon-bone junction. A study in rabbits." *Journal of Orthopaedic Research* **21**(6): 984–9. doi:10.1016/S0736-0266(03)00104-9.

Wang, Y., G. Wang, X. Luo, J. Qiu, and C. Tang. 2012. "Substrate stiffness regulates the proliferation, migration, and differentiation of epidermal cells." *Burns* **38**(3): 414–20. doi:10.1016/j.burns.2011.09.002.

Webb, K., R. W. Hitchcock, R. M. Smeal, W. Li, S. D. Gray, and P. A. Tresco. 2006. "Cyclic strain increases fibroblast proliferation, matrix accumulation, and elastic modulus of fibroblast-seeded polyurethane constructs." *Journal of Biomechanics* **39**(6): 1136–44. doi:10.1016/j.jbiomech.2004.08.026.

Weiss, D. S., R. Kirsner, and W. H. Eaglstein. 1990. "Electrical stimulation and wound healing." *Archives of Dermatology* **126**(2): 222–5. PMID:2405781.

Whitney, N. P., A. C. Lamb, T. M. Louw, and A. Subramanian. 2012. "Integrin-mediated mechanotransduction pathway of low-intensity continuous ultrasound in human chondrocytes." *Ultrasound in Medicine & Biology* **38**(10): 1734–43. doi:0.1016/j.ultrasmedbio.2012.06.002.

Winter, G. D. 1962. "Formation of the scab and the rate of epithelization of superficial wounds in the skin of the young domestic pig." *Nature* **193**: 293–4. PMID:14007593.

Yang, G., C. Luo, X. Yan, L. Cheng, and Y. Chai. 2011. "Extracorporeal shock wave treatment improves incisional wound healing in diabetic rats." *Tohoku Journal of Experimental Medicine* **225**(4): 285–92. PMID:22104424.

Young, S. R. and M. Dyson. 1990. "Macrophage responsiveness to therapeutic ultrasound." *Ultrasound in Medicine & Biology* **16**(8): 809–16. PMID:2095011.

Zhao, M., H. Bai, E. Wang, J. V. Forrester, and C. D. McCaig. 2004. "Electrical stimulation directly induces pre-angiogenic responses in vascular endothelial cells by signaling through VEGF receptors." *Journal of Cell Science* **117**(Pt. 3): 397–405. doi:10.1242/jcs.00868.

Zhao, M., B. Song, J. Pu, T. Wada, B. Reid, G. Tai, et al. 2006. "Electrical signals control wound healing through phosphatidylinositol-3-OH kinase-gamma and PTEN." *Nature* **442**(7101): 457–60. doi:10.1038/nature04925.

Zhao, M., J. Penninger, and R. R. Isseroff. 2010. "Electrical activation of wound-healing pathways." *Advances in Skin and Wound Care* **1**: 567–73. doi:10.1089/9781934854013.567.

Zhao, H., X. Li, S. Zhao, Y. Zeng, L. Zhao, H. Ding, et al. 2014. "Microengineered in vitro model of cardiac fibrosis through modulating myofibroblast mechanotransduction." *Biofabrication* **6**(4): 045009. doi:10.1088/1758-5082/6/4/045009.

16

Mechanobiology and Mechanotherapy for Cutaneous Scarring

Rei Ogawa[1] *and Chenyu Huang*[2]

[1] *Department of Plastic, Reconstructive, and Aesthetic Surgery, Nippon Medical School, Tokyo, Japan*
[2] *Department of Plastic, Reconstructive, and Aesthetic Surgery, Beijing Tsinghua Changgung Hospital, Tsinghua University, Beijing, China*

16.1 Introduction

In infancy and childhood, the skin expands to cover the growing skeleton and soft tissues. During this stage and thereafter, the skin is constantly subjected to intrinsic mechanical forces: these forces arise from the continuous tension of the skin and the cyclically applied skin tension that arises from body movements. The skin is also constantly subjected to extrinsic forces, namely, compression force and shear force. As humans age, the intrinsic mechanical forces start to decrease, as indicated by the development of skin creases. This change is partnered with a drop in skin thickness, which causes the skin to be more susceptible to damage induced by extrinsic forces. As a result, the skin of older people tears readily.

All of these changes during the normal course of life are mediated by the responses of cells that perceive the intrinsic and extrinsic mechanical forces on the skin (Chin et al. 2010). A similar principle applies when the skin is injured: the injury drastically changes the mechanophysiological conditions of the skin, which galvanizes the skin cells to produce a complex biological reaction that ultimately leads to wound-healing (Gurtner et al. 2008).

16.2 Cutaneous Wound-Healing and Mechanobiology

In cutaneous wound-healing, the gaps in the skin are normally closed by the generation of granulation tissue and the re-establishment of an effective epidermal barrier. This phenomenon associates with a cascade of complex biochemical events that can be categorized into four general processes: coagulation, inflammation, proliferation, and remodeling. Coagulation begins immediately after injury, upon which the platelets release various growth factors and cytokines. A few days later, the inflammatory and proliferative phases start: due to increased vascular permeability, inflammatory cells and soluble factors arrive in the wound from the blood vessels. Thereafter, fibroblasts

Mechanobiology: Exploitation for Medical Benefit, First Edition. Edited by Simon C. F. Rawlinson.

migrate into the wound, secrete collagen, and promote angiogenesis, thereby closing the gap in the skin surface. The remodeling phase, which generates the scar, starts within a week of injury and continues over months.

All phases of wound-healing, including granulation tissue formation, wound contraction, and epithelialization, are influenced by both intrinsic and extrinsic mechanical forces (Van De Water et al. 2013). For example, the formation of granulation tissue is driven by intrinsic and extrinsic mechanical stimulation of fibroblasts, myofibroblasts, endothelial cells, and epithelial cells in and near the wound. The wound itself contracts via forces produced by myofibroblasts; this process is also shaped by many extrinsic forces, including the natural tension in the skin. During remodeling, fibroblasts secrete collagen and fibronectin, which are key components of the extracellular matrix (ECM). These cells then regulate the volume of the ECM by secreting collagenase. This sequential synthesis and enzymatic breakdown of ECM proteins remodels the three-dimensional (3D) structure of the ECM. If the balance between collagen synthesis and degradation is not carefully maintained, scars can become either hypertrophic or atrophic (Ogawa 2011). These scar aberrations are induced by mechanical forces on the ECM, which are transmitted to the resident fibroblasts because they are bound to the affected matrix proteins. Even very small forces on the cells can cause scar deformation. The wound-healing phases are also affected by the increased volume and flow of extracellular fluid (ECF) in the wound, which result from the increased blood-vessel permeability after wounding. Thus, the ECM and ECF produce intrinsic mechanical forces, such as tension, pressure, shear, osmotic pressure, and hydrostatic pressure. The skin cells detect these mechanical stimuli via mechanosensors and convert them into electrical signals that induce their proliferation, angiogenesis, and epithelization (Chin et al. 2010). In addition to these cellular responses, the mechanosensitive nociceptors of the nervous system induce important macroscale tissue responses (Akaishi et al. 2008a; Chin et al. 2009).

16.3 Cutaneous Scarring and Mechanobiology

The last phase of wound-healing produces the scar. Scars mainly consist of collagens that are covered by epidermis. The early stage of scarring is marked by inflammation, whose purpose is to close the wound gap. The inflammation causes inflammatory cells, blood vessels, nerve fibers, and collagen-secreting fibroblasts to accumulate in the damaged area, resulting in an immature scar that is red, elevated, hard, and painful. Normally, this inflammation decreases naturally over months. This is known as the "scar maturation process" and is characterized by a decrease in the number of blood vessels, collagen fibers, and fibroblasts (Huang et al. 2013).

When the scar maturation process is excessive, it can result in an atrophic scar (e.g., the typical scar that is left by chickenpox), which is a depression in the skin. By contrast, when the maturation process is not properly engaged because inflammation continues in the scar, the immature scar stage is prolonged. This results in the pathological scars known as hypertrophic scars and keloids. Many lines of evidence in recent years have suggested that mechanical force can be an important cause of such pathological scar development (Ogawa 2011). This notion is further supported by the empirical understanding that in order to prevent noticeable scarring, surgical wounds must be stabilized (see later). In other words, though an appropriate amount of intrinsic tension is needed to induce wound closure, it is also important to suppress the extrinsic mechanical forces on the scar edges. When the mechanical forces on and in the scar are imbalanced, a heavy scar can result.

16.4 Cellular and Tissue Responses to Mechanical Forces

Mechanical forces, including stretching tension, shear force, scratch, compression, hydrostatic pressure, and osmotic pressure, are perceived by cellular mechanosensors and/or nerve fiber mechanoreceptors (including mechanosensitive nociceptors), which produce the somatic sensation of mechanical force (Ogawa and Hsu 2013). The cellular mechanosensors include mechanosensitive ion channels such as the Ca^{2+}, K^{+}, Na^{2+}, and Mg^{2+} ion channels, cytoskeleton components such as actin filaments, and cell-adhesion molecules such as integrins (Martinac 2014). Skin resident cells are attached to the ECM via cell-adhesion molecules, while the cytoskeleton is connected to mechanosensitive ion channels and cell-adhesion molecules. When the ECM is distorted by mechanical forces such as skin tension, the cytoskeleton is altered and the mechanosensitive ion channels are activated. However, not all mechanical forces activate mechanosensitive ion channels by altering the cytoskeleton: for example, ECF-based pressure activates these channels directly, because hydrostatic pressure impacts ion inflow but not cell shape. The mechanosensing cells then convert the mechanical stimuli into electrical signals, which employ various mechanotransduction pathways to regulate cell proliferation, angiogenesis, and epithelialization (Huang et al. 2012).

The key mechanotransduction pathways involved in scarring at the cellular level appear to be the integrin, mitogen-activated protein kinase (MAPK)/G-protein, tumor necrosis factor alpha (TNF-α)/nuclear factor κB (NF-κB), Wnt/β-catenin, interleukin (IL), calcium ion, and transforming growth factor beta (TGF-β)/Smad pathways (Huang et al. 2012). The latter seems to play an especially important role in the way scar tissue reacts to mechanical forces. Supporting this is that when keloid-derived fibroblasts are subjected to mechanical force in the form of equibiaxial strain, they produce more TGF-β1 and -β2 than normal skin-derived fibroblasts (Wang et al. 2006). Another study has shown that stretching a myofibroblast-derived ECM in the presence of mechanically apposing stress fibers immediately activates latent TGF-β1, and that compared to relaxed tissues, stressed tissues exhibit increased activation of Smad2 and 3, which are the downstream targets of TGF-β1 signaling (Wipff et al. 2007). Other membrane proteins that modulate cellular mechanosignaling pathways include the G-proteins (Silver et al. 2003): mechanical stimulation alters G-protein conformation, leading to growth factor-like changes that initiate secondary messenger cascades and initiate cell growth. Similarly, calcium ion-mechanosensitive channels are involved in phospholipase C activation, which can lead to protein kinase C activation and subsequent epidermal growth factor (EGF) activation.

At the tissue level, sensory fibers act as mechanoreceptors in the skin (Akaishi et al. 2008a). When mechanical stimuli are received by mechanosensitive nociceptors somewhere on the body, the signals are transmitted to the dorsal root ganglia, which contain neuronal cell bodies in the afferent spinal nerves. This results in neuropeptide release from the peripheral terminals of the primary afferent sensory neurons, which innervate the skin and are often in physical contact with epidermal and dermal cells. These neuropeptides include substance P, calcitonin gene-based peptide (CGRP), neurokinin A, vasoactive intestinal peptide, and somatostatin, and all of them can directly modulate the functions of keratinocytes, fibroblasts, and Langerhans, mast, dermal, microvascular, endothelial, and infiltrating immune cells. By this mechanism, the triggering of the mechanosensitive nociceptors alters cell proliferation, cytokine production, antigen presentation, sensory neurotransmission, mast cell degradation, and vasodilation. This mechanism is known especially for increasing vascular permeability in both physiological and pathophysiological conditions. These proinflammatory responses are termed "neurogenic inflammatory responses." Substance P and CGRP act through the neurokinin 1 and the CGRP1 receptor, respectively, and both are synthesized during nerve growth factor regulation. It has been suggested that neurogenic inflammation/neuropeptide activity plays a role in burn and abnormal scars, such as keloids and hypertrophic scars.

16.5 Keloids and Hypertrophic Scars and Mechanobiology

Traditionally, hypertrophic scars and keloids are diagnosed as separate clinical and pathological entities, though they are both characterized by prolonged and aberrant ECM accumulation. The so-called "typical keloids" grow beyond the confines of their original wounds and exhibit accumulation of dermal hyalinized collagens under the microscope (Figure 16.1). By contrast, hypertrophic scars generally grow within the boundaries of wounds and appear histologically as dermal nodules. However, even senior clinicians sometimes have difficulty in differentiating between the two conditions, particularly with atypical cases. Pathologists traditionally rely on the thick eosinophilic

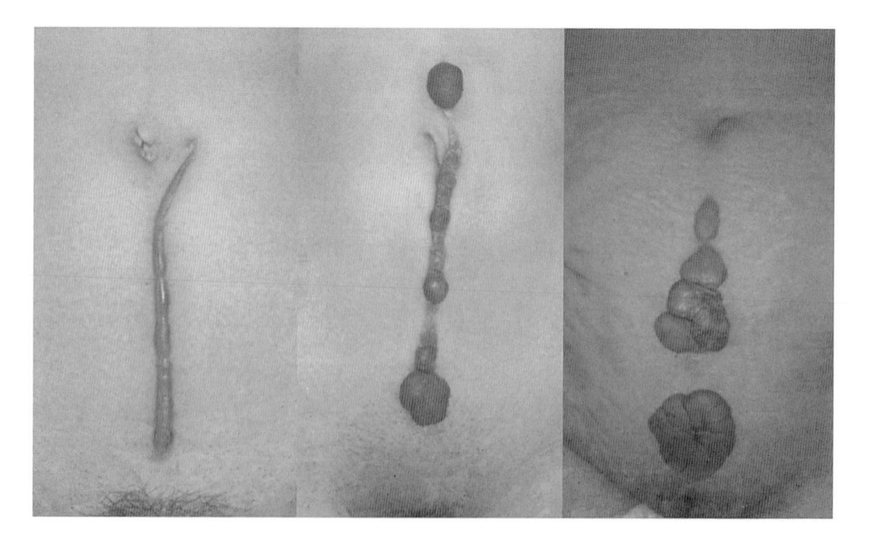

Figure 16.1 Keloids, hypertrophic scars, and gray area. Typical hypertrophic scars generally grow within the boundaries of wounds, and typical keloids grow beyond the confines of their original wounds. However, even senior clinicians sometimes have difficulty in differentiating between the two conditions, particularly with atypical cases.

collagen bundles in keloids to distinguish them from hypertrophic scars, but the diagnosis is not always straightforward. Indeed, we currently believe that hypertrophic scars and keloids represent successive or alternative stages of the same underlying fibroproliferative pathological lesion, and that the progression into one or the other classical form may be determined by a variety of proinflammatory risk factors, including mechanical forces (Figure 16.2). This notion is supported by a number of observations, including the fact that keloids can grow from mature scars and that hypertrophic scars can be generated by mechanical forces in experimental animal models (Huang et al. 2014).

We also believe that mechanical forces not only promote keloid and hypertrophic scar growth, but may also be a primary trigger for their generation. This is supported by a statistical study of 1500 anatomic regions in Asian patients, which showed that keloids tend to occur at specific sites, including the anterior chest, shoulder, scapular, and lower abdomen–suprapubic regions (Ogawa 2011) (Figure 16.3). All of these sites are constantly or frequently subjected to mechanical forces, including skin stretching as a result of daily body movements. Thus, the anterior chest skin is regularly stretched horizontally by upper-limb movements, the shoulder and scapula skin is constantly stretched by upper-limb movements and body-bending motions, and the lower abdomen and suprapubic skin regions are stretched hundreds of times a day by sitting and standing motions. By contrast, heavy scars on the scalp, upper eyelid, and anterior lower leg are rare, even in patients with extensive keloids or hypertrophic scars that cover much of the body. This pattern is likely to reflect the absence of tension on the skin in these regions, even in cases of deep wounding: the skin on the scalp and anterior lower leg is stabilized by the bones that lie directly under it, while there is little tension on the upper eyelid during the opening and closing of the eyes.

These observations together suggest that keloids and hypertrophic scars largely differ because the inflammation in keloids is more prolonged and stronger than that in hypertrophic scars. It is possible that the greater inflammation in keloids is due to pronounced

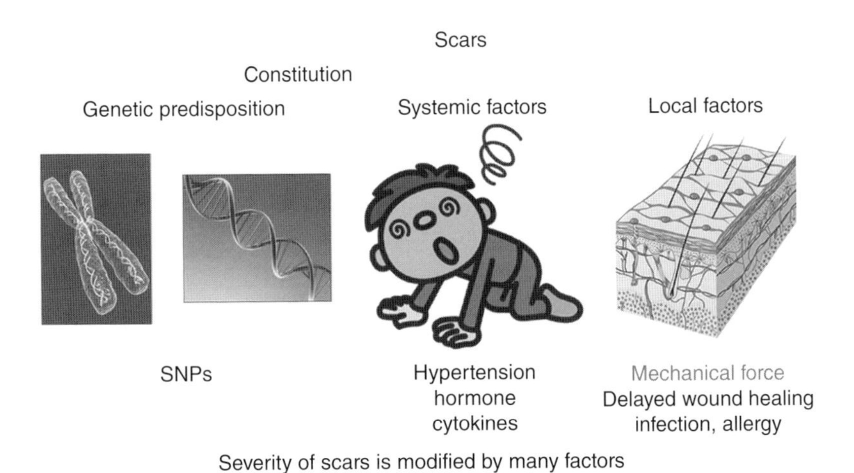

Figure 16.2 Severity of scars is modified by many factors. It is likely that the inflammatory status in scars is modified by many other risk factors, including genetic, systemic, and local factors, such as hypertension (high blood pressure).

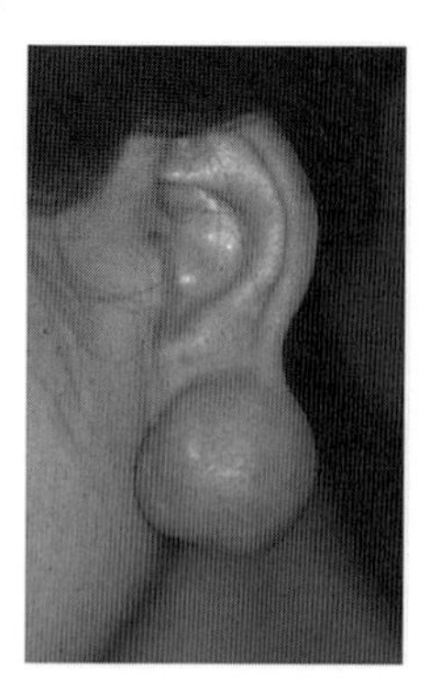
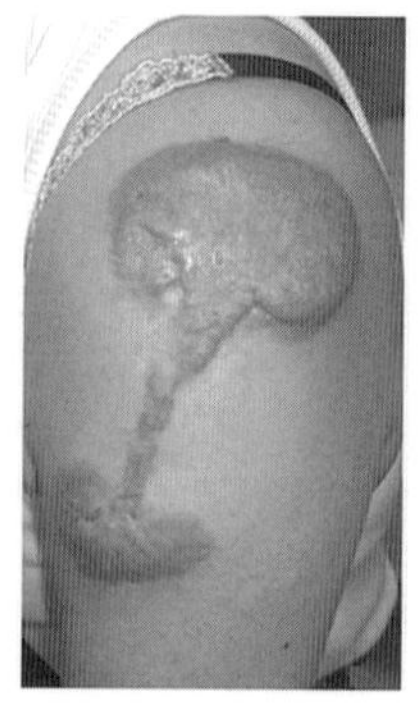
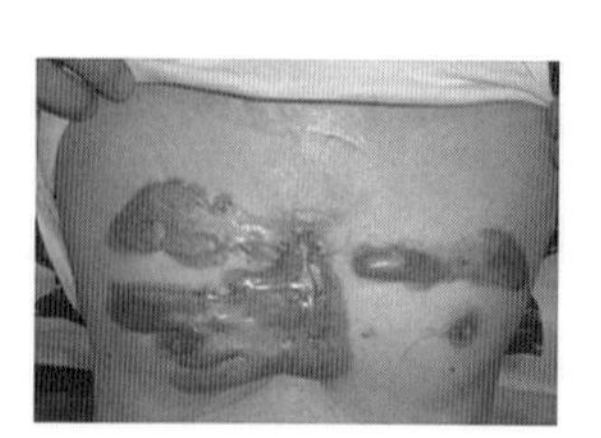
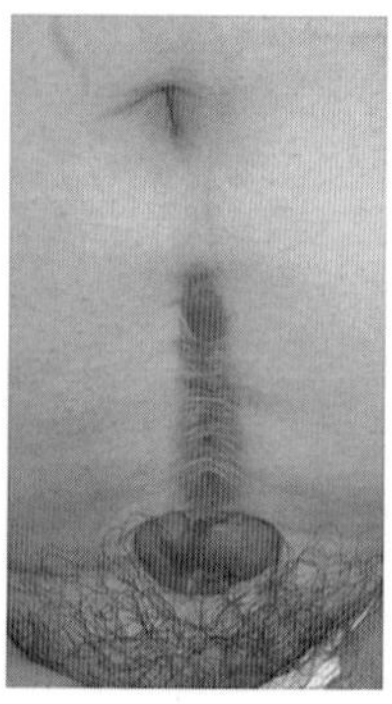

Figure 16.3 Typical keloids. Keloids tend to occur at specific sites, including the anterior chest, shoulder, scapular, and lower abdomen–suprapubic regions. *Source:* Ogawa (2011). Reproduced with permission of John Wiley and Sons.

skin tension on the wound, whereas the inflammation in hypertrophic scars is the result of less strong or different mechanical forces. However, it is also likely that the inflammatory status in heavy scars is modified by many other risk factors, including genetic, systemic, and local factors, such as hypertension (high blood pressure) (Arima et al. 2015) (Figure 16.2).

16.6 Relationship Between Scar Growth and Tension

Keloids grow and spread both vertically and horizontally, and are thus similar in many respects to slowly growing malignant tumors. The direction of their horizontal growth results in characteristic shapes, which depend on their location. For example, keloids on the anterior chest grow in a "crab's claw"-like pattern, whereas shoulder keloids grow in a "butterfly" shape. These patterns may reflect the predominant directions of skin tension at these sites (Akaishi et al. 2008b).

Our previous finite-element analysis of the mechanical force distribution around keloids showed that there is high skin tension at the keloid edges and lower tension at the center (Akaishi et al. 2008b; Ogawa et al. 2011) (Figure 16.4). This observation may explain why keloids generally stop growing in their central regions. This is supported by the fact that histological keloid sections that move from the periphery to the center exhibit a gradual decrease in the intensity of key features of inflammation, such as the presence of microvessels, fibroblasts, and inflammatory cells. This indicates that there is reduced inflammation in the center. Our finite-element analysis also showed that keloid expansion occurs in the direction of dominant skin-pulling and that the stiffness of the skin at the keloid circumference correlates directly with the degree of skin tension. These observations together strongly support the notion that skin tension associates closely with both the pattern and the degree of keloid growth. We speculate that individual (genetic) variation in scar responsiveness to skin tension may explain why keloids have a different growth pattern to hypertrophic and normal scars.

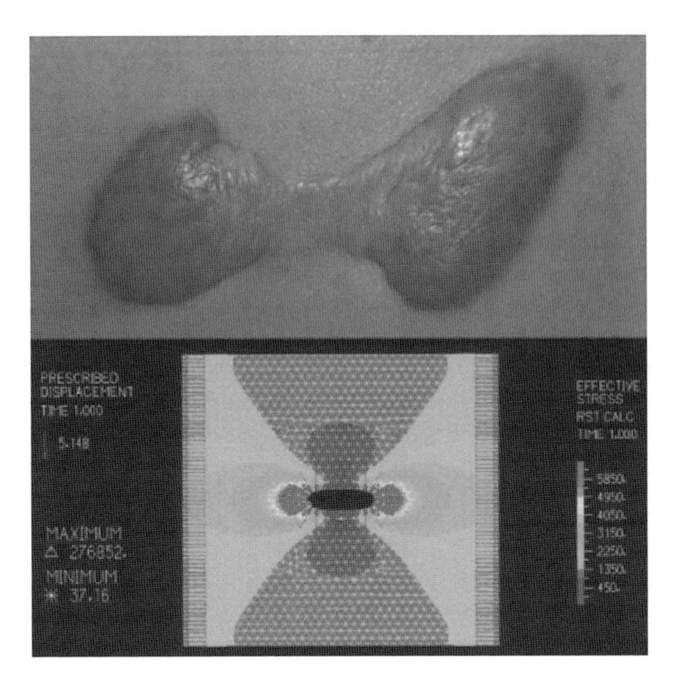

Figure 16.4 Finite-element analysis of the mechanical force distribution around keloids, showing that there is high skin tension at the edges. This observation strongly supports the notion that skin tension is closely associated with the pattern and degree of keloid growth. *Source:* Ogawa (2011). Reproduced with permission of John Wiley and Sons.

16.7 A Hypertrophic Scar Animal Model Based on Mechanotransduction

Many researchers have sought to develop animal models of heavy scars by using mice, rats, and rabbits. However, all of these models, especially the keloid models, seem to be driven more by an acute inflammatory response than by chronic inflammation, and they thus largely generate immature scars. Nevertheless, one of these models – a hypertrophic scar mouse model in which heavy scars are induced by mechanical force loading – shows that scars subjected to tension exhibit less apoptosis and that inflammatory cells and mechanical forces promote fibrosis (Aarabi et al. 2007). These findings support the now well-established notion that mechanical forces strongly modulate cellular behavior in the scar. Analyses of fibroblasts harvested from human or animal scar tissues have elucidated the mechanosignaling pathways that may participate in the formation and growth of cutaneous scars. The TGF-β/Smad, integrin, and calcium ion pathways have been definitively proven to play this role, while the MAPK and G-protein, Wnt/β-catenin, TNF-α/NF-κB, and IL pathways play a possible but as yet unproven role. During scar development, these cellular mechanosignaling pathways interact actively with the ECM and crosstalk extensively with the hypoxia, inflammation, and angiogenesis pathways. The elucidation of scar mechanosignaling pathways provides a new platform for understanding scar development. This better understanding will facilitate research into this promising field, and may help to promote the development of pharmacological interventions that could ultimately prevent, reduce, or even reverse scar formation and progression.

16.8 Mechanotherapy for Scar Prevention and Treatment

16.8.1 Wound and Scar Stabilization Materials

To limit skin-stretching and external mechanical stimuli during wound-healing/scarring, wounds or scars should be covered by fixable materials such as tape, bandages, garments, or silicone gel sheets. A randomized controlled trial (RCT) showed that tape fixation helped to prevent hypertrophic scar formation after cesarean section in 70 subjects: the scar volume was significantly lower when paper tape was used (Atkinson et al. 2005). Other RCTs showed that silicone gel sheeting significantly reduces the incidence of hypertrophic scars or keloids (Gold 1994). Our computer analysis of the mechanical force conditions around scars also showed that silicone gel sheeting reduces the tension at the scar edges (Akaishi et al. 2010).

16.8.2 Sutures

The preceding observations, and the fact that keloids and hypertrophic scars arise from the dermis, led us to speculate that we could reduce the risk of keloid and hypertrophic scar formation after surgery (and the tendency for keloids to recur after flap treatment) by using sutures that place little tension on the wound dermis (Ogawa et al. 2011). In general surgery (e.g., cardiac, abdominal, and gynecological surgery), the epidermis and dermis are usually sutured together after subcutaneous sutures have been placed. By contrast, three-layered sutures consisting of separate subcutaneous, dermal, and superficial sutures are used in plastic surgery; these sutures clearly decrease the risk of both surgical-site infections and hypertrophic scars. However, further modifications of suture techniques are needed to prevent the development of severe hypertrophic scars and keloids, because even three-layered sutures place tension on the dermis. Consequently, we have started to use subcutaneous/fascial tensile reduction sutures that place the tension on the deep fascia and superficial fascia layer. This means that the use of dermal sutures is minimized; indeed, dermal sutures can be avoided altogether if the wound edges can be joined naturally under very low tension. We prefer 2-0 polydioxanone sutures (PDSII: Ethicon Japan, Tokyo), or 3-0 PDSII for subcutaneous/fascial sutures, 4-0 or 5-0 PDSII for dermal sutures (if they are necessary), or 6-0 or 7-0 polypropylene or nylon sutures (Proline or Ethilon: Ethicon Japan, Tokyo) for superficial sutures. The consequence of this suturing strategy is that the wound edges are elevated smoothly, with minimal tension on the dermis. Our preliminary evidence suggests that this approach prevents the development of large scars (Ogawa et al. 2011).

Sometimes, after suturing, there are small nodules under the skin that can be sensed when the wound surface is touched. These probably reflect surgical damage to the dermis. Since we have noticed that keloids and hypertrophic scars tend to recur from these nodules (indeed, it seems that keloid recurrence usually starts from the suture marks, rather than the sutured surfaces), we ensure that we do not nick the dermal layer during surgery.

16.8.3 Skin Grafting, Flaps, and Z-Plasty

Keloids can be treated surgically in two ways: they can either be radically resected or they can undergo mass reduction. For both types of approach, skin grafting or flap transfer with postoperative radiotherapy may be required if the keloid is difficult to excise completely and suture directly. However, skin grafting associates with two problems: keloid recurrence at the margins of the skin graft and depigmentation of the center of the skin graft. Since our computer simulation studies suggest that the tension on the edge of the keloid will be reduced if there is soft tissue under the keloid (data not shown), we are currently attempting to reduce the tension on skin flaps for keloid reconstruction by using flaps with fat under the skin. Our preliminary study of 20 patients with huge keloids who were treated with such flaps and with postoperative radiotherapy in our facility shows that all cases had uneventful postoperative courses and did not exhibit keloid recurrence (Ogawa et al. 2011). While a longer follow-up period is needed to confirm these observations, they are encouraging.

In the past, keloid reconstruction using flaps was discouraged because it was thought that the donor site could develop keloids. However, the development of keloids on the donor site can be prevented by using subcutaneous/fascial tensile reduction sutures and postoperative radiotherapy. Thus, we believe flap surgery is suitable as a treatment of choice when dealing with severe keloids. In particular, since perforator flaps (especially the perforator-pedicled propeller flap) associate with little donor-site morbidity, we often use these sophisticated flaps to reconstruct keloids.

Z-plasty is also used to decouple long keloids and hypertrophic scars, especially those located on a joint. When scars are segmented into short sections, they tend to mature in a relatively short time. We can effectively use z-plasty for limb parts; namely, the shoulder joint, axilla, elbow joint, wrist joint, finger joints, groin region, knee joints, and foot joint (Figure 16.5).

16.9 Conclusion

Wound-healing and scarring involve some of the most complicated biological reactions in our body. Nevertheless, we believe that these phenomena can be viewed more simply, thereby helping to identify key drivers that can be exploited in the clinical setting.

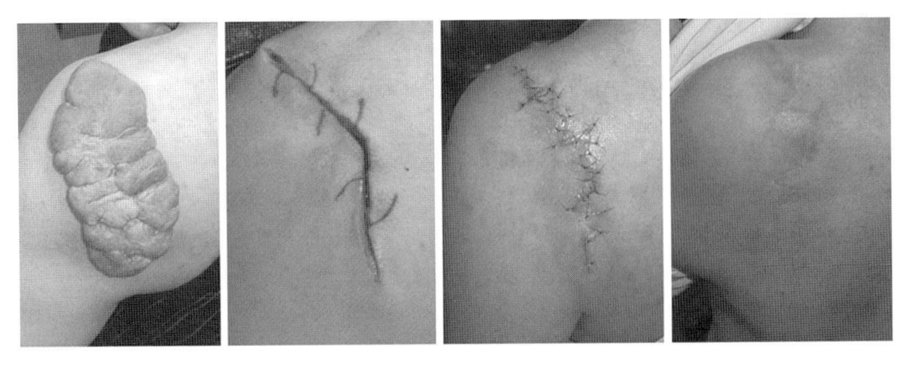

Figure 16.5 Z-plasty for shoulder-joint keloids. Z-plasty is effective in tension reduction and in decoupling long keloids and hypertrophic scars, especially those located on a joint.

Basic mechanobiology research and our own studies have shown that mechanical force, together with hypersensitivity to mechanical force, is a major cause of abnormal scarring. More specifically, it has been shown that cyclically applied mechanical forces may induce continuous skin inflammation and mechanosignaling pathway activation. These observations together suggest that reducing skin tension may prevent the development and recurrence of abnormal scars. Further research into mechanosignaling pathways may aid the development of effective therapies for abnormal scars.

References

Aarabi, S., K. A. Bhatt, Y. Shi, J. Paterno, E. I. Chang, S. A. Loh, et al. 2007. "Mechanical load initiates hypertrophic scar formation through decreased cellular apoptosis." *FASEB Journal* **21**: 3250–61. doi:10.1096/fj.07-8218com.

Akaishi, S., R. Ogawa, and H. Hyakusoku. 2008a. "Keloid and hypertrophic scar: neurogenic inflammation hypotheses." *Medical Hypotheses* **71**: 32–8. doi:10.1016/j.mehy.2008.01.032.

Akaishi, S., M. Akimoto, R. Ogawa, and H. Hyakusoku. 2008b. "The relationship between keloid growth pattern and stretching tension: visual analysis using the finite element method." *Annals of Plastic Surgery* **60**: 445–51. doi:10.1097/SAP.0b013e3181238dd7.

Akaishi, S., M. Akimoto, H. Hyakusoku, and R. Ogawa. 2010. "The tensile reduction effects of silicone gel sheeting." *Plastic and Reconstructive Surgery* **126**: 109–11e. doi:10.1097/PRS.0b013e3181df7073.

Arima, J., C. Huang, B. Rosner, S. Akaishi, and R. Ogawa. 2015. "Hypertension: a systemic key to understanding local keloid severity." *Wound Repair and Regeneration* **23**: 213–21. doi:10.1111/wrr.12277.

Atkinson, J. A., K. T. McKenna, A. G. Barnett, D. J. McGrath, and M. Rudd. 2005. "A randomized, controlled trial to determine the efficacy of paper tape in preventing hypertrophic scar formation in surgical incisions that traverse Langer's skin tension lines." *Plastic and Reconstructive Surgery* **116**: 1648–56; disc. 1657–8. PMID:16267427.

Chin, M. S., L. Lancerotto, D. L. Helm, P. Dastouri, M. J. Prsa, M. Ottensmeyer, et al. "Analysis of neuropeptides in stretched skin." *Plastic and Reconstructive Surgery* **124**: 102–13. doi:10.1097/PRS.0b013e3181a81542.

Chin, M. S., R. Ogawa, L. Lancerotto, G. Pietramaggiori, K. T. Schomacker, J. C. Mathews, et al. 2010. "In vivo acceleration of skin growth using a servo-controlled stretching device." *Tissue Engineering. Part C, Methods* **16**: 397–405. doi:10.1089/ten.TEC.2009.0185.

Gold, M. H. 1994. "A controlled clinical trial of topical silicone gel sheeting in the treatment of hypertrophic scars and keloids." *Journal of the American Academy of Dermatology* **30**: 506–7. PMID:8113473.

Gurtner, G., C S. Werner, Y. Barrandon, and M. T. Longaker. 2008. "Wound repair and regeneration." *Nature* **453**(7193): 314–21. doi:10.1038/nature07039.

Huang, C., S. Akaishi, and R. Ogawa. 2012. "Mechanosignaling pathways in cutaneous scarring." *Archives of Dermatological Research* **304**: 589–97. doi:10.1007/s00403-012-1278-5.

Huang, C., G. F. Murphy, S. Akaishi, and R. Ogawa. 2013. "Keloids and hypertrophic scars: update and future directions." *Plastic and Reconstructive Surgery. Global Open* **1**(4): e25. doi:10.1097/GOX.0b013e31829c4597.

Huang, C., S. Akaishi, H. Hyakusoku, and R. Ogawa. 2014. "Are keloid and hypertrophic scar different forms of the same disorder? A fibroproliferative skin disorder hypothesis based on keloid findings." *International Wound Journal* **11**: 517–22. doi:10.1111/j.1742-481X.2012.01118.x.

Martinac, B. 2014. "The ion channels to cytoskeleton connection as potential mechanism of mechanosensitivity." *Biochimica et Biophysica Acta* **1838**: 682–91. doi:10.1016/j.bbamem.2013.07.015.

Ogawa, R. 2011. "Mechanobiology of scarring." *Wound Repair and Regeneration* **19**(Suppl. 1): s2–9. doi:10.1111/j.1524-475X.2011.00707.x.

Ogawa, R. and C. K. Hsu. 2013. "Mechanobiological dysregulation of the epidermis and dermis in skin disorders and in degeneration." *Journal of Cellular and Molecular Medicine* **17**(7): 817–22. doi:10.1111/jcmm.12060.

Ogawa, R, S. Akaishi, C. Huang, T. Dohi, M. Aoki, Y. Omori, et al. 2011. "Clinical applications of basic research that shows reducing skin tension could prevent and treat abnormal scarring: the importance of fascial/subcutaneous tensile reduction sutures and flap surgery for keloid and hypertrophic scar reconstruction." *Journal of Nippon Medical School* **78**(2): 68–76. PMID:21551963.

Silver, F. H., L. M. Siperko, and G. P. Seehra. 2003. "Mechanobiology of force transduction in dermal tissue." *Skin Research and Technology* **9**(1): 3–23. PMID:12535279.

Van De Water, L., S. Varney, and J. J. Tomasek. 2013. "Mechanoregulation of the myofibroblast in wound contraction, scarring, and fibrosis: opportunities for new therapeutic intervention." *Advances in Wound Care* **2**(4): 122–41. doi:10.1089/wound.2012.0393.

Wang, Z., K. D. Fong, T. T. Phan, I, J. Lim, M. T. Longaker, and G. P. Yang 2006. "Increased transcriptional response to mechanical strain in keloid fibroblasts due to increased focal adhesion complex formation." *Journal of Cell Physiology* **206**: 510–17. doi:10.1002/jcp.20486.

Wipff, P. J., D. B. Rifkin, J. J. Meister, and B. Hinz. 2007. "Myofibroblast contraction activates latent TGF-β1 from the extracellular matrix." *Journal of Cell Biology* **179**: 1311–23. doi:10.1083/jcb.200704042.

17

Mechanobiology and Mechanotherapy for the Nail

Hitomi Sano and Rei Ogawa

Department of Plastic, Reconstructive, and Aesthetic Surgery, Nippon Medical School, Tokyo, Japan

17.1 Introduction

Our bodies are constantly being subjected to mechanical forces that directly affect cellular functions (Aronson et al. 1990; Duncan and Turner 1995; Carter et al. 1998; Kim et al. 2010). Mechanobiology is an emerging field of study that looks at the responses to mechanical forces which shape development, physiology, and disease. Several therapies, including tissue expanders (Lantieri et al. 1998) and negative-pressure wound therapy (NPWT) (Orgill et al. 2009), utilize the biological responses of tissues to mechanical forces to effect desired tissue changes. These approaches can be defined as "mechanotherapies" (Huang et al. 2013).

Nails are defined as the smooth hard layers on the ends of fingers and toes. They have numerous functions, including protecting the distal phalanges, enhancing tactile discrimination, performing fine manipulation, and contributing to pedal biomechanics (Ashbell et al. 1967; Russell and Casas 1989; Drake et al. 1998; Salazard et al. 2004). Thus, nails are continuously exposed to physical stimulation. Nail configuration is known to be influenced by numerous factors, including genetic factors, the shape of the distal phalangeal bone (Tosti and Piraccini 2008), mechanical force (Sano and Ogawa 2013, 2015; Sano et al. 2014), malnutrition (Al-Dabbagh and Al-Abachi 2005), neurogenic factors (Horowitz 1993; Andersen 2012), blood flow (Alemany 2012; Loenneke et al. 2012), and factors that cause the thinning and softening of nails, such as pharmaceutical use (Baron 1999). Several recent studies have shown that of these factors, mechanical forces have a particularly pronounced effect on nail configuration and thus may be involved in the development of nail deformity. This chapter introduces the effect of mechanical force on nail configuration from a mechanobiological point of view.

17.2 Nail Anatomy

To understand the effect of mechanical force on nails, it is important to have an overview of the nail anatomy (Figure 17.1).

Mechanobiology: Exploitation for Medical Benefit, First Edition. Edited by Simon C. F. Rawlinson.

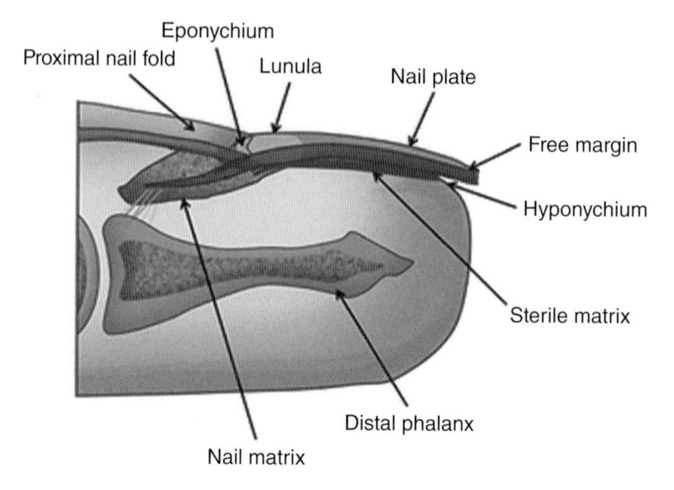

Figure 17.1 Nail anatomy.

17.2.1 Nail Plate

The nail plate is a keratinized structure that is continuously produced throughout life. It results from the maturation and keratinization of the nail matrix epithelium and is firmly attached to the sterile matrix. In transverse section, the nail plate consists of three parts: the dorsal, intermediate, and ventral nail plates (Perrin 2008). The dorsal and intermediate plates are produced by the nail matrix, whereas the ventral plate is produced by the sterile matrix.

17.2.2 Nail Matrix

The nail matrix is a specialized epithelial structure that lies above the mid portion of the distal phalanx. In longitudinal section, the matrix consists of a proximal (dorsal) and a distal (ventral) portion. Keratinization of the proximal nail matrix cells produces the dorsal nail plate, while keratinization of the distal nail matrix cells produces the intermediate nail plate. The distal matrix is visible through the nail plate as a white half-moon-shaped area called the lunula.

17.2.3 Sterile Matrix

The sterile matrix epidermis extends from the distal margin of the lunula to the hyponychium. The ventral plate that is produced by the sterile matrix constitutes approximately one-fifth of the terminal nail thickness and mass (Johnson et al. 1991). The dermis of the sterile matrix contains numerous sensory nerve endings, including Merkel endings and Meissner's corpuscles (Standring 2005).

17.3 Role of Mechanobiology in Nail Morphology

Our recent research shows that excessively convex nails are associated with limited or absent upward mechanical forces on the nail, such as in the case of bedridden patients. By contrast, nails with a flattened or concave configuration are associated with excessive upward mechanical forces. An example is the thumb nail of the working hand of a carpenter who frequently uses a hammer (Sano and Ichioka 2012; Sano and Ogawa 2013, 2015). These observations have led us to speculate that normal nails reflect standard mechanical forces and that nail deformities are the result of unusually small or large mechanical forces on the nail.

The example of hair may provide some clues to the mechanism by which the nail changes its configuration in response to the degree of mechanical force. Like nails, hair is composed of cornified keratinocytes and its morphology can be shaped by mechanical forces (Xu and Chen 2011). Hair consists of a central layer and a surrounding layer, and the mechanical force that results from a disparity in the growth rates of these layers is responsible for the curling hair phenotype (Bernard 2003; Krause and Foitzik 2006; Xu and Chen 2011). Since nails also consist of several layers (i.e., plates) (Zaias 1980), we hypothesize that the ratio of the individual growth rates of the three nail plates is determined in large part by the habitual strength of the mechanical forces on the nail. This ratio in turn determines whether the nail develops a curved, normal, or convex shape. For example, if the ventral nail plate grows faster than the dorsal and intermediate plates, the nail will tend to flatten; if the ventral plate grows more slowly than the other two plates, the nail will develop an excessive curved configuration. This hypothesis is supported by the fact that the dorsal and intermediate layers are produced by the nail matrix, whereas the ventral layer is produced by the sterile matrix (Tosti and Piraccini 2008).

There are multiple possible mechanisms by which mechanical forces might induce changes in nail layer growth rate. One possibility is that the mechanical stimuli directly affect the proliferation of the sterile matrix cells and their production of the ventral plate. For example, hard mechanical forces could increase this proliferation, whereas lack of stimulation could blunt it. Since the dermis of the sterile matrix contains numerous mechanoreceptors (Standring 2005), it is also possible that the effect of mechanical stimuli on sterile matrix proliferation is mediated indirectly by these mechanoreceptors. Thus, inadequate or excessive mechanical stimulation of the nail would alter ventral plate growth, while the dorsal and intermediate layers would continue to grow at a normal rate. The result would be a deformed nail.

17.4 Nail Diseases and Mechanical Forces

Our proposed mechanism is best illustrated by examining two representative nail deformities: koilonychia and pincer nail.

17.4.1 Koilonychia

Koilonychia is characterized by spooning of the nails. Its etiology is often associated with thinning and softening of the nails (Baron 1999) as a result of iron deficiency, chronic renal failure (Salem et al. 2008), malnutrition (Al-Dabbagh and Al-Abachi 2005), or pharmaceutical use. Our recent study showed that finger- and toenails that undergo substantial physical stimulation tend to have a flat or concave shape (Sano and Ichioka 2012; Sano and Ogawa 2013; Sano et al. 2014). Koilonychia can also be observed in the nails of healthy adults that are exposed to frequent strong force (Bentley-Phillips and Bayles 1971) and in the toenails of normal children, which are still too immature to withstand normal mechanical forces (Tosti and Piraccini 2008). These observations support the notion that mechanical forces exceeding the automatic curvature force will cause the nail to curve outward, resulting in koilonychia.

17.4.2 Pincer Nail

Pincer nail is defined as the transverse overcurvature of the nail plate, which increases along the longitudinal axis (Cornelius and Shelley 1968). Plausible causes of pincer nail include heredity (Chapman 1973; de Berker and Carmichael 1995), inappropriate nail-cutting, and ill-fitting shoes (Baran and Dawber 2001; Baran et al. 2001). All of these etiologies may have the same underlying mechanism, namely, a mismatch between the mechanical forces on the nail and the automatic curvature force (Sano and Ichioka 2012;Sano and Ogawa 2013; Sano et al. 2014). This notion is supported by a recent study which showed that when finger or toenails receive less physical stimulation than normal, they tend to curve inward. For example, the degree of toenail curvature in bed-ridden patients tends to increase as the duration of the bedridden state increases (Sano and Ogawa 2013). At a certain point, when the nail curvature has progressed beyond the range of normal, the nail is called a "pincer nail." Thus, pincer nails may be caused either by a lack of upward mechanical forces or by an inherent increase in the automatic curvature force.

17.5 Current Nail Treatment Strategies

Various conservative (Effendy et al. 1993; Kim and Sim 2003) and surgical (DuVries 1959; Suzuki et al. 1979; Brown et al. 2000; Plusje 2000, 2001; Aksakal et al. 2001; Haneke 2001; Hatoko et al. 2003; Zook et al. 2005) treatments of pincer nail have been reported. Conservative procedures are usually employed first. These include the use of cotton, elastic tape (Nishioka et al. 1985), polyacryl sculptured nails (Nishioka et al. 1985), flexible tube splinting (Schulte et al. 1998; Arai et al. 2004), an elastic wire (Moriue et al. 2008), or a plastic device (Harrer et al. 2005; Di Chiacchio et al. 2006). Though these treatments are relatively noninvasive, they demand frequent care, and the recurrence rate is high.

Surgical procedures can be deployed as an alternative, including total or partial excision of the sterile matrix (Pettine et al. 1988; Umeda et al. 1992), phenolization (Robb and Murray 1982), and carbon dioxide laser matricectomy (Leshin and Whitaker 1988). However, these procedures have several disadvantages, including surgical complexity, pain, cosmetic deformity, and the need for local anesthesia.

17.6 Mechanotherapy for Nail Deformities

Our mechanobiology-based hypothesis suggests that those therapies that control the degree of mechanical stimulus on the nail (thereby improving the balance between the automatic nail curvature force and the upward mechanical forces) may be particularly effective for nail deformities. Such therapies include several current treatments, namely, those employing an elastic wire or plastic device to reinforce the daily upward mechanical force.

In addition, several new therapies that reduce the automatic curvature force of the nail have been reported. One of these is the external use of a thioglycolic acid preparation that softens the nail. This method has been shown to be a useful pincer nail therapy (Okada and Okada 2012). Another approach that effectively reduces the automatic curvature force in overcurved nails is to thin the nails with a nail rasp (Sano 2015). These therapies, which essentially harness our proposed mechanobiology-based mechanism, have both been shown to prevent recurrence after the conventional therapies described in the previous section.

Though treatments for koilonychia have rarely been documented, we speculate that this nail condition would be improved by reducing the physical stimulus on the nails and/or by supplying nutrients that reinforce the hardness of the nails.

We hope that this review of the mechanisms that may underlie nail deformities will promote the future development of innovative methods to prevent and treat nail deformities. Future experiments such as isolation of cells from nail/sterile matrices and measurement of responses to mechanical force should be verified in order to increase our understanding of the pathogenesis of automatic nail curvature force.

17.7 Conclusion

Mechanical forces may affect nail configuration and participate in the development of nail deformities. Such deformities can be treated by improving the balance between automatic nail curvature force and the upward mechanical forces arising from the finger/toepad.

References

Aksakal, A. B., A. Akar, H. Erbil, and M. Onder. 2001. "A new surgical therapeutic approach to pincer nail deformity." *Dermatologic Surgery* **27**(1): 55–7. PMID:11231245.

Al-Dabbagh, T. Q. and K. G. Al-Abachi. 2005. "Nutritional koilonychia in 32 Iraqi subjects." *Annals of Saudi Medicine* **25**(2): 154–7. PMID:15977696.

Alemany, M. 2012. "Regulation of adipose tissue energy availability through blood flow control in the metabolic syndrome." *Free Radical Biology & Medicine* **52**(10): 2108–19. doi:10.1016/j.freeradbiomed.2012.03.003.

Andersen, H. 2012. "Motor dysfunction in diabetes." *Diabetes/Metabolism Research and Reviews* **28**(Suppl. 1): 89–92. doi:10.1002/dmrr.2257.

Arai, H., T. Arai, H. Nakajima, and E. Haneke. 2004. "Formable acrylic treatment for ingrowing nail with gutter splint and sculptured nail." *International Journal of Dermatology* **43**(10): 759–65. doi:10.1111/j.1365-4632.2004.02342.x.

Aronson, J., B. Good, C. Stewart, B. Harrison, and J. Harp. 1990. "Preliminary studies of mineralization during distraction osteogenesis." *Clinical Orthopaedics and Related Research* **250**: 43–9. PMID:2293943.

Ashbell, T. S., H. E. Kleinert, S. M. Putcha, and J. E. Kutz. 1967. "The deformed finger nail, a frequent result of failure to repair nail bed injuries." *Journal of Trauma* **7**(2): 177–90. PMID:5335262.

Baran, R. and R. P. R. Dawber. 2001. *Diseases of the Nail and their Management*, 3rd edn. Boston, MA: Blackwell Science.

Baran, R., E. Haneke, and B. Richert. 2001. "Pincer nails: definition and surgical treatment." *Dermatologic Surgery* **27**(3): 261–6. PMID:11277894.

Baron, R. B. 1999. "Nutrition." In: L. M. Tierney, S. J. McPhee, and M. A. Papadakis (eds.). *Current Medical Diagnosis and Treatment*. New York: Appleton and Lange. pp. 1174–82.

Bentley-Phillips, B. and M. A. Bayles. 1971. "Occupational koilonychia of the toe nails." *British Journal of Dermatology* **85**(2): 140–4. PMID:5571027.

Bernard, B. A. 2003. "Hair shape of curly hair." *Journal of the American Academy of Dermatology* **48**(6 Suppl.): S120–6. doi:10.1067/mjd.2003.279.

Brown, R. E., E. G. Zook, and J. Williams. 2000. "Correction of pincer-nail deformity using dermal grafting." *Plastic and Reconstructive Surgery* **105**(5): 1658–61. PMID:10809094.

Carter, D. R., G. S. Beaupre, N. J. Giori, and J. A. Helms. 1998. "Mechanobiology of skeletal regeneration." *Clinical Orthopaedics and Related Research* **355**(Suppl.): S41–55. PMID:9917625.

Chapman, R. S. 1973. "Letter: Overcurvature of the nails – an inherited disorder." *British Journal of Dermatology* **89**(3): 317–18. PMID:4743435.

Cornelius, C. E. 3rd and W. B. Shelley. 1968. "Pincer nail syndrome." *Archives of Surgery* **96**(2): 321–2. PMID:5212470.

de Berker, D. and A. J. Carmichael. 1995. "Congenital alternate nail dystrophy." *British Journal of Dermatology* **133**(2): 336–7. PMID:7547416.

Di Chiacchio, N., B. V. Kadunc, A. R. Trindade de Almeida, and C. L. Madeira. 2006. "Treatment of transverse overcurvature of the nail with a plastic device: measurement of response." *Journal of the American Academy of Dermatology* **55**(6): 1081–4. doi:10.1016/j.jaad.2006.05.045.

Drake, L. A., R. K. Scher, E. B. Smith, G. A. Faich, S. L. Smith, J. J. Hong, and M. J. Stiller. 1998. "Effect of onychomycosis on quality of life." *Journal of the American Academy of Dermatology* **38**(5 Pt. 1): 702–4. PMID:9591814.

Duncan, R. L. and C. H. Turner. 1995. "Mechanotransduction and the functional response of bone to mechanical strain." *Calcified Tissue International* **57**(5): 344–58. PMID:8564797.

DuVries, H. L. 1959. "Diseases and deformities of the toenails." In: V. T. Inman (ed.). *Surgery of the Foot*. St. Louis, MO: CV Mosby. pp. 204–22.

Effendy, I., B. Ossowski, and R. Happle. 1993. "[Pincer nail. Conservative correction by attachment of a plastic brace]." *Hautarzt* **44**(12): 800–2. PMID:8113046.

Haneke, E. 2001. "Pincer nail." E. A. Krull, E. G. Zook, R. Baran and E. Haneke (eds.). In: *Nail Surgery: A Text and Atlas*. Philadelphia, PA: Lippincott Williams and Wilkins. pp. 168–71.

Harrer, J., V. Schoffl, W. Hohenberger, and I. Schneider. 2005. "Treatment of ingrown toenails using a new conservative method: a prospective study comparing brace treatment with Emmert's procedure." *Journal of the American Podiatric Medical Association* **95**(6): 542–9. PMID:16291845.

Hatoko, M., H. Iioka, A. Tanaka, M. Kuwahara, S. Yurugi, and K. Niitsuma. 2003. "Hard-palate mucosal graft in the management of severe pincer-nail deformity." *Plastic and Reconstructive Surgery* **112**(3): 835–9. doi:10.1097/01.prs.0000070178.32975.9e.

Horowitz, S. H. 1993. "Diabetic neuropathy." *Clinical Orthopaedics and Related Research* **296**: 78–85. PMID:8222454.

Huang, C., J. Holfeld, W. Schaden, D. Orgill, and R. Ogawa. 2013. "Mechanotherapy: revisiting physical therapy and recruiting mechanobiology for a new era in medicine." *Trends in Molecular Medicine* **19**(9): 555–64. doi:10.1016/j.molmed.2013.05.005.

Johnson, M., J. S. Comaish, and S. Shuster. 1991. "Nail is produced by the normal nail bed: a controversy resolved." *British Journal of Dermatology* **125**(1): 27–9. PMID:1873199.

Kim, K. D. and W. Y. Sim. 2003. "Surgical pearl: nail plate separation and splint fixation – a new noninvasive treatment for pincer nails." *Journal of the American Academy of Dermatology* **48**(5): 791–2. doi:10.1067/mjd.2003.196.

Kim, I. S., Y. M. Song, and S. J. Hwang. 2010. "Osteogenic responses of human mesenchymal stromal cells to static stretch." *Journal of Dental Research* **89**(10): 1129–34. doi:10.1177/0022034510375283.

Krause, K. and K. Foitzik. 2006. "Biology of the hair follicle: the basics." *Seminars in Cutaneous Medicine and Surgery* **25**(1): 2–10. doi:10.1016/j.sder.2006.01.002.

Lantieri, L. A., N. Martin-Garcia, J. Wechsler, M. Mitrofanoff, Y. Raulo, and J. P. Baruch. 1998. "Vascular endothelial growth factor expression in expanded tissue: a possible mechanism of angiogenesis in tissue expansion." *Plastic and Reconstructive Surgery* **101**(2): 392–8. PMID:9462772.

Leshin, B. and D. C. Whitaker. 1988. "Carbon dioxide laser matricectomy." *Journal of Dermatologic Surgery and Oncology* **14**(6): 608–11. PMID:3372844.

Loenneke, J. P., T. Abe, J. M. Wilson, R. S. Thiebaud, C. A. Fahs, L. M. Rossow, and M. G. Bemben. 2012. "Blood flow restriction: an evidence based progressive model (Review)." *Acta Physiologica Hungarica* **99**(3): 235–50. doi:10.1556/APhysiol.99.2012.3.1.

Moriue, T., K. Yoneda, J. Moriue, Y. Matsuoka, K. Nakai, I. Yokoi, et al. 2008. "A simple therapeutic strategy with super elastic wire for ingrown toenails." *Dermatologic Surgery* **34**(12): 1729–32. doi:10.1111/j.1524-4725.2008.34360.x.

Nishioka, K., I. Katayama, Y. Kobayashi, and C. Takijiri. 1985. "Taping for embedded toenails." *British Journal of Dermatology* **113**(2): 246–7. PMID:4027192.

Okada, K. and E. Okada. 2012. "Novel treatment using thioglycolic acid for pincer nails." *Journal of Dermatology* **39**(12): 996–9. doi:10.1111/j.1346-8138.2012.01670.x.

Orgill, D. P., E. K. Manders, B. E. Sumpio, R. C. Lee, C. E. Attinger, G. C. Gurtner, and H. P. Ehrlich. 2009. "The mechanisms of action of vacuum assisted closure: more to learn." *Surgery* **146**(1): 40–51. doi:10.1016/j.surg.2009.02.002.

Perrin, C. 2008. "The 2 clinical subbands of the distal nail unit and the nail isthmus. Anatomical explanation and new physiological observations in relation to the nail growth." *American Journal of Dermatopathology* **30**(3): 216–21. doi:10.1097/DAD.0b013e31816a9d31.

Pettine, K. A., R. H. Cofield, K. A. Johnson, and R. M. Bussey. 1988. "Ingrown toenail: results of surgical treatment." *Foot & Ankle* **9**(3): 130–4. PMID:3229700.

Plusje, L. G. 2000. "Correction of pincer-nail deformity using dermal grafting." *Dermatologic Surgery* **105**(5): 1658–61.

Plusje, L. G. 2001. "Pincer nails: a new surgical treatment." *Dermatologic Surgery* **27**(1): 41–3.

Robb, J. E. and W. R. Murray. 1982. "Phenol cauterization in the management of ingrowing toenails." *Scottish Medical Journal* **27**(3): 236–9. PMID:7112084.

Russell, R. C. and L. A. Casas. 1989. "Management of fingertip injuries." *Clinics in Plastic Surgery* **16**(3): 405–25. PMID:2673624.

Salazard, B., F. Launay, C. Desouches, P. Samson, J. L. Jouve, and G. Magalon. 2004. "[Fingertip injuries in children: 81 cases with at least one year follow-up]." *Revue de chirurgie orthopédique et réparatrice de l'appareil moteur* **90**(7): 621–7. PMID:15625512.
Salem, A., S. Al Mokadem, E. Attwa, S. Abd El Raoof, H. M. Ebrahim, and K. T. Faheem. 2008. "Nail changes in chronic renal failure patients under haemodialysis." *Journal of the European Academy of Dermatology and Venereology* **22**(11): 1326–31. doi:10.1111/j.1468-3083.2008.02826.x.
Sano, H. 2015. "A novel nonsurgical treatment for pincer nail that involves mechanical force control." **3**(2): e311. doi:10.1097/gox.0000000000000220.
Sano, H. and S. Ichioka. 2012. "Influence of mechanical forces as a part of nail configuration." *Dermatology* **225**(3): 210–14. doi:10.1159/000343470.
Sano, H. and R. Ogawa. 2013. "Role of mechanical forces in hand nail configuration asymmetry in hemiplegia: an analysis of four hundred thumb nails." *Dermatology* **226**(4): 315–18. doi:10.1159/000350260.
Sano, H. and R. Ogawa. 2015. "A novel nonsurgical treatment for pincer nail that involves mechanical force control." *Plastic and Reconstructive Surgery. Global Open* **3**(2): e311. doi:10.1097/gox.0000000000000220.
Sano, H., K. Shionoya, and R. Ogawa. 2014. "Effect of mechanical forces on finger nail curvature: an analysis of the effect of occupation on finger nails." *Dermatologic Surgery* **40**(4): 441–5. doi:10.1111/dsu.12439.
Schulte, K. W., N. J. Neumann, and T. Ruzicka. 1998. "Surgical pearl: nail splinting by flexible tube--a new noninvasive treatment for ingrown toenails." *Journal of the American Academy of Dermatology* **39**(4 Pt. 1): 629–30. PMID:9777771.
Standring, S. 2005. "Skin and skin appendages." In: S. Standring (ed.). *Grey's Anatomy: The Anatomical Basis of Clinical Practice.* Edinburgh: Elsevier Churchill Livingston. pp. 157–78.
Suzuki, K., I. Yagi, and M. Kondo. 1979. "Surgical treatment of pincer nail syndrome." *Plastic and Reconstructive Surgery* **63**(4): 570–3. PMID:424467.
Tosti, A. and B. M. Piraccini. 2008. "Biology of nails and nail disorders." In: K. Wolff, L. A. Goldsmith, S. I. Katz, B. Gilchrest, A. Paller, and D. Leffell (eds.). *Fitzpatrick's Dermatology in General Medicine.* New York: McGraw-Hill. pp. 778–94.
Umeda, T., K. Nishioka, and K. Ohara. 1992. "Ingrown toenails: an evaluation of elevating the nail bed-periosteal flap." *Journal of Dermatology* **19**(7): 400–3. PMID:1401497.
Xu, B. and X. Chen. 2011. "The role of mechanical stress on the formation of a curly pattern of human hair." *Journal of the Mechanical Behavior of Biomedical Materials* **4**(2): 212–21. doi:10.1016/j.jmbbm.2010.06.009.
Zaias, N. 1980. *The Nail in Health and Disease,* 2nd edn. New York: Appleton and Lange.
Zook, E. G., C. P. Chalekson, R. E. Brown, and M. W. Neumeister. 2005. "Correction of pincer-nail deformities with autograft or homograft dermis: modified surgical technique." *Journal of Hand Surgery* **30**(2): 400–3. doi:10.1016/j.jhsa.2004.09.005.

18

Bioreactors

Recreating the Biomechanical Environment *In Vitro*

James R. Henstock[1] *and Alicia J. El Haj*[2]

[1] *Institute for Ageing and Chronic Disease, University of Liverpool, Liverpool, UK*
[2] *Institute for Science and Technology in Medicine, University of Keele, Staffordshire, UK*

18.1 The Mechanical Environment: Forces in the Body

The human body is exposed to a diverse range of forces throughout life, as it interacts with a complex and ever-changing physical environment. Contractions of skeletal muscle result in the compressive loading of bones and joints and stretching forces within tendons and ligaments, while contractions of cardiac muscle pump blood around the body, perfusing tissues with a pressurized flowing fluid. Cells therefore reside in a very dynamic environment, in which mechanical cues from their surroundings combine with biochemical signals to regulate all aspects of their behavior, from the initial growth and development of a tissue to maintenance of homeostasis and response to injury.

The generation of simultaneous, multiple forces by the whole body leads to their effects being distributed throughout the entire organism, and as a result tissues across the body experience different levels of these forces, depending on a wide variety of mediating factors. For example, forces that are applied in fluid-filled environments generate dynamic hydrostatic pressures as compressive loads are converted into a transitory increase in local pressure. In the load-bearing joints of the lower limb, these forces can be extraordinarily high, reaching 15 MPa in the knee synovium and 18 MPa in the hip during strenuous exercise, such as running (Morrell et al. 2005) –this is 60–80 times the pressure in an average car tire. It might be expected that these very high loads must be somehow endured by cells throughout the whole of the skeleton in the lower limbs, but the rigid extracellular matrix (ECM) serves to absorb and redirect the majority of these forces, such that osteocytes in their fluid-filled lacunae detect a much lower amount of pressure, estimated to be around 0.3 MPa (Zhang et al. 1998). Similarly, the unique biological, chemical, and electrostatic properties of cartilage serve to protect chondrocytes from the extremes of compressive loading in the joints.

The hydrodynamic pressure generated by arterial blood flow is much lower – up to 0.016 MPa. Pressures significantly higher than this result in vessel wall rupture (aneurism) and can prove fatal (McDonald 2001), but in these tissues, cells can sense and respond to the shear forces caused by the flowing fluid. Therefore, it is reasonable to suspect that cells in each tissue type have evolved sensitive mechanisms for detecting

Mechanobiology: Exploitation for Medical Benefit, First Edition. Edited by Simon C. F. Rawlinson.

and responding to forces within the actual physiological range experienced by that particular tissue, allowing them to adapt to the naturally occurring changes in their mechanical environment during development and adult life. Research has shown that this is indeed the case, and cells isolated from a range of tissues are extremely diverse in their specific sensitivity to applied mechanical loading through either compressive, tensile, or shear forces. In order to develop a greater understanding of how these forces influence cell behavior, it is necessary to be able to recreate the mechanical elements of the body's internal environment under testable laboratory conditions, to allow for experimentation and a detailed investigation of the mechanotransduction process. For this, we need bioreactors.

18.2 Bioreactors: A Short History

Bioreactors can be described simply as artificial culture environments for living cells; by this definition, the scope for determining "What is a bioreactor?" is extremely large. It can be argued that the earliest bioreactors were developed several thousand years ago for the brewing of fermented beverages (beer, wine, and mead) using environmentally sourced yeast cells. However, it was the pharmaceutical industry's use of biotechnology for the commercial manufacture of bacterial and fungal-derived products, such as antibiotics, vitamins, and recombinant human proteins, that drove the expansion of scalable, controlled, artificial environments for living cells (Zhou et al. 2010).

The rapid increase in understanding of the environmental factors required to grow single-celled organisms under optimal conditions did not immediately translate into a similar understanding of the culture of mammalian cells. Eukaryotic cell culture was pioneered in the late 19th century by both Sydney Ringer and Willhelm Roux, who developed salt solutions which allowed tissue explants to survive and function "*in vitro*" for several days. The methodology of tissue and cell culture was further developed during the 20th century, particularly during the 1940s and 1950s, in support of research in virology and the production of vaccines (Norrby 2008). Thus, the principles underlying human cell culture were first defined over 100 years ago and remain similar today: cells are grown in a single monolayer culture on a flat surface in an incubator containing 5% CO_2 and 20% O_2, surrounded with an isotonic solution that both provides essential nutrients and dilutes cell waste, but which exists without any perturbation aside from periodic changes of the culture media. This highly artificial situation has proved more than satisfactory for culturing virtually all of the cell types found within the human body, facilitating an intricately detailed understanding of how cells function and behave, and enabling researchers to find and test cures for diseases.

Nevertheless, many researchers are now coming to realize that the limits of traditional cell culture may be approaching, and that for a more detailed and realistic investigation into how native cells exist in the dynamic mechanical environment of the body, this environment must somehow be replicated *in vitro* through the development of more advanced cell culture systems (Figure 18.1). Fortunately, advances in technology have been synergistic with advances in biology, so an array of mechanical devices (bioreactors) has been constructed in which dynamic compressive, tensile, or shear forces can be applied to cells *in vitro*, mimicking aspects of the physical environment found *in vivo*. As has been mentioned, cells experience specific types and combinations

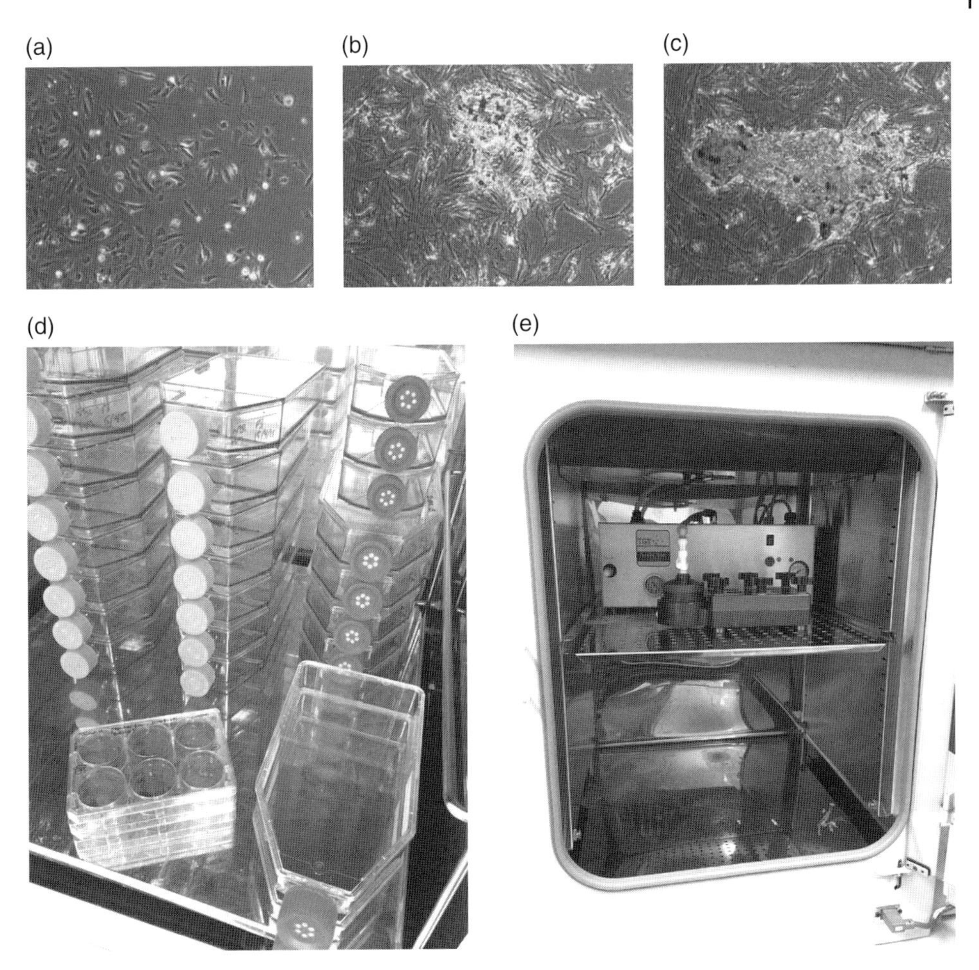

Figure 18.1 Cell culture has evolved significantly since its origins in the 19th century, but is still generally performed in two-dimensional (2D) monolayers cultured under static conditions, which do not reflect the rich structural and mechanical environment of the native tissue. Certain cell types, such as osteoblasts, change their morphology *in vitro* as they differentiate from (a) mesenchymal stem cells (MSCs) into terminally differentiated osteocytes, which become encased in mineralizing matrix and form characteristic (b) three-dimensional (3D) "nodule" structures. Mechanically stimulating these cells (in this case, by using magnetic nanoparticles directly against the mechanically gated ion channel, TREK1) shows that (c) mechanical cues can have a significant enhancement on tissue formation. Reproducible biomaterial scaffolds have now elevated cell culture into the third dimension, mimicking many of the structural and chemical aspects of the tissue; however, without mechanical cues the environment is still incomplete. The purpose of bioreactors, therefore, is to build a mechanical apparatus that integrates with existing cell culture, thereby facilitating a straightforward transition from (d) static culture in plates and flasks to dynamic microenvironmental culture, such as in (e) a hydrostatic bioreactor, which provides compressive forces to cells cultured in standard plates, augmenting the standard incubator with a mechanical component and bridging the gap between traditional and advanced cell culture.

of forces depending on their physiological environment within the body, so it is necessary to consider what type of loading is most relevant to a particular cell before we can select an appropriate bioreactor.

18.3 Bioreactor Types

Many different types of bioreactor have been developed for research into mechanobiology, ranging from simple systems built "in house" by researchers to extremely advanced commercial systems. In order to test the wide range of forces and tissue types of interest, bioreactors have been specifically designed to provide mechanical stimulation via hydrostatic pressurization, tensile strain, compressive loading, shearing fluid flow, or a combination of these elements. As has been mentioned, the forces experienced by tissues in the body vary both in type and magnitude depending on the physiological location, and so both the amount and the frequency of loading can usually be varied within the bioreactor design.

In order to design a bioreactor, it is necessary to consider (and ideally measure and mathematically model) the tissue within the body. Bioreactors can then be engineered to generate forces *in vitro* similar to those the cells would normally experience in their native environmental niche. As a result, there is no single ideal bioreactor that suits all purposes for all cells (except for the body itself), but advanced combination bioreactors are now able to replicate many of the physiological forces experienced *in vitro* and allow for experimental testing, mechanical conditioning, and even online monitoring of living cells in dynamic culture environments. A checklist of criteria to consider when designing a bioreactor for a specific purpose is given in Table 18.1, while a summary of the effects of example bioreactors of each type is given in Table 18.2.

18.3.1 Perfusion

Perfusion bioreactors are perhaps the most common type of dynamic environment used for cell culture. They are designed to allow cells in three-dimensional (3D) scaffolds to obtain nutrients and eliminate waste products. Research has shown that the diffusion "mass-transport" issue is a key limiting factor in the growth of 3D constructs and tissues *in vitro*, restricting the size of construct that can be created and cultured to 100–200 μm (Lovett et al. 2009). In the body, the perfusion of tissues with nutrients and the carrying away of waste are performed by the vasculature, with capillaries and tissue fluid perfusing the tissue with low-flow, low-pressure fluid.

In theory, therefore, this is a relatively straightforward engineering problem to solve, but in practice the challenges prove to be more subtle and multifactorial. For example, creating a uniform fluid flow through a complex natural or engineered scaffold is extremely challenging, with areas of excessively high flow (usually around the inlet port and exterior or interior of the scaffold, depending on its porosity and position; see Figure 18.2) and "dead spots" of extremely restricted flow occurring (Hidalgo-Bastida et al. 2012). Either of these regions can prove fatal for cells, via simple mechanical removal, induced apoptosis, or necrosis. Consequently, the flow rates within perfusion reactors and porous scaffolds are a common theme in hydrodynamic mathematical modeling exercises, and have resulted in a synergistic evolution of bioreactors, scaffold

Table 18.1 Generalized considerations for determining bioreactor conditions.

Type	Loading forces can be compressive (direct platens (Concaro et al. 2009) or hydrostatic (Elder and Athanasiou 2009)), tensile (Benhardt and Cosgriff-Hernandez 2009), shear (McCoy and O'Brien 2010), perfusion (Gaspar et al. 2012), or a combination.
Monitoring	Online monitoring and recording of data allows for complete information to be obtained over the culture and stimulation period (Lourenço et al. 2012). Direct imaging of constructs under loading requires early consideration in order to attain the correct focal lengths (Mather et al. 2007).
Control	Programmable software can react to changing conditions (e.g., stiffness) and is generally more sensitive (Burdge and Libourel 2014), while simple mechanical control is robust and effective.
Modeling	Detailed mathematical modeling of the system allows for rapid, synergistic analysis and interpretation of the experimental data (Pearson et al. 2015).
Duration	Samples can be cultured in the bioreactor for short periods (e.g., 1 hour per day) or established permanently in the system for the duration of the experiment (e.g., 4 weeks) (Puetzer et al. 2012).
Substrate	Cells can be seeded into biomaterials (Murphy and Atala 2013) on standard culture plates or cultured scaffold-free, supported by increasing amounts of their ECM (Gauvin et al. 2011).
Environment	In addition to mechanical simulation of the *in vitro* environment, cells can be grown in bioreactor chambers under traditional atmospheric oxygen concentration (20%) or in more physiologically appropriate hypoxic (≤1%) conditions (Malda et al. 2007).

designs, and modeling tools for the optimization of 3D cell culture in these systems (Melchels et al. 2011). In addition to considering perfusion bioreactors as independent systems, it is valuable to understand that the effects of shear forces within these systems act as a key mechanical component in the bioreactor environment.

18.3.2 Shear

As with hydrostatic forces, shear (or hydrodynamic) forces act on a wide range of cell types in many different tissues. Vascular endothelial cells are commonly used to study sensation and response to flow, but other cell types have also been shown to be shear-responsive, including nephrite cells of the kidney, hair cells of the inner ear, and osteocytes. The sensation mechanisms in these cells are linked to the deformation of cell shape, which results in changes to the cytoskeleton and activation of membrane stretch-activated ion channels. Researchers have demonstrated that shear forces induce movement and sensation in the primary cilia (Ferrell et al. 2010; Hoey et al. 2012).

Fluid mechanics in biological systems is highly complex and is influenced by blood vessel diameter, curvature, rigidity, obstructions (e.g., atherosclerotic plaques), and damage, in addition to the presence of biological material ranging from viscosity-affecting biomolecules to freely flowing erythrocytes or neutrophils, which use the vascular conduits, yet attach and migrate through the endothelial walls. Clearly, there are significant unanswered research questions in this field in terms of the biological interactions with vessel walls, and mathematical modeling is proving an extremely useful tool for understanding how these complex systems operate in the body (Lawford et al. 2008).

Table 18.2 Experimental bioreactors designed to research the effects of mechanical forces on cells and tissues *in vitro*.

Compressive (indirect/hydrostatic)	Pressure (kPa)	Frequency	Effects
Fetal mouse bone (Klein-Nulend et al. 1993)	13	0.3	Release of soluble growth factors
Osteocytes (MLO-Y4) (Liu et al. 2010)	68	0.5	Increase in intracellular [Ca]; microtubule reorganization; increase in [COX-2 mRNA] and RANKL/OPG mRNA ratio; decrease in apoptosis
Chick fetal femur (Henstock et al. 2013)	0–280	0–2	Increase in bone density directly proportional to frequency of pressure
Calf bone explants (Takai et al. 2004)	3000	0.3	Increase in osteoblast proliferation/osteocyte viability at day 8; effect lost thereafter
Dental pulp cells (Yu et al. 2009)	3000	0.5	Slight differences: cells detached, decrease in adherin experession
Pluripotent mouse cells (Elder et al. 2005)	5000	1	Doubled sGAG and collagen synthesis
Compressive (direct)	**Strain (%)**	**Frequency**	**Effects**
Juvenile bovine chondrocytes (Lima et al. 2007)	10 > 2	1	Increase in mechanical properties, but only after TGF-β supplementation ends
Porcine intervertebral disk cells (Fernando et al. 2011)	15	0; 0.1; 1	Promotion of ATP production via glycolysis by dynamic compression
Tensile	**Strain (%)**	**Frequency**	**Effects**
Decellularized equine tendon and equine MSC (Youngstrom et al. 2015)	3–5	0.33	Cells integrated into scaffolds; altered ECM composition; tendonlike gene expression; increased elastic modulus and strength
Rabbit Achilles tendon (Wang et al. 2013b)	0–9	0.25	Matrix deterioration at 3% and elevated MMP-1, -3, and -12; 6% maintained structural integrity and cell function; collagen rupture at 9%
Scaffold-free dermal fibroblasts (Gauvin et al. 2011)	10	0; 1	Cytoskeletal alignment and ECM parallel to the strain axis; increased ECM content; increased tensile strength and modulus
Murine RAW264.7 osteoclasts (Li et al. 2015)	10–15	-	Increased resorption area; decreased apoptosis; increased Bcl-2/Bax ratio; inhibited caspase-3; downregulated cytochrome C
Shear	**Flow (dynes/cm^2)**	**Frequency**	**Effects**
Peridontal ligament cells (Martinez et al. 2013)	1 or 5–6	Constant	Differentiation to smooth muscle and, particularly, endothelial cells; 5–6 dynes/cm^2 augments the synthesis of ECM collagen

Human MSC (Glossop and Cartmell 2009)	1, 5, or 10	Constant	MAP3K8-induced activation of signaling; shear-induced IL1B expression
Murine embryonic stem cells (mESCs) (Wolfe and Ahsan 2013)	1.5–15.0	-	Induced endothelial phenotype; FLK1 protein (VEGF receptor) as critical mediator
Adipose-derived human stem cells (McIlhenny et al. 2010)	Increased to 9	-	Increased attachment within vascular graft due to upregulation of the $\alpha5\beta1$-integrin

Designing bioreactors to precisely replicate the dynamic complexity is therefore more challenging than it initially appears, yet it offers huge scope for innovation in design (Figure 18.2). It is relatively straightforward to achieve basic flow even in standard cell monolayer cultures: one must simply place cell-seeded microtiter plates on apparatus designed for biochemical mixing (e.g., orbital shakers, tilting or rocking mixers); these common lab tools can be conveniently placed within incubators, which have minimal potential for introducing infection. While this is a good starting point for experimenting with hydrodynamic flow in cell cultures, the relative lack of control or measurement of the shear forces sets limits on how well the data from simple experiments can be interpreted. Research has shown that some cell types are critically sensitive to specific ranges of shear force, and so specifying these values in a bespoke system that can be modeled is much preferred (Hutmacher and Singh 2008).

Bioreactors can provide quantifiable flow rates in various ways, which can be separated into two main types: microfluidic systems and microcariers. In microfluidics, cells are grown on a porous scaffold or a flat surface (Glossop and Cartmell 2009; Martinez et al. 2013) and a nutrient solution is pumped across the cell layer. Microcarriers, on the other hand, make use of closed systems in which the cells are cultured on free-floating scaffolds in an agitated suspension (stirred spinner flasks and rotating-wall bioreactors (see Figure 18.2a,b); Gaspar et al. 2012). In both spinner flasks and rotating-wall bioreactors, the motion of the cells within the media depends on a complex relationship between fluid velocity, viscosity, and the buoyancy of the cell constructs, which changes due to cell proliferation and matrix deposition: ultimately, constructs will sink and may become physically damaged by the spinning impeller or by the rotating walls of the bioreactor. By contrast, shear forces over the cell layer in microfluidic systems can be very precisely controlled, but the physical requirements for connective tubing present significant challenges in operating the system across multiple wells.

18.3.3 Compression

Compressive forces acting on hydrated tissues in the body result in the generation of shear forces as fluid flows from the compressed area through the interstitial spaces, while a significant portion of the force is experienced as hydrostatic pressure. It is thought that this compound input serves to inform cells such as osteocytes and chondrocytes how the surrounding tissue is being compressed, and osteocytes in particular

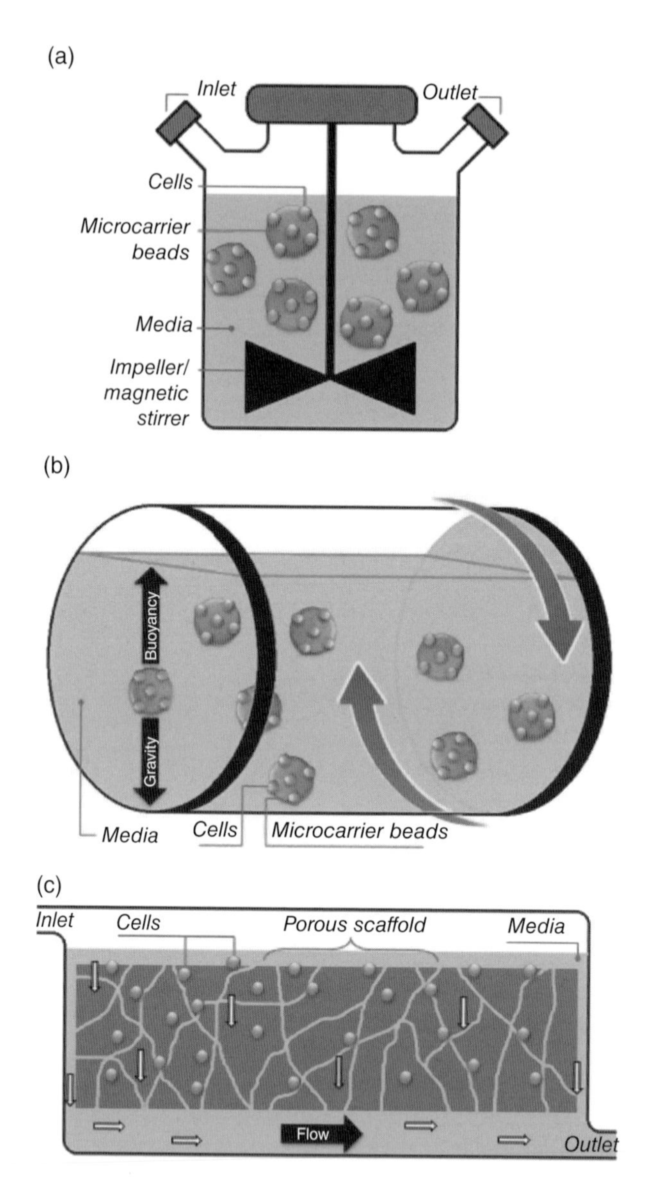

Figure 18.2 Bioreactors can deliver shear forces in many different ways. (a) Stirred spinner flask. (b) Rotating-wall bioreactor. These were initially developed for the industrial expansion of microorganisms, but adherent mammalian cells must first be seeded on to microcarriers or biomaterial scaffolds or cultured as tissue explants. (c) A simple method for providing shear is to perfuse the cell-seeded scaffold with fluid, which can be pressurized; the nonuniform effects on flow rate and the subsequent cell responses are a common theme of mathematical modeling.

have sensitive dendritic processes for detecting pressurized flow (Bonewald 2011; Klein-Nulend et al. 2012). Evidence suggests that the mechanotransduction of pressure is via specific pathways, and so cells may be capable of sensing pressure and flow separately, combining inputs intracellularly to form a coherent response (Elder and Athanasiou 2009).

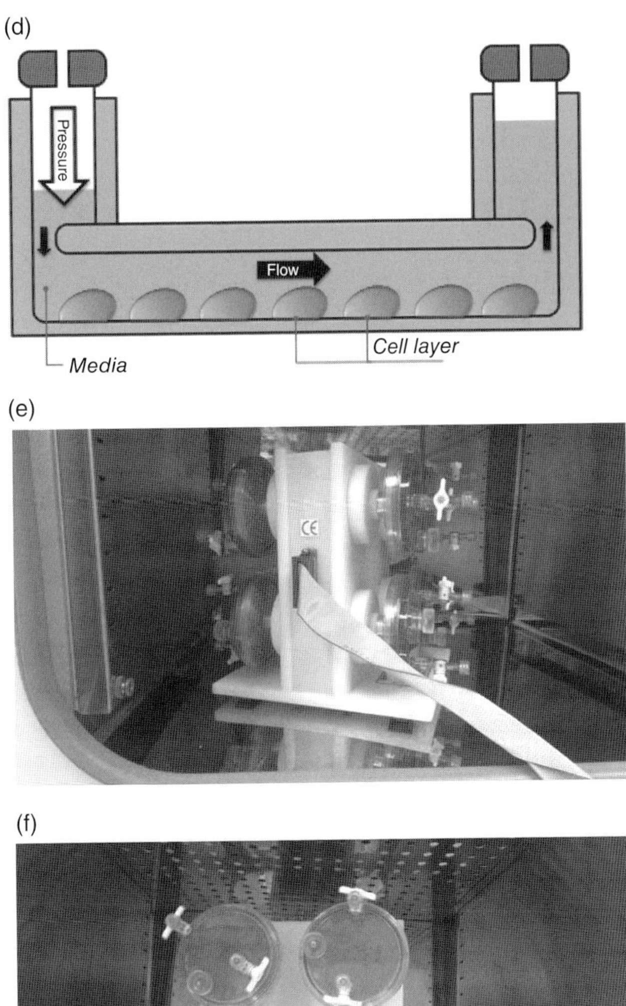

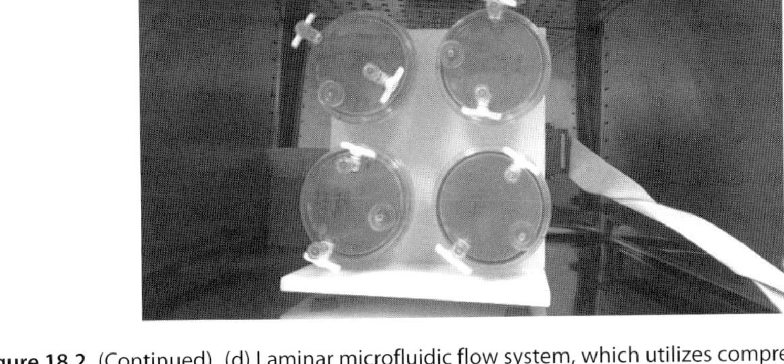

Figure 18.2 (Continued) (d) Laminar microfluidic flow system, which utilizes compressed air to drive media flow across an adherent cell monolayer. *Source:* Adapted from Martinez et al. (2013). (e,f) Commercial version of a rotating-wall bioreactor, marketed by Synthecon. *Source:* Courtesy of Dr. Yvonne Reinwald. All of these types of bioreactor deliver shear forces in different ways, and each has unique challenges in delivering consistent, measurable flow rates. (*See insert for color representation of the figure.*)

Compressive hydrostatic forces are shown to be crucially important in regulating the development of many tissues, particularly those in the musculoskeletal system, such as bone and cartilage. Research has shown that compression of the developing limb (due to resistance from the uterine wall or chorioallantoic membrane) is essential to correct joint formation (Nowlan et al. 2010, 2014), and recreating these forces in developmental models of endochondral ossification has been shown to be effective in optimizing their culture (Foster et al. 2015).

Direct compression of tissue explants and cell-seeded scaffolds between solid plates is an inherently challenging technique, as the interface between platen and tissue is highly nonphysiological in terms of force distribution. To accommodate variability within the tissue, constructs are often placed under an initial strain or creep, and confined or nonconfined compression is then applied (Kelly et al. 2013). A major issue with having a moving upper platen is that of sterility: having moving parts above a[n open] cell culture markedly increases opportunities for infection, while having the moving platen under the culture increases both the complexity and the inherent cost of the bioreactor design (Lujan et al. 2011). Direct compression therefore presents nontrivial engineering challenges, but allows relatively large forces to be directly applied to cultures, and importantly also allows immediate responses to recorded, such as changes in compressive modulus over long culture periods. As the software running these devices is often adapted from compression-testing equipment, the tools for running the bioreactor and analyzing the effects are conveniently integrated (Kelly et al. 2013).

By contrast, hydrostatic pressure acts uniformly and nondirectionally, and is therefore considered a general mechanical signal to cells. Several researchers have experimented with using hydrostatic-pressure bioreactors to apply forces to cells and tissues in culture (see Table 18.2 for examples). A commercial example of a hydrostatic bioreactor, shown in Figure 18.3, uses compressed, recycled incubator air to apply a pressure to the gaseous head space above cells cultured in a standard well plate (Henstock et al. 2013; Reinwald et al. 2015). This approach has been shown to have many advantages, as no bespoke substrates or cultureware are required to grow the cells, which allows for broad comparisons to traditional static culture.

Creating a novel bioreactor presents both opportunities and challenges in determining the effectiveness of the design, and validation is paramount in order to demonstrate that the bioreactor is not generating side effects that might confound experiments. A common problem with pressurization in bioreactors is that the partial pressures of the gases in the incubator are linked to the atmospheric pressure, and so the resulting altered chemical composition and pH of the media may have a large effect on cell behavior, independently of mechanical loading. Coupling experimental measurements of the bioreactor environment with mathematical modeling allows many of the questions regarding this dynamic environment to be answered (Reinwald et al. 2015). In order to validate the stimulatory effectiveness of this bioreactor, mechanically sensitive fetal femurs were isolated from developing chicks and cultured either in static conditions or under dynamic hydrostatic pressure (0–280 kPa at 1 Hz for 1 hour/day). Analysis by X-ray microtomography (μCT) and whole-mount histology showed significant increases in the volume of bone formation by endochondral ossification compared to static, unstimulated femurs. These changes were found to be directly proportional to the frequency of applied stimulation and were most effective on femurs containing more mature cells (Henstock et al. 2013).

It is interesting to note that over the last 3 decades, the magnitude of compression in bioreactors has generally been revised down, as research has shown that cells are very

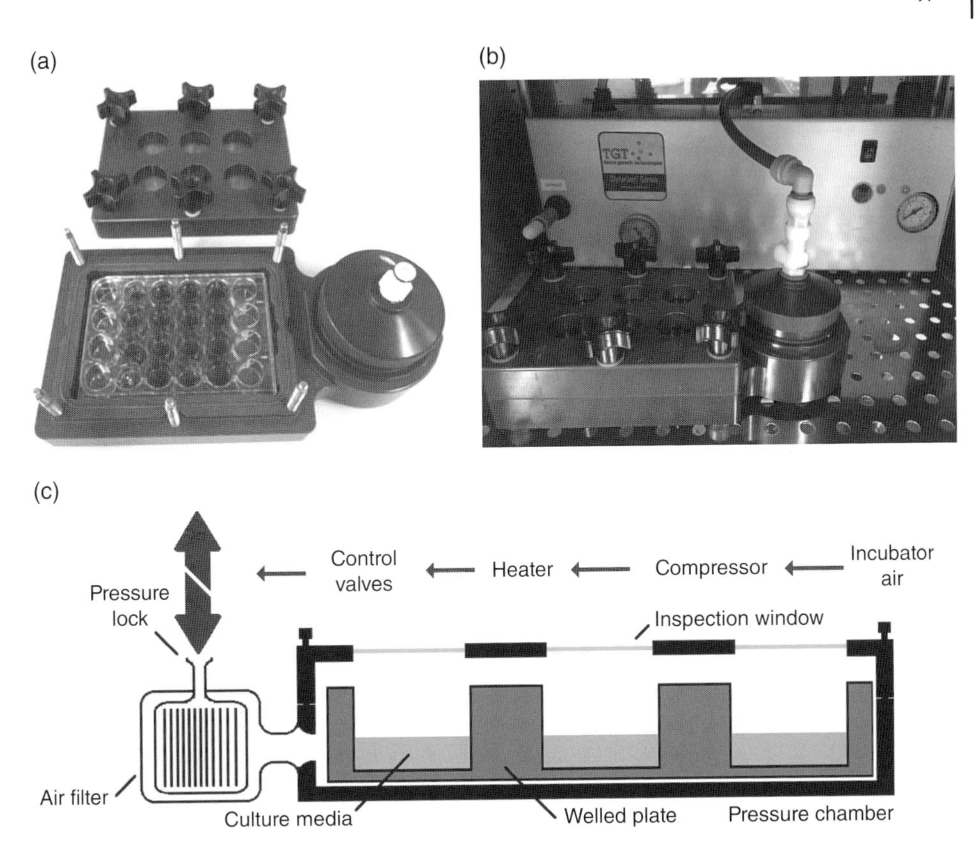

Figure 18.3 Hydrostatic bioreactor designed and created in partnership between Keele University (Professor Alica El-Haj) and TissueGrowthTechnologies (now a part of Instron). The key features of this bioreactor have been optimized to enhance high laboratory throughput and maintain maximum sterility, with the majority of the bioreactor and most of the moving hardware being located outside of the incubator environment, including the compressor and the computer control system. The bioreactor chamber itself exists as a "mini-incubator": a sealed, autoclavable chamber accommodating a standard multiwelled plate, allowing for standard experiments to be run under conventional, static incubator cultures and under dynamic hydrostatic loading. Separation from the outside space is provided by a replaceable and autoclavable filter, and is maintained by wide flanges to the chamber and a gasket, which helps reduce microbial infiltration when the chamber is opened in order to replace culture plates. The addition of inspection windows allows the cultures to be viewed, but reflections from the glass and the height above the culture restrict opportunities to derive measurable data or high-resolution images – this will be a point of optimization of future models in this range, which has now been commercialized as "CartiGen HP" (hydrostatic pressure). (a) Bioreactor chamber, containing cells in a standard culture plate. (b) Chamber and valve-control box – the only parts of the bioreactor placed within the incubator. (c) Schematic of the bioreactor. (*See insert for color representation of the figure.*)

sensitive to specific ranges of applied pressure and that the huge forces applied to the skeleton do not translate to similar loads on cells (Takai et al. 2004; Elder et al. 2005; Yu et al. 2009; Henstock et al. 2013). Being able to recreate accurate physiological forces at much lower pressures is therefore an enormous advantage for biologists and engineers, and permits a range of finely tuned bioreactors that are appropriate for many cell types to be developed.

18.3.4 Tensile

Tensile forces are commonly experienced in cells in tendon, ligament, and muscle, and each of these tissues has structural adaptations in the ECM which enable them to withstand these types of loading. Transitory tensile loads in large, active mammals can be extreme – for example, forces in the horse forelimb can exceed 5000 N in the deep digital flexor tendons and approach 12 000 N in the suspensory ligament under normal locomotion (Takahashi et al. 2014). Obviously, failure of these tissues under such extreme loads is not unknown, and repairs to tendon and ligament are a common veterinary and clinical surgical procedure – albeit, one which is hampered by a lack of available autologous donor sites for the repair of lost or severely damaged tissue.

Attempts to grow replacement tendon, ligament, and muscle tissues *in vitro* have revealed the specific requirement for strain in order to align cell growth along the appropriate axis, during which both the intracellular cytoskeleton and deposited ECM are aligned parallel to the strain axis. Research has shown that static strain is sufficient to generate alignment, while dynamic loading of up to 10% strain results in increased tensile modulus, due to the deposition of mechanically resilient ECM (Gauvin et al. 2011).

Growing tissues under tensile loading generally requires that the cells be cultured on an appropriate scaffold – one that is elastic, biocompatible, and strong enough to withstand the tension, such as collagen-coated aligned fibers (Shah et al. 2013) or decellularized animal tendon (Wang et al. 2013a; Youngstrom et al. 2015), though some researchers have managed to load tissues formed *de novo* entirely of their own ECM (Gauvin et al. 2011). The other requirement for tensile bioreactors is a durable interface with the grips (i.e., the point at which the tissue culture is attached to the actuator providing the strain). There are several ways to achieve this (outlined in Figure 18.4), but the interface remains a critical point in the practical development of the bioreactor design. An additional consideration is that many tendons and ligaments in the body are contoured around joints, such that their natural flexor tension is also combined with torsion. Several research groups have therefore included rotation of the actuator to reflect this more complex loading (Lee et al. 2013).

As with direct-compression systems, many of the bioreactors designed to apply tension to cultures use existing mechanical testing equipment to provide very precise loads to the tissue and to reciprocally measure changes in tensile modulus over the culture period. One of the most successful is the Electroforce system, developed by Bose and now marketed by TA Instruments (Figure 18.4). A number of design improvements were implemented in this system following a productive dialogue between the manufacturers and researchers using the equipment (Cartmell et al. 2011). The Electroforce system is a closed vertical chamber and is directly connected to the vertical testing station, but most other systems are horizontal, allowing for a more standard approach to cell culture and media changing. In all cases, the engineered tissue is fixed at one or both ends, and tension is applied by moving one anchor point (manually or via a motorized actuator). Depending on the approach, tensile strain can be either externally applied or allowed to build internally through the natural process of cell contraction (Dennis et al. 2009). A unique feature of muscle tissue is that it generates its own tensile forces via contraction in response to electrical stimulation by nerves, a process that can be recreated *in vitro* by controlled electrical stimulation (Donnelly et al. 2010). This provides a novel method for delivering appropriate environmental signals to drive tissue formation – the ultimate goal of bioreactor design.

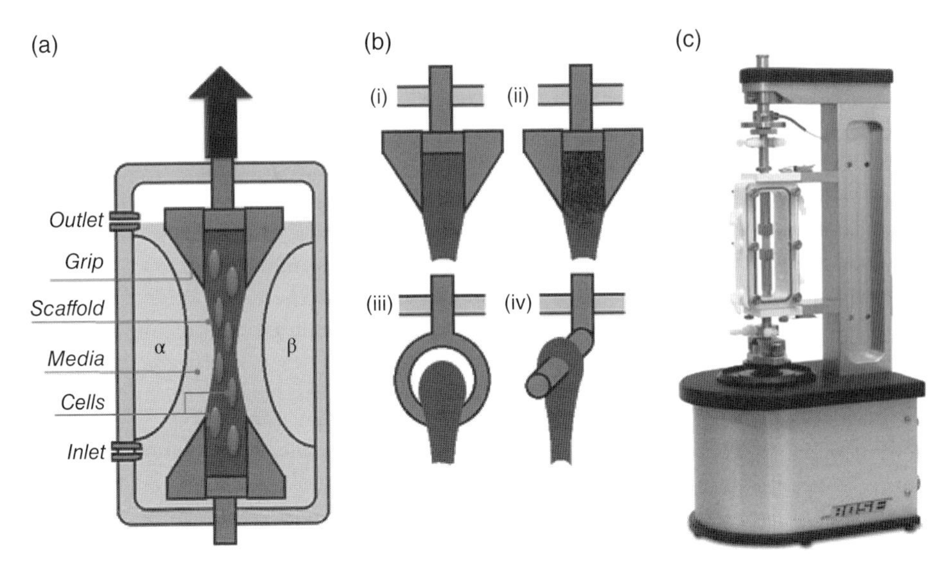

Figure 18.4 Schematic of a typical tension bioreactor, in which cells in a biomaterial scaffold such as collagen are cultured between two grips, which can be pulled apart to generate strain within the cell-seeded construct. (a) A useful adaptation of the second-generation Bose Electroforce series is the addition of space-filling solid baffles (α and β), which require less medium in the bioreactor chamber, and thus provide substantial savings on the expensive growth factors used in culture. (b) A particular engineering challenge is (i) the grip–biomaterial interface, where a great deal of mechanical failure occurs. Researchers have thus developed various ways of reinforcing this region. Options include (ii) the use of composite materials with an integrated solid scaffold or sacrificial zone, which is mechanically more resilient than the region under tensile load. Alternatively, the cell-seeded scaffold can be fabricated as a circular band and (iii) connected to the tension actuator via a loop or (iv) pinned on to the grips. Often, researchers will use a combination of these approaches, depending on the application (see Table 18.1). (c) The Electroforce series (now owned by TA Instruments) has been extensively used by researchers, as it conveniently attaches to existing mechanical testing systems, which include adaptable commercial software. (*See insert for color representation of the figure.*)

18.3.5 Combination

Compression, shear, and perfusion are shown to have combined effects on the macromolecular composition of many tissues, and cartilage in particular has an interesting relationship with compression that is responsible for providing nutrition in the deep layers of the tissue and for mechanically extracting cellular waste (Mardones et al. 2015). Additionally, the compression of cartilage, in concert with shear forces, is responsible for cleaving the inactive, latent form of transforming growth factor beta (TGF-β) into its active form in the synovial fluid (but not in the cartilage itself), and the transduction pathway of this family of growth factors is shown to be strongly influenced by mechanotransduction (Albro et al. 2012, 2013; Kopf et al. 2012).

A growing appreciation of the compound forces involved in most tissues has led to the development of new bioreactors that can deliver a range of forces at different times and in different magnitudes. An example of this synergistic development is the evolution of testing rigs used to analyze the lifetime wear of lower-limb joint prostheses. In addition to analyzing composite artificial joints, these can be adapted to deliver physiological loading comprising complex high-impact or sustained compressive and shear forces, which more accurately replicate the loads experienced during locomotion (Nugent-Derfus et al. 2007).

Combination bioreactors are extremely useful for recreating the complex, dynamic environment within the body, but they are intrinsically more complex to understand. Nevertheless, in order to generate physiologically accurate forces across tissue constructs, the range of magnitudes, frequencies, and types of stimulation should be considered in total. As has been discussed, many bioreactor–biomaterial scaffolds integrate their forces either by accident or by design (e.g., combined pressure and shear in some perfusion bioreactors, or tension and torque in the actuator design of tensile bioreactors). These design considerations lead to tremendous insights into cellular mechanotransduction processes and reveal how tissues and cells adapt to the continuous and changing mechanical demands placed upon them during life.

18.3.6 Unconventional Bioreactors

Amidst research into the direct replication of physiological environments encountered in biological processes, some research groups are also using technological methods to provide stimuli not encountered under native circumstances, but which nonetheless act on cells in culture. Perhaps the most interesting are techniques which deliver electrical or magnetic fields and ultrasonic vibration to affect cell mechanosensors at the molecular level. These bioreactors, which can employ micro- or nanoparticles to amplify signaling, allow the culture of novel combinations of biomaterials and cells that can still experience mechanical forces acting on their receptors, subverting the usual mechanics of the tissue to deliver stimuli and allowing bone cells to be "mechanically stimulated" in hydrogels, for example (Walker et al. 2007; Balint et al. 2013; Henstock et al. 2014; Henstock and El Haj 2015).

These technologies are interesting, and their applications are highly translational – providing novel methods for delivering stimuli to mechanoreceptors on therapeutic cells injected or implanted into regenerating wound sites. For a more detailed overview of electrical, ultrasound, and magnetic nanoparticle targeting of mechanoreceptors, see Henstock and El Haj (2015), Balint et al. (2013), and Walker et al. (2007).

18.4 Commercial versus Homemade Bioreactors

As has been mentioned, the early types of bioreactor were designed and built by researchers using readily available components, often with the assistance of biomedical engineering departments with expertise in clinical materials engineering. This approach led to fascinating innovations, but the increasing complexity of these systems caused the ultimate failure of some designs – often due to infection of the cell cultures. As most cell biologists are uncomfortably aware, sterility in tissue culture incubators is a primary concern, and introducing devices which have moving parts, require lubrication (often generating vaporized oil droplets), include electrical power and control connections to the outside, or contain parts, corners, and cavities that are difficult to clean and sterilize markedly increases the opportunities to infect cell cultures. Furthermore, the obvious and perhaps even unavoidable position for a stimulatory device, platen, or actuator is directly above the exposed culture – a situation that can compromise sterility unless painstaking efforts are made to sterilize the apparatus before each use.

These lessons have been learned, and more recent bioreactor designs are streamlined such that the elements which directly interface with cells are compact, smooth, and often autoclavable, while control systems are located outside of the incubator. This more complex engineering is of course well suited to commercial manufacturers, whose skills in design and mass-production can be employed to generate reliable systems. Furthermore, a design used across multiple labs allows for comparison of experimental data between groups – a proposition that was difficult in the past, due to the unique mechanical loading features of individual, bespoke bioreactors and the particular parameters used by different research groups.

There are now a number of companies making bioreactors for research use, including enterprises that offer bioreactors as their sole product (often spin-out companies from university research groups) and larger, multinational companies, such as Bose and Instron, whose expertise in generating mechanical testing systems has naturally evolved into mechanical-stimulation devices.

The evolution of bioreactors has now completed a full circle, as the original mechanical testing of tissues which informed the initial bioreactor designs has led to bioreactors being designed by mechanical-testing companies. At each stage, there have been significant discoveries, and our understanding of the complex nature of the mechanical environment in the body has been revolutionized. Unnatural static culture methods do replicate this *in vivo* environment, and so they are gradually being replaced by bioreactor culture systems. Researchers now have a large choice of bioreactor designs commercially available at competitive prices, while innovation continues to drive the designs forward to create more well-characterized *in vitro* mechanical environments. The future of bioreactors may therefore be in general cell culture, providing modular systems that are a standard feature of routine *in vitro* culture, turning our artificial incubators into more accurate recreations of physiological environments, and facilitating the next generation of biological research.

18.5 Automated Cell-Culture Systems

A recent drive in the application of stem cells in regenerative medicine has been to lower the production cost of therapeutic cells to a level where they can conceivably be competitive against conventional drugs and other treatments for clinical problems. As part of this trend, research institutes and certain companies have invested heavily in technology that allows the creation of automated, GMP (Good Manufacturing Practice), cost-effective, and scalable production methods in which most of the tasks traditionally performed by technicians (such as feeding, passaging, and harvesting cells) can be automated. In effect, these closed systems are bioreactors, and with the reproducibility of automation it has become possible to dramatically standardize protocols for the effective manufacture of cells. Some of these mass-production systems have evolved from bacterial cell-culture bioreactors, and they contain methods for ensuring adequate nutrition of the cells being cultured. This usually involves culturing the cells on microcarriers using a wave- or stirred-tank bioreactor, or else on surrounding membranous hollow fibers, allowing perfusion and flow of nutrient fluid past the cells.

As has been discussed already, these methods for providing nutrients also inevitably provide mechanical forces, which may stimulate (or damage) the cells being grown.

Indeed, the great majority of the mathematical and experimental modeling performed on these systems has been carried out in order to characterize the negative effects on cell proliferation and survival of excessive shear forces on microcarriers within impelled, stirred-tank bioreactors. It is therefore likely that the emerging evidence surrounding the positive influences of certain mechanical cues on cell growth and their potential activity may lead to the deliberate inclusion of mechanical stimulation within the next generation or two of these automated systems, in which appropriate (rather than inadvertent) mechanical forces can be applied to cells in culture, leading to enhanced growth and differentiation. As it is now known that mechanical forces serve to regulate the transduction of growth factors in stem cells, getting these environmental, physical signals correct may allow a dramatic increase in the efficiency of the desired outputs (highly active stem cells) and a huge saving in the cost of expensive recombinant growth factors used in such substantial volumes.

18.6 The Future of Bioreactors in Research and Translational Medicine

The purpose of bioreactors is twofold: first, to culture native tissues and cells *in vitro* in order to gain unique insights into how they respond to mechanical forces; and, second, to provide anabolic environments for the generation of tissue-engineered replacements for damaged or diseased parts of the body. Substantial progress has been made in both of these disciplines, and bioreactors have proven essential in understanding the complex interactions that underpin cell behavior under normal and pathological conditions. Bioreactors providing mechanical environments in culture have also been instrumental in elucidating the mediation of mechanotransduction in growth factor signal transduction (Kim et al. 2009; Kopf et al. 2012), cancer metastasis (Yates et al. 2007), and the enhancement of drug transport (Ginai et al. 2013). A wealth of data has demonstrated that the addition of mechanical cues to growing tissues assists in the differentiation and growth of stem cells and the production of more appropriate, functional ECM, such as aligned tendon or mineralized bone.

Perfusion bioreactors have enabled tissue engineers to generate lab-grown tissues much larger than can be constructed without a dynamic nutrient supply. A variety of these scalable bioreactors have been developed for applications in stem cell manufacture, including hollow-fiber bioreactors (Shipley and Waters 2012), rotating-wall bioreactors (Kasper et al. 2007; Radtke and Herbst-Kralovetz 2012), wave-tank bioreactors, and modular stirred-tank systems (Heathman et al. 2015); these different systems are summarized in Figure 18.5.

Over the last several decades, bioreactor development has evolved from rudimentary DIY apparatus into a mainstream commercial enterprise in which major manufacturers are investing significantly in well-characterized, mass-produced, automated devices. The future of bioreactors therefore seems to be in providing additive environments to existing cell culture, both as generally available lab equipment for research use and as an element within automated systems used to create commercial medical cell products.

Bioreactors therefore point toward the next generation in cell culture, in which increasingly easy-to-use systems will allow a greater number of researchers to apply

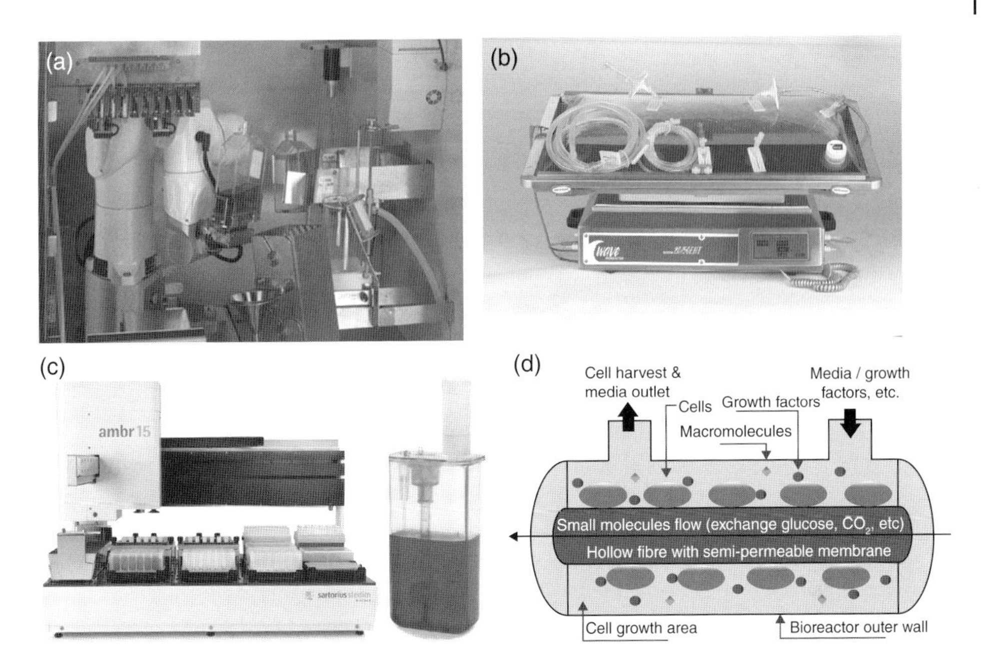

Figure 18.5 Commercial bioreactors are often designed principally to improve bioproduction methods – increasing cell yields and reducing costs through the extensive use of automation, in order to achieve consistent, reliable cell growth *in vitro*. Most of these systems use an element of perfusion to provide nutrients to cells, though the effects of this as a mechanical stimulus are generally a secondary consideration and are often limited to ameliorating the negative effects of excessive shear forces on cell viability. Nevertheless, this is likely to be the direction in which cells become commercialized as manufactured therapeutic products. Cell factories may be relatively easy to re-engineer with more appropriate mechanical characteristics which support and enhance cell growth, ultimately leading to more effective cell products. (a) Robotic T-flask handling (CompacT SelecT TAP Biosystems). (b) GE Healthcare's WAVE Bioreactor 2050. *Source:* Courtesy of GE Healthcare. (c) Ambr 15 microscale (10–15 mL) disposable microbioreactors for cells cultured on microcarriers and stirred by an impeller (TAP Biosystems). (d) Generalized schematic of a hollow-fiber bioreactor that allows cell culture with nutrient perfusion and a degree of shear/flow through the cell growth area while keeping expensive growth factors and important biological macromolecules within the culture. (*See insert for color representation of the figure.*)

appropriate mechanical loading to their cell-culture experiments. Similarly, optimizing the physical environment of cell manufacture may be instrumental in commercializing stem cell therapy, providing fully optimized cell products for clinical applications. Obtaining a deeper understanding of how bioreactors operate and can best be optimized remains a fascinating challenge at the interface of engineering and biology.

References

Albro, M. B., A. D. Cigan, R. J. Nims, K. J. Yeroushalmi, S. R. Oungoulian, C. T. Hung, and G. A. Ateshian. 2012. "Shearing of synovial fluid activates latent TGF-beta." *Osteoarthritis and Cartilage* **20**(11): 1374–82. doi:10.1016/j.joca.2012.07.006.

Albro, M. B., R. J. Nims, A. D. Cigan, K. J. Yeroushalmi, J. J. Shim, C. T. Hung, and G. A. Ateshian. 2013. "Dynamic mechanical compression of devitalized articular cartilage does not activate latent TGF-beta." *Journal of Biomechanics* **46**(8): 1433–9. doi:10.1016/j.jbiomech.2013.03.006.

Balint, R., N. J. Cassidy, and S. H. Cartmell. 2013. "Electrical stimulation: a novel tool for tissue engineering." *Tissue Engineering. Part B, Reviews* **19**(1): 48–57. doi:10.1089/ten.TEB.2012.0183.

Benhardt, H. A. and Cosgriff-Hernandez, E. M. 2009. "The role of mechanical loading in ligament tissue engineering." *Tissue Engineering. Part B, Reviews* **15**: 467–75. doi:10.1089/ten.TEB.2008.0687.

Bonewald, L. F. 2011. "The amazing osteocyte." *Journal of Bone Mineral Research* **26**(2): 229–38. doi:10.1002/jbmr.320.

Burdge, D. A. and Libourel, I. G. L. 2014. "Open source software to control bioflo bioreactors." *PLoS One* **9**(3): e92108. doi:10.1371/journal.pone.0092108.

Cartmell, S. H., S. Rathbone, G. Jones, and L. A. Hidalgo-Bastida. 2011. "3D sample preparation for orthopaedic tissue engineering bioreactors." *Methods in Molecular Biology* **695**: 61–76. doi:10.1007/978-1-60761-984-0_5.

Concaro, S., Gustavson, F., and Gatenholm, P. 2009. "Bioreactors for tissue engineering of cartilage." *Advances in Biochemical Engineering/Biotechnology* **112**: 125–43. doi:10.1007/978-3-540-69357-4_6.

Dennis, R. G., B. Smith, A. Philp, K. Donnelly, and K. Baar. 2009. "Bioreactors for guiding muscle tissue growth and development." *Advances in Biochemical Engineering/Biotechnology* **112**: 39–79. doi:10.1007/978-3-540-69357-4_3.

Donnelly, K., A. Khodabukus, A. Philp, L. Deldicque, R. G. Dennis, and K. Baar. 2010. "A novel bioreactor for stimulating skeletal muscle in vitro." *Tissue Engineering. Part C, Methods* **16**(4): 711–18. doi:10.1089/ten.TEC.2009.0125.

Elder, B. D. and K. A. Athanasiou. 2009. "Hydrostatic pressure in articular cartilage tissue engineering: from chondrocytes to tissue regeneration." *Tissue Engineering. Part B, Reviews* **15**(1): 43–53. doi:10.1089/ten.teb.2008.0435.

Elder, S. H., K. S. Fulzele, and W. R. McCulley. 2005. "Cyclic hydrostatic compression stimulates chondroinduction of C3H/10T1/2 cells." *Biomechanics and Modeling in Mechanobiology* **3**(3): 141–6. doi:10.1007/s10237-004-0058-3.

Fernando, H. N., Czamanski, J., Yuan, T. Y., Gu, W., Salahadin, A., and Huang, C. Y. 2011. "Mechanical loading affects the energy metabolism of intervertebral disc cells." *Journal of Orthopaedic Research* **29**: 1634–41. doi:10.1002/jor.21430.

Ferrell, N., R. R. Desai, A. J. Fleischman, S. Roy, H. D. Humes, and W. H. Fissell. 2010. "A microfluidic bioreactor with integrated transepithelial electrical resistance (TEER) measurement electrodes for evaluation of renal epithelial cells." *Biotechnology and Bioengineering* **107**(4): 707–16. doi:10.1002/bit.22835.

Foster, N. C., J. R. Henstock, Y. Reinwald, and A. J. El Haj. 2015. "Dynamic 3D culture: models of chondrogenesis and endochondral ossification." *Birth Defects Research. Part C, Embryo Today* **105**(1): 19–33. doi:10.1002/bdrc.21088.

Gaspar, D. A., V. Gomide, and F. J. Monteiro. 2012. "The role of perfusion bioreactors in bone tissue engineering." *Biomatter* **2**(4): 167–75. doi:10.4161/biom.22170.

Gauvin, R., R. Parenteau-Bareil, D. Larouche, H. Marcoux, F. Bisson, A. Bonnet, et al. 2011. "Dynamic mechanical stimulations induce anisotropy and improve the tensile properties of engineered tissues produced without exogenous scaffolding." *Acta Biomaterialia* **7**(9): 3294–301. doi:10.1016/j.actbio.2011.05.034.

Ginai, M., R. Elsby, C. J. Hewitt, D. Surry, K. Fenner, and K. Coopman. 2013. "The use of bioreactors as in vitro models in pharmaceutical research." *Drug Discovery Today* **18**(19–20): 922–35. doi:10.1016/j.drudis.2013.05.016.

Glossop, J. R. and S. H. Cartmell. 2009. "Effect of fluid flow-induced shear stress on human mesenchymal stem cells: differential gene expression of IL1B and MAP3K8 in MAPK signaling." *Gene Expression Patterns* **9**(5): 381–8. doi:10.1016/j.gep.2009.01.001.

Heathman, T. R., A. W. Nienow, M. J. McCall, K. Coopman, B. Kara, and C. J. Hewitt. 2015. "The translation of cell-based therapies: clinical landscape and manufacturing challenges." *Regenerative Medicine* **10**(1): 49–64. doi:10.2217/rme.14.73.

Henstock, J. and A. El Haj. 2015. "Controlled mechanotransduction in therapeutic MSCs: can remotely controlled magnetic nanoparticles regenerate bones?" *Regenerative Medicine* **10**(4): 377–80. doi:10.2217/rme.15.23.

Henstock, J. R., M. Rotherham, J. B. Rose, and A. J. El Haj. 2013. "Cyclic hydrostatic pressure stimulates enhanced bone development in the foetal chick femur in vitro." *Bone* **53**(2): 468–77. doi:10.1016/j.bone.2013.01.010.

Henstock, J. R., M. Rotherham, H. Rashidi, K. M. Shakesheff, and A. J. El Haj. 2014. "Remotely activated mechanotransduction via magnetic nanoparticles promotes mineralization synergistically with bone morphogenetic protein 2: applications for injectable cell therapy." *Stem Cells Translational Medicine* **3**(11): 1363–74. doi:10.5966/sctm.2014-0017.

Hidalgo-Bastida, L. A., S. Thirunavukkarasu, S. Griffiths, S. H. Cartmell, and S. Naire. 2012. "Modeling and design of optimal flow perfusion bioreactors for tissue engineering applications." *Biotechnology and Bioengineering* **109**(4): 1095–9. doi:10.1002/bit.24368.

Hoey, D. A., S. Tormey, S. Ramcharan, F. J. O'Brien, and C. R. Jacobs. 2012. "Primary cilia-mediated mechanotransduction in human mesenchymal stem cells." *Stem Cells* **30**(11): 2561–70. doi:10.1002/stem.1235.

Hutmacher, D. W. and H. Singh. 2008. "Computational fluid dynamics for improved bioreactor design and 3D culture." *Trends in Biotechnology* **26**(4): 166–72. doi:10.1016/j.tibtech.2007.11.012.

Kasper, C., K. Suck, F. Anton, T. Scheper, S. Kall, and M. van Griensven. 2007. "A newly developed rotating bed bioreactor for bone tissue engineering." In: N. Ashammakhi, R. Reis and E. Chiellini (eds.). *Topics in Tissue Engineering*. Oulu, Finland: University of Oulu.

Kelly, T. A., B. L. Roach, Z. D. Weidner, C. R. Mackenzie-Smith, G. D. O'Connell, E. G. Lima, et al. 2013. "Tissue-engineered articular cartilage exhibits tension-compression nonlinearity reminiscent of the native cartilage." *Journal of Biomechanics* **46**(11): 1784–91. doi:10.1016/j.jbiomech.2013.05.017.

Kim, I. S., Y. M. Song, T. H. Cho, J. Y. Kim, F. E. Weber, and S. J. Hwang. 2009. "Synergistic action of static stretching and BMP-2 stimulation in the osteoblast differentiation of C2C12 myoblasts." *Journal of Biomechanics* **42**(16): 2721–7. doi:10.1016/j.jbiomech.2009.08.006.

Klein-Nulend, J., Semeins, C. M., Veldhuijzen, J. P., and Burger, E. H. 1993. "Effect of mechanical stimulation on the production of soluble bone factors in cultured fetal mouse calvariae." *Cell and Tissue Research* **271**: 513–17. PMID:8472308.

Klein-Nulend, J., R. G. Bacabac, and A. D. Bakker. 2012. "Mechanical loading and how it affects bone cells: the role of the osteocyte cytoskeleton in maintaining our skeleton." *European Cells & Materials* **24**: 278–91.

Kopf, J., A. Petersen, G. N. Duda, and P. Knaus. 2012. "BMP2 and mechanical loading cooperatively regulate immediate early signalling events in the BMP pathway." *BMC Biology* **10**: 37. doi:10.1186/1741-7007-10-37.

Lawford, P. V., Y. Ventikos, A. W. Khir, M. Atherton, D. Evans, D. R. Hose, et al. 2008. "Modelling the interaction of haemodynamics and the artery wall: current status and future prospects." *Biomedicine & Pharmacotherapy* **62**(8): 530–5. doi:10.1016/j.biopha.2008.07.054.

Lee, K. I., J. S. Lee, J. G. Kim, K. T. Kang, J. W. Jang, Y. B. Shim, and S. H. Moon. 2013. "Mechanical properties of decellularized tendon cultured by cyclic straining bioreactor." *Journal of Biomedical Materials Research. Part A* **101**(11): 3152–8. doi:10.1002/jbm.a.34624.

Li, F., Sun, X., Zhao, B., Ma, J., Zhang, Y., Li, S., et al. 2015. "Effects of cyclic tension stress on the apoptosis of osteoclasts in vitro." *Experimental and Therapeutic Medicine* **9**: 1955–61. doi:10.3892/etm.2015.2338.

Lima, E. G., Bian, L., Ng, K. W., Mauck, R. L., Byers, B. A., Tuan, R. S., et al. 2007. "The beneficial effect of delayed compressive loading on tissue-engineered cartilage constructs cultured with TGF-beta3." *Osteoarthritis and Cartilage* **15**: 1025–33. doi:10.1016/j.joca.2007.03.008.

Liu, C., Zhao, Y., Cheung, W. Y., Gandhi, R., Wang, L., and You, L. 2010. "Effects of cyclic hydraulic pressure on osteocytes." *Bone* **46**: 1449–56. doi:10.1016/j.bone.2010.02.006.

Lourenço, N. D., Lopes, J. A., Almeida, C. F., Sarraguça, M. C., and Pinheiro, H. M. 2012. "Bioreactor monitoring with spectroscopy and chemometrics: a review." *Analytical and Bioanalytical Chemistry* **404**: 1211–37. doi:10.1007/s00216-012-6073-9.

Lovett, M., K. Lee, A. Edwards, and D. L. Kaplan. 2009. "Vascularization strategies for tissue engineering." *Tissue Engineering. Part B, Reviews* **15**(3): 353–70. doi:10.1089/ten.TEB.2009.0085.

Lujan, T. J., K. M. Wirtz, C. S. Bahney, S. M. Madey, B. Johnstone, and M. Bottlang. 2011. "A novel bioreactor for the dynamic stimulation and mechanical evaluation of multiple tissue-engineered constructs." *Tissue Engineering. Part C, Methods* **17**(3): 367–74. doi:10.1089/ten.TEC.2010.0381.

Malda, J., Klein, T. J., and Upton, Z. 2007. "The roles of hypoxia in the in vitro engineering of tissues." *Tissue Engineering* **13**: 2153–62. doi:10.1089/ten.2006.0417.

Mardones, R., C. M. Jofre, and J. J. Minguell. 2015. "Cell therapy and tissue engineering approaches for cartilage repair and/or regeneration." *International Journal of Stem Cells* **8**(1): 48–53. doi:10.15283/ijsc.2015.8.1.48.

Martinez, C., S. Rath, S. Van Gulden, D. Pelaez, A. Alfonso, N. Fernandez, et al. 2013. "Periodontal ligament cells cultured under steady-flow environments demonstrate potential for use in heart valve tissue engineering." *Tissue Engineering. Part A* **19**(3–4): 458–66. doi:10.1089/ten.TEA.2012.0149.

Mather, M. L., Morgan, S.P., and Crowe, J. A. 2007. "Meeting the needs of monitoring in tissue engineering." *Regenerative Medicine* **2**: 145–60. doi:10.2217/17460751.2.2.145.

McCoy, R. J. and O'Brien, F. J. 2010. "Influence of shear stress in perfusion bioreactor cultures for the development of three-dimensional bone tissue constructs: a review." *Tissue Engineering. Part B, Reviews* **16**: 587–601. doi:10.1089/ten.TEB.2010.0370.

McDonald, A. G. 2001. "Effects of high pressure on cellular processes." In: N. Sperelakis (ed.). *Cell Physiology Sourcebook: A Molecular Approach*. Philadelphia, PA: Elsevier. pp. 1003–23.

McIlhenny, S. E., Hager, E. S., Grabo, D. J., DiMatteo, C., Shapiro, I. M., et al. 2010. "Linear shear conditioning improves vascular graft retention of adipose-derived stem cells by upregulation of the α5β1 integrin." *Tissue Engineering. Part A* **16**(1): 245–55. doi:10.1089/ten.TEA.2009.0238.

Melchels, F. P., B. Tonnarelli, A. L. Olivares, I. Martin, D. Lacroix, J. Feijen, et al. 2011. "The influence of the scaffold design on the distribution of adhering cells after perfusion cell seeding." *Biomaterials* **32**(11): 2878–84. doi:10.1016/j.biomaterials.2011.01.023.

Morrell, K. C., W. A. Hodge, D. E. Krebs, and R. W. Mann. 2005. "Corroboration of in vivo cartilage pressures with implications for synovial joint tribology and osteoarthritis causation." *Proceedings of the National Academy of Sciences of the United States of America* **102**(41): 14 819–24. doi:10.1073/pnas.0507117102.

Murphy, S. V. and Atala, A. 2013. "Organ engineering – combining stem cells, biomaterials, and bioreactors to produce bioengineered organs for transplantation." *Bioessays* **35**: 163–72. doi:10.1002/bies.201200062.

Norrby, E. 2008. "Nobel Prizes and the emerging virus concept." *Archives of Virology* **153**(6): 1109–23. doi:10.1007/s00705-008-0088-8.

Nowlan, N. C., J. Sharpe, K. A. Roddy, P. J. Prendergast, and P. Murphy. 2010. "Mechanobiology of embryonic skeletal development: Insights from animal models." *Birth Defects Research. Part C, Embryo Today* **90**(3): 203–13. doi:10.1002/bdrc.20184.

Nowlan, N. C., V. Chandaria, and J. Sharpe. 2014. "Immobilized chicks as a model system for early-onset developmental dysplasia of the hip." *Journal of Orthopaedic Research* **32**(6): 777–85. doi:10.1002/jor.22606.

Nugent-Derfus, G. E., T. Takara, K. O'Neill J, S. B. Cahill, S. Gortz, T. Pong, et al. 2007. "Continuous passive motion applied to whole joints stimulates chondrocyte biosynthesis of PRG4." *Osteoarthritis and Cartilage* **15**(5): 566–74. doi:10.1016/j.joca.2006.10.015.

Pearson, N. C., Waters, S. L., Oliver, J. M., and Shipley, R. J. 2015. "Multiphase modelling of the effect of fluid shear stress on cell yield and distribution in a hollow fibre membrane bioreactor." *Biomechanics and Modeling in Mechanobiology* **14**: 387–402. doi:10.1007/s10237-014-0611-7.

Puetzer, J. L., Ballyns, J. J., and Bonassar, L. J. 2012. "The effect of the duration of mechanical stimulation and post-stimulation culture on the structure and properties of dynamically compressed tissue-engineered menisci." *Tissue Engineering. Part A* **18**: 1365–75. doi:10.1089/ten.TEA.2011.0589.

Radtke, A. L. and M. M. Herbst-Kralovetz. 2012. "Culturing and applications of rotating wall vessel bioreactor derived 3D epithelial cell models." *Journal of Visualized Experiments* **62**: pii: 3868. doi:10.3791/3868.

Reinwald, Y., K. H. Leonard, J. R. Henstock, J. P. Whiteley, J. M. Osborne, S. L. Waters, et al. 2015. "Evaluation of the growth environment of a hydrostatic force bioreactor for preconditioning of tissue-engineered constructs." *Tissue Engineering. Part C, Methods* **21**(1): 1–14. doi:10.1089/ten.tec.2013.0476.

Shah, R., J. C. Knowles, N. P. Hunt, and M. P. Lewis. 2013. "Development of a novel smart scaffold for human skeletal muscle regeneration." *Journal of Tissue Engineering and Regenerative Medicine* **10**(2): 162–71. doi:10.1002/term.1780.

Shipley, R. J. and S. L. Waters. 2012. "Fluid and mass transport modelling to drive the design of cell-packed hollow fibre bioreactors for tissue engineering applications." *Mathematical Medicine and Biology* **29**(4): 329–59. doi:10.1093/imammb/dqr025.

Takahashi, T., K. Mukai, H. Ohmura, H. Aida, and A. Hiraga. 2014. "In vivo measurements of flexor tendon and suspensory ligament forces during trotting using the thoroughbred forelimb model." *Journal of Equine Science* **25**(1): 15–22. doi:10.1294/jes.25.15.

Takai, E., R. L. Mauck, C. T. Hung, and X. E. Guo. 2004. "Osteocyte viability and regulation of osteoblast function in a 3D trabecular bone explant under dynamic hydrostatic pressure." *Journal of Bone Mineral Research* **19**(9): 1403–10. doi:10.1359/jbmr.040516.

Walker, N. A., C. R. Denegar, and J. Preische. 2007. "Low-intensity pulsed ultrasound and pulsed electromagnetic field in the treatment of tibial fractures: a systematic review." *Journal of Athletic Training* **42**(4): 530–5. PMID:18174942.

Wang, Z., R. S. Lakes, J. C. Eickhoff, and N. C. Chesler. 2013a. "Effects of collagen deposition on passive and active mechanical properties of large pulmonary arteries in hypoxic pulmonary hypertension." *Biomechanics and Modeling in Mechanobiology* **12**(6): 1115–25. doi:10.1007/s10237-012-0467-7.

Wang, T., Lin, Z., Day, R.E., Gardiner, B., Landao-Bassonga, E., Rubenson, J., et al. 2013b. "Programmable mechanical stimulation influences tendon homeostasis in a bioreactor system." *Biotechnology and Bioengineering* **10**: 495–507. doi:10.1002/bit.24809.

Wolfe, R.P. and T. Ahsan. 2013. "Shear stress during early embryonic stem cell differentiation promotes hematopoietic and endothelial phenotypes." *Biotechnology and Bioengineering* **110**: 1231–42. doi:10.1002/bit.24782.

Yates, C., C. R. Shepard, G. Papworth, A. Dash, D. Beer Stolz, S. Tannenbaum, et al. 2007. "Novel three-dimensional organotypic liver bioreactor to directly visualize early events in metastatic progression." *Advances in Cancer Research* **97**: 225–46. doi:10.1016/s0065-230x(06)97010-9.

Youngstrom, D. W., I. Rajpar, D. L. Kaplan, and J. G. Barrett. 2015. "A bioreactor system for in vitro tendon differentiation and tendon tissue engineering." *Journal of Orthopaedic Research* **33**(6): 911–18. doi:10.1002/jor.22848.

Yu, V., M. Damek-Poprawa, S. B. Nicoll, and S. O. Akintoye. 2009. "Dynamic hydrostatic pressure promotes differentiation of human dental pulp stem cells." *Biochemical and Biophysical Research Communications* **386**(4): 661–5. doi:10.1016/j.bbrc.2009.06.106.

Zhang, D., S. Weinbaum, and S. C. Cowin. 1998. "Estimates of the peak pressures in bone pore water." *Journal of Biomechanical Engineering* **120**(6): 697–703. PMID:10412451.

Zhou, T.-C., W.-W. Zhou, W. Hu, and J.-J. Zhong. 2010. "Bioreactors, cell culture, commercial production." In: M. C. Flickinger (ed.). *Encyclopedia of Industrial Biotechnology*. Chichester: John Wiley and Sons. pp. 1–18.

19

Cell Sensing of the Physical Properties of the Microenvironment at Multiple Scales

Julien E. Gautrot

Institute of Bioengineering and School of Engineering and Materials Science, Queen Mary University of London, London, UK

19.1 Introduction

Biomaterials play an important role in the development of tissue-engineering platforms, as they provide a physical scaffold capable of holding the shape of an implant or construct and sustaining repetitive mechanical loading. In addition, biomaterials used for the regeneration of tissues should ideally allow a complete degradation and replacement with natural components of the extracellular matrix (ECM), enabling the preservation of the physical structure of the healing tissue in the long term. Finally, biomaterials provide important biochemical cues controlling cell behavior and function, allowing them to accomplish the remodeling of the scaffold and the tissue being regenerated. In particular, cell adhesion to biomaterials is essential for the control of cell phenotype and the remodeling of the healing tissue. This process not only controls key functions such as proliferation, apoptosis, and differentiation, but is integrated with matrix remodeling and the long-term mechanical stability of the regenerated tissue. In addition, it is becoming clear that cell adhesion is not simply based on cell receptor–ligand binding events and is modulated by a number of physical signals and substrate mechanics, as well as topography and morphology (Lutolf and Hubbell 2005; Discher et al. 2009; Guilak et al. 2009). Therefore, an understanding of the adhesion of cells to biomaterials requires the study of the mechanisms by which such physical properties are sensed by cells.

Though cell adhesion mediated by integrins is initially driven by the binding of a few molecules, meaning it is inherently a nanoscale phenomenon, it results in the assembly of large complexes of proteins, termed "focal adhesions" (FAs) (see Figure 19.1). Often only several microns in size, the coupling of FAs to the contractile actin cytoskeleton allows mechanical forces to be generated and transmitted over long distances (tens of microns at the single-cell level, and millimeters and above for multicellular assemblies). The initial binding of integrin heterodimers to ECM molecules is followed by their clustering and the recruitment of several key cytoplasmic adapter proteins, such as talin (Zhang et al. 2008), vinculin (Humphries et al. 2007; Thievessen et al. 2013), and zyxin (Petit and Thiery 2000). These molecules allow the coupling of these initial integrin clusters, termed "nascent adhesions," to the actin cytoskeleton. Other proteins, such as

Mechanobiology: Exploitation for Medical Benefit, First Edition. Edited by Simon C. F. Rawlinson.

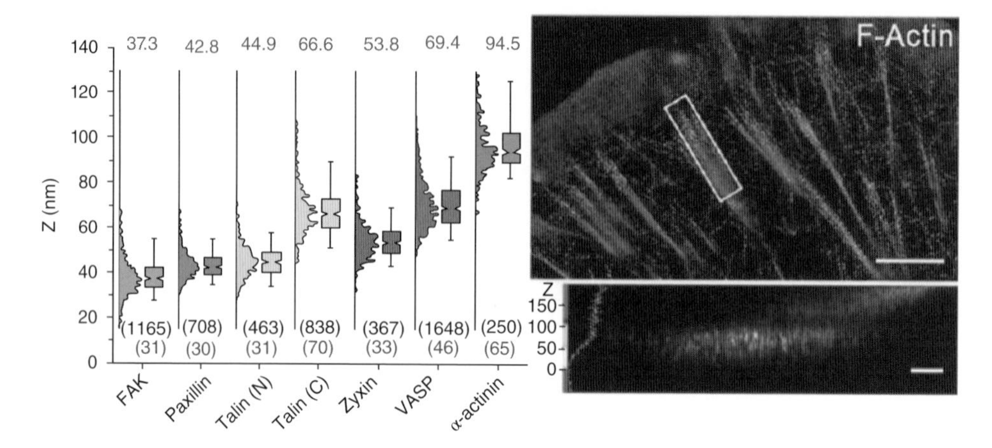

Figure 19.1 Nanoscale architecture of focal adhesions (FAs). Left: average z-position of different FA-associated proteins (Liu et al. 2015). Right: super-resolution image (top: top view; bottom: side view) of the actin network, with color-coded z-spatial information, at cell protrusions and FAs (scale bars top: 2 μm; bottom: 250 nm). *Source:* Reproduced with permission from Liu et al. (2015). (*See insert for color representation of the figure.*)

formins (Pollard et al. 2000) and α-actinin (Roca-Cusachs et al. 2013), then allow the crosslinking of actin bundles, and myosin couples them mechanically to generate contractile forces (Choi et al. 2008; Pasapera et al. 2015).

These molecular assembly processes allow the maturation of nascent integrin clusters into microscale FAs, highly structured in the z-direction at the nanoscale (with the cell membrane/integrin binding at one end and the actin network at the other end; see Figure 19.1) (Kanchanawong et al. 2010; Liu et al. 2015) and capable of applying strong shear forces and deforming their ECMs (Gardel et al. 2008). Such forces are generated at the cell front by the strong shear associated with actin flow and along stress fibers through myosin-based contractility, and are coupled to the ECM through integrins. The mechanisms by which such forces are established and the geometry and dynamics of FAs contribute to making these complexes of proteins the primary sensors of the nano- to microscale physical properties of the cell microenvironment. This chapter will detail some of the mechanisms by which FAs sense the physical environment, how this environment impacts on cell phenotype, and some of the platforms developed to study and control these phenomena.

19.2 Cells Sense their Mechanical Microenvironment at the Nanoscale Level

19.2.1 Impact of Substrate Mechanics on Cell Phenotype

Reports that cells can sense the mechanical properties of their microenvironment and deform inert materials date back to the 1970s and 80s. Harris et al. (1980) reported the wrinkling pattern that single cells and groups of cells generate when spreading at the surface of soft silicone substrates, and noted the implications of these observation for the possible role of mechanical tension in the direction of cell locomotion and the remodeling of matrices. Keese and Giaever (1991) demonstrated that cells can spread at the surface of ultrathin protein layers and that their proliferative behavior and ability to form colonies are affected by the mechanical behavior of these substrates (Giaever and Keese 1983). These observations were quickly followed by the development of techniques allowing the quantification and study of cell-generated traction forces. Traction force microscopy (TFM) was widely used to study local forces generated by cells and the FAs that they develop in order to adhere to the substrate (Oliver et al. 1995). Magnetic bead twisting assays were developed in order to study the transmission of forces from the cell membrane to the cytoskeleton, cytoplasm, and nucleus (Wang et al. 1993). Soft silicone-based microposts displaying tunable flexural moduli in the range of 2.7–1600.0 nN/μm (depending on their height) were used to quantify traction forces exerted by cells on adhesion sites (Tan et al. 2003). These experiments provided clear evidence for the large local mechanical deformation exerted by cells on their substrate and coincided with the discovery of the role of integrin heterodimers in the formation of cell adhesions (Albelda and Buck 1990).

Following these initial reports and observations, the use of poly(acrylamide) (PAAm) hydrogels and similar biomaterials with controlled mechanical properties (with Young's moduli ranging from sub-kPa to MPa) allowed the systematic study of the impact of matrix stiffness on cell behavior. Pelham and Wang (1997) showed that the spreading and locomotion of rat kidney cells and fibroblasts were controlled by the mechanical properties of PAAm gels. Cells seeded on soft substrates were unable to form stable FAs, as evidenced by vinculin and phosphotyrosine immunostainings. Similar observations were made for endothelial cells and fibroblasts spreading on PAAm gels coated with fibronectin and collagen (Yeung et al. 2005). Cells spreading on soft substrates were unable to assemble a stable actin cytoskeleton and spread. Other cell adhesive structures were also shown to be sensitive to matrix stiffness. For example, the dynamics of podosome formation was shown to depend on the stiffness of the substrate on which cells spread (Collin et al. 2006). As a result of such changes in cell adhesions, cell motility was found to be affected by matrix stiffness and guided by rigidity gradients (durotaxis) (Lo et al. 2000), the motility of smooth-muscle cells (SMCs) was found to be regulated by matrix rigidity in a biphasic fashion (increasing at intermediate moduli and decreasing at higher moduli) (Peyton and Putnam 2005), and the position of the maximal cell velocity was dependent on the density of integrin ligands (fibronectin concentration) but did not fully correlate with cell spreading or FA maturation (which displayed a monophasic behavior).

Considering the importance of matrix adhesion in the initiation and mediation of signaling pathways, it is not surprising that the impact of matrix stiffness on cell adhesions often also correlates with a change in phenotype (Discher et al. 2005, 2009; Guilak et al. 2009); hence, cell proliferation has been correlated to matrix stiffness (Georges and Janmey 2005; Assoian and Klein 2008; Wells 2008). The proliferation of chondrocytes was found to be increased on crosslinked collagen scaffolds with higher moduli, though the interplay of other parameters, such as matrix contraction and differences in crosslinking chemistry, might play a role here (Lee et al. 2001). The stiffness of acrylamide hydrogels was found to regulate the incorporation of BrdU in mouse embryonic fibroblasts, vascular SMCs, and MCF10A cells (Klein et al. 2009); this behavior was integrin-dependent and mediated by FAK, Rac, and cyclin D1. The proliferation of neural stem cells (NSCs) was also found to be regulated by matrix stiffness. The growth rate of

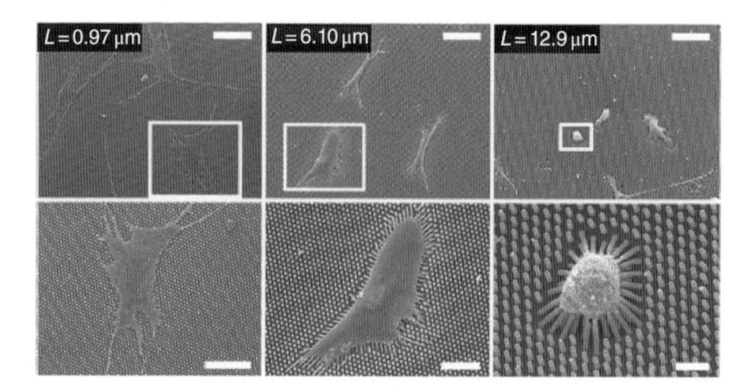

Figure 19.2 MSCs spreading on polydimethylsiloxane (PDMS) microposts with controlled flexural moduli. *Source:* Adapted from Fu et al. (2010). With permission of Macmillan Publishers Ltd., copyright 2010.

NSCs was found to be optimal on matrices with moderately stiff moduli (above 100 Pa) expressing high levels of the progenitor cell marker nestin and low levels of differentiation marker β-tubulin III (Saha et al. 2008).

The differentiation of stem cells was also shown to be controlled by the rigidity of the substrates on which they were cultured. Engler et al. (2006) showed that mesenchymal stem cells (MSCs) differentiate into defined lineages according to the mechanical properties of PAAm gels on which they are cultured. MSCs differentiated into neural lineages on soft (0.1–1.0 kPa) hydrogels, expressed myogenic markers on matrices of intermediate stiffness (8–17 kPa), and differentiated into bone lineages on more rigid substrates (25–40 kPa). Similarly, MSC differentiation was found to be controlled by the flexural modulus of microposts (see Figure 19.2) (Fu et al. 2010). When exposed to bipotential differentiation medium, they committed more frequently toward adipogenic lineages on long and soft microposts, whereas they differentiated more into osteoblasts on rigid, short patterns. Though FA formation and cell spreading are often correlated with such a response to matrix stiffness, a multiple range of signaling pathways seems to be involved, and the mechanotransduction mechanisms are emerging as intricate and complex. Hence, yes-associated protein (YAP) localization to the nucleus has been found to control MSC fate decision in response to matrix stiffness (Dupont et al. 2011). Nuclear lamin A expression correlates with tissue stiffness and enhances matrix-directed MSC differentiation (Swift et al. 2013); integrin internalization and endocytosis mediated by a caveolae pathway control the neurogenic differentiation of MSCs in response to soft matrices (Du et al. 2011); and keratinocyte differentiation of soft PAAm hydrogels is controlled by a MEK/ERK/AP1 pathway (Trappmann et al. 2012).

The phenotype of other stem cells was also shown to be controlled by matrix compliance. Hence, the self-renewal of muscle stem cells was preserved when these cells were cultured on matrices with moduli close to that of muscle tissues (near 12 kPa) (Gilbert et al. 2010). When muscle stem cells were cultured in such optimal conditions *in vitro*, their engraftment in mice was significantly improved, compared to control cells cultured on rigid plastic substrates. Similarly, the expansion of hemopoietic stem cells on compliant tropoelastin substrates was improved by two- to threefold compared to tissue culture plastic controls, though it may be difficult to distinguish matrix-associated biochemical signals from the contribution of substrate mechanics in this study (Holst et al. 2010). Keratinocytes spreading on soft PAAm hydrogels were unable to form stable FAs or to spread, and committed to differentiation, expressing high levels of the cornified envelop marker involucrin (Trappmann et al. 2012). The differentiation of NSCs spreading on PAAm-based interpenetrated networks into glial cells was modulated by the stiffness of the substrates (Saha et al. 2008). On soft gels (500 Pa), NSCs formed neurons, whereas on stiffer matrices (10 kPa), they differentiated into glial cells.

Beyond the control of decisions on stem cell fate, matrix stiffness has been reported to play an important role in the development of pathologies such as cancer (Ingber 2002), though its contribution is mediated, directly or indirectly, by a variety of processes. For example, through its impact on cell adhesion dynamics, matrix stiffness has been found to regulate the motility of tumor cells in two dimensions (2D) (Tzvetkova-Chevolleau et al. 2008; Ulrich et al. 2009; Tilghman et al. 2010). Studying such processes in three dimensions (3D) is considerably more difficult, due to the intricate effects of local heterogeneity in matrix stiffness, geometry, and porosity typical of the 3D matrices used for cell culture. The migration of DU-145 human prostate carcinoma cells through matrigel matrices of different densities and stiffnesses has been found to be strongly correlated with the composition of the matrix (Zaman et al. 2006). When the integrin ligand density was kept constant, cell speeds were higher on softer matrices, perhaps suggesting that pore size and matrix degradation are important limiting factors in such 3D locomotion. However, competing interactions contribute to make a full understanding of tumor progression through 3D matrices complex and difficult to predict. Breast cancer progression was found to be enhanced by matrix crosslinking, as a result of increased mechanical feedback and FA formation (Levental et al. 2009). In addition, in cohesive epithelial tissues, cell–cell interactions, often mediated by cadherins, are also important contributors to migratory and invasive behaviors (Hegedus et al. 2006). Hence, matrix stiffness was shown to promote epithelial–mesenchymal transition (EMT) in a breast cancer 3D model, through the control of the nuclear translocation of TWIST1 (Wei et al. 2015). Similarly, EMT was found to be promoted when normal murine mammary gland epithelial cells and Madin–Darby canine kidney (MDCK) epithelial cells were cultured on stiff matrices, through a switch in transforming growth factor beta (TGF-β)-induced cell function (Leight et al. 2012). Interestingly, though this effect seems in some cases to be driven by increased integrin ligation and FA formation on stiffer matrices (Levental et al. 2009; Leight et al. 2012), in other systems, particularly in 3D, integrin clustering dominates at lower stiffnesses (Chaudhuri et al. 2014). It has been proposed that soft matrices allow greater remodeling of the nanoscale spatial distribution of ligands and the clustering of integrins, resulting in increased signaling.

19.2.2 Nano- to Microscale Sensing of Mechanical Properties of the Matrix

As demonstrated in the previous subsection, the mechanism by which cells sense the mechanical properties of their microenvironment is complex and depends on both the nanoscale heterogeneity of the matrix and the dimensionality of the system. The local sensing of the mechanical properties of the ECM is clearly evidenced by recent reports showing that cell adhesion, spreading, and phenotype do not always correlate with matrix stiffness. Hence, the spreading and differentiation of human primary keratinocytes cultured on polydimethylsiloxane (PDMS) substrates remained unchanged by variation in matrix stiffness in the range of 1 kPa to 1 MPa (Trappmann et al. 2012), while the behavior of keratinocytes cultured on PAAm hydrogels was strongly affected by the stiffness of the matrix. This behavior was correlated with changes in FA formation on PAAm hydrogels (keratinocytes were unable to form stable adhesions on soft gels), whereas cells formed mature FAs on both soft and stiff PDMS substrates. It has been proposed that these contrasting responses to matrix stiffness arise from differences in the tethering of ECM proteins used to biofunctionalize these materials and the associated change in mechanical coupling. Similarly, fibroblasts cultured on crosslinked collagen gels showed clear changes in spreading and motility as a function of gel density and porosity, rather than gel stiffness (Miron-Mendoza et al. 2010). In good agreement with these observations, fibroblast spreading was unaffected by the stiffness of PDMS substrates (5 kPa vs. 2 MPa), but cell polarization was found to be influenced by substrate mechanics, through a mechanism controlled by FA formation and involving protein tyrosine kinases (Prager-Khoutorsky et al. 2011). It was recently proposed that the ability of fibroblasts to spread on soft matrices can be explained by their ability to remodel their matrix locally (Chaudhuri et al. 2015). Hence, whereas cells were unable to spread on soft (1.4 kPa) covalently crosslinked alginate gels, they couold spread and form FAs on ionically crosslinked (through Ca^{2+} ions) alginate gels of similar stiffness. In addition, this phenomenon was only observed at high ligand density. These results imply that noncovalently crosslinked matrices, which allow greater local nanoscale remodeling and ligand clustering, enable the formation of stable FAs and thus the assembly of a well-structured actin cytoskeleton.

In order to explore the respective roles of bulk stiffness and ligand tethering (and therefore local mechanical properties) on cell spreading and the differentiation of adipose-derived stem cells, Wen et al. (2014) varied the apparent porosity of matrices independently of their bulk stiffness. They found that cell response was more strongly affected by bulk stiffness than by changes in apparent porosity, though they observed changes in cell differentiation as a function of porosity. These results thus seem to conflict with the hypothesis that ligand tethering and local nanoscale mechanical coupling are key factors contributing to the sensing of matrix mechanics (Trappmann et al. 2012). However, it is possible that varying the apparent porosity does not lead to control of ligand tethering; indeed, the density of ligand tethering sites is controlled by the concentration of coupling agent (in the case of these studies, sulfo-SANPAH) and the concentration of polymer chains within the hydrogel matrix – varying the coupling agent clearly affected cell spreading and differentiation (Trappmann et al. 2012). However, the strategy used to vary the apparent porosity of PAAm hydrogels independently of their bulk stiffness requires an increase in polymer chain density to compensate for lower crosslinking levels. Hence, though the apparent porosity of hydrogels was decreased, the concentration of polymer chains within the matrix was high, resulting in high tethering densities (see Figure 19.3). In contrast, PDMS substrates of varying moduli (either all hydrophobic and nonporous or swollen in aqueous conditions) allowed high tethering densities to be achieved (see Figure 19.3). It should be pointed out that, in these different studies, the tethering densities for PAAm gels and PDMS substrates were not directly measured. Hence, the debate over the respective roles of bulk mechanical properties, apparent porosity, and ligand tethering is still open. However, it is clear that local nano- to microscale mechanical properties of matrices play an important role in regulating cell behavior.

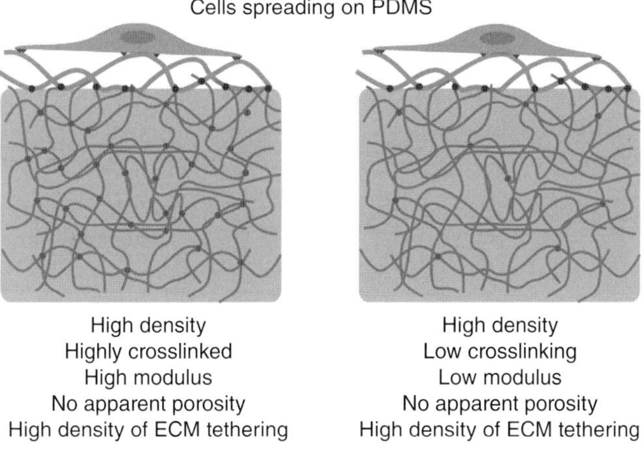

Figure 19.3 Proposed model to account for the distinct cell response to the bulk moduli of ECM protein-functionalized PAAm gels and PDMS substrates. Key parameters influencing the bulk modulus and cell response are summarized. (*See insert for color representation of the figure.*)

The notion that, at the single-cell level, mechanical sensing of the matrix occurs locally at the nano- to microscale is in good agreement with the current understanding of molecular mechanisms sustaining cell adhesion. Indeed, cell adhesion is initiated by membrane protrusions driven by actin polymerization and resulting in large shear forces (Ponti et al. 2004; Gardel et al. 2008). These forces are transmitted to the matrix via clusters of integrins connected to the cytoskeleton by adapter proteins such as talin (Zhang et al. 2008) and vinculin (Humphries et al. 2007; Thievessen et al. 2013), which grow in size and composition as they are reinforced. This process results in the structuring of the actin cytoskeleton (Xu et al. 2012) through bundling of individual fibers, mediated by molecules such as α-actinin (Roca-Cusachs et al. 2013) and formins (Courtemanche et al. 2013), and through myosin-based contractility (Laakso et al. 2008; Pasapera et al. 2015). Hence, primary sensing of matrix stiffness, at the single-cell level, relies on local deformations based on dynamic molecular assemblies. Further evidence of the importance of local mechanical properties of the matrix on cell adhesion can be found in the various reports highlighting the role of nanoscale remodeling of the matrix in the clustering of cell membrane receptors. TFM has provided direct evidence of the strong stresses that cells exert on the ECM. Interestingly, the stresses exerted on 2D matrices are relatively complex and involve local wrinkling of the matrix, with traction forces oriented both normal and parallel to the adhesion plane (Legant et al. 2013). Fluorescence resonance energy transfer (FRET) microscopy has also provided evidence of ligand clustering as a result of cell-mediated traction forces (Kong et al. 2005). This clustering was found to be optimal on gels of moderate stiffness (near 60 kPa) and was strongly dependent on ligand density and the stress-relaxation behavior of the matrix (Chaudhuri et al. 2015). Other systems based on flexible ligands tethered to rigid matrices have also highlighted the importance of ligand mobility in sustaining the clustering processes required for the development of FAs (Thid et al. 2007). However, other reports have highlighted that higher ligand mobility can equally be detrimental to FA assembly and cell spreading, presumably through a lack of mechanical engagement (Lautscham et al. 2014).

Local matrix remodeling seems to be equally crucial in 3D microenvironments. Though the formation of FAs in 3D matrices is still disputed, proteins associated with these structures in 2D (talin, paxillin, vinculin, α-actinin, zyxin, VASP, FAK, and p130Cas) seem to be involved in the regulation of membrane protrusion in 3D environments, albeit through very different structures, with different dynamics and mechanisms resulting in the generation of stress (Cukierman et al. 2001; Fraley et al. 2010). In addition, cell protrusions also form in 3D, despite displaying different sizes than their 2D counterparts, and are regulated by proteins found in FAs on planar substrates. Despite these differences in the structure and dynamics of adhesion complexes and the local topography of the matrix, cells exert strong forces on 3D matrices, too. In stable protrusions, forces localize at the tips of protrusive structures. During cellular extension, however, the growing tip is not associated with strong contractile forces, and stresses localize several microns behind the leading edge (Legant et al. 2010). In addition to sensing the local mechanical properties of the matrix, cell protrusions in 3D have also been shown to sense the local morphology of the microenvironment. Cell branching has been found to correlate with the local curvature of the matrix and to be controlled by myosin II activity through the minimization of the local cell membrane curvature (Elliott et al. 2015). Disentangling which parameters control the generation of these forces is difficult, and matrix stiffness, degradability, and morphology are likely to combine in dictating the mechanism by which protrusions are generated and exert tension on the microenvironment on order to sustain cell spreading and motility.

The complexity of the 3D microenvironment perhaps explains why the impact of matrix stiffness on cell phenotype is difficult to predict. Hence, the proliferation of fibroblasts has been found to be regulated by a combination of matrix stiffness, density, and degradability, as well as the density of the integrin ligands they present to cells (Bott et al. 2010). The differentiation of MSCs in covalently crosslinked hyaluronic acid hydrogels is controlled by matrix degradation rather than matrix mechanics *per se* (see Figure 19.4) (Khetan et al. 2013). Cells that degrade their 3D environment locally (within soft matrices) are able to spread further and to generate traction forces that regulate the signaling pathways controlling osteogenic differentiation. On the other hand, cells that are unable to degrade their matrix commit to adipogenesis. These observations

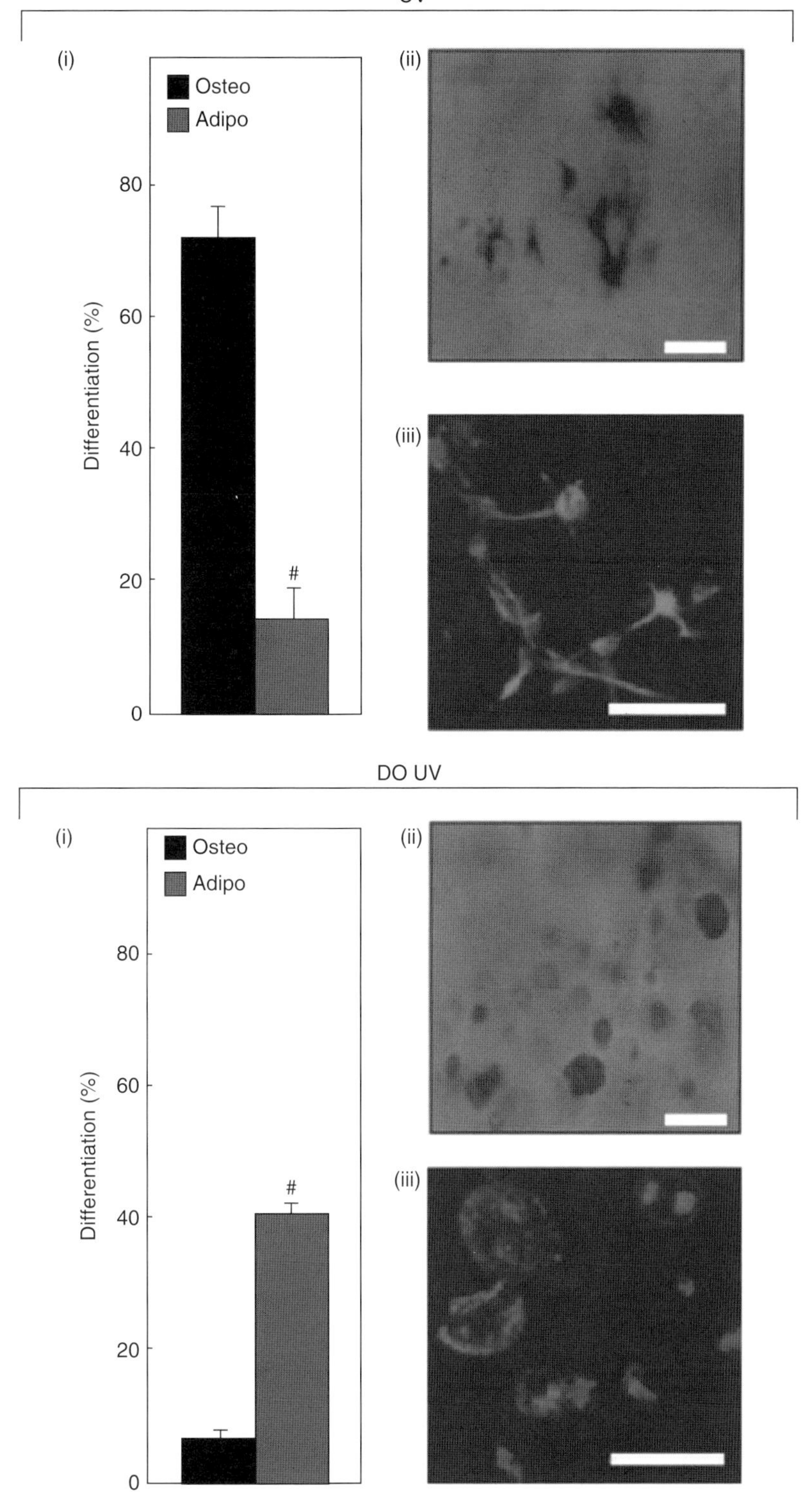

Figure 19.4 The differentiation of MSCs is regulated by matrix degradation in 3D. Top: osteogenic differentiation of MSCs in degradable hyaluronic acid gels. Bottom: adipogenic differentiation in nondegradable gels. *Source:* Adapted from Khetan et al. (2013). With permission of Macmillan Publishers, copyright 2013.

contrast with the differentiation of MSCs in 3D alginate gels, which commit to osteogenesis when their microenvironment is stiff (11–30 kPa) but to adipogenesis when the matrix is soft (2.5–5.0 kPa) (Huebsch et al. 2010). In this latter, non-cell-degradable system, changes in ligand clustering with matrix compliance have been proposed to control the cell behaviors observed.

Overall, these observations suggest that the cell response to substrate mechanics is complex and is often influenced by other biochemical and physical parameters, such as ligand density, matrix degradability, and morphology, especially in 3D microenvironments. However, whether in 2D or 3D systems, mechanical sensing seems to be local, at the single-cell level, and to rely on the direct coupling of the mechanical environment to molecular complexes and the cell cytoskeleton.

19.3 Cell Sensing of the Nanoscale Physicochemical Landscape of the Environment

Not surprisingly, other nanoscale physicochemical properties of the matrix have been shown to impact on cell phenotype. Often, these effects are mediated by the modulation of molecular processes underlying cell adhesion, such as integrin clustering, vinculin stabilization of FAs, or the bundling of actin filaments into contractile stress fibers. These physicochemical properties of the matrix can be subdivided into matrix topography (the quasi-3D geometry of a substrate) and geometry (the quasi-2D arrangement of adhesive ligands or biochemical cues).

19.3.1 Impact of Nanotopography

As discussed in the previous section, the cell response to substrate mechanics, especially in 3D, is complicated by the morphology of the environment, which not only contributes to the local mechanical heterogeneity of the matrix, but also can impact on the geometrical boundary imposed on the cell. In many cases, *in vivo*, the ECM displays a fibrillar morphology, based on the assembly of molecules such as fibronectin, collagen, and laminin. Understanding how such fibrillar morphology can impact cell behavior is therefore essential for the development of biomaterials for tissue engineering.

Beyond the chemical and biochemical composition of fibrillar substrates, the dimensions of the fibers seem to play an important role in controlling cell adhesion and behavior. Hence, the differentiation and proliferation of NSCs has been found to be influenced by the diameter of electrospun nanofibers. In differentiation medium, NSCs differentiated more frequently into oligodendrocytes on thinner fibers (300 nm), compared to larger fibers (750 or 1500 nm) or flat tissue culture plastic (Christopherson et al. 2009). In normal culture medium, proliferation was higher on tissue culture plastic and thin fibers. In addition, changes in cell phenotype as a function of fiber size correlated with changes in cell spreading. Whereas cells spread well on thin fibers, they remained more rounded and aggregated on larger fibers. NSCs initially spread more on small (300 nm), randomly oriented fibers than on larger (1000 nm) fibers. On oriented fibers, this effect on cell shape was not as pronounced, and cells were observed to align and stretch in the direction of the fibers. In addition, NSCs formed longer primary neurites on smaller fibers, especially when they were oriented (He et al. 2010). In the case of osteoblasts, it was found that, via changes in FA formation, the diameter of nanofibers controlled the activation of the MAPK pathway, through ERK, p38, and JNK – potentially explaining the role of fiber dimension in the regulation of cell phenotype (Jaiswal and Brown 2012).

Not only does the size of fibers impact on the size of FAs formed at their surface, but it can also impact the curvature of the membrane at its contact. This change in membrane curvature can be sensed by membrane proteins such as POR1, the binding of which depends on the membrane curvature. On small fibers (100 nm), which maximize membrane curvature, POR1 binding increases, activating the RAC1 pathway (Higgins et al. 2015), while on larger fibers (1000 nm), POR1 binding decreases, and associated RAC1 signaling activates alkaline phosphatase (ALP) and triggers osteogenic differentiation of MSCs. The increase in FA formation at nanofibers (with their inherent curvature) also results in an increase in myosin-based cell contractility and RhoA activity (Ozdemir et al. 2013). This in turn regulates the osteogenic differentiation of osteoprogenitor cells. Monte Carlo simulations (computer-based simulations relying on iterative calculations and optimization of convergence) also suggest that integrin clustering (even in the absence of binding forces between two adjacent integrins) is strongly affected by substrate curvature and membrane stiffness (which may vary with the composition of phospholipids, abundance and nature of membrane proteins, and morphology of the cortical actin network) (Lepzelter et al. 2012). Hence, integrin clustering, in combination with other curvature-sensing molecules, may act as a direct sensor of surface topography.

In addition to curvature, size, and density, other properties of nanofibers may also impact on cell adhesion and phenotype. Hence, it was found that the shape of glass ribbons generated via self-assembly (helical vs. twisted, with periodicities of 63 and 100 nm, respectively) influenced the formation of FAs and cell spreading in MSCs (Das et al. 2013). MSCs cultured on helical nanoribbons displayed larger FAs, decreased STRO-1 expression, and increased Osx expression, suggesting that nanofiber shape regulates osteogenic differentiation of MSCs. Other properties of fibrous networks also impact on cellular adhesion, motility, and phenotype. Hence, the mechanics of self-assembled peptide nanofibers have been found to control neuron development (Sur et al. 2013). The density of neurons adhering from hippocampal cultures is higher on soft than on stiffer fibers, with fewer astrocytes. In addition, neurons cultured on soft nanofibers display fewer neurites but longer primary neurites. This behavior has been proposed to result from the facilitated cell polarization due to mechanosensing of the matrix. The porosity of fibrous mats also plays an important role in regulating cell proliferation, spreading, and motility. Cells seeded on polycaprolactone (PCL) mats proliferated when the pore size of the mats was larger than 6 μm, but struggled to spread when pore size increased above 20 μm (Lowery et al. 2010). For such large pore size, cells adhered to single fibers and did not bridge across the pores of the mats. In addition, the size of pores formed by fibrillar mats or gels doesn't only regulate cell adhesion, but also cell infiltration and migration through the 3D structure of the matrix (Wolf and Friedl 2011). The migration of cells through collagen gels is controlled by the size of pore formed by the fibrillar network, as well as the ability of cells to locally degrade the matrix and exert traction forces on it, and to deform their nuclei (Wolf et al. 2013). In turn, such parameters impact on the ability of cells to invade the matrix and proliferate, affecting tissue-engineering applications (Mandal and Kundu 2009; Rnjak-Kovacina et al. 2011). Therefore, the cellular response to fibrous mats is complex and results from the combination of chemical, biochemical, and a number of physical properties of these substrates.

The nanotopography of continuous surfaces also strongly influences cell adhesion and phenotype. The dimensions, geometries, and regularities of the associated patterns or textures have been reported to control such responses. In the case of nanogrooves, cell alignment (along the main axis of the grooves) is typically observed, depending on the size and depth of the patterns, as well as their spacing (Kim et al. 2010). Membrane protrusions within grooves are restricted on narrow features (400 vs. 800 nm), presumably by membrane curvature and stiffness, resulting in more extensive FAs. This behavior correlates with the abundance and orientation of FAs and actin stress fibers. Such changes in cell adhesion, spreading, and cytoskeletal organization can impact on fate decisions, as in the control of osteoblast phenotype (Cassidy et al. 2014). For example, large grooves (15 μm wide and 2 μm deep) resulted in adipogenesis in MSCs, whereas smaller grooves (2 μm wide and 2 μm deep) led to osteogenesis (Abagnale et al. 2015).

Other types of nanotopography have also been shown to control cell adhesion and the differentiation of stem cells. In particular, substrates presenting nanopits with controlled dimensions and regularity (in which nanopits are either aligned into arrays or disordered) have been developed by Dalby et al. (2007). Typical dimensions for these nanopits are 120 nm diameter and 100 nm depth, organized in hexagonal or square arrays (with spacing between features ranging from 20 to 50 nm). When nanopits are disordered, the fate of MSCs adhering to their surface is pushed toward osteogenesis (osteopontin and osteocalcin expression), correlating with stronger cell adhesion (see Figure 19.5) (Dalby et al. 2007; Biggs et al. 2009; Tsimbouri et al. 2014). Interestingly, though the number and size of FAs formed on disordered arrays was higher than for ordered arrays, cell spreading was higher on substrates presenting ordered nanopits. This contrasts with correlations between matrix stiffness, FA formation, and cell spreading. In addition, MSCs cultured on ordered arrays retained stem cell markers and could be expanded to higher passage levels than those cultured on disordered substrates (McMurray et al. 2011). In addition, the density of nanoposts (and therefore the distance between two features, varying from 1.2 to 5.6 μm) was found to control the differentiation of human MSCs (Ahn et al. 2014). On sparse patterns, cells differentiated toward osteogenic lineages, whereas on denser patterns, they committed to adipogenic differentiation. These phenotypes were associated with marked changes in cell contractility and stiffness. Similarly, the height of titania nanopillars (ranging from 15 to 100 nm) was

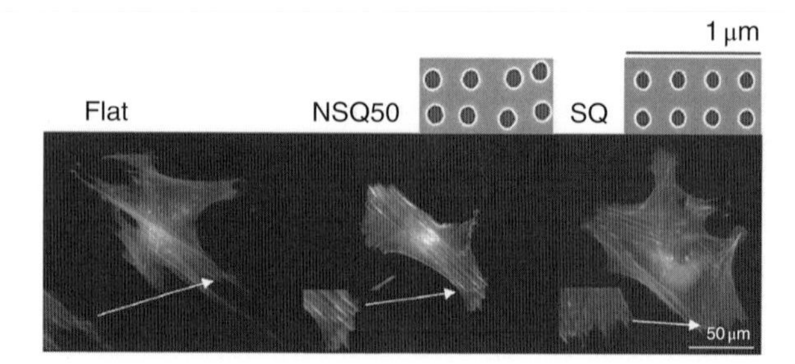

Figure 19.5 Impact of substrate topography on FA formation and cell spreading. *Source:* Reprinted with permission from Tsimbouri et al. (2014). Copyright 2014 John Wiley and Sons. (*See insert for color representation of the figure.*)

found to control the spreading and osteogenic differentiation of human MSCs (Sjostrom et al. 2009). Finally, as a result of changes in nanotopography, the mechanical behavior of substrates was also found to be affected (Kuo et al. 2014). Therefore, the effects of nanotopography and substrate mechanics can combine to control cell phenotype. Cells were cultured on soft pillars displaying sizes in the range of 200–680 nm and flexural rigidities between 0.26 and 94 nN/nm, respectively. Though FA maturation and cell spreading gradually increased as a function of nanopattern size, on the smallest pillars (200–300 nm), cells were able to deform the substrates, and FAs bridging was observed across several patterns.

More generally, surface roughness controls cell adhesion and phenotype and is an important design parameter (often simple to engineer) for biomaterials used in tissue-engineering applications. Hence, human embryonic stem cells (hESCs) generate fewer but larger FAs on smooth glass surfaces compared to nanorough silica surfaces (root-mean-square roughness of 100 nm) (Chen et al. 2012). This correlates with an improved proliferation and retention of the pluripotent marker Oct3/4 on smooth surfaces. However, as with cell sensing of nanopits and grooves, the mechanism by which FA formation and dynamics are controlled by the roughness of such silica substrates is not fully clear.

The nanotopography of biomaterials is therefore an important parameter controlling cell adhesion, spreading, motility, and fate, and the nanotexturing of surfaces constitutes a simple method for controlling cell phenotype independently of other bulk properties and is important for the design of implants and scaffolds for use in tissue engineering and regenerative. However, the underlying mechanism by which cell adhesions sense such nanoscale physical properties and modulate the signaling pathways controlling cell fate is not always clear, and important research efforts are required to improve our understanding of this topic.

19.3.2 Impact of Nanoscale Geometry

In contrast to nanotextured surfaces displaying continuous chemistry, though with 3D nanotopography, other biomaterials display nanoscale domains with well-defined chemistry. Such geometrical patterns with controlled chemistry often result from self-assembly processes and constitute simple model systems by which to dissociate geometrical properties from other morphological and mechanical parameters. Substrates with controlled nanoscale geometries have thus been designed in order to study the processes underlying adhesion formation, from initial integrin clustering (for sub-100 nm patterns) to the maturation of FA (for nanopatterns with dimensions in the range of 100 nm to 1 μm).

Nanopatterned substrates presenting gold clusters, each capable of accommodating the binding of one integrin heterodimer, and separated by defined distances in the range of 20–250 nm, were generated via the self-assembly of block copolymers (Cavalcanti-Adam et al. 2006; Huang et al. 2009). This technique – block-copolymer micelle nanolithography – allows the study of the influence of binding geometry and ligand density on integrin clustering and the development of stable FAs. A threshold distance of 60 nm between integrin ligands (RGD peptides) was identified as a key mediator of integrin clustering and FA assembly (Cavalcanti-Adam et al. 2006). Cells cultured on substrates displaying ligands with larger spacing than this critical distance were unable to spread. More specifically, clusters of four or five integrins were identified as the minimal cluster size capable of sustaining cell adhesion in fibroblasts, using a different nanopatterning approach (Schvartzman et al. 2011). Similarly, the level of order of the pattern was found to impact on cell adhesion, as it resulted in a local decrease in ligand spacing (Huang et al. 2009). In addition, such local nanoscale effects dominated the global distribution of adhesive ligands (microscale geometry of the ECM; see Figure 19.6) (Deeg et al. 2011). When the global ligand density of gold clusters (and therefore RGD peptides) spaced by

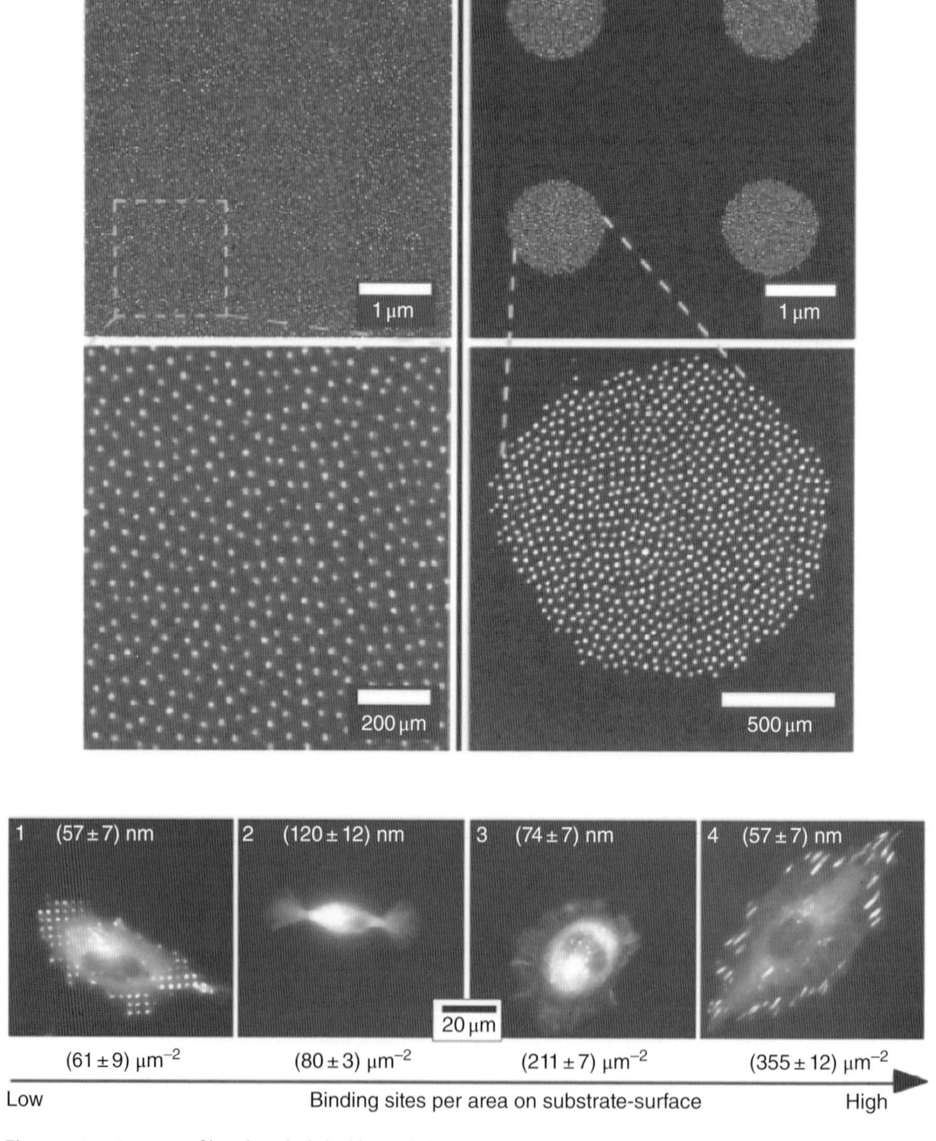

Figure 19.6 Impact of local and global ligand density on cell adhesion. Top: examples of nanopatterned substrates. Bottom: impact of ligand density on cell adhesion and spreading. *Source:* Deeg et al. (2011). Copyright 2011 American Chemical Society.

57 nm was adjusted to be below that of clusters spaced by 120 nm (using micropatterning to define 1 μm patches of dense clusters), cell spreading and FA formation were found to be similar to those for substrates displaying high local and global ligand densities (Figure 19.6).

To probe the relationship between integrin clustering and force generation, this self-assembly platform was modified to allow gold clusters to be generated at the tops of hydrogel micropillars (Rahmouni et al. 2013). Forces generated by cells were measured by monitoring the displacement of these pillars. Using this system, the α5β1-integrin heterodimer was identified as an important player in the generation of forces on the matrix. In addition, the ability of integrin clustering to translate into traction forces at the nanoscale was studied using ligands presenting fluorophores that are activated when under tension (quenched by the close proximity of the underlying gold cluster) (Liu et al. 2014). It was proposed that the sensing of ligand density occurred via a two-stage process: the first stage driven by actin polymerization, with forces generated on ligands in the range of 1–3 pN; the second controlled by ligand density, integrin clustering, and myosin contractility, allowing forces per ligand in the range of 6–8 pN.

Beyond the initial ligation and clustering of integrin, FA maturation and signaling involve the recruitment of a large number of proteins and the maturation of nascent adhesions into large, microscale FAs (Geiger et al. 2009; Harburger and Calderwood 2009). Nano- to micropatterning techniques have been developed to allow the control of such maturation and the study of this process and its relationship with the generation of forces and cell spreading.

In order to study the geometrical maturation of FAs, colloidal lithography was used to generate arrays of gold islands in the range of 100 nm to 1 μm, surrounded by protein- and cell-resistant polymer coatings (Malmstrom et al. 2010). These islands were coated with ECM proteins to promote integrin binding and control the maximum size of the adhesions formed. Cell spreading was gradually impaired through decreasing adhesive patch size, correlating with the disruption of the actin cytoskeleton. The type of ECM protein deposited on to these nanoislands impacted on their behavior: vitronectin allowed some level of bridging of adhesions between islands, presumably through matrix deposition, but fibronectin did not (Malmstrom et al. 2011). Keratinocyte spreading on these patterns was also gradually impaired, correlating with an increase in cell differentiation (expression of involucrin) (see Figure 19.7) (Gautrot et al. 2014). In addition, it was found that the recruitment of FA-associated proteins (talin, vinculin, paxilin) and the phosphorylation of proteins such as FAK and cortactin were not impaired on the smallest nanoislands. Laminin deposition was also observed, further suggesting that evolution of the composition of FAs is not directly coupled to their geometrical maturation – the size of adhesions correlates better with actin cytoskeleton assembly and force generation (Balaban et al. 2001; Galbraith et al. 2002; van Hoorn et al. 2014). This lack of mechanical coupling can in turn influence both cell spreading and motility (Slater et al. 2015).

In contrast to these studies, patterned substrates developed using microcontact printing identified a size threshold of approximately 300 nm, below which adhesion formation was impaired (Coyer et al. 2012). This behavior was controlled by talin- and vinculin-mediated stabilization of FAs, and it was proposed that small patches are unable to sustain the large forces generated by actin threadmilling and myosin-based contractility. The occurrence of this size threshold in this system may result from the

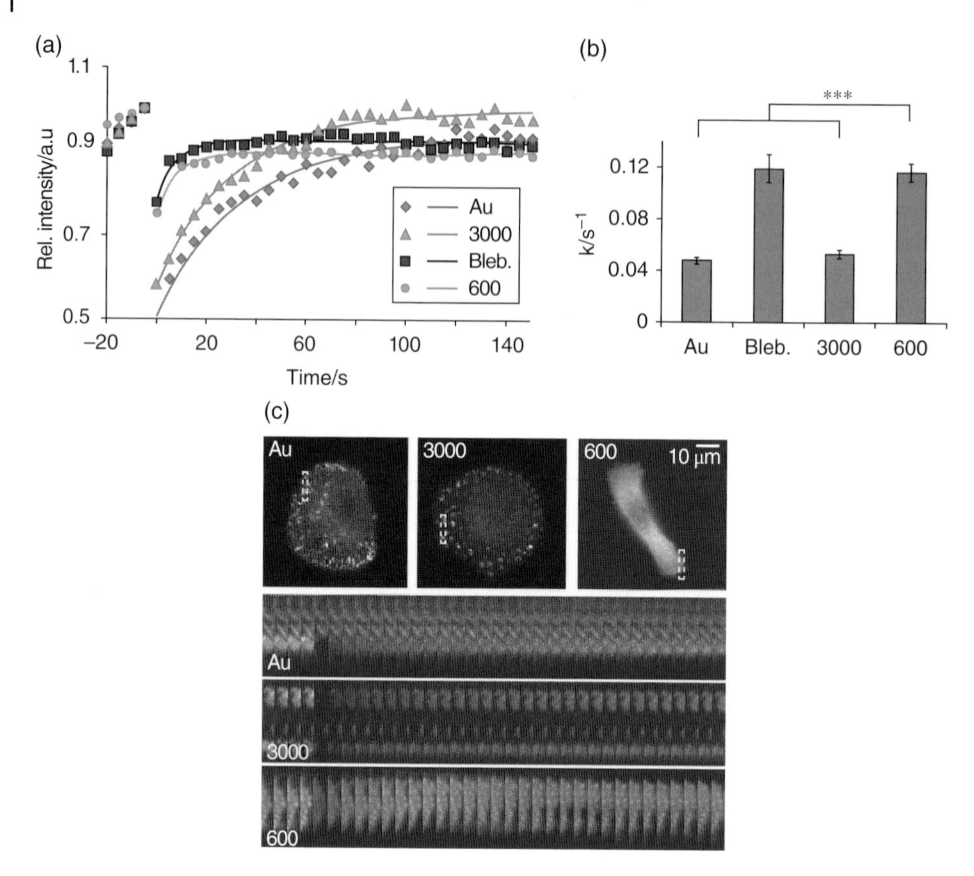

Figure 19.7 The dynamics of recruitment of vinculin is controlled by adhesive patch size. *Source:* Gautrot et al. (2014). Copyright 2014 American Chemical Society.

presence of a larger adhesive patch at the center of each cell pattern (and surrounded by the nanopatches studied), which may preferentially segregate adhesion complexes and associated molecules. In addition, super-resolution microscopy provided evidence that the large adhesion complexes observed by epifluorescence or confocal microscopy may result from the close proximity of several smaller adhesions. Stresses generated at such adhesions were calculated to be in the range of 10–300 nN/μm^2 – one order of magnitude higher than previously thought (van Hoorn et al. 2014). These observations (and others based on the study of FA dynamics), correlated with force generation (Thievessen et al. 2013), suggest that the relationship between adhesion composition, geometrical maturation, and force generation is complex and does not necessarily follow simple linear laws.

19.4 Cell Sensing of the Microscale Geometry and Topography of the Environment

Moving up in the scale of the geometry of biological systems, the microstructure of the cell environment plays a key role in cellular organization, cell phenotype, and tissue architecture and function. Cells *in vivo*, within one cell type and in a given tissue, display very reproducible sizes and shapes. In addition, they are not randomly distributed throughout the structure of a tissue, but often occupy well-defined locations with respect to ECM components and other cells, which in the case of stem cells are designated as "stem cell niches." The well-defined geometry of this microenvironment is typically lost *in vitro*, and cells display altered and less reproducible shapes and sizes in tissue culture dishes. Hence, the control and study of cell shape and geometry at the microscale *in vitro* is an important focus in the field of tissue engineering and has led to the development of micropatterned platforms that allow control of the physical and biochemical landscape to which cells are exposed in culture.

19.4.1 Control of Cell Spreading by ECM Geometry

Several micropatterning systems have been developed to control cell adhesion and spreading at the microscale. In particular, microcontact printing of self-assembled monolayers (Kane et al. 1999), proteins (Thery et al. 2006a), or polymer brushes (Gautrot et al. 2010) has been shown to be a simple, efficient, and reproducible approach to the control of cell adhesion and spreading. These techniques highlight several rules according to which the ECM geometry controls the formation of cell adhesions and the organization of the cytoskeleton. FAs are primarily stabilized at corners or sharp edges of ECM geometries, though they may form and migrate in other areas where the cell is in contact with the matrix (Thery et al. 2006a). This strengthening mechanism allows the cell to maximize its spreading area. Stress fibers are then found to connect these stable adhesions; these fibers are particularly reinforced when bridging over nonadhesive areas, working against membrane tension, and reducing membrane curvature. The biophysical principle underlying these observations has been further explored using biomechanical computational models that are able to capture the distributions of FAs and stress fibers (Deshpande et al. 2008; Pathak et al. 2008). The results of these studies highlight some of the important strengthening mechanisms that couple FA formation and stabilization to the generation of contractile forces sustained by the actin cytoskeleton. Similarly, the assembly of other components of the cytoskeleton, such as microtubules, is also directed by the ECM geometry, through the localization of FAs (Thery et al. 2006b; Huda et al. 2012).

An important question that is directly addressed by studies using ECM micropatterns is: What is the minimum set of biochemical entities (proteins, enzymes, and small molecules) that allows the sensing of geometrical cues, the assembly of a cytoskeleton, and the ability of the cytoskeleton to generate coherent mechanical forces. To further test the concept that ECM geometry directly controls cytoskeleton assembly and the dynamics of contractile forces, micropatterned substrates were used to direct the nucleation and formation of actin fibers. It was found that the orientations, interactions, and lengths of the resulting actin filaments were controlled by the geometry of the nucleation pattern (Reymann et al. 2010). In addition, the resulting networks were dynamically controlled by myosin-based contractile forces and the crosslinking of actin fibers (Reymann et al. 2012). Hence, when filaments assembled in an antiparallel fashion, they deformed and ultimately disassembled, whereas when they interacted in a parallel fashion, they bundled and stabilized. This simple process, through which the geometry of nucleation directs the architecture and stability of the actin network, constitutes a very direct sensing mechanism, allowing the geometry of the ECM to be sensed by cells and control their spreading. Similar simple geometry-sensing mechanisms may also direct the assembly of cytoskeletal structures in other biological processes, such as the assembly of the actin ring observed during mitosis at the equatorial plane of a dividing cell. Such a phenomenon was recreated in cell-sized droplets, taken from free actin molecules, in the presence of bundling factors, without external geometrical parameters others than the spherical geometry of the droplets (Miyazaki et al. 2015).

Beyond the reproduction of cell shape, localization of adhesion sites, and architecture of the cytoskeleton, micropatterned platforms can direct the internal organization of other cell compartments and organelles. Hence, it has been found that the asymmetry of the ECM directs cell polarity in retinal pigment epithelial (RPE) cells (Thery et al. 2006b). Cortactin accumulated at ECM-coated regions and at the leading edge of cells was induced to polarize on arrow-shaped adhesive islands. This, together with the assembly and polarization of the cytoskeleton, resulted in the respective positioning of the Golgi and the centrosome "in front" of the nucleus (in the direction of the arrow). Similar observations were made in the case of the organization of the cytoskeleton of keratinocytes (Gautrot et al. 2010), though the positioning of other organelles was not as well defined as for RPE cells (unpublished data). Such cell-dependent variability is perhaps a reflection of differences in integrin expression, as heterodimers such as $\alpha v\beta 3$ and $\alpha 5\beta 1$ seem to play distinct and synergistic roles in mediating cytoskeleton organization and the generation of myosin-based contractile forces (Schiller et al. 2013). In addition, this polarization of cells in response to the asymmetry of the ECM is functional, and cells released from such patterns begin migrating in the direction in which the pattern induced their orientation (Jiang et al. 2005). Finally, in extreme conditions that force cells to elongate, the various cellular compartments and organelles, including the nucleus, must also accommodate the change in cell shape. This gives rise to striking nuclear elongations, due to the balance of compressive and tensile forces arising from the deformation of the cytoskeleton (Versaevel et al. 2012; Kim and Wirtz 2015).

19.4.2 Directing the Phenotype of Single Cells via Matrix Geometry

Not surprisingly, such important changes in cytoskeletal architecture and the positioning of organelles can have important effects on cell phenotypes. Perhaps one of the most striking examples of such phenomena is the impact of ECM geometry on the orientation of cell division. Cell division is known to be oriented in cells that are slightly elongated just before the onset of mitosis. Using ECM micropatterns of different geometries but which conferred identical shapes to cells, it was found that the geometry of ECM, rather than cell shape, controls the orientation of the spindle (see Figure 19.8) (Thery et al. 2005). The geometry of ECM and the external mechanical forces applied to the cell microenvironment just before the onset of mitosis control the direction and localization of retraction fibers formed during cell retraction in the early stages of mitosis. These retraction fibers bias the organization of the subcortical actin network and the orientation of the mitotic spindle through interactions with astral microtubules (Fink et al. 2011). The importance of such geometrical and mechanical constraints on cell division may play a role in tissue development and asymmetric cell division, contributing to the shape of tissues and the maintenance of their homeostasis. Hence, these interactions have been found to control the partitioning of DNA in daughter cells and asymmetric cell fate in mouse skeletal muscle stem cells (Freida et al. 2013; Yennek et al. 2014).

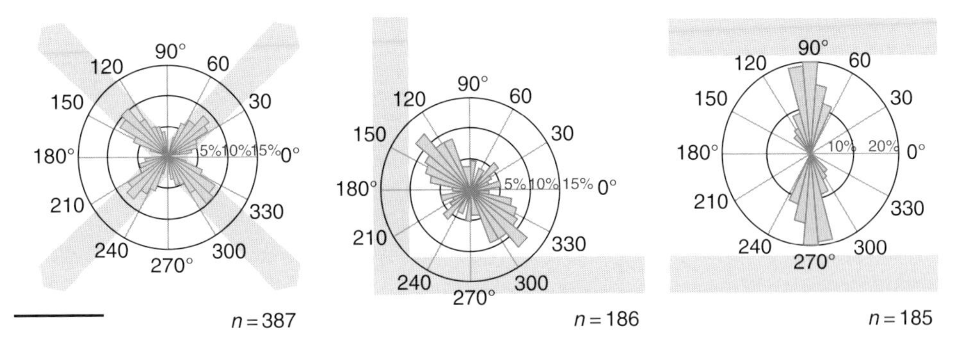

Figure 19.8 ECM guides the orientation of cell division. *Source:* Thery et al. (2005). Reproduced with permission of *Nature*.

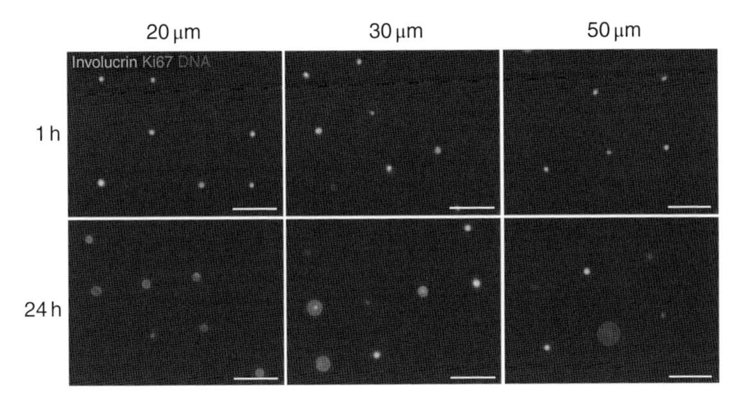

Figure 19.9 Cell shape controls the differentiation of keratinocytes. *Source:* Connelly et al. (2010). Reproduced with permission of *Nature*. (*See insert for color representation of the figure.*)

Therefore, local asymmetry of the cell microenvironment, whether purely geometrical in origin or coupled to biochemical asymmetry, may play an important role in directing the fate and partitioning of daughter cells.

Other phenotypes have been connected to the matrix geometry, independent of cell division. Hence, cell death has been found to be controlled by adhesion areas and cell spreading (Chen et al. 1997). DNA synthesis and apoptosis are differentially regulated by the geometry of the ECM and the ability of cells to spread on the surface of adhesive islands. Stem cell differentiation has also been found to be regulated by such geometrical cues: the differentiation of human primary keratinocytes was prevented on large adhesive islands on which cells were able to fully spread, whereas the expression of differentiation markers (involucrin) increased on small islands on which cells were rounded (Watt et al. 1988). This report, by Fiona Watt et al. (1988), constitutes the first experiment in which cell spreading and phenotype were controlled using micropatterned substrates. Other markers, such as transglutaminase, were also found to be upregulated on rounded cells adhering to small adhesive islands, and cell commitment was found to be regulated through a G-actin, MAL, SRF, AP1 pathway (see Figure 19.9) (Connelly et al. 2010). Interestingly, this pathway allows the direct control of gene transcription through cytoskeleton rearrangement, as it is controlled by the balance of

G- and F-actin, and hence represents an extremely fast mechanosensitive mechanism. This platform was also used to identify the role of p38 and histone acetylation in mediating terminal differentiation of human keratinocytes (Connelly et al. 2011). Similar behaviors, in terms of nuclear localization of MAL and histone acetylation, were also reported in fibroblast cell lines (Jain et al. 2013). Hence, cell-based assays are emerging as powerful tools for interrogating molecular pathways that regulate fate decision in stem cells. Similar control of stem cell fate was reported in the case of human MSCs, which committed to adipogenesis when seeded as single cells on small islands, but differentiated into osteoblasts on large islands, maximizing cell spreading (McBeath et al. 2004). This phenotype was regulated by the assembly of the cytoskeleton and its contractility in a RhoA-dependent process. Interestingly, fine differences in adhesive geometry (e.g., star shapes vs. flower shapes) were found to mediate similar phenotypes, as a result of cytoskeleton assembly and of differences in the recruitment of FAs (Kilian et al. 2010; Song et al. 2011). Therefore, as with the control of cell division, beyond cell shape, it is the regulation of the adhesion–cytoskeleton machinery and its underlying biomechanical activity that seems to regulate fate decision.

Considering the role of the actin cytoskeleton and its remodeling in mediating mechanotransduction cues, it is not surprising that other cell phenotypes have been found to depend on ECM geometry. Cell shape enables the EMT of mouse mammary epithelial cells (Nelson et al. 2008); this effect is potentiated by MMP3 treatment but is independent of exposure to TGF-β. Ciliogenesis, the process of formation of a primary cilium, is mediated by rearrangements in the cytoskeleton (Pitaval et al. 2010): at low cell density, for which cells in culture are typically proliferative, few cells present a clear cilium, but the proportion of ciliated cells gradually increases as the density does. Whether this effect is mediated by cell contact and associated growth inhibition or whether it results from the restricted cell spreading typical of dense cultures is not clear. Using a micropattern assay, it was found that cells cultured on small adhesive islands (750 μm^2) presented cilia more frequently than cells allowed to spread on large islands (3000 μm^2). The length of the cilia was also found to increase for rounded cells. When cells were treated with drugs perturbing cytoskeletal assembly (cytochalasin or ROCK inhibitor Y27632) or contractility (blebbistatin), the incidence of ciliated cells and the average cilium length increased, even though the cells were seeded on large islands.

Finally, important crosstalks and synergies between geometrical cues and other parameters of the microenvironment are likely to occur *in vitro* as well as *in vivo*, and studying such interactions requires the development of novel high-throughput platforms. An example of such crosstalks was evidenced by a micropatterned polymer brush platform, in which cell shape was controlled independently of the chemistry of the surrounding coating defining the ECM pattern (with a range of neutral, hydrophilic, and negatively charged brushes) (Tan et al. 2013). Interestingly, it was found that matrix geometry was the primary factor directing the fate of primary keratinocytes (differentiation, assessed via the expression of the cornified envelop marker involucrin), but that negatively charged coatings also had an impact on cell differentiation, correlating with a destabilization of FAs, independent of cell shape. The combination of biochemical, geometrical, morphological, and mechanical parameters that can be explored increases rapidly, and becomes impossible to probe with conventional biomaterials platforms. The development of novel systems allowing high-throughput screening of multiple physicochemical parameters, together with appropriate imaging and analysis tools, is therefore required.

19.4.3 Sensing of Microenvironment Geometry by Cell Clusters and Organoids

At the multicellular level, geometrical cues have been shown to control the organization, structure, and functioning of biological systems. As with single cells, tissues and organs usually display well-defined shapes and architectures *in vivo* – information that is lost *in vitro*. Micropatterning has therefore been used to control and recreate some of these structures in multicellular assemblies. Interestingly, similar rules seem to apply to the organization of the cytoskeleton of cell clusters as for single cells. Hence, FAs accumulate at the boundary between nonadhesive and adhesive areas and stress fibers form in these regions, particularly when bridging nonadhesive areas (Gautrot et al. 2012). Obviously, cell–cell adhesions, such as those mediated by cadherins, also contribute to the overall cluster architecture, and the two types of adhesion (cell–matrix and cell–cell) crosstalk to define the precise structure of the multicellular cluster. Such crosstalks between different adhesion types have been reported for cardiomyocytes (McCain et al. 2012) as well as for endothelial cells (Liu et al. 2010) and result in cooperative mechanical coupling of cell–cell adhesions to cell–matrix anchoring sites through the cytoskeleton. E-cadherin-mediated junctions in MCF10A cells are directed by the ECM geometry, highlighting the importance of the balance of adhesive forces for the overall organization of cell clusters (Tseng et al. 2012). Such a balance of forces doesn't just structure multicellular assemblies, but also results in the polarization of subcellular components and organelles. In primary rat astrocytes, it was found that the nuclei of cells at the periphery of small clusters (total area 11 000 μm^2, capable of accommodating 5–10 cells) were localized close to cell–cell junctions rather than the cluster edge (Dupin et al. 2009). This process was found to be mediated by the assembly and architecture of the intermediate filament network (Dupin et al. 2011). In addition, the centrosome was positioned close to the nucleus and directed toward the cluster edge, as in wound beds. This behavior was found to be controlled by the level of calcium present in the medium and by the formation of N-cadherin-mediated junctions.

Several types of cell–cell junction may also contribute to maintaining the architecture of cell clusters. Hence, primary keratinocytes form compact cell clusters on 100 μm^2 adhesive islands, with differentiated cells found in the clusters' centers (see Figure 19.10) (Gautrot et al. 2012). This cellular self-partitioning process, like the partitioning of cells in embryoid bodies (Steinberg 2007), has been found to result from the balance of matrix-adhesive forces and cell–cell adhesions. In addition, two distinct types of adhesions – adherens junctions and desmosomes – play redundant roles in maintaining this architecture. When the center of such ECM patterns was replaced by nonadhesive areas (coated with protein-resistant polymer brushes), the partitioning process was improved, as differentiating cells typically lose integrin expression and upregulate the expression of cadherins. Hence, this platform allows the simultaneous study and quantification of molecules and pathways controlling differentiation and partitioning in the epidermis, thereby mimicking some of the architectural features of this tissue *in vitro* in a human context. As a proof of concept, the partitioning of cancer cells isolated from patients presenting different grades of squamous cell carcinoma (SCC) was studied using this assay and was found to closely match the level of perturbation of the diseased tissue from which cells had been derived (Gautrot et al. 2012).

Human micro-epidermis

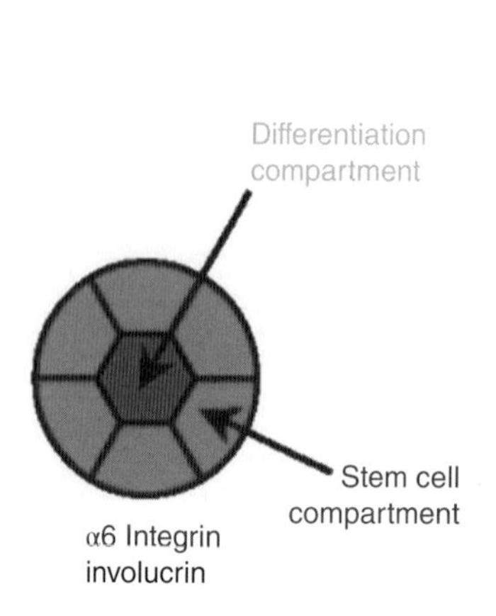

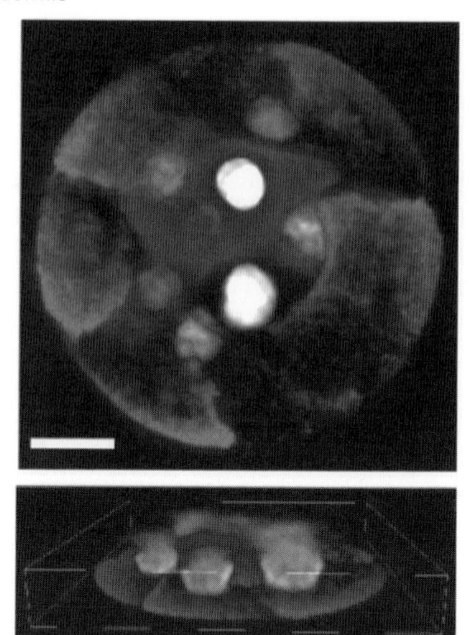

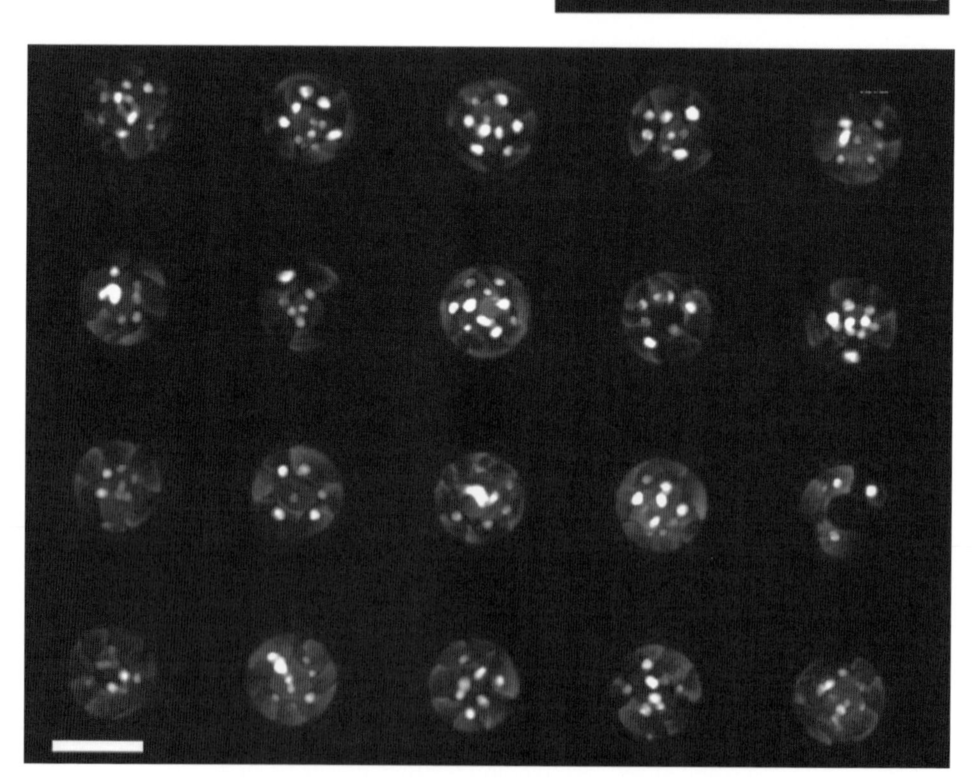

Figure 19.10 Arrays of microepidermis with controlled partitioning of differentiated keratinocytes. *Source:* Gautrot et al. (2012). Reproduced with permission of Elsevier. (*See insert for color representation of the figure.*)

Other mechanisms by which cell partitioning may occur have also been highlighted. MSCs spreading on large micropatterned adhesive islands differentiate into distinct lineages depending on the local geometry of the pattern (Ruiz and Chen 2008). Local geometries that are found to result in contractile cell phenotypes lead to osteogenic differentiation, whereas those that result in less contractile, more rounded morphologies lead to adipogenesis. Similarly, mammary epithelial cells located at regions of micropatterns that induce higher cell contractility express markers of EMT such as smooth-muscle alpha-actin (α-SMA) and vimentin (Gomez et al. 2010). Interestingly, this behavior also correlates with higher nuclear localization of the factor MTRFA (also known as MAL). The local geometry of multicellular assemblies is often associated with variations in local cell densities and the resulting changes in concentrations of cytokines or autocrine signaling molecules. This has been found to control the occurrence of mammary branching: mammary epithelial cells located at apices of patterns at which the local concentration of inhibitory morphogenetic factors was low initiated branching and expressed markers such as vimentin and keratin 8 (Nelson et al. 2006).

19.5 Conclusion

To broaden their use in the field of regenerative medicine, biomaterials should capture the complex interplay of biochemical, mechanical, topographical, and morphological properties of the *in vivo* cell microenvironment in order to improve our control of cell phenotype and the generation of structures resembling tissues to be regenerated. Our understanding of such intricate interactions has improved greatly in the last few decades, but significant questions remain open. A detailed understanding of how cells sense mechanical properties at the nanoscale remains to be eluciated. Though clustering of ligands has been reported in several cases, this is insufficient to account for the behavior of cells in many situations. In particular, the molecular structure and dynamics of the “strain sensor” that allows cells to either engage mechanically with their substrate (at least in 2D systems) or to disassemble the growing protrusion and “look” somewhere else remains to be clearly identified. The response of cells to topography, though widely documented, is also poorly understood. Beyond a few mechanistic studies, little is known of the mechanism via which FAs (and potentially other structures and proteins) sense the topography of a substrate that is homogenous from a chemical and biochemical point of view. Finally, a detailed understanding of how cells sense the nano- to microscale geometry of the adhesive landscape is still required. Importantly, this question touches on the detailed nature and role of FAs and the relationship between their composition, size, and shape. Addressing these important questions will require novel biological tools and a biomaterials platform that allows the control of cell adhesion and cell–biomaterials interactions. However, this challenge should be rewarding, as such an understanding underpins the development of novel biomaterials for tissue engineering and regenerative medicine.

References

Abagnale, G., M. Steger, V. H. Nguyen, N. Hersch, A. Sechi, S. Joussen, et al. 2015. “Surface topography enhanced differentiation of mesenchymal stem cells towards osteogenic and adipogenic lineages.” *Biomaterials* **61**: 316–26. doi:10.1016/j.biomaterials.2015.05.030.

Ahn, E. H., Y. Kim, Kshitiz, S. S. An, J. Afzal, S. S. C. Lee, et al. 2014. “Spatial control of adult stem cell fate using nanotopographic cues.” *Biomaterials* **35**: 2401–10. doi:10.1016/j.biomaterials.2013.11.037.

Albelda, S. M. and C. A. Buck. 1990. “Integrins and other cell adhesion molecules.” *FASEB Journal* **4**(11): 2868–80. PMID:2199285.

Assoian, R. K. and E. A. Klein. 2008. “Growth control by intracellular tension and extracellular stiffness.” *Trends in Cell Biology* **18**(7): 347–52. doi:10.1016/j.tcb.2008.05.002.

Balaban, N. Q., U. S. Schwarz, D. Riveline, P. Goichberg, G. Tzur, I. Sabanay, et al. 2001. “Force and focal adhesion assembly: a close relationship studied using elastic micropatterned substrates.” *Nature Cell Biology* **3**: 466–72. doi:10.1038/35074532.

Biggs, M. J. P., R. G. Richards, N. Gadegaard, C. D. W. Wilkinson, R. O. C. Oreffo, and M. J. Dalby. 2009. “The use of nanoscale topography to modulate the dynamics of adhesion formation in primary osteoblasts and ERK/MAPK singalling in STRO-1+ enriched skeletal stem cells.” *Biomaterials* **30**: 5094–103. doi:10.1016/j.biomaterials.2009.05.049.

Bott, K., Z. Upton, K. Schrobback, M. Ehrbar, J. A. Hubbell, M. P. Lutolf, and S. C. Rizzi. 2010. “The effect of matrix characteristics on fibroblasts proliferation in 3D gels.” *Biomaterials* **31**: 8454–64. doi:10.1016/j.biomaterials.2010.07.046.

Cassidy, J. W., J. N. Roberts, C.-A. Smith, M. Robertson, K. White, M. Biggs, et al. 2014. “Osteogenic lineage restriction by osteoprogenitors cultured on nanometric grooved surfaces: the role of focal adhesion maturation.” *Acta Biomaterialia* **10**: 651–60. doi:10.1016/j.actbio.2013.11.008.

Cavalcanti-Adam, E. A., A. Micoulet, J. Blummel, J. Auernheimer, H. Kessler, and J. P. Spatz. 2006. “Lateral spacing of integrin ligands influences cell spreading and focal adhesion assembly.” *European Journal of Cell Biology* **85**: 219–24. doi:10.1016/j.ejcb.2005.09.011.

Chaudhuri, O., S. T. Koshy, C. Branco da Cunha, J.-W. Shin, C. S. Verbeke, et al. 2014. “Extracellular matrix stiffness and composition jointly regulate the induction of malignant phenotypes in mammary epithelium.” *Nature Materials* **13**: 970–8. doi:10.1038/nmat4009.

Chaudhuri, O., L. Gu, M. Darnell, D. Klumpers, S. A. Bencherif, J. C. Weaver, et al. 2015. “Substrate stress relaxation regulates cell spreading.” *Nature Communications* **6**: 6364. doi:10.1038/ncomms7365.

Chen, C. S., M. Mrksich, S. Huang, G. M. Whitesides, and D. E. Ingber. 1997. “Geometric control of cell life and death.” *Science* **276**: 1425–8. PMID:9162012.

Chen, W., L. G. Villa-Diaz, Y. Sun, S. Weng, J. K. Kim, R. H. W. Lam, et al. 2012. “Nanotopography influences adhesion, spreading, and self-renewal of human embryonic stem cells.” *ACS Nano* **6**(5): 4094–103. doi:10.1021/nn3004923.

Choi, C. K., M. Vicente-Manzanares, J. Zareno, L. A. Whitmore, A. Mogilner, and A. R. Horwitz. 2008. “Actin and a-actinin orchestrate the assembly and maturation of nascent adhesions in a myosin II motor-independent manner.” *Nature Cell Biology* **10**: 1039–50. doi:10.1038/ncb1763.

Christopherson, G. T., H. Song, and H.-Q. Mao. 2009. “The influence of fiber diameter of electrospun substrates on neural stem cell differentiation and proliferation.” *Biomaterials* **30**: 556–64. doi:10.1016/j.biomaterials.2008.10.004.

Collin, O., P. Tracqui, A. Stephanou, Y. Usson, J. Clement-Lacroix, and E. Planus. 2006. “Spatiotemporal dynamics of actin-rich adhesion microdomains: influence of substrate flexibility.” *Journal of Cell Science* **119**(9): 1914–25. doi:10.1242/jcs.02838.

Connelly, J., J. E. Gautrot, B. Trappmann, D. W. M. Tan, G. Donati, W. T. S. Huck, and F. M. Watt. 2010. "Actin and SRF transduce physical cues from the microenvironment to regulate epidermal stem cell fate decisions." *Nature Cell Biology* **12**: 711–18. doi:10.1038/ncb2074.

Connelly, J., A. Mishra, J. E. Gautrot, and F. M. Watt. 2011. "Shape-induced terminal differentiation of human epidermal stem cells requires p38 and is regulated by histone acetylation." *PLoS ONE* **6**(11): e27259. doi:10.1371/journal.pone.0027259.

Courtemanche, N., J. Y. Lee, T. D. Pollard, and E. C. Greene. 2013. "Tension modulates actin filament polymerization mediated by formin and profilin." *Proceedings of the National Academy of Sciences of the United States of America* **110**(24): 9752–7. doi:10.1073/pnas.1308257110.

Coyer, S. R., A. Singh, D. W. Dumbauld, D. A. Calderwood, S. W. Craig, E. Delamarche, and A. Garcia. 2012. "Nanopatterning reveals an ECM area threshold for focal adhesion assembly and force transmission that is regulated by integrin activation and cytoskeleton tension." *Journal of Cell Science* **125**: 5110–23. doi:10.1242/jcs.108035.

Cukierman, E., R. Pankov, D. R. Stevens, and K. M. Yamada. 2001. "Taking cell-matrix adhesions to the third dimension." *Science* **294**: 1708–12. doi:10.1126/science.1064829.

Dalby, M. J., N. Gadegaard, R. Tare, A. Andar, M. O. Riehle, P. Herzyk, et al. 2007. "The control of human mesenchymal cell differentiation using nanoscale symmetry and disorder." *Nature Materials* **6**: 997–1003. doi:10.1038/nmat2013.

Das, R. K., O. F. Zouani, C. Labrugere, R. Oda, and M.-C. Durrieu. 2013. "Influence of nanohelical shape and periodicity on stem cell fate." *ACS Nano* **7**(4): 3351–61. doi:10.1021/nn4001325.

Deeg, J. A., I. Louban, D. Aydin, C. Selhuber-Unkel, H. Kessler, and J. P. Spatz. 2011. "Impact of local versus global ligand density on cellular adhesion." *Nano Letters* **11**: 1469–76. doi:10.1021/nl104079r.

Deshpande, V. S., M. Mrksich, R. M. McMeeking, and A. G. Evans. 2008. "A bio-mechanical model for coupling cell contractility with focal adhesion formation." *Journal of the Mechanics and Physics of Solids* **56**: 1484–510. doi:10.1016/j.jmps.2007.08.006.

Discher, D. E., P. Janmey, and Y.-L. Wang. 2005. "Tissue cells feel and respond to the stiffness of their substrate." *Science* **310**: 1139–43. doi:10.1126/science.1116995.

Discher, D. E., D. J. Mooney, and P. W Zandstra. 2009. "Growth factors, matrices, and forces combine and control stem cells." *Science* **324**: 1673–7. doi:10.1126/science.1171643.

Du, J., X. Chen, X. Liang, G. Zhang, J. Xu, L. He, et al. 2011. "Integrin activation and internalization on soft ECM as a mechanism of induction of stem cell differentiation by ECM elasticity." *Proceedings of the National Academy of Sciences of the United States of America* **108**(23): 9466–71. doi:10.1073/pnas.1106467108.

Dupin, I., E. Camand, and S. Etienne-Manneville. 2009. "Classical cadherins control nucleus and centrosome position and cell polarity." *Journal of Cell Biology* **185**(5): 779–86. doi:10.1083/jcb.200812034.

Dupin, I., Y. Sakamoto, and S. Etienne-Manneville. 2011. "Cytoplasmic intermediate filaments mediate actin-driven positioning of the nucleus." *Journal of Cell Science* **124**(6): 865–72. doi:10.1242/jcs.076356.

Dupont, S., L. Morsut, M. Aragona, E. Enzo, S. Giulitti, M. Cordenonsi, et al. 2011. "Role of YAP/TAZ in mechanotransduction." *Nature* **474**: 179–83. doi:10.1038/nature10137.

Elliott, H., R. S. Fischer, K. A. Myers, R. A. Desai, L. Gao, C. S. Chen, et al. 2015. "Myosin II controls cellular branching morphogenesis and migration in three dimensions by minimizing cell-surface curvature." *Nature Cell Biology* **17**(2): 137–47. doi:10.1038/ncb3092.

Engler, A. J., S. Sen, H. L. Sweeney, and D. E. Discher. 2006. "Matrix elasticity directs stem cell lineage specification." *Cell* **126**: 677–89. doi:10.1016/j.cell.2006.06.044.

Fink, J., N. Carpi, T. Betz, A. Betard, M. Chebah, A. Azioune, et al. 2011. "External forces control mitotic spindle positioning." *Nature Cell Biology* **13**(7): 771–8. doi:10.1038/ncb2269.

Fraley, S. I., Y. Feng, R. Krishnamurthy, D.-H. Kim, A. Celedon, G. D. Longmore, and D. Wirtz. 2010. "A distinctive role for focal adhesion proteins in three-dimensional cell motility." *Nature Cell Biology* **12**(6): 598–604. doi:10.1038/ncb2062.

Freida, D., S. Lecourt, A. Cras, V. Vanneaux, G. Letort, X. Gidrol, et al. 2013. "Human bone marrow mesenchymal stem cells regulate biased DNA segregation in response to cell adhesion asymmetry." *Cell Reports* **5**(3): 601–60. doi:10.1016/j.celrep.2013.09.019.

Fu, J., Y.-K. Wang, M. T. Yang, R. Desai, X. Yu, Z. Liu, and C. S. Chen. 2010. "Mechanical regulation of cell function with geometrically modulated elastomeric substrates." *Nature Methods* **7**(9): 733–6. doi:10.1038/nmeth.1487.

Galbraith, C. G., K. M. Yamada, and M. P. Sheetz. 2002. "The relationship between force and focal complex development." *Journal of Cell Biology* **159**(4): 695–705. doi:10.1083/jcb.200204153.

Gardel, M. L., B. Sabass, L. Ji, G. Danuser, U. S. Schwarz, and C. M. Waterman. 2008. "Traction stress in focal adhesions correlates biphasically with actin retrograde flow speed." *Journal of Cell Biology* **183**(6): 999–1005. doi:10.1083/jcb.200810060.

Gautrot, J. E., B. Trappmann, F. Oceguera-Yanez, J. Connelly, X. He, F. M. Watt, and W. T. S. Huck. 2010. "Exploiting the superior protein resistance of polymer brushes to control single cell adhesion and polarisation at the micron scale." *Biomaterials* **31**: 5030–41. doi:10.1016/j.biomaterials.2010.02.066.

Gautrot, J. E., C. Wang, X. Liu, S. J. Goldie, B. Trappmann, W. T. S. Huck, and F. M. Watt. 2012. "Mimicking normal tissue architecture and perturbation in cancer with engineered micro-epidermis." *Biomaterials* **33**(21): 5221–9. doi:10.1016/j.biomaterials.2012.04.009.

Gautrot, J. E., J. Malmstrom, M. Sundh, C. Margadant, A. Sonnenberg, and D. S. Sutherland. 2014. "The nanoscale geometrical maturation of focal adhesions controls stem cell differentiation and mechanotransduction." *Nano Letters* **14**: 3945–52. doi:10.1021/nl501248y.

Geiger, B., J. P. Spatz, and A. D. Bershadsky. 2009. "Environmental sensing through focal adhesions." *Nature Reviews. Molecular Cell Biology* **10**: 21–33. doi:10.1038/nrm2593.

Georges, P. C. and P. A. Janmey. 2005. "Cell type-specific response to growth on soft materials." *Journal of Applied Physiology* **98**: 1547–53. doi:10.1152/japplphysiol.01121.2004.

Giaever, I. and C. R. Keese. 1983. "Behavior of cells at fluid interfaces." *Proceedings of the National Academy of Sciences of the United States of America* **80**: 219–22. PMID:6571995.

Gilbert, P. M., K. L. Havenstrite, K. E. G. Magnusson, A. Sacco, N. A. Leonardi, P. Kraft, et al. 2010. "Substrate elasticity regulates skeletal muscle stem cell self-renewal in culture." *Science* **329**: 1078–81. doi:10.1126/science.1191035.

Gomez, E. W., Q. K. Chen, N. Gjorevski, and C. M. Nelson. 2010. "Tissue geometry patterns epithelial-mesenchymal transition via intercellular mechanotransduction." *Journal of Cellular Biochemistry* **110**: 44–51. doi:10.1002/jcb.22545.

Guilak, F., D. M. Cohen, B. T. Estes, J. M. Gimble, W. Liedtke, and C. S. Chen. 2009. "Control of stem cell fate by physical interactions with the extracellular matrix." *Cell Stem Cell* **5**: 17–26. doi:10.1016/j.stem.2009.06.016.

Harburger, D. S. and D. A. Calderwood. 2009. "Integrin signalling at a glance." *Journal of Cell Science* **122**: 159–63. doi:10.1242/jcs.018093.

Harris, A. K., P. Wild, and D. Stopak. 1980. "Silicone rubber substrata: a new wrinkle in the study of cell locomotion." *Science* **208**(4440): 177–9. PMID:6987736.

He, L., S. Liao, D. Quan, K. Ma, C. Chan, S. Ramakrishna, and J. Lu. 2010. "Synergistic effects of electrospun PLLA fiber dimension and pattern on neonatal mouse cerebellum C17.2 stem cells." *Acta Biomaterialia* **6**: 2960–9. doi:10.1016/j.actbio.2010.02.039.

Hegedus, B., F. Marga, K. Jakab, K. L. Sharpe-Timms, and G. Forgacs. 2006. "The interplay of cell-cell and cell-matrix interactions in the invasive properties of brain tumors." *Biophysics Journal* **91**: 2708–16. doi:10.1529/biophysj.105.077834.

Higgins, A. M., B. L. Banik, and J. L. Brown. 2015. "Geometry sensing through POR1 regulates Rac1 activity controlling early osteoblast differentiation in response to nanofiber diameter." *Integrative Biology* **7**: 229–36. doi:10.1039/c4ib00225c.

Holst, J., S. Watson, M. S. Lord, S. S. Eamegdool, D. V. Bax, L. B. Nivison-Smith, et al. 2010. "Substrate elasticity provides mechanical signals for the expansion of hemopoietic stem and progenitor cells." *Nature Biotechnology* **28**(10): 1123–8. doi:10.1038/nbt.1687.

Huang, J., S. V. Grater, F. Corbellini, S. Rinck, E. Bock, R. Kemkemer, et al. 2009. "Impact of order and disorder in RGD nanopatterns on cell adhesion." *Nano Letters* **9**(3): 1111–16. doi:10.1021/nl803548b.

Huda, S., S. Soh, D. Pilans, M. Byrska-Bishop, J. Kim, G. Wilk, et al. 2012. "Microtubule guidance tested through controlled cell geometry." *Journal of Cell Science* **125**: 5790–9. doi:10.1242/jcs.110494.

Huebsch, N., P. R. Arany, P. R. Mao, D. Shvartsman, O. A. Ali, S. A. Bencherif, et al. 2010. "Harnessing traction-mediated manipulation of the cell/matrix interface to control stem-cell fate." *Nature Materials* **9**: 518–26. doi:10.1038/nmat2732.

Humphries, J. D., P. Wang, C. Streuli, B. Geiger, M. J. Humpries, and C. Ballestrem. 2007. "Vinculin controls focal adhesion formation by direct interactions with talin and actin." *Journal of Cell Biology* **179**: 1043–57. doi:10.1083/jcb.200703036.

Ingber, D. E. 2002. "Cancer as a disease of epithelial-mesenchymal interactions and extracellular matrix regulation." *Differentiation* **70**: 547–60. doi:10.1046/j.1432-0436.2002.700908.x.

Jain, N., K. V. Iyer, A. Kumar, and G. V. Shivashankar. 2013. "Cell geometric constraints induce modular gene-expression patterns via redistribution of HDAC3 regulated by actomyosin contractility." *Proceedings of the National Academy of Sciences of the United States of America* **110**(28): 11 349–54. doi:10.1073/pnas.1300801110.

Jaiswal, D. and J. L. Brown. 2012. "Nanofiber diameter-dependent MAPK activity in osteoblasts." *Journal of Biomedical Materials Research. Part A* **100**(11): 2921–8. doi:10.1002/jbm.a.34234.

Jiang, X., D. A. Bruzewicz, A. P. Wong, M. Piel, and G. M. Whitesides. 2005. "Directing cell migration with asymmetric micropatterns." *Proceedings of the National Academy of Sciences of the United States of America* **102**(4): 975–8. doi:10.1073/pnas.0408954102.

Kanchanawong, P., G. Shtengel, A. M. Pasapera, E. B. Ramko, M. W. Davidson, H. F. Hess, and C. M. Waterman. 2010. "Nanoscale architecture of integrin-based cell adhesions." *Nature* **468**: 580–4. doi:10.1038/nature09621.

Kane, R. S., S. Takayama, E. Ostuni, D. E. Ingber, and G. M. Whitesides. 1999. "Patterning proteins and cells using soft lithography." *Biomaterials* **20**: 2363–76. PMID:10614942.

Keese, C. R. and I. Giaever. 1991. "Substrate mechanics and cell spreading." *Experimental Cell Research* **195**: 528–32. PMID:2070833.

Khetan, S., M. Guvendiren, W. R. Legant, D. M. Cohen, C. S. Chen, and J. A. Burdick. 2013. "Degradation-mediated cellular traction directs stem cell fate in covalently crosslinked three-dimensional hydrogels." *Nature Materials* **12**: 458–65. doi:10.1038/nmat3586.

Kilian, K. A., B. Bugarija, B. T. Lahn, and M. Mrksich. 2010. "Geometric cues for directing the differentiation of mesenchymal stem cells." *Proceedings of the National Academy of Sciences of the United States of America* **107**(11): 4872–7. doi:10.1073/pnas.0903269107.

Kim, D.-H. and D. Wirtz. 2015. "Cytoskeletal tension induces the polarized architecture of the nucleus." *Biomaterials* **48**: 161–72. doi:10.1016/j.biomaterials.2015.01.023.

Kim, D.-H., E. A. Lipke, P. Kim, R. Cheong, S. Thompson, M. Delannoy, et al. 2010. "Nanoscale cues regulated the structure and function of macroscopic cardiac tissue constructs." *Proceedings of the National Academy of Sciences of the United States of America* **107**(2): 565–70. doi:10.1073/pnas.0906504107.

Klein, E. A., L. Yin, D. Kothapalli, P. Castagnino, F. J. Byfield, T. Xu, et al. 2009. "Cell-cycle control by physiological matrix elasticity and in vivo tissue stiffening." *Current Biology* **19**: 1511–18. doi:10.1016/j.cub.2009.07.069.

Kong, H. J., T. R. Polte, E. Alsberg, and D. J. Mooney. 2005. "FRET measurements of cell-traction forces and nano-scale clustering of adhesion ligands varied by substrate stiffness." *Proceedings of the National Academy of Sciences of the United States of America* **102**: 4300–5. doi:10.1073/pnas.0405873102.

Kuo, C. W., D.-Y. Chueh, and P. Chen. 2014. "Investigation of size-dependent cell adhesion on nanostructured interfaces." *Journal of Nanobiotechnology* **12**: 54. doi:10.1186/s12951-014-0054-4.

Laakso, J. M., J. H. Lewis, H. Shuman, and E. M. Ostap. 2008. "Myosin I can act as a molecular force sensor." *Science* **321**: 133–6. doi:10.1126/science.1159419.

Lautscham, L. A., C. Y. Lin, V. Auernheimer, C. A. Naumann, W. H. Goldmann, and B. Fabry. 2014. "Biomembrane-mimicking lipid bilayer system as a mechanically tunable cell substrate." *Biomaterials* **35**: 3198–207. doi:10.1016/j.biomaterials.2013.12.091.

Lee, C. R., A. J. Grodzinsky, and M. Spector. 2001. "The effects of cross-linking of collagen-glycosaminoglycan scaffolds on compressive stiffness, chondrocyte-mediated contraction, proliferation and biosynthesis." *Biomaterials* **22**: 3145–54. PMID:11603587.

Legant, W. R., J. S. Miller, B. L. Blakely, D. M. Cohen, G. M. Genin, and C. S. Chen. 2010. "Measurement of mechanical tractions exerted by cells in three-dimensional matrices." *Nature Methods* 7(12): 969–71. doi:10.1038/nmeth.1531.

Legant, W. R., C. K. Choi, J. S. Miller, L. Shao, L. Gao, E. Betzig, and C. S. Chen. 2013. "Multidimensional traction force microscopy reveals out-of-plane rotational moments about focal adhesions." *Proceedings of the National Academy of Sciences of the United States of America* **110**(3): 881–6. doi:10.1073/pnas.1207997110.

Leight, J. L., M. A. Wozniak, S. Chen, M. L. Lynch, and C. S. Chen. 2012. "Matrix rigidity regulates a switch between TGF-b1-induced apoptosis and epithelial-mesenchymal transition." *Molecular Biology of the Cell* **23**(5): 781–91. doi:10.1091/mbc.E11-06-0537.

Lepzelter, D., O. Bates, and M. Zaman. 2012. “Integrin clustering in two and three dimensions.” *Langmuir* **28**: 5379–86. doi:10.1021/la203725a.

Levental, K. R., H. Yu, L. Kass, J. N. Lakins, M. Egeblad, J. Erler, et al. 2009. “Matrix crosslinking forces tumor progression by enhancing integrin signalling.” *Cell* **139**: 891–906. doi:10.1016/j.cell.2009.10.027.

Liu, Z., J. L. Tan, D. M. Cohen, M. T. Yang, N. J. Sniadecki, S. A. Ruiz, et al. 2010. “Mechanical tugging force regulates the size of cell-cell junctions.” *Proceedings of the National Academy of Sciences of the United States of America* **107**(22): 9944–9. doi:10.1073/pnas.0914547107.

Liu, Y., R. Medda, Z. Liu, K. Galior, K. Yehl, J. P. Spatz, et al. 2014. “Nanoparticle tension probes patterned at the nanoscale: impact of integrin clustering on force transmission.” *Nano Letters* **14**: 5539–46. doi:10.1021/nl501912g.

Liu, J., Y. Wang, W. I. Goh, H. Goh, M. A. Baird, S. Ruehland, et al. 2015. “Talin determines the nanoscale architecture of focal adhesions.” *Proceedings of the National Academy of Sciences of the United States of America* **112**(35): E4864–73. doi:10.1073/pnas.1512025112.

Lo, C.-M., H.-B. Wang, M. Dembo, and Y.-L. Wang. 2000. “Cell movement is guided by the rigidity of the substrate.” *Biophysics Journal* **79**: 144–52. doi:10.1016/S0006-3495(00)76279-5.

Lowery, J. L., N. Datta, and G. C. Rutledge. 2010. “Effect of fiber diameter, pore size and seeding method on growth of human dermal fibroblasts in electrospun poly(e-caprolactone) fibrous mats.” *Biomaterials* **31**: 491–504. doi:10.1016/j.biomaterials.2009.09.072.

Lutolf, M. P. and J. A. Hubbell. 2005. “Synthetic biomaterials as instructive extracellular microenvironments for morphogenesis in tissue engineering.” *Nature Biotechnology* **23**: 47–55. doi:10.1038/nbt1055.

Malmstrom, J., B. Christensen, H. P. Jakobsen, J. Lovmand, R. Foldbjerg, E. S. Sorensen, and D. S. Sutherland. 2010. “Large area protein patterning reveals nanoscale control of focal adhesion development.” *Nano Letters* **10**: 686–94. doi:10.1021/nl903875r.

Malmstrom, J., J. Lovmand, S. Kristensen, M. Sundh, M. Duch, and D. S. Sutherland. 2011. “Focal complex maturation and bridging on 200 nm vitronectin but not fibronectin patches reveal different mechanisms of focal adhesion formation.” *Nano Letters* **11**: 2264–71. doi:10.1021/nl200447q.

Mandal, B. B. and S. C. Kundu. 2009. “Cell proliferation and migration in silk fibroin 3D scaffolds.” *Biomaterials* **30**: 2956–65. doi:10.1016/j.biomaterials.2009.02.006.

McBeath, R., D. M. Pirone, C. M. Nelson, K. Bhadriraju, and C. S. Chen. 2004. “Cell shape, cytoskeletal tension and RhoA regulate stem cell lineage commitment.” *Developmental Cell* **6**: 483–95. PMID:15068789.

McCain, M. L., H. Lee, Y. Aratyn-Schaus, A. G. Kleber, and K. K. Parker. 2012. “Cooperative coupling of cell-matrix and cell-cell adhesions in cardiac muscle.” *Proceedings of the National Academy of Sciences of the United States of America* **109**: 9881–6. doi:10.1073/pnas.1203007109.

McMurray, R. J., N. Gadegaard, P. M. Tsimbouri, K. V. Burgess, L. E. McNamara, R. Tare, et al. 2011. “Nanoscale surfaces for the long term maintenance of mesenchymal stem cell phenotype and multipotency.” *Nature Materials* **10**: 637–44. doi:10.1038/nmat3058.

Miron-Mendoza, M., J. Seemann, and F. Grinnell. 2010. “The differential regulation of cell motile activity through matrix stiffness and porosity in three dimensional collagen matrices.” *Biomaterials* **31**: 6425–35. doi:10.1016/j.biomaterials.2010.04.064.

Miyazaki, M., M. Chiba, H. Eguchi, T. Ohki, and S. Ishiwata. 2015. "Cell-sized spherical confinement induces the spontaneous formation of contractile actomyosin rings in vitro." *Nature Cell Biology* **17**(4): 480–9. doi: 10.1038/ncb3142.

Nelson, C. M., M. M. VanDuijn, J. L. Inman, D. A. Fletcher, and M. J. Bissell. 2006. "Tissue geometry determines sites of mammary branching morphogenesis in organotypic cultures." *Science* **314**: 298–300. doi:10.1126/science.1131000.

Nelson, C. M, D. Khauv, M. J. Bissell, and D. C. Radisky. 2008. "Change in cell shape is required for matrix metalloproteinas-induced epithelial-mesenchymal transition of mammary epithelial cells." *Journal of Cellular Biochemistry* **105**: 25–33. doi:10.1002/jcb.21821.

Oliver, T., M. Dembo, and K. Jacobson. 1995. "Traction forces in locomoting cells." *Cell Motility and the Cytoskeleton* **31**(3): 225–40. doi:10.1002/cm.970310306.

Ozdemir, T., L.-C. Xu, C. Siedlecki, and J. L. Brown. 2013. "Substrate curvature sensing through myosin IIa upregulates early osteogenesis." *Integrative Biology* **5**: 1407–16. doi:10.1039/c3ib40068a.

Pasapera, A. M., S. V. Plotnikov, R. S. Fischer, L. B. Case, T. T. Egelhoff, and C. M. Waterman. 2015. "Rac1-dependent phosphorylation and focal adhesion recruitment of myosin IIA regulates migration and mechanosensing." *Current Biology* **25**: 175–86. doi:10.1016/j.cub.2014.11.043.

Pathak, A., V. S. Deshpande, R. M. McMeeking, and A. G. Evans. 2008. "The simulation of stress fibre and focal adhesion development in cells on patterned substrates." *Journal of the Royal Society, Interface* **5**: 507–24. doi:10.1098/rsif.2007.1182.

Pelham, R. J. and Y.-L. Wang. 1997. "Cell locomotion and focal adhesions are regulated by substrate flexibility." *Proceedings of the National Academy of Sciences of the United States of America* **94**: 13 661–5. PMID:9391082.

Petit, V. and J. P. Thiery. 2000. "Focal adhesions: structure and dynamics." *Biology of the Cell* **92**: 477–94. PMID:11229600.

Peyton, S. R. and A. J. Putnam. 2005. "Extracellular matrix rigidity governs smooth muscle cell motility in a biphasic fashion." *Journal of Cell Physiology* **204**: 198–209. doi:10.1002/jcp.20274.

Pitaval, A., Q. Tseng, M. Bornens, and M. Thery. 2010. "Cell shape and contractility regulate ciliogenesis in cell cycle-arrested cells." *Journal of Cell Biology* **191**(2): 303–12. doi:10.1083/jcb.201004003.

Pollard, T. D., L. Blanchoin, and R. D. Mullins. 2000. "Molecular mechanisms controlling actin filament dynamics in nonmuscle cells." *Annual Reviews in Biophysics and Biomolecular Structure* **29**: 545–76. doi:10.1146/annurev.biophys.29.1.545.

Ponti, A., M. Machacek, S. Gupton, C. Waterman-Storer, and G. Danuser. 2004. "Two distinct actin networks drive the protrusion of migrating cells." *Science* **305**: 1782–6. doi:10.1126/science.1100533.

Prager-Khoutorsky, M., A. Lichtenstein, R. Krishnan, K. Rajendran, A. Mayo, Z. Kam, et al. 2011. "Fibroblast polarization is a matrix-rigidity-dependent process controlled by focal adhesion mechanosensing." *Nature Cell Biology* **13**: 1457–65. doi:10.1038/ncb2370.

Rahmouni, S., A. Lindner, F. Rechenmacher, S. Neubauer, T. R. A. Sobahi, H. Kessler, et al. 2013. "Hydrogel micropillars with integrin selective peptidomimetic functionalized nanopatterned tops: a new tool for the measurement of cell traction forces transmitted through avb3- or a5b1-integrins." *Advanced Materials* **25**: 5869–74. doi:10.1002/adma.201301338.

Reymann, A.-C., J.-L. Martiel, T. Cambier, L. Blanchoin, R. Boujemaa-Paterski, and M. Thery. 2010. "Nucleation geometry governs ordered actin networs structures." *Nature Materials* **9**: 827–32. doi:10.1038/nmat2855.

Reymann, A.-C., R. Boujemaa-Paterski, J.-L. Martiel, C. Guerin, W. Cao, H. F. Chin, et al. 2012. "Actin network architecture can determine myosin motor activity." *Science* **336**: 1310–14. doi:10.1126/science.1221708.

Rnjak-Kovacina, J., S. G. Wise, Z. Li, P. K. M. Maitz, C. J. Young, Y. Wang, and A. S. Weiss. 2011. "Tailoring the porosity and pore size of electrospun synthetic human elastin scaffolds for dermal tissue engineering." *Biomaterials* **32**: 6729–36. doi:10.1016/j.biomaterials.2011.05.065.

Roca-Cusachs, P., A. del Rio, E. Pulkin-Faucher, N. C. Gauthier, N. Biais, and M. P. Sheetz. 2013. "Integrin-dependent force transmission to the extracellular matrix by alpha-actinin triggers adhesion maturation." *Proceedings of the National Academy of Sciences of the United States of America* **110**: E1361–70. doi:10.1073/pnas.1220723110.

Ruiz, S. A. and C. S. Chen. 2008. "Emergence of patterned stem cell differentiation within multicellular structures." *Stem Cells* **26**(11): 2921–7. doi:10.1634/stemcells.2008-0432.

Saha, K., A. J. Keung, E. F. Irwin, Y. Li, L. Little, D. V. Schaffer, and K. E. Healy. 2008. "Substrate modulus directs neural stem cell behavior." *Biophysics Journal* **95**: 4426–38. doi:10.1529/biophysj.108.132217.

Schiller, H. B., M.-R. Hermann, J. Polleux, T. Vignaud, S. Zanivan, C. C. Friedel, et al. 2013. "β1- and αv-class integrins cooperate to regulate myosin II during rigidity sensing of fibronectin-based microenvironments." *Nature Cell Biology* **15**(6): 625–36. doi:10.1038/ncb2747.

Schvartzman, M., M. Palma, J. Sable, J. Abramson, X. Hu, M. P. Sheetz, and S. J. Wind. 2011. "Nanolithographic control of the spatial organization of cellular adhesion receptors at the single-molecule level." *Nano Letters* **11**: 1306–12. doi:10.1021/nl104378f.

Sjostrom, T., M. J. Dalby, A. Hart, R. Tare, R. O. C. Oreffo, and B. Su. 2009. "Fabrication of pillar-like titania nanostructures on titanium and their interactions with human skeletal stem cells." *Acta Biomaterialia* **5**: 1433–41. doi:10.1016/j.actbio.2009.01.007.

Slater, J. H., P. J. Boyce, M. P. Jancaitis, H. E. Gaubert, A. L. Chang, M. K. Markey, and W. Frey. 2015. "Modulation of endothelial cell migration via manipulation of adhesion site growth using nanopatterned surfaces." *ACS Applied Materials & Interfaces* **7**: 4390–400. doi:10.1021/am508906f.

Song, W., H. Lu, N. Kawazoe, and G. Chen. 2011. "Adipogenic differentiation of individual mesenchymal stem cell on different geometric micropatterns." *Langmuir* **27**: 6155–62. doi:10.1021/la200487w.

Steinberg, M. S. 2007. "Differential adhesion in morphogenesis: a modern view." *Current Opinions in Genetics & Development* **17**: 281–6. doi:10.1016/j.gde.2007.05.002.

Sur, S., C. J. Newcomb, M. J. Webber, and S. I. Stupp. 2013. "Tuning supramolecular mechanics to guide neuron development." *Biomaterials* **34**: 4749–57. doi:10.1016/j.biomaterials.2013.03.025.

Swift, J., I. L. Ivanovska, A. Buxboim, T. Harada, P. C. D. P. Dingal, J. Pinter, et al. 2013. "Nuclear lamin-A scales with tissue stiffness and enhances matrix-directed differentiation." *Science* **341**: 1240104. doi:10.1126/science.1240104.

Tan, J. L., J. Tien, D. M. Pirone, D. S. Gray, K. Bhadriraju, and C. S. Chen. 2003. "Cells lying on a bed of microneedles: an approach to isolate mechanical force." *Proceedings of the National Academy of Sciences of the United States of America* **100**(4): 1484–9. doi:10.1073/pnas.0235407100.

Tan, K. Y., H. Lin, M. Ramstedt, F. M. Watt, W. T. S. Huck, and J. E. Gautrot. 2013. "Decoupling geometrical and chemical cues directing epidermal stem cell fate on polymer brush-based cell micro-patterns." *Integrative Biology* **5**: 899–910. doi:10.1039/c3ib40026c.

Thery, M., V. Racine, A. Pepin, M. Piel, Y. Chen, J.-B. Sibarita, and M. Bornens. 2005. "The extracellular matrix guides the orientation of the cell division axis." *Nature Cell Biology* **7**(10): 947–53. doi:10.1038/ncb1307.

Thery, M., A. Pepin, E. Dressaire, Y. Chen, and M. Bornens. 2006a. "Cell distribution of stress fibres in response to the geometry of the adhesive environment." *Cell Motility and the Cytoskeleton* **63**: 341–55. doi:10.1002/cm.20126.

Thery, M., V. Racine, M. Piel, A. Pepin, A. Dimitrov, Y. Chen, et al. 2006b. "Anisotropy of cell adhesive microenvironment governs cell internal organization and orientation of polarity." *Proceedings of the National Academy of Sciences of the United States of America* **103**(52): 19 771–6. doi:10.1073/pnas.0609267103.

Thid, D., M. Bally, K. Holm, S. Chessari, S. Tosatti, M. Textor, and J. Gold. 2007. "Issues of ligand accessibility and mobility in initial cell attachment." *Langmuir* **23**: 11 693–704. doi:10.1021/la701159u.

Thievessen, I., P. M. Thompson, S. Berlemont, K. M. Plevock, S. V. Plotnikov, A. Zemljic-Harpf, et al. 2013. "Vinculin-actin interaction couples actin retrograde flow to focal adhesions, but is dispensable for focal adhesion growth." *Journal of Cell Biology* **202**(1): 163–77. doi:10.1083/jcb.201303129.

Tilghman, R. W., C. R. Cowan, J. D. Mih, Y. Koryakina, D. Gioeli, J. K. Slack-Davis, et al. 2010. "Matrix rigidity regulates cancer cell growth and cellular phenotype." *PLoS One* **5**(9): e12905. doi:10.1371/journal.pone.0012905.

Trappmann, B., J. E. Gautrot, J. Connelly, D. G. T. Strange, Y. Li, M. L. Oyen, et al. 2012. "Extracellular matrix tethering regulates stem cell fate." *Nature Materials* **11**: 642–9. doi:10.1038/nmat3339.

Tseng, Q., E. Duchemin-Pelletier, A. Deshiere, M. Balland, H. Guillou, O. Filhol, and M. Thery. 2012. "Spatial organization of the extracellular matrix regulates cell-cell junction positioning." *Proceedings of the National Academy of Sciences of the United States of America* **109**(5): 1506–11. doi:10.1073/pnas.1106377109.

Tsimbouri, P., N. Gadegaard, K. Burgess, K. White, P. Reynolds, P. Herzyk, et al. 2014. "Nanotopographical effects on mesenchymal stem cell morphology and phenotype." *Journal of Cellular Biochemistry* **115**: 380–90. doi:10.1002/jcb.24673.

Tzvetkova-Chevolleau, T., A. Stephanou, D. Fuard, J. Ohayon, P. Schiavone, and P. Tracqui. 2008. "The motility of normal and cancer cells in response to the combined influence of the substrate rigidity and anisotropic microstructure." *Biomaterials* **29**: 1541–51. doi:10.1016/j.biomaterials.2007.12.016.

Ulrich, T. A., E. M. de Juan Pardo, and S. Kumar. 2009. "The mechanical rigidity of the extracellular matrix regulates the struture, motility, and proliferation of glioma cells." *Cancer Research* **69**(10): 4167–74. doi:10.1158/0008-5472.CAN-08-4859.

van Hoorn, H., R. Harkes, E. M. Spiesz, C. Sotrm, D. van Noort, B. Ladoux, and T. Schmidt. 2014. "The nanoscale architecture of force-bearing focal adhesions." *Nano Letters* **14**: 4257–62. doi:10.1021/nl5008773.

Versaevel, M., T. Grevesse, and S. Gabriele. 2012. "Spatial coordination between cell and nuclear shape within micropatterned endothelial cells." *Nature Communications* **14**(3): 671. doi:10.1038/ncomms1668.

Wang, N., J. P. Butler, and D. E. Ingber. 1993. "Mechanotransduction across the cell surface and through the cytoskeleton." *Science* **260**(5111): 1124–7. PMID:7684161.

Watt, F. M., P. W. Jordan, and C. H. O'Neill. 1988. "Cell shape controls terminal differentiation of human epidermal keratinocytes." *Proceedings of the National Academy of Sciences of the United States of America* **85**: 5576–80. PMID:2456572.

Wei, S., L. Fattet, J. H. Tsai, Y. Guo, V. H. Pai, H. E. Majeski, et al. 2015. "Matrix stiffness drives epithelial-mesenchymal transition and tumour metastasis through a TWIST-G3BP2 mechanotransduction pathway." *Nature Cell Biology* **17**(5): 678–88. doi:10.1038/ncb3157.

Wells, R. G. 2008. "The role of matrix stiffness in regulating cell behavior." *Hepatology* **47**(4): 1394–400. doi:10.1002/hep.22193.

Wen, J. H., L. G. Vincent, A. Fuhrmann, Y. S. Choi, K. C. Hribar, H. Taylor-Weiner, et al. 2014. "Interplay of matrix stiffness and protein tethering in stem cell differentiation." *Nature Materials* **13**: 979–87. doi:10.1038/nmat4051.

Wolf, K. and P. Friedl. 2011. "Extracellular matrix determinants of proteolytic and non-proteolytic cell migration." *Trends in Cell Biology* **21**(12): 736–44. doi:10.1016/j.tcb.2011.09.006.

Wolf, K., M. te Lindert, M. Krause, S. Alexander, J. te Riet, A. L. Willis, et al. 2013. "Physical limits of cell migration: control by ECM space and nuclear deformation and tuning by proteolysis and traction force." *Journal of Cell Biology* **201**(7): 1069–84. doi:10.1083/jcb.201210152.

Xu, K., H. P. Babcock, and X. Zhuang. 2012. "Dual-objective STORM reveals three-dimensional filament organization in the actin cytoskeleton." *Nature Methods* **9**(2): 185–8. doi:10.1038/nmeth.1841.

Yennek, S., M. Burute, M. Thery, and S. Tajbakhsh. 2014. "Cell adhesion geometry regulates non-random DNA segregation and asymmetric cell fates in mouse skeletal muscle stem cells." *Cell Reports* **7**: 961–70. doi:10.1016/j.celrep.2014.04.016.

Yeung, T., P. C. Georges, L. A. Flanagan, B. Marg, M. Ortiz, M. Funaki, et al. 2005. "Effect of substrate stiffness on cell morphology, cytoskeletal structure, and adhesion." *Cell Motility and the Cytoskeleton* **60**: 24–34. doi:10.1002/cm.20130.

Zaman, M. H., L. M. Trapani, A. L. Sieminski, D. MacKellar, H. Gong, R. D. Kamm, et al. 2006. "Migration of tumor cells in 3D matrices is governed by matrix stiffness along with cell-matrix adhesion and proteolysis." *Proceedings of the National Academy of Sciences of the United States of America* **103**(29): 10 889–94. doi:10.1073/pnas.0604460103.

Zhang, X., G. Jiang, Y. Cai, S. J. Monkley, D. R. Critchley, and M. P. Sheetz. 2008. "Talin depletion reveals independence of initial cell spreading from integrin activation and traction." *Nature Cell Biology* **10**: 1062–8. doi:10.1038/ncb1765.

20

Predictive Modeling in Musculoskeletal Mechanobiology

Hanifeh Khayyeri[1], *Hanna Isaksson*[1], *and Patrick J. Prendergast*[2]

[1] *Department of Biomedical Engineering, Lund University, Lund, Sweden*
[2] *Trinity Centre for Bioengineering, School of Engineering, Trinity College Dublin, Dublin, Ireland*

20.1 What is Mechanobiology? Background and Concepts

Mechanobiology is the study of how mechanical forces influence biological processes over time at the molecular, cellular, tissue, and organ levels. It is a relatively new multidisciplinary scientific field compared to its "big brother," biomechanics (the study of the structure and function of biology using the principles of mechanics), which has been studied by many prominent anatomists, mathematicians, and engineers since the time of Aristotle (Fung 1993). It was not until the 17th century, however, when Darwin presented his theory of evolution, that mechanobiology – specifically, skeletal adaptation – began to receive the attention of scientists.

The vertebrate skeleton exists in different forms, with each form adapted for its unique function. The earliest evidence of vertebrates consists in fishlike fossils from about 500 million years ago with external bony skeletons (exoskeleton) instead of the internal bony skeleton (endoskeleton) that is most common in vertebrates today (Carter and Beaupré 2001). Researchers believe that these adaptations in skeletal structure occurred because of the ability of skeletal cells to modulate their synthetic activities in response to local mechanical stimuli (Hall 2005). This same evolutionary process has led to new skeletal forms that are even more suited for their function (i.e., the new load-bearing environment).

The fact that the structure of bone in the body is arranged to withstand loads was first discovered by Willhem Roux, while studying the evolutionary theory of Darwin. Roux called his observation *the principle of functional adaptation*, which described what seemed to be a tissue-based self-organizing process. Not long after came Wolff's Law, describing how bone adapts its external shape and internal structure in response to mechanical forces. As suggested by Darwin and Roux, Wolff's concept has grounds in that form follows function, where it seems like the shape and structure of the skeleton are adapted for its required mechanical function. This type of mechanobiological response can be observed in the arms of tennis players, for example, where the bone in the racket arm is thicker than that in the other arm as a result of the higher forces to which it is regularly exposed (Figure 20.1). Another well-known example is the skeletons

Mechanobiology: Exploitation for Medical Benefit, First Edition. Edited by Simon C. F. Rawlinson.

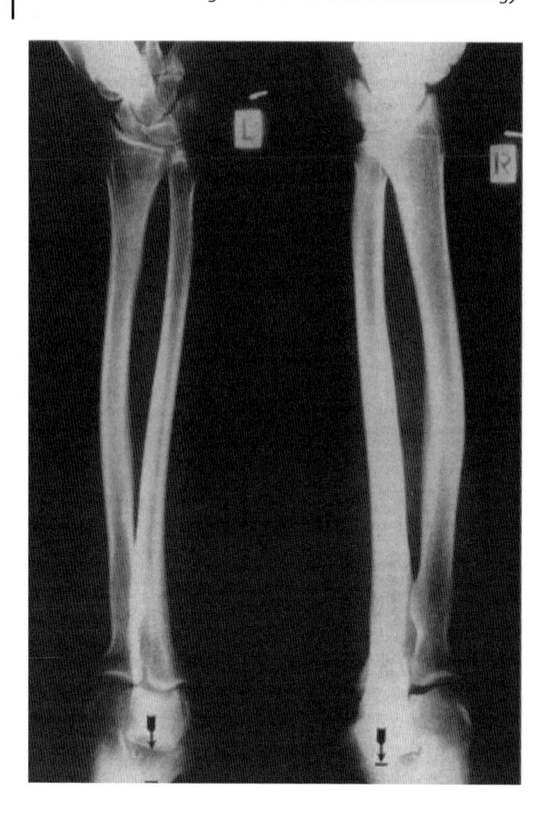

Figure 20.1 Forearms of a professional tennis player (right-hander). The image illustrates both widening and lengthening of the right arm as an adaptive response to mechanical loading through exercise. *Source:* Krahl et al. (1994). Reproduced with permission of SAGE.

of astronauts, which become less dense during longer space flights due to the low-gravity environment (if the bone is subjected to low mechanical forces, then the bone mass is reduced).

The question that arises is: If the musculoskeletal system in the body adapts to loading demands due to the mechanical milieu created within the tissues, can the adaptive response be controlled by regulating the mechanical environment? In theory, at least, this is possible. However, in order for us to successfully do so, we would need to know which mechanical cues are most important for regulating the mechanoresponse of different tissues. This question is fundamental to understanding the musculoskeletal system in order to develop treatments for musculoskeletal disorders, as well as to regeneration of skeletal tissues.

All musculoskeletal tissues in the body originate from the differentiation of pluripotent stem cells, specifically mesenchymal stem cells (MSCs), into various cell lineages. In living organisms, this differentiation process is preprogrammed in the DNA and takes place during embryonic development (Hall 2005). MSC differentiation is a central concept in mechanobiological tissue formation. The cells differentiate into different cell phenotypes, which synthesize extracellular matrix (ECM) to form tissues such as bone, cartilage, fat, muscles, and tendon when exposed to appropriate mechanical stimuli (van der Meulen and Huiskes 2002). The hierarchical development of MSCs into functional tissues is reversible and highly complex, as the cells can also dedifferentiate and transdifferentiate into other lineages. In fact, the discovery of dedifferentiation and

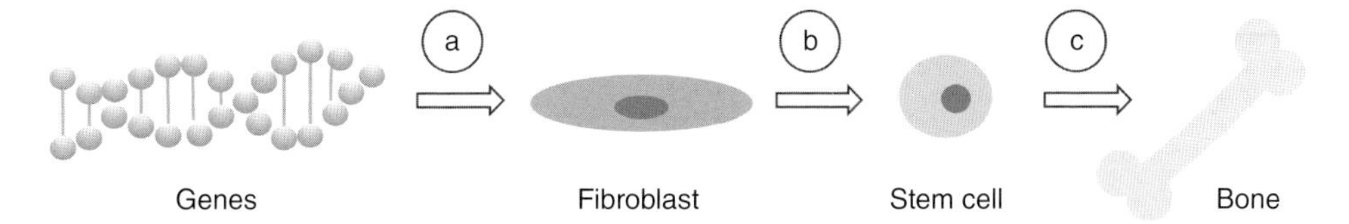

Figure 20.2 Discoveries of Dr. Shinya Yamanaka that were awarded the Nobel Prize. He studied genes that regulate stem cell function and (a) transferred such genes into fibroblasts from a mouse. These fibroblasts (b) dedifferentiated into stem cells, which could (c) redifferentiate into, for example, musculoskeletal cells.

redifferentiation by Dr. John B. Gurdon and Dr. Shinya Yamanaka led to them being awarded the Nobel Prize in Physiology in 2012 (Figure 20.2).

During regeneration and repair, quiescent and uncommitted MSCs are activated in response to external chemical and mechanical signals. These control cells' proliferation, migration, and differentiation activities. The cells residing in the tissue matrix are sensitive to stimulation (mechanosensitive) and can undergo apoptosis (programmed cell death) to give way to another, "more suited" cell phenotype and matrix if the environment is altered. Mathematically, this can be explained as a negative feedback loop. For example, a bone injury creates an environment with high mechanical strains. Invading cells such as MSCs fill the site of injury and produce soft-tissue ECM. This new matrix reduces the mechanical strains (negative feedback) and enables the differentiation of other cell phenotypes, such as cartilage and bone cells, that produce stiffer tissue matrices. Eventually, the stimulus at the injury site is sufficiently reduced to allow viable bone to restore homeostatic conditions (Prendergast 2007) (Figure 20.3a). On the same note, Weinans et al. (1992) have proposed that the system for stem cell differentiation could be turned into a positive feedback loop if the mechanoresponsiveness of the cells were targeted (Figure 20.3b): altering cells' mechanosensors with the aid of pharmaceutical drugs can heighten their response to the mechanical environment. This additional sensitization creates cells that more efficiently produce ECM, for example, which could lead to rapid bone formation, or in other words, to ever more sensitized cells. Thus, the concept of mechanobiological stem cell differentiation is one of the most powerful for understanding how to regulate the response of cells and tissues to mechanical conditions in order to ultimately enable (self-)regeneration of healthy load-bearing skeletal tissues.

20.2 Examples of Mechanobiological Experiments

Investigations of mechanobiology are carried out on many levels simultaneously in order to understand the influence of mechanical forces on the skeleton. Mechanobiological studies on the organ level involve *in vivo* experimentation, where the response to mechanical loading is studied. The vast extent of studies in mechanobiology cannot be covered in this chapter. Instead, we have decided to focus on the topic of mechanobiological tissue differentiation, in particular on the cues of mechanobiology in bone fracture healing. Fracture-healing experiments are a very common type of mechanobiological organ experiment and have shown that mechanical loading has a profound effect on the bone regeneration process. These experiments exhibit the

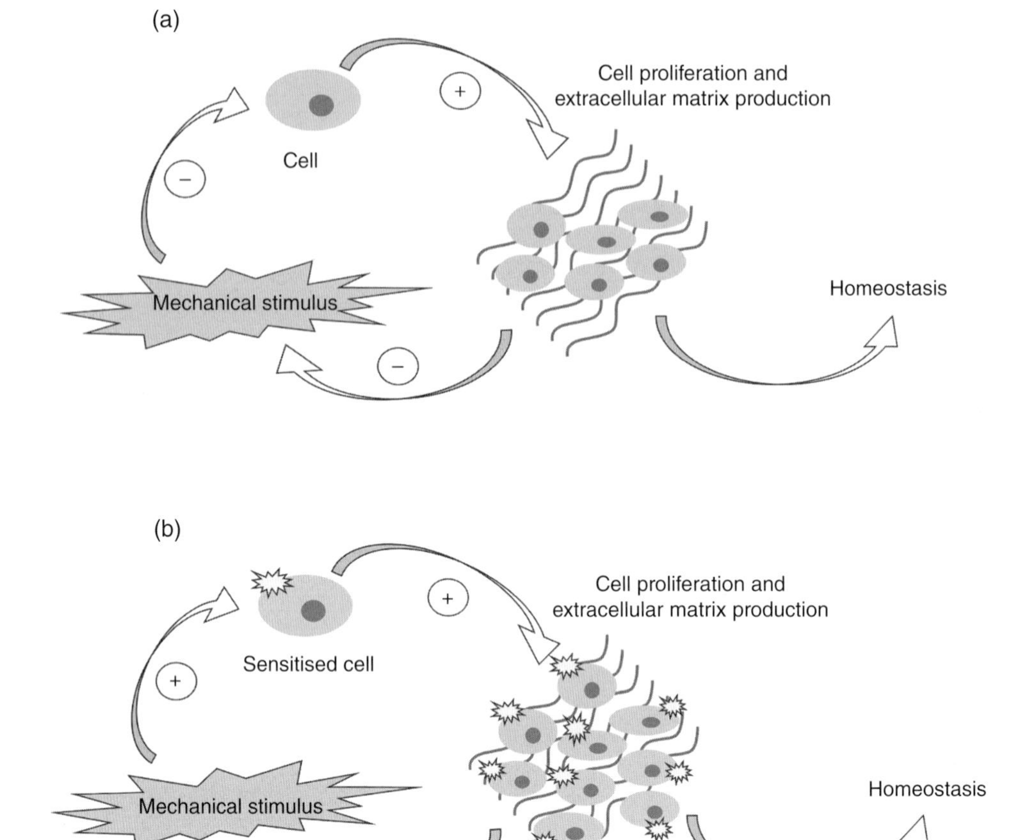

Figure 20.3 (a) Negative feedback loop, where the formation of new tissue matrix and cell phenotypes reduces the pericellular mechanical strains detected by cells. (b) Positive feedback loop, where sensitized cells respond more efficiently to the mechanical environment. New tissue matrices and cell phenotypes are formed to reduce the stimuli sensed by cells, but as the cells are more mechanosensitive, their response to loading is stronger than that of nonsensitized cells. Thus, tissue formation can become more effective.

fundamental concept of mechanobiology, that *form follows function*, meaning that tissues are formed, structured, and shaped to function most favorably in a given mechanical environment. The normal bone-healing process can be divided into three phases: (i) inflammatory phase; (ii) reparative phase; and (iii) remodeling phase (Frost 1989). Mechanical stimulation plays a key role in both the reparative and the remodeling phases. Briefly, the inflammatory phase (also known as the reactive phase) starts immediately after the fracture and is the body's response to the trauma. It results in bleeding and the formation of a hematoma. This hematoma consists primarily of inflammatory cells and granulation tissue. During the reparative phase, the cells (MSCs and fibroblasts) in the callus migrate and proliferate, and form soft tissue (e.g., fibrous tissue and cartilage in the fracture gap). Cells distal to the gap form bone (woven bone formation through intramembranous ossification). Next, the cartilage is replaced with woven

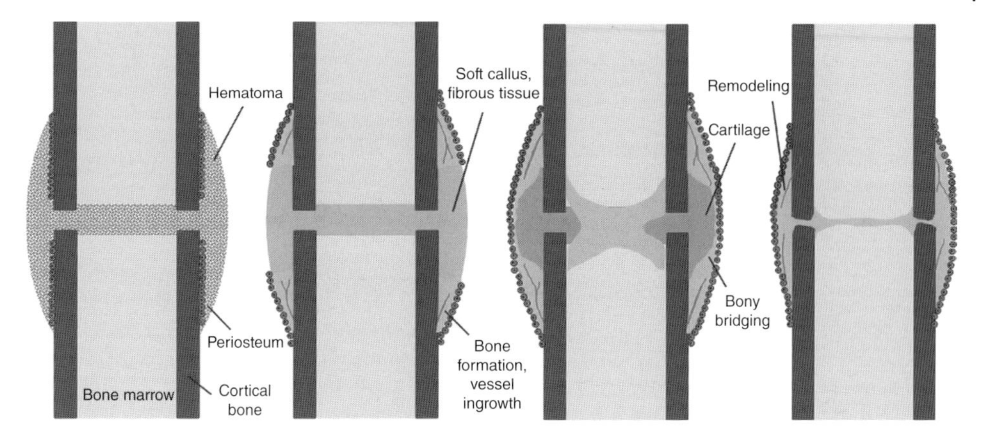

Figure 20.4 The phases of fracture healing. Inflammatory phase: a hematoma, consisting of platelets, inflammatory cells, and signaling molecules, is formed. MSCs migrate from the periosteum or the bone-lining surface to the gap, which organizes into connective granulation tissue. Reparative phase: bone formation begins from the periosteal surface and grows to form the callus. Stem cells start to differentiate, which results in soft-tissue production (e.g., fibrous tissue and cartilage) in the gap. The cartilage is replaced by bone through endochondral ossification. Remodeling phase: after maturation of the bone, the tissue remodels and becomes more organized. The fracture callus gradually resorbs and restores the bone's original shape.

bone through a process called "endochondral ossification" (Einhorn 1998). The bone is usually considered clinically healed when the callus has bridged with it, and this defines the end of the reparative phase; it takes about 3–6 months (depending on the bone) for human long bones to reach this stage after a fracture. Finally, in the remodeling phase, woven bone is replaced by lamellar bone, which is more highly organized and largely restores the bone's load-bearing strength (Marsh and Li 1999). The callus is gradually resorbed and replaced by compact bone (Figure 20.4).

The phases of fracture healing involve the differentiation of many different musculoskeletal cell phenotypes and tissue-formation pathways. It has long been recognized that mechanical stimulation can induce fracture healing or alter its biological pathway (Claes et al. 1998). Therefore, numerous fracture-healing studies have been conducted by, for example, examining various loading magnitudes and modes. Such studies report that moderate axial cyclic compression has a positive effect on rate of healing, as compared to distraction or static forces (Goodship and Kenwright 1985; Kenwright et al. 1991; Augat et al. 2003; Hente et al. 2004). Moreover, cyclic bending stimulates cartilage differentiation, indicating that bending of the fracture callus creates mechanical stimuli more suited to cartilage than to bone formation (Palomares et al. 2009). Shear movements, however, have produced contradictory results. Some studies have found them detrimental to healing, while others find them to be positive simulators (Augat et al. 2003; Bishop et al. 2006). Hence, the effect of shear versus axial motion appears to be sensitive to timing, magnitude, and gap size (Augat et al. 2005).

Studies on the cellular and molecular level are best performed using *in vitro* experiments, where the cells are cultured from their natural environment and subjected to a controlled mechanical loading in the laboratory. Cell culture experiments are simplified models of the complexity of mechanobiology and enable better understanding of the

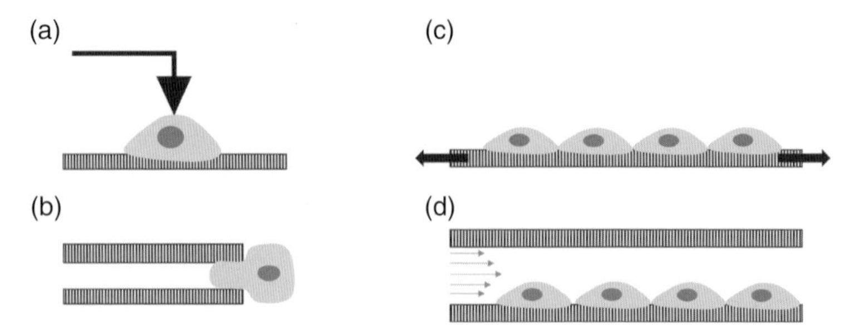

Figure 20.5 Examples of experimental techniques for the investigation of cell mechanobiology. (a) Nanoindentation: a technique whereby the cell is carefully indented. The indenter records force and deformation data to investigate cell material properties. (b) Micropipette aspiration: a technique whereby the cell is sucked into a pipette with a known diameter, which deforms it. (c) Substrate stretch: a monolayer of cells is stretched by stretching the substrate on which it lies. Based on the stiffness of the substrate and the magnitude of stretch, the cells are exposed to mechanical strain. (d) Flow chambers: confluent layer of cells in a bioreactor are exposed to fluid flow. This technique is often used to investigate the effect of fluid shear on cells.

underlying mechanisms by which mechanical forces influence biological cellular activities. The mechanobiology of a cell is primarily tested by conducting (see Figure 20.5 for illustrative examples):

- *Single-cell experiments:* Cells are deformed (mechanical stimuli) using various techniques, including nanoindentation, micropipette aspiration, and magnetic bead twisting (MacQueen et al. 2013; Rodriguez et al. 2013).
- *Monolayer experiments:* Confluent layer of cells are exposed to stretch, fluid flow, and hydrostatic pressure (e.g., in bioreactors) (MacQueen et al. 2013; Rodriguez et al. 2013).

In these experiments, the mechanosensitive cell response can be measured and visualized in the form of protein expression and cytoskeletal changes in the cell. Moreover, researchers have discovered that cells can "feel" the stiffness of the substrate on which they reside (Buxboim et al. 2010; Fletcher and Mullins 2010), up to several microns below the surface, depending on the hardness of the substrate (cells can feel deeper on softer substrate). This finding has become well established in the field, making this sensation mechanism another way of regulating the cell response to the pericellular environment and of studying cell mechanobiology (Iskratsch et al. 2014).

An example of how findings from experimental cell biology can be used in mechanobiological tissue regeneration is provided by a study of *in vivo* bone formation around a micro motion-loaded implant. Leucht et al. (2013) investigated how the mechanoregulated bone-formation process is affected by the deletion of an important mechanosensory organelle (primary cilium) on pre-bone cells in mice. This organelle has been shown to be very sensitive to mechanical loading in cell-culture experiments (Malone et al. 2007; Anderson et al. 2008), so the cells were expected to become less sensitive to their *in vivo* pericellular environment. Leucht et al. (2013) reported that upon deletion of the primary cilium, the cells in these mice could not proliferate in response to mechanical stimulus, deposit and orient tissue matrix to strain fields, or differentiate into bone-forming osteoblasts.

20.3 Modeling Mechanobiological Tissue Regeneration

In the early days, concepts of mechanoregulation could only be tested by conducting animal experiments. Scientists proposed theories based on their experimental observations of, for example, embryonic bone development and fracture healing. This was problematic, of course, since it was difficult to conduct quantitative measurements of either the mechanical environment within the tissue or the loads to which the animals were subjected. Moreover, it was impossible to measure the local mechanical environment in the tissue where the cell was residing. With the advent of computer power and advances in numerical tools, the science of mechanobiology was developed in a whole new direction of experimentation – namely, computational analysis.

Computational methods enabled constitutive modeling of biological tissues and improved understanding of their biomechanical behavior, which is subject to changes according to the biophysical environment. The use of finite element analysis was one of the largest breakthroughs, allowing modeling of complex anatomical geometries and more accurate computations of mechanical quantities (such as stresses, strains, and fluid flows) with advanced constitutive descriptions (Taylor and Prendergast 2015). This enabled the performance of computational experiments, which calculate the mechanical environment more accurately than traditional animal experiments can do. Computer models have the advantage that they can be developed to analyze complex processes on all levels of biology. Their ability to consider a high degree of complexity enables them to be deconstructed into their constituents, providing insight into how the mechanobiological responses on the molecular, cellular, tissue, and organ levels are integrated and synergized in a single system. Mechanobiological models for regeneration have been developed with consideration of the following aspects:

- *Geometry of the site being studied:* Image-analysis techniques are often used to create models of organ, tissue structure, and cell geometries. Anatomically more accurate models can be created, including patient-specific (rather than generic) models (Isaksson et al. 2009).
- *Constitutive behavior:* Tissues can be described as a continuum with a biomechanical behavior that is represented in material models. Improving material models can give a more accurate description of the local mechanical environment, by, for example, including the poroelastic nature of the tissue (Prendergast et al. 1996).
- *Cell activities:* The cells are the constituent parts of the model. They proliferate, migrate, apoptose, and differentiate. Modeling these activities involves consideration of the temporal and spatial aspects of the regeneration process (Isaksson et al. 2008). Cells can also be models of their own (single-cell models), where biophysical stimuli promote structural changes inside the cell and induce biochemical reactions (Ingber 2010; Barreto et al. 2013; Khayyeri et al. 2015a; Xue et al. 2015).
- *Biochemical expressions:* Cells under specific mechanical conditions synthesize biochemical signals that influence the differentiation pathway (Geris et al. 2009a). Identifying and modeling the most essential biochemical signals can not only improve mechanobiological predictions but also help develop new drug therapies that enhance the overall process.

These modeling components must all be combined to create a computer simulation of a mechanobiological process. In addition, the simulation has to be based on a mechanoregulation theory. Theories of mechanoregulation describe how mechanical loads modulate tissue formation and maintenance. They have a key role in modern mechanobiology, and provide the basis for predicting how tissues are formed and adapted to their local mechanical environment.

20.4 Mechanoregulation Theories for Bone Regeneration

Pauwels' (1960) mechanoregulation theory is often recognized as one of the first hypotheses about mechanoregulated stem cell differentiation. By studying the process of fracture healing, Pauwels, who was an orthopaedic surgeon, could suggest that hydrostatic pressure and shear strain guide MSCs into different differentiation pathways to form musculoskeletal tissues. He proposed that fibrous tissue forms in regions of tension, since collagen fibers are highly resistant to exclusively tensile stress, while cartilage tissue forms under hydrostatic pressure, since cartilage tissue forms fluid-filled spherical structures around chondrocytes, which swell osmotically. Thus, Pauwels suggested that biophysical stimulation would guide the differentiation of soft tissues such as fibrocartilage and fibrous tissue, which when matured and stabilized would induce formation of bone. During this period in time, however, the computational tools required to calculate the proposed biophysical stimuli at specific regeneration sites and under specific loading conditions were not available. However, his novel approach has inspired many scientists to present their own theories of mechanoregulated stem cell differentiation, aimed at unraveling the complex interactions between mechanical forces and biology (see Figure 20.6). The best-known theories belong to:

- Carter et al. (1988), who suggest that hydrostatic stress history and principal strain history guide tissue differentiation.
- Claes and Heigele (1999), who propose that hydrostatic pressure and strain determine stem cell fate.
- Prendergast et al. (1997), who present fluid flow and octahedral shear strain as regulators of tissue formation.
- Geris et al. (2010), who suggest fluid flow is only the main mechanoregulatory stimulus, and that when it is combined with bioregulatory stimulus, they can guide stem cell differentiation.
- Gomez-Benito et al. (2005), who propose that stem cells abide by the mechanical quality of deviatoric strain during the differentiation process.

A more recent theory of stem cell differentiation by Burke and Kelly (2012) suggests that the cell differentiation process is not directly regulated by the mechanical environment; rather, mechanical forces influence revascularization of tissues under healing, and stem cell fate is primarily guided by the availability of oxygen in combination with substrate stiffness (neighboring tissue stiffness).

Many mechanobiological computer models have been developed to test different mechanoregulation theories. However, in order to say anything about the predictive capacity of a given theory, it is important to differentiate between the theory and the method of implementation, which includes numerical schemes that describe biological activity such as cell motility and death. In fact, the existing mechanoregulation theories for stem cell differentiation have not changed since they were first introduced by their

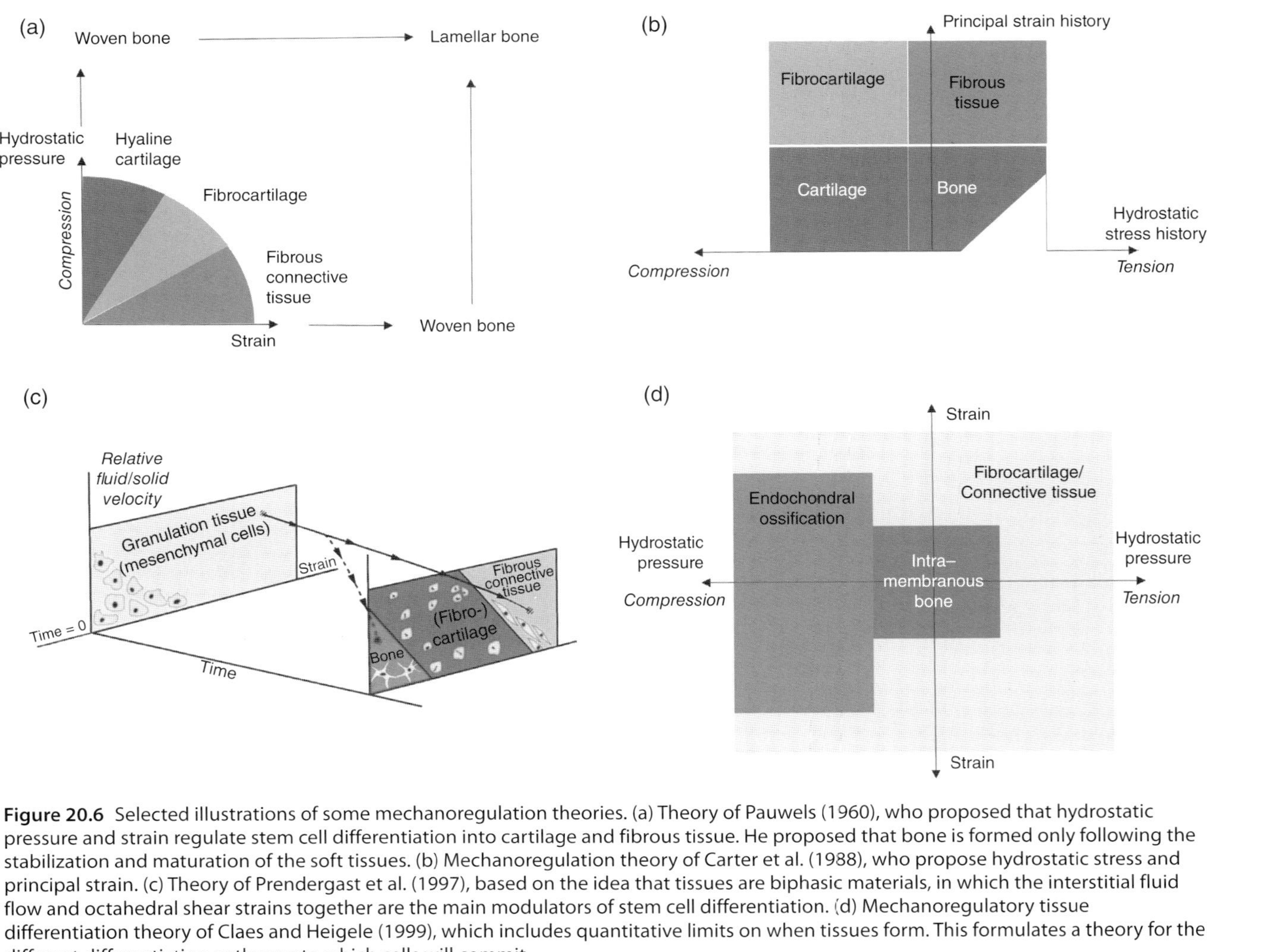

Figure 20.6 Selected illustrations of some mechanoregulation theories. (a) Theory of Pauwels (1960), who proposed that hydrostatic pressure and strain regulate stem cell differentiation into cartilage and fibrous tissue. He proposed that bone is formed only following the stabilization and maturation of the soft tissues. (b) Mechanoregulation theory of Carter et al. (1988), who propose hydrostatic stress and principal strain. (c) Theory of Prendergast et al. (1997), based on the idea that tissues are biphasic materials, in which the interstitial fluid flow and octahedral shear strains together are the main modulators of stem cell differentiation. (d) Mechanoregulatory tissue differentiation theory of Claes and Heigele (1999), which includes quantitative limits on when tissues form. This formulates a theory for the different differentiation pathways to which cells will commit.

authors, but the computational implementation algorithms have undergone rapid developments. Experimental findings still support many of these theories (Miller et al. 2015).

20.5 Use of Computational Modeling Techniques to Corroborate Theories and Predict Experimental Outcomes

The first computational analyses of mechanoregulation were only single time-point analyses, with finite-element models created for the regenerated domain and a single static load applied to determine the mechanical environment and the predicted tissue differentiation. Dennis Carter et al. (1988) at Stanford University suggested fracture healing was guided by mechanical factors such as hydrostatic pressures and shear strains. They calculated these factors in the finite-element domain and compared the distribution of stresses and strains was to the distribution of tissues formed *in vivo*; if the proposed mechanical stimuli followed the pattern of the observed tissue phenotype, then the mechanoregulation theory was regarded as corroborated. These first models had major simplifications and very limited predictive capacity, as they used idealized geometries of the tissue regeneration domain and could not consider any temporal or spatial changes during the healing process (Khayyeri et al. 2015b). Moreover, they did not include quantified thresholds of when each tissue type would form. Despite these limitations, they were ground-breaking when first proposed and were applied to many different mechanobiological scenarios.

Later, mechanobiological computations developed into computer simulations with an initial starting condition, from which the process of differentiation and adaptation over time was simulated. These early simulations did not explicitly consider biological activities and used only finite-element tools to compute the mechanical environment; they determined the tissue-differentiation process in an iterative fashion, without explicit consideration of cell activities. For example, a simulation model of an implant interface, conducted by Rik Huiskes et al. (1997), assumed the interface region to be filled with fibrous connective tissues at the beginning of the simulation. In each iteration, the mechanical environment was computed and the regenerative domain was updated with the new tissue phenotypes predicted by the theory of Prendergast et al. (1997), which were gradually replaced with cartilage and bone. The simulations were run until they reached equilibrium, which was when no more tissue transitions occurred. This type of early simulation was able to capture the overall features of mechanoregulated stem cell differentiation, but it lacked the ability to give detailed information about the temporal behavior of the process. Still, it captured gradual spatial changes in the regeneration domain.

The bioengineering research group at Trinity College Dublin, Ireland, suggested that the simulations should include cellular activities such as migration, proliferation, and apoptosis in the models, such that the models could capture greater biological complexity, including temporal effects of the mechanoregulatory process. Damien Lacroix (who was a PhD student at the time) was the first to work on this new development in mechanoregulated fracture-healing simulations. Expanding the models in this direction opened up a whole new set of studies, in which simulations could test the influence of the source of precursor cells, which was shown to have a significant effect on healing pattern and rate (Lacroix et al. 2002). Further, spatial tissue patterns, in combination with reorientation of collagen fibers in the newly formed matrix (Nagel and Kelly 2010) and disruptive healing events, are best captured with models that can consider cell phenotype-specific activities (Isaksson et al. 2008). Today, many other models have been presented, but the inclusion of cell activities is primary implemented by the following three modeling techniques:

1) Introducing differential equations to model cell activities, where cell dispersal and proliferation are modeled as diffusion. The cells are assumed to advance to areas of lower cell density at a rate determined by the diffusion coefficient (Isaksson et al. 2006).
2) Introducing differential equations to model bioregulatory pathways. The equations are used to describe cells' production of biological factors, such as vascular endothelial growth factor (VEGF) and bone morphogenetic proteins (BMPs), where access to such growth factors is assumed to influence stem cell fate (Geris et al. 2009b).
3) Considering the cells as single discrete units that move stochastically in a given space – the lattice approach. Random walk and occupation of free neighboring space are used to describe cell motility and proliferation (Perez and Prendergast 2007).

These methods have also enabled the modeling of blood vessel formation (angiogenesis) (Geris et al. 2008; Checa and Prendergast 2009; Burke et al. 2015) and thereby of the influence of oxygen tension on cell differentiation. Mapping out the blood vessel distribution in an injury site is essential, since it gives information about the cells' distance from a source of oxygen in the region. This information entails knowledge of whether the local cell environment fulfils both the biomechanical and the biochemical requirements for the formation of bone cells, since cells in low-oxygen regions (hypoxic regions) tend to differentiate into chondrocytes despite suitable mechanical stimuli for bone cell differentiation.

Computational mechanobiological models have come a long way since they were first introduced as single time-point analyses. Currently, they can predict principal tissue differentiation phenomena observed during experiments on fracture healing, distraction osteogenesis, and implant interfaces. For example, they can predict the non-unions during a fracture-healing process due to excessive loading (Byrne et al. 2011) and can capture the formation of fibrous tissue around bone implants subjected to high external forces (Checa and Prendergast 2009). These types of simulation hold great potential for using mechanobiology in health applications.

20.6 Horizons of Computational Mechanobiology

Existing mechanobiological computer models are phenomenological and often rely on organ-level finite-element analysis, in which a continuum-level quantity is computed. Stimuli that the cell senses or responds to are not computed directly. When a tissue is loaded, the strains transmitted to the cells in musculoskeletal tissues are primarily determined by the composition of the ECM: in bone, mainly hydroxyapatite and collagen, and in cartilage, proteoglycans and collagen. As a result, biophysical stimuli are very heterogeneous at the tissue level, and it is likely that the mechanical stimuli applied on tissues are different to those transmitted to the individual cells through the ECM. Therefore, modeling the hierarchical tissue levels as one unified continuum can introduce constraints into the simulations. Hence, mechanobiological simulations need to consider the biomechanical behavior of the ECM using advanced constitutive equations and model the strain transmission on to single cells residing within the matrix. This was highlighted in bone cells by Mullen et al. (2013), among others, who showed that the biomechanical behavior of the ECM affects differentiation of osteoblasts into osteocytes.

The need for multiscale modeling has been underlined by many researchers, and is a challenging task for mechanobiologists. Musculoskeletal tissues are hierarchical and mechanosensitive where the cells are the modulators of the mechanoresponse. Therefore, if multiscale models are to be used in mechanobiology, one must be able to explain how these different levels of biology communicate (biologically *and* computationally in algorithms) back and forth and how they affect one another. For example, we must figure out how a single-cell simulation, producing computational results as strains, can be linked to biological responses such as protein expression and matrix production – and how, in turn, these affect the cells' pericellular environment and the ECM of the tissue. These mechanobiological changes in the environment influence the continuum behavior of the tissue (e.g., anisotropy and stiffness, which are common parameters for evaluating the tissue's healing process and the organ's functional load-bearing capacity).

Nonetheless, predictive mechanobiology holds great potential for improving patient health and lifestyles (with and without multiscaling). *In silico* techniques (e.g., computer simulations) have become the new experimental tools for testing scientific hypotheses and capturing the complex dynamics of biological systems. Marco Viceconti (2015) suggests that by formulating our existing knowledge in biology and physiology in mathematical terms, we can build computer models that encapsulate the physiology of humans. These models can be corroborated against experimental and clinical data in order to obtain reliable predictive models of human health. In general, mechanobiological research is rapidly advancing toward clinical applications, with more and more patient-specific models and patient databases that can be used for modeling purposes. Imaging techniques enable the modeling of patient-specific geometries (e.g., of hip and knee joints). Electronic health care data repositories allow for patient-specific data storage, which can be used for follow-up therapies or progression of diseases, providing modelers with data across space and time. The concept of e-Health (tools and services that use information technology to improve prevention, diagnosis, treatment, monitoring, and management of public health issues; http://ec.europa.eu/health/ehealth) is becoming more grounded. Mechanobiological simulations can play an essential part in developing e-Health technologies.

The blue-sky research on computational mechanobiology could ultimately allow us to design hip prostheses suited to patients based on their age, lifestyle, and bone morphology, such that the implant can endure a lifetime of activities. Mechanobiology will enable the prescription of elderly therapies that target bone mechanobiology and help modulate or ameliorate the effects of osteoporosis (bone fragility), enabling patients to maintain an active lifestyle in old age and not risk fracturing their bones when they fall. Mechanobiological knowledge can be used to develop biomaterials that help injured athletes recover from fractures and tears rapidly, without damaging their sports career.

These are only a few examples of the great promise of the field. To realize this, we need to better understand the exact mechanisms by which mechanical stimulation is sensed (mechanosensation, mechanotransduction) and which mechanical cues are most important for manipulation of the desired tissue-formation response (mechanoregulation).

References

Anderson, C. T., A. B. Castillo, S. A. Brugmann, J. A. Helms, C. R. Jacobs, and T. Stearns. 2008. "Primary cilia: cellular sensors for the skeleton." *Anatomical Record – Advances in Integrative Anatomy and Evolutionary Biology* **291**(9): 1074–8. doi:10.1002/Ar.20754.

Augat, P., J. Burger, S. Schorlemmer, T. Henke, M. Peraus, and L. Claes. 2003. "Shear movement at the fracture site delays healing in a diaphyseal fracture model." *Journal of Orthopaedic Research* **21**(6): 1011–7. doi:10.1016/S0736-0266(03)00098-6.

Augat, P., U. Simon, A. Liedert, and L. Claes. 2005. "Mechanics and mechano-biology of fracture healing in normal and osteoporotic bone." *Osteoporosis International* **16**(Suppl. 2): S36–43. doi:10.1007/s00198-004-1728-9.

Barreto, S., C. H. Clausen, C. M. Perrault, D. A. Fletcher, and D. Lacroix. 2013. "A multi-structural single cell model of force-induced interactions of cytoskeletal components." *Biomaterials* **34**(26): 6119–26. doi:10.1016/j.biomaterials.2013.04.022.

Bishop, N. E., M. van Rhijn, I. Tami, R. Corveleijn, E. Schneider, and K. Ito. 2006. "Shear does not necessarifly inhibit bone healing." *Clinical Orthopaedics and Related Research* **443**:307–14. doi:10.1097/01.blo.0000191272.34786.09.

Burke, D. P. and D. J. Kelly. 2012. "Substrate stiffness and oxygen as regulators of stem cell differentiation during skeletal tissue regeneration: a mechanobiological model." *PLoS One* 7(7): e40737. doi:10.1371/journal.pone.0040737.

Burke, D. P., H. Khayyeri, and D. J. Kelly. 2015. "Substrate stiffness and oxygen availability as regulators of mesenchymal stem cell differentiation within a mechanically loaded bone chamber." *Biomechanics and Modeling in Mechanobiology* **14**(1): 93–105. doi:10.1007/s10237-014-0591-7.

Buxboim, A., I. L. Ivanovska, and D. E. Discher. 2010. "Matrix elasticity, cytoskeletal forces and physics of the nucleus: how deeply do cells 'feel' outside and in?" *Journal of Cell Science* **123**(Pt. 3): 297–308. doi:10.1242/jcs.041186.

Byrne, D. P., D. Lacroix, and P. J. Prendergast. 2011. "Simulation of fracture healing in the tibia: mechanoregulation of cell activity using a lattice modeling approach." *Journal of Orthopaedic Research* **29**(10): 1496–503. doi:10.1002/jor.21362.

Carter, D. R. and G. Beaupré (eds.). 2001. *Skeletal Function and Form: Mechanobiology of Skeletal Development, Aging, and Regeneration*, 1st edn. Cambridge: Cambridge University Press.

Carter, D. R., P Blenman, and G. S. Beaupré. 1988. "Correlations between mechanical stress history and tissue differentiation in initial fracure healing." *Journal of Orthopedic Research* **6**(5): 736–48. doi:10.1002/jor.1100060517.

Checa, S. and P. J. Prendergast. 2009. "A mechanobiological model for tissue differentiation that includes angiogenesis: a lattice-based modeling approach." *Annals of Biomedical Engineering* **37**(1): 129–45. doi:10.1007/s10439-008-9594-9.

Claes, L. E. and C. A. Heigele. 1999. "Magnitudes of local stress and strain along bony surfaces predict the course and type of fracture healing." *Journal of Biomechanics* **32**(3): 255–66. doi:S0021-9290(98)00153-5.

Claes, L. E., C. A. Heigele, C. Neidlinger-Wilke, D. Kaspar, W. Seidl, K. J. Margevicius, and P. Augat. 1998. "Effects of mechanical factors on the fracture healing process." *Clinical Orthopaedics and Related Research* **355**(Suppl.): S132–47. PMID:9917634.

Einhorn, T. A. 1998. "The cell and molecular biology of fracture healing." *Clinical Orthopaedics and Related Research* **355**(Suppl.): S7–21. PMID:9917622.

Fletcher, D. A. and D. Mullins. 2010. "Cell mechanics and the cytoskeleton." *Nature* **463**(7280): 485–92. doi:10.1038/Nature08908.

Frost, H. M. 1989. "The biology of fracture-healing. An overview for clinicians. Part I." *Clinical Orthopaedics and Related Research* **248**: 283–93. PMID:2680202.

Fung, Y. C. 1993. *Biomechanics: Mechanical Properties of Living Tissues*, 2nd edn. New York: Springer-Verlag.

Geris, L., A. Gerisch, J. V. Sloten, R. Weiner, and H. V. Oosterwyck. 2008. "Angiogenesis in bone fracture healing: a bioregulatory model." *Journal of Theoretical Biology* **251**(1): 137–58. doi:10.1016/j.jtbi.2007.11.008.

Geris, L., J. V. Sloten, and H. Van Oosterwyck. 2009a. "In silico biology of bone modelling and remodelling: regeneration." *Philosophical Transactions of the Royal Society A. Mathematical Physical and Engineering Sciences* **367**(1895): 2031–53. doi:10.1098/rsta.2008.0293.

Geris, L., K. Vandamme, I. Naert, J. Vander Sloten, J. Duyck, and H. Van Oosterwyck. 2009b. "Numerical simulation of bone regeneration in a bone chamber." *Journal of Dental Research* **88**(2): 158–63. doi:10.1177/0022034508329603.

Geris, L., J. Vander Sloten, and H. Van Oosterwyck. 2010. "Connecting biology and mechanics in fracture healing: an integrated mathematical modeling framework for the study of nonunions." *Biomechanics and Modeling in Mechanobiology* **9**(6): 713–24. doi:10.1007/s10237-010-0208-8.

Gomez-Benito, M. J., J. M. Garcia-Aznar, and M. Doblare. 2005. "Finite element prediction of proximal femoral fracture patterns under different loads." *Journal of Biomechanical Engineering – Transactions of the ASME* **127**(1): 9–14. doi:10.1115/1.1835347.

Goodship, A. E. and J. Kenwright. 1985. "The influence of induced micromovement upon the healing of experimental tibial fractures." *Journal of Bone and Joint Surgery. British Volume* **67**(4): 650–5. PMID:4030869.

Hall, B. K. 2005. *Bones and Cartilage: Developmental and Evolutionary Skeletal Biology*. San Diego, CA: Elsevier.

Hente, R., B. Fuchtmeier, U. Schlegel, A. Ernstberger, and S. M. Perren. 2004. "The influence of cyclic compression and distraction on the healing of experimental tibial fractures." *Journal of Orthopaedic Research* **22**(4): 709–15. doi:10.1016/j.orthres.2003.11.007.

Huiskes, R., W. D. Van Driel, P. J. Prendergast, and K. Søballe. 1997. "A biomechanical regulatory model for periprosthetic fibrous-tissue differentiation." *Journal of Materials Science: Materials in Medicine* **8**(12): 785–8. PMID:15348791.

Ingber, D. E. 2010. "From cellular mechanotransduction to biologically inspired engineering: 2009 Pritzker Award Lecture, BMES Annual Meeting October 10, 2009." *Annals of Biomedical Engineering* **38**(3): 1148–61. doi:10.1007/s10439-010-9946-0.

Isaksson, H., C. C. van Donkelaar, R. Huiskes, and K. Ito. 2006. "Corroboration of mechanoregulatory algorithms for tissue differentiation during fracture healing: comparison with in vivo results." *Journal of Orthopaedic Research* **24**(5): 898–907. doi:10.1002/jor.20118.

Isaksson, H., C. C. van Donkelaar, R. Huiskes, and K. Ito. 2008. "A mechano-regulatory bone-healing model incorporating cell-phenotype specific activity." *Journal of Theoretical Biology* **252**(2): 230–46. doi:10.1016/j.jtbi.2008.01.030.

Isaksson, H., I. Gröngröft, W. Wilson, C. C. van Donkelaar, B. van Rietbergen, A. Tami, R. Huiskes, K. Ito. 2009. "Remodeling of fracture callus in mice is consistent with mechanical loading and bone remodeling theory." *Journal of Orthopaedic Research* **27**(5): 664–72. doi:10.1002/jor.20725.

Iskratsch, T., H. Wolfenson, and M. P. Sheetz. 2014. "Appreciating force and shape – the rise of mechanotransduction in cell biology." *Nature Reviews. Molecular Cell Biology* **15**(12): 825–33. doi:10.1038/nrm3903.

Kenwright, J., J. B. Richardson, J. L. Cunningham, S. H. White, A. E. Goodship, M. A. Adams, et al. 1991. "Axial movement and tibial fractures. A controlled randomised trial of treatment." *Journal of Bone and Joint Surgery. British Volume* **73**(4): 654–9. PMID:2071654.

Khayyeri, H., S. Barreto, and D. Lacroix. 2015. "Primary cilia mechanics affects cell mechanosensation: a computational study." *Journal of Theoretical Biology* **379**: 38–46. doi:10.1016/j.jtbi.2015.04.034.

Khayyeri, H., H. Isaksson, and P. J. Prendergast. 2015. "Corroboration of computational models for mechanoregulated stem cell differentiation." *Computer Methods in Biomechanics and Biomedical Engineering* **18**(1): 15–23. doi:10.1080/10255842.2013.774381.

Krahl, H., U. Michaelis, H. G. Pieper, G. Quack, and M. Montag. 1994. "Stimulation of bone-growth through sports – a radiologic investigation of the upper extremities in professional tennis players." *American Journal of Sports Medicine* **22**(6): 751–7. doi:10.1177/036354659402200605.

Lacroix, D., P. J. Prendergast, G. Li, and D. Marsh. 2002. "Biomechanical model to simulate tissue differentiation and bone regeneration: application to fracture healing." *Medical and Biological Engineering and Computing* **40**(1): 14–21. PMID:11954702.

Leucht, P., S. D. Monica, S. Temiyasathit, K. Lenton, A. Manu, M. T. Longaker, et al. 2013. "Primary cilia act as mechanosensors during bone healing around an implant." *Medical Engineering & Physics* **35**(3): 392–402. doi:10.1016/j.medengphy.2012.06.005.

MacQueen, L., Y. Sun, and C. A. Simmons. 2013. "Mesenchymal stem cell mechanobiology and emerging experimental platforms." *Journal of the Royal Society, Interface* **10**(84): 20130179. doi:10.1098/rsif.2013.0179.

Malone, A. M. D., C. T. Anderson, P. Tummala, R. Y. Kwon, T. R. Johnston, T. Stearns, and C. R. Jacobs. 2007. "Primary cilia mediate mechanosensing in bone cells by a calcium-independent mechanism." *Proceedings of the National Academy of Sciences of the United States of America* **104**(33): 13 325–30. doi:10.1073/pnas.0700636104.

Marsh, D. R. and G. Li. 1999. "The biology of fracture healing: optimising outcome." *British Medical Bulletin* **55**(4): 856–69. PMID:10746335.

Miller, G. J., L. C. Gerstenfeld, and E. F. Morgan. 2015. "Mechanical microenvironments and protein expression associated with formation of different skeletal tissues during bone healing." *Biomechanics and Modeling in Mechanobiology* **14**(6): 1239–53. doi:10.1007/s10237-015-0670-4.

Mullen, C. A., M. G. Haugh, M. B. Schaffler, R. J. Majeska, and L. M. McNamara. 2013. "Osteocyte differentiation is regulated by extracellular matrix stiffness and intercellular separation." *Journal of the Mechanical Behavior of Biomedical Materials* **28**: 183–94. doi:10.1016/j.jmbbm.2013.06.013.

Nagel, T. and D. J. Kelly. 2010. "Mechano-regulation of mesenchymal stem cell differentiation and collagen organisation during skeletal tissue repair." *Biomechanics and Modeling in Mechanobiology* **9**(3): 359–72. doi:10.1007/s10237-009-0182-1.

Palomares, K. T., R. E. Gleason, Z. D. Mason, D. M. Cullinane, T. A. Einhorn, L. C. Gerstenfeld, and E. F. Morgan. 2009. "Mechanical stimulation alters tissue differentiation and molecular expression during bone healing." *Journal of Orthopaedic Research* **27**(9): 1123–32. doi:10.1002/jor.20863.

Pauwels, F. 1960. "[A new theory of the influence of mechanical stimuli on the differentiation of supporting tissue. The tenth contribution to the functional anatomy and causal morphology of the supporting structure.]" *Zeitschrift für Anatomie und Entwicklungsgeschichte* **121**: 478–515. PMID:14431062.

Perez, M. A. and P. J. Prendergast. 2007. "Random-walk models of cell dispersal included in mechanobiological simulations of tissue differentiation." *Journal of Biomechanics* **40**(10): 2244–53. doi:10.1016/j.jbiomech.2006.10.020.

Prendergast, P.J. 2007. "Computational modelling of cell and tissue mechanoresponsiveness." *Gravitational and Space Biology* **20**(2): 43–50.

Prendergast, P. J., W. D. van Driel, and J. H. Kuiper. 1996. "A comparison of finite element codes for the solution of biphasic poroelastic problems." *Proceedings of the Institution of Mechanical Engineers. Part H, Journal of Engineering in Medicine* **210**(2): 131–6. PMID:8688118.

Prendergast, P. J., R. Huiskes, and K. Søballe. 1997. "ESB Research Award 1996. Biophysical stimuli on cells during tissue differentiation at implant interfaces." *Journal of Biomechanics* **30**(6): 539–48. doi:S0021929096001406.

Rodriguez, M.L., P.J. McGarry, and N.J. Sniadecki. 2013. "Review on cell mechanics: experimental and modeling approaches." *Applied Mechanics Reviews* **65**: 060801. doi: 10.1115/1.4025355.

Taylor, M. and P. J. Prendergast. 2015. "Four decades of finite element analysis of orthopaedic devices: where are we now and what are the opportunities?" *Journal of Biomechanics* **48**(5): 767–78. doi:10.1016/j.jbiomech.2014.12.019.

van der Meulen, M. C. H. and R. Huiskes. 2002. "Why mechanobiology? A survey article." *Journal of Biomechanics* **35**(4): 401–14. PMID:11934410.

Weinans, H., R. Huiskes, and H. J. Grootenboer. 1992. "The behavior of adaptive bone-remodeling simulation-models." *Journal of Biomechanics* **25**(12): 1425–41. doi:10.1016/0021-9290(92)90056-7.

Viceconti, M. 2015. "Biomechanics-based in silico medicine: the manifesto of a new science." *Journal of Biomechanics* **48**(2): 193–4. doi:10.1016/j.jbiomech.2014.11.022.

Xue, F., A. B. Lennon, K. K. McKayed, V. A. Campbell, and P. J. Prendergast. 2015. "Effect of membrane stiffness and cytoskeletal element density on mechanical stimuli within cells: an analysis of the consequences of ageing in cells." *Computer Methods in Biomechanics and Biomedical Engineering* **18**(5): 468–76. doi:10.1080/10255842.2013.811234.

21

Porous Bone Graft Substitutes

When Less is More

Charlie Campion and Karin A. Hing

Institute of Bioengineering and School of Engineering and Materials Science, Queen Mary University of London, London, UK

21.1 Introduction

Bone is a multifaceted living tissue with a complex hierarchical structure that performs several key functions within the body, from providing structural support and protection to bodily organs, to enabling mineral homeostasis and the supply of mesenchymal and hematopoietic stem cells (MSCs and HSCs). As a living tissue, bone requires a constant supply of oxygen and nutrients and is subject to infection and to degenerative, metabolic and metastatic disease. Moreover, like any structural material, bone will fracture spontaneously when overloaded. However, one of bone's key attributes is its ability to undergo self-repair and to adaptively respond to gradual changes in mechanical demand.

While bone possesses the ability to repair small fractures or defects without external intervention, its ability to self-heal is limited in the size of fracture or defect it is able to restore to healthy tissue. Alternatively, where there is an imbalance in the body's normal hormonal regulatory system, resulting in metabolic bone disease and either depletion (osteoporosis) or overproduction (Paget's disease) of bone, spontaneous restoration to healthy tissue is unlikely. For these reasons, clinical intervention is sometimes necessary. This can take two quite different approaches: either external or internal stabilization to facilitate "natural" bone regeneration or excision and complete or partial replacement of a portion of the bone with a medical device.

The considerable disparity in the approaches and materials used to replace or repair damaged or diseased bone tissue reflects the fact that development of these repair strategies has been studied from different perspectives – as a biomedical engineering challenge aimed at efficiently restoring load-bearing function and rapidly restoring quality of life to the patient, and as the restoration of structurally or metabolically "normal" tissue that retains its physiologically and biomechanically responsive characteristics. While both of these approaches can exploit bone's natural capacity for adaptation and repair, when taken to extremes they can result in very different recommendations for treatment, ranging from whole or partial limb amputation and the fitting of an osteointegrated state-of-the-art limb prosthesis to the targeted delivery of powerful growth factors and stem cells in order to augment bone's natural self-healing capacity.

Mechanobiology: Exploitation for Medical Benefit, First Edition. Edited by Simon C. F. Rawlinson.

In the clinical setting, bone regeneration is generally accepted as preferable to bone replacement, reflecting the fact that despite the best efforts of many leading researchers and clinicians, the materials and devices available for bone replacement still fall short of the ideal. One of the biggest hurdles in bone replacement is the mechanical mismatch between traditional orthopedic biomedical materials (e.g., stainless steel, cobalt-chrome, and titanium alloys) and natural bone tissue, leading to a phenomenon known as "stress shielding." Stress shielding occurs when a high-modulus prosthesis (100–250 GPa, depending on the orthopedic alloy used) is attached to or implanted in bone: initially, the metal component stabilizes motion or channels load, providing either a protective environment to allow bone to heal around the prosthesis or an alternative mechanism which enables the bone or joint to perform its normal load-bearing or locomotive function; however, with time, the fact that the stiffness of the prosthesis is greater than that of natural bone (the modulus for cortical bone is typically reported to range from 10 to 25 GPa) results in the local tissue receiving a reduced biomechanical stimulus, leading ultimately (through mechanobiological pathways) to its resorption. Bone loss then results in loosening of the prosthesis, patient discomfort, implant failure, and the need for further surgical intervention in an environment now depleted of bone stock. This is a situation in which mechanobiology has been ignored to medical detriment, and while the need for materials with matched mechanics to work alongside the host tissue is clear, provision of such a biocompatible, low-modulus material with sufficient fatigue resistance still presents a technical challenge to medical engineers and biomedical materials scientists alike.

Bone regeneration, on the other hand, makes a virtue of the fact that bone is an adaptive living tissue, and harnesses bone's natural ability to heal itself and to undergo mechanical adaptation. The hurdles to this approach are the limited volume of tissue that bone can spontaneously regenerate and the rate at which the regeneration of fully functional load-bearing tissue occurs. These hurdles place an interim requirement for a support or scaffold that, at worst, does not hinder the regeneration process and, at best, assists or stimulates it.

The obvious candidate "material" for such a scaffold is fresh autologous cancellous bone itself, and in modern medicine, autografting – the procedure of replacing missing bone with fresh bone from the patient's own body (Czitrom and Gross 1992; Meeder and Eggers 1994) – is still regarded by many as the "gold standard" in bone repair. Autografting is utilized in severe trauma cases, oncology, primary total hip revisions, and the correction of large "bony defects," where a significant piece of bone is missing or damaged. However, the amount of bone that can be safely harvested from any one patient is limited, while the additional surgical procedure may itself be complicated by donor site pain and morbidity. These factors have a significant impact on individual patient recovery and global health care socioeconomics – with patients who undergo an autograft procedure from a separate surgical site averaging an extra hour in surgery and an extra day in hospital as a consequence. Allografting (Czitrom and Gross 1992) – using donor tissue stored within regulated bone banks – overcomes these difficulties, but the demand far outstrips the supply, there is no assurance of freedom from disease (Barriga et al. 2004; McCann et al. 2004), and healing can be inconsistent (Togawa et al. 2004). Consequently, there is an increasing demand for alternative approaches that can avoid these complications. This has led to the development of two sometimes overlapping approaches: the development of osteogenic bone grafts, which can either be wholly synthetic in origin or may be processed from natural tissue (collectively known as "synthetics"), and treatment with osteoinductive growth factors (collectively known as "biologics"), which are often delivered from a form of synthetic bone graft.

The development of synthetic bone graft substitutes (BGSs) was initially driven by a desire among materials engineers to replicate the chemistry and/or structure of bone matrix so as to encourage "safe" bone formation within a structurally protective graft material, whereas the development of biologics was inspired by the discovery by biologists (notably Marshall Urist) that certain natural proteins possessed the ability to stimulate bone formation when implanted in muscle tissue *in vivo*, as well as the ability to stimulate osteoblastic cell differentiation *in vitro* (Urist 1965).

Both of these routes have had their setbacks. Initial preoccupation with the strength of the synthetic materials (there being initially a general consensus that strength should be maximized) and a lack of sufficient control of the graft chemistry led to poor clinical outcomes and surgical disillusionment with early synthetic graft materials, while off-label use of powerful growth factors such as BMP-2 at the recommended on-label "supraphysiological" doses resulted in serious clinical complications and subsequent expensive litigation issues (Carragee et al. 2011). This has left many surgeons understandably cautious in their selection of an alternative product to autograft.

Moreover, the choice of products available to the surgeon is vast in terms of the combinations of graft chemistries, pore structures, and growth factors (Hing 2005; Chau and Mobbs 2009). Considering the pore structure alone, while it is well recognized that both the rate of integration and the final volume of regenerated bone may be primarily dependent on the structural characteristics of the porosity, there is still controversy as to which ones are key. Well-studied characteristics include total porosity volume fraction and features of the so-called "macroporosity" (i.e., pores typically in the range of 50–1000 μm), such as the modal macropore diameter and the modal pore interconnection size. Additionally, there are now a number of additional claims being made regarding the importance of much smaller (1–50 μm scale) pores located within the "dense" struts of the graft, known as "strut-porosity" or "microporosity" (Figure 21.1).

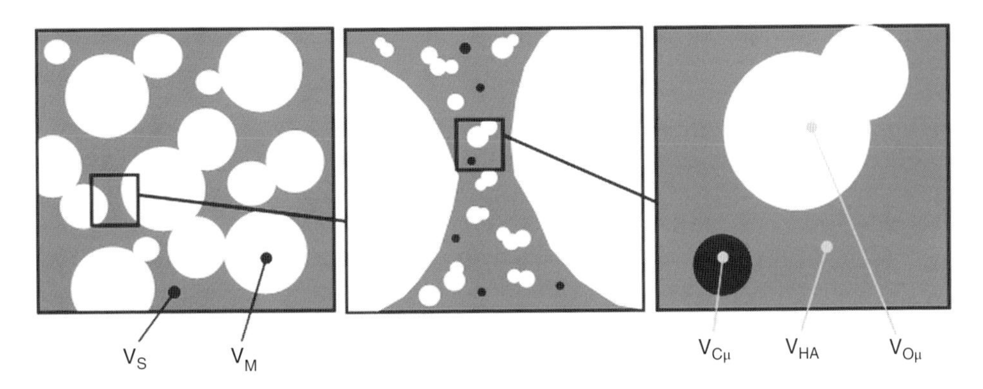

Figure 21.1 Schematic diagram of the different material and porosity volume fractions within a porous hydroxyapatite scaffold. V_S, total strut volume fraction; V_M, total macropore volume fraction; $V_{C\mu}$, total closed micropore volume fraction; V_{HA}, total hydroxyapatite volume fraction; $V_{O\mu}$, total open micropore volume fraction. $V_S = V_{HA} + V_{C\mu} + V_{O\mu}$. Total porosity $(P_T) = 1 - [V_{HA}/(V_S + V_M)]$. Strut porosity $(P_S) = (V_{C\mu} + V_{O\mu})/V_S$. If ρ_{HA} = theoretical density of hydroxyapatite, then apparent density $= \rho_{HA}[V_{HA}/(V_S + V_M)]$ and real density $= \rho_{HA}[V_{HA}/(V_{C\mu} + V_{HA})]$.

Furthermore, the optimum feature size or volume fraction for both scales of parameters will be dependent on the graft chemistry.

This chapter aims to review some of the recent developments in bioceramic BGSs, and while graft chemistry clearly plays a significant role, it will concentrate on understanding how mechanobiology might be behind the biological sensitivity to alteration in various structural characteristics and the apparent ability of some of these synthetic graft structures to behave "osteoinductively" *in vivo*.

21.2 Bone: The Ultimate Smart Material

Apart from being capable of self-repair and self-regulation, bone also has a highly remarkable complex structure, or more correctly a series of complex structures. From a materials engineering perspective, bone is a multiphase composite, hierarchically ordered from the nanoscale right through to the macroscale, where the precise structure and the proportion of the components of any piece of bone vary in response to local hormonal, mechanical, and nutritional conditions, resulting in local differences in structural anisotropy and density, and thus many different classifications of bone that exhibit very different mechanical and functional characteristics.

At a macrostructural level, the two principle classifications of bone are cancellous (or spongy) and cortical (or compact). Generally, cortical bone is found on the surface (and the thickness of this protective skin increases in mechanically demanding regions, such as the shafts of long bones), while cancellous bone is found in the interior, such as within the femoral head and vertebra.

The density of cortical bone is relatively consistent, typically varying by about 3%, and ranging from 1.85–2.05 g/cm^3 in human bone. The basic functional unit of mature compact bone is the osteon or *Haversian system*. The Haversian system consists of canals (*Haversian canals*) surrounded by concentric rings of mineralized lamellar bone, in which bone cells are sandwiched between layers of calcified matrix (*lamellae*) in small, semi-isolated spaces called *lacunae*. Canaliculi penetrate the lamellae, radiating through the matrix and connecting lacunae with one another and with nutrient sources. Perforating canals, the *canals of Volkmann*, extend perpendicular to the osteons and supply blood to osteons deeper in the bone and to tissues of the marrow cavity (Martini 1998). This "osteonal" microstructure is highly anisotropic: when traced longitudinally, osteons have no definite end, being part of a dense three-dimensional (3D) network of branching and converging tubular or cylindrical elements. In long bones, such as the humerus or femur, most osteons have their longitudinal axis approximately parallel to the shaft of the bone, so that when sectioned in a transverse direction, they appear with an essentially circular (as opposed to elliptical or parallel-sided) profile. Secondary Haversian systems are easier to discern, as they as larger (~100 μm in diameter; Epker et al. 1964) and are bounded by a clear "cement" line, and their organization stops abruptly at the cement line and does not conform with the organization of adjacent osteons. The Haversian canal in a secondary Haversian system will usually contain only a single capillary or blood vessel. Haversian systems are separated by irregular angular wedges of "interstitial bone," which are the remains of older osteons that have been partially resorbed during remodeling. Primary Haversian systems are smaller but may contain two or more vessels. They have no cement lines, their organization fits in better with their neighbors, and there is no interstitial bone between them.

Cancellous bone is characterized by its foam- or spongelike appearance and can be considered to be made up from a series of interconnected bony struts, ranging from 100 to 300 μm in width and framing a series of interconnected pores with spacings of 300–1500 μm between adjacent trabeculae. The porosity accounts for approximately 75–95% of the total volume of cancellous bone (Athanasiou et al. 2000). Nutrient transfer to the osteocytes occurs via diffusion along canaliculi that open on to the surfaces of the trabeculae. In response to local mechanical demands, the architecture and relative volume of pores and struts varies significantly: in some regions, the pores dominate and the struts resemble a network of thin interconnected rods, often with considerable anisotropy, due to the orientation of major trabeculae along lines of principle stress, whereas in others, the pores are embedded in a significant volumes of bone matrix and are only connected by small channels. This leads to a location-dependent variation in the apparent density of cancellous bone of up to 30%, typically ranging from 0.21 to 0.39 g/cm^3 in healthy tissue.

While changes in the mineralization of the trabeculae have been reported to have little effect (Hodgskinson and Currey 1990), apparent density and trabecular architecture strongly influence the mechanical properties of cancellous bone tissue (Carter and Hayes 1976; Gibson 1985; Hodgskinson and Currey 1990; Linde et al. 1991). The structural or apparent density of cancellous bone is the mass per unit bulk volume of the tissue, whereas the matrix, material, or real density is the mass per unit volume of the bone trabeculae. The strength and moduli of cancellous bone are related to its apparent density by a power-law function (Gibson 1985; Keller 1994). The strength of cancellous bone varies widely over the range of observed densities – unlike the density of cortical bone, the density of cancellous bone can vary by an order of magnitude, so this is quite a significant variation potential; it is related to the square of the apparent density. Its modulus varies as either the square or the cube of the apparent density. Both strength and modulus are sensitive to the rate of loading, as well as the loading direction, though the effect of loading rate is less sensitive than the effect of density (Ouyang et al. 1997). Midrange values for the strength and modulus of cancellous bone are 2.00–5.00 MPa and 90.0–400 MPa, respectively (Rohl et al. 1991). As would be expected, the mechanical strength of cancellous bone is also highly dependent upon architecture, reflecting anatomical site (Goldstein 1987), and on the age of the individual, due to age-related variation in mineral content according to changes in diet, health, and lifestyle (Burstein et al. 1976; McCalden et al. 1997).

There are, additionally, subclassifications of both cancellous and cortical bone, based on age and developmental history. Cancellous bone can be classed as either coarse or fine, where coarse cancellous is characteristic of healthy adult mammalian skeleton, while fine cancellous is characteristic of the fetal skeleton or early fracture callus and comes in two further forms, depending on the route of osteogenesis: fine cancellous membranous bone (bone formed *de novo*) and fine cancellous endochondral bone (bone formed from a cartilaginous template). Cortical bone can be considered as surface, primary, or secondary osteonal: as with cancellous bone, the distinction depends on features of the bone's osteonal microstructure, which vary with location, age, and bone origin. Furthermore, the microstructure of both cancellous and cortical bone also varies as a function of its formation history. Rapidly formed bone, such as that formed initially in fracture callus, often has a disordered, less dense structure, and is known as "woven bone," while bone formed more sedately during normal growth or callus remodeling has a more ordered dense structure and is known as "lamellar bone," due to its striated structure. This variation in microstructure also contributes to considerable variation in stiffness, strength, and toughness in both cortical and cancellous bone.

At a microstructural level, bone can be considered a multiphase porous composite, containing a grid of bone cells located within a network of lacunae, all interconnected by micron-scale channels (cannalicunae) embedded in bone matrix. The bone matrix is itself a composite at the nanoscale, comprising an organic fibrous network reinforced with inorganic nanocrystallites (Cameron 1972).

The organic phase of the bone matrix makes up roughly 20 wt% of the bone's wet mass and consists of a number of different glycoproteins, polysaccharides, and citrates, but it is dominated by a highly elastic protein: collagen. Collagen is the most abundant protein in the body, accounting for 70–90% of the nonmineralized component of the bone matrix and varying from an almost random network of coarse bundles to a highly organized system of parallel-fibered sheets or helical bundles. Collagen consists of carefully arranged arrays of tropocollagen molecules, which are long, rigid molecules (300 nm long, 1.5 nm wide) comprising three left-handed helices of peptides (known as α-chains) bound together in a right-handed triple helix. Though all α-chains contain the glycine-X-Y sequence, different types of collagen may be produced via the combination of different amounts and sequences of other amino acids within the tropocollogen molecule. To date, over 20 different types of collagen have been identified, but bone contains mostly collagen type I (containing two identical and one dissimilar α-chain ($\alpha1(I)_2\alpha2(I)$) within its tropocollogen molecule), which is the body's most abundant form, accounting for 90% of its total collagen. Collagen types I and V are organized into collagen fibrils, which are formed by the assembly of tropocollagen molecules in a three-quarter stagger, parallel array. As a result of this assembly, the fibrils exhibit characteristic cross–striations, or banding, which occur in a repeating pattern approximately every 64 nm (Robinson and Watson 1952). The fibrils are stabilized by inter- and intramolecular crosslinks (the number and distribution of which determine the tissue's mineralization state), and have average individual diameters of 100 nm. In collagen I, the fibrils are wound into bundles to form collagen fibers that range from 0.2 to 12.0 μm in diameter. The arrangement, packing texture, and density of these fibers impact directly on the local mechanics of the collagen matrix, providing a mechanism for sensitizing the local tissue properties to the manner in which the matrix was laid down by collagen-synthesizing cells (typically, osteoblasts). *De novo* bone matrix can be formed through two principle routes – endochondral and membraneous – often resulting in a loosely organized form of matrix known as "woven bone." Much of this woven bone tissue is subsequently remodeled into a more ordered, stronger tissue known as "lamellar bone."

The inorganic phase of bone matrix is usually referred to as "bone mineral" and accounts for 70–75 wt% of the mass of dry bone. However, it is worth noting that due to bone mineral's significantly greater density (as compared to that of the inorganic phase) and bone's naturally hydrated state, this equates to less than 40 vol% of the bone matrix (i.e., in volume terms, the organic component is the dominant phase). Bone mineral is often incorrectly referred to as "hydroxyapatite" (HA), which has a chemical formula of $Ca_{10}(PO_4)_6(OH)_2$ and a Ca : P ratio of 5 : 3 (1.66). Hydroxyapatite is a hydrated calcium phosphate ceramic, with a similar (but not identical) crystallographic structure to bone mineral (de Jong 1926). Bone mineral is additionally characterized by calcium, phosphate, and hydroxyl deficiency (reported Ca : P ratios of 1.37–1.87) (Posner 1969; Mc Connel 1973), internal crystal disorder, and ionic substitution within the apatite lattice, resulting in the presence of significant levels of additional key trace elements, including carbonate, sodium, magnesium, zinc, silicate, and fluoride. Bone mineral is not a direct analog of hydroxyapatite, as is commonly believed, but is more closely related to an A-B-type carbonate-substituted apatite (Le Geros and Le Geros 1993; Elliot 1994). These factors all contribute to an apatite that is sufficiently insoluble for stability, yet sufficiently reactive to allow the *in vivo* submicroscopic (5–100 nm) crystallites to be constantly resorbed and reformed, as required by the body, meeting bone's additional role as the body's mineral store.

It is the nanostructurally ordered combination of these highly ordered elastic collagen fibers, reinforced by hard submicroscopic inorganic crystallites, that enables bone to simultaneously possess stiffness, elasticity, hardness, and toughness. Moreover, variation in the nanostructural, microstructural, and macrostructural arrangement of these components gives bone matrix the capacity to display a wide range of mechanical properties, enabling local tissue to be tailored to the local mechanical environment for a minimal weight, maximizing its efficiency. There is currently no artificial engineering material that can match the performance or adaptability of bone gram for gram. Researchers have long recognized that it is the composite nature of bone matrix that is key to its success; however, we are only just beginning to understand how to manipulate composite materials at the submicroscopic "nano" level required for the degree of control found in bone structure, and the concept of smart adaptive self-building materials is in its infancy.

21.3 Bone-Grafting Classifications

Bone grafts are required for a wide range of situations in which the bone cannot naturally regenerate. They are typically used to fill small voids (such as small bone tumors, following bone fracture reduction, or in osteotomies and plastic surgery). Bone grafts are also applied clinically to enhance and stabilize instrumentation over time. With a variety of sources of graft, both natural and synthetic, come a wide range of inherent properties. An entire nomenclature exists for these properties, which can be used to differentiate between these grafts in terms of their clinical utility. Some of the most commonly used terms are as follows:

- *Osteoconductive:* Provides a physical structure into and along which bone may grow (Miyazaki et al. 2009).
- *Osteoinductive:* Capable of inducing bone formation in a non-bony site by recruiting and inducing (pluripotent) stem cells to become osteoblasts (Miyazaki et al. 2009).
- *Osteogenic:* Contains the cells required to produce bone (Miyazaki et al. 2009).
- *Osteostimulative:* Has the ability to signal or activate cells in order to enhance bone growth (Miyazaki et al. 2009).
- *Bioactive:* Has the propensity to form a bonelike mineral layer on the surface following submersion in simulated body fluid (Kokubo et al. 1990a,b).

21.3.1 Autografts

Autologous bone, otherwise known as "autograft," is bone harvested from the patient's own body (Meeder and Eggers 1994; Giannoudis et al. 2005; Dinopoulos et al. 2012). Autologous bone is most frequently harvested from the iliac crest, as it provides access to relatively good-quality and -quantities of cancellous bone. It can also be collected during surgery and applied in a defect local to the harvest site. As such, this autograft is known as "local bone." Local autologous bone can typically be collected during surgeries such as posterolateral and interbody lumbar fusions in the spine, during which the spinous processes, facets, and laminae may be expended in order to provide procedural access or local graft, or both.

Autograft, whether from the iliac crest or from local sources, has been the standard of care for bone grafting for many years, and is thus considered the gold standard by clinicians (Giannoudis et al. 2005; Dinopoulos et al. 2012). The advantages of autografting are that there are no direct acquisition costs relating to sourcing the graft (unlike with other bone grafts) and, as it is harvested from and implanted in the same patient, there is no chance for disease transmission and a very low risk of immunogenic response (Giannoudis et al. 2005; Dinopoulos et al. 2012). According to the definitions provided at the start of this section, autograft can be considered osteoconductive, osteogenic, and osteoinductive.

Despite the wide use of autografting, there are several known clinical and logistical limitations to the use of this approach. The most commonly understood include: a wide range of complication rates (9–49%), donor-site pain, scarring, increased risk of infection, poor-quality donor bone, elongated surgery time, additional need for post-operative analgesia, and limited supply (Giannoudis et al. 2005; Kim et al. 2009; Dinopoulos et al. 2012). Furthermore, additional adverse clinical effects include: hematoma formation, increased blood loss, nerve damage, hernia formation, arterial injury, ureteral injury, pelvic instability, cosmetic defects, and tumor transplantation (Giannoudis et al. 2005; Dinopoulos et al. 2012).

21.3.2 Allografts

An alternative option is the use of allograft human bone, which is bone donated either from a live patient (e.g., a femoral head from a hip procedure) or from the deceased (Czitrom and Gross 1992; Giannoudis et al. 2005; Miyazaki et al. 2009; Dinopoulos et al. 2012). The advantage of using an allograft compared to an autograft is that there is no need for a second procedure to harvest the graft material. On top of the reduced surgery time, this eliminates the problems of donor-site morbidity and pain. Allografts are claimed to retain their osteoinductive potential, since the noncollagenous proteins remain in the graft prior to implantation. However, *in vitro* studies have demonstrated variability in the osteoinductivity of allografts, arising from the method used to process them for clinical use (Giannoudis et al. 2005; Miyazaki et al. 2009; Dinopoulos et al. 2012). Other disadvantages of allografting include: variable quality of donor bone, risk of disease transmission, infection, low therapeutic dose of osteoinductive factors, and the need for donor consent (Giannoudis et al. 2005; Miyazaki et al. 2009; Dinopoulos et al. 2012).

21.3.3 Demineralized Bone Matrix

Often considered the modern allograft, demineralized bone matrix (DBM) was developed to provide clinicians with an alternative to traditional strut allograft. The manufacture of such grafts typically involves submerging and rinsing the donated bone in acid solutions in order to remove the mineral content of the bone. This process leaves behind all the collagen and noncollagenous proteins in the source bone. The resultant material is typically admixed to natural or synthetic biomaterials, which provide cohesivity, moldability, and adhesiveness to the graft. This provides surgical utility to the graft during a grafting procedure. The advantages of DBMs are that they have been demonstrated to be osteoinductive through *in vitro* and *in vivo* tests, they provide some degree of osteoconductive scaffold through the provision of collagen particles and fibers, and they have advantageous handling properties (Giannoudis et al. 2005; Lee et al. 2005; Miyazaki et al. 2009; Bae et al. 2010; Dinopoulos et al. 2012). The disadvantages of DBM include the lack of mineral scaffold for osteoconduction, the potential for migration due to compression or irrigation, the variable quality (according to donor), the cost, and the risk of infection (Giannoudis et al. 2005; Lee et al. 2005; Miyazaki et al. 2009; Bae et al. 2010; Dinopoulos et al. 2012).

21.3.4 Synthetic Bone Grafts

Synthetic bone grafts, such as hydroxyapatite, tricalcium phosphate (TCP), calcium sulfate (CaS), and bioglass, have been used in clinical practice for several decades. We have already outlined the clinical need for synthetic replacements of autograft and allograft. Carbonate-substituted hydroxyapatite with the approximate formula $(Ca,Mg,Na)_{10}(PO_4HPO_4CO_3)_6(OH)_2$ makes up approximately 5% of the human body mass (Ravaglioli 1992), so the development of pure synthetic hydroxyapatite forms that were readily available and compatible with sterilization led to a considerable interest in the material in the 20th century as a biomaterial for the treatment, augmentation, and replacement of osseous tissue. The term “apatite” was derived from the Greek word “apato,” meaning deceit, by a mineralogist named Werner in the 1970s. Apatites can undergo numerous ionic substitutions into their crystalline structure and can assume many identities, which was Werner’s basis for naming them in this way. Biologically, this phenomenon is observed in the numerous substitutions, such as the Fe^{2+}, Mg^{2+}, Zn^{2+}, F^-, Cl^- carbonate, silicate, and citrate ionic substitutions, that occur in tissues such as bone, dentine, and enamel (Carlisle 1970; LeVier 1975; Aoki 1994; LeGeros 2002). Importantly, the ability of bone mineral to perform ionic substitutions to this extent allows it to act as a calcium and mineral reservoir, which gives it a significant role in mineral homeostasis. Synthetic hydroxyapatite, however, is generally manufactured in the pure stoichiometric form, but it is an attractive biomaterial as it closely resembles bone mineral chemically and has been proven to be compatible with bone tissues *in vitro* and *in vivo* (Jarcho 1981; Winter and Griss 1981).

The perceived clinical disadvantage of phase-pure hydroxyapatite is its slow rate of resorption *in vivo*, due to its high crystallinity and stability (Suchanek et al. 1996; Greenspan 1999; Laurencin 2003). The fact that the hydroxyapatite remains *in situ* may not offer any significant biological disadvantages, but clinicians have anecdotally expressed a desire to see such bone grafts resorb in order that clinical X-rays can be used to identify the rate and progress of remodeling of new bone. Hydroxyapatite is radiopaque and hence masks – to a certain extent – the presence of new bone on X-ray. The technical solution to this is to select a more resorbable phase of calcium phosphate (Barrere et al. 2003), which will dissolve *in situ* and thus be less radiopaque. TCP, with the chemical formula $Ca_3(PO_4)_6$ and a calcium/phosphorus ratio of 1.5, has been investigated for this purpose in preclinical studies and has been used in clinical practice (Van Blitterswijk et al. 1991; Hunter et al. 1995; Boyan et al. 1996; Cao and Hench 1996; Healy et al. 1996; Ozawa and Kasugai 1996; Anselme 2000; Park and Bronzino 2003), as has CaS, otherwise known as plaster of Paris. The challenge with highly resorbable phases such as TCP and CaS is that due to their relative instability compared to hydroxyapatite, they can dissolve too quickly for complete osteointegration into a treated bone defect (Hing et al. 2007).

21.3.5 Bone Morphogenetic Proteins

In recent years, recombinant human bone morphogenetic proteins (rhBMPs) have found a place in the bone-grafting treatment paradigm, as they can direct MSCs to differentiate into an osteogenic bone line. The advantages of rhBMPs in clinical application are that they are osteoinductive, effective at promoting bone growth, and supported by a substantial amount of clinical data. The adverse event profile of these therapies is becoming better understood. Reported disadvantages include ectopic bone growth, swelling, and excessive resorption, among many others (Vaidya et al. 2007; Wong et al. 2008).

21.4 Synthetic Bone Graft Structures

The ideal synthetic graft should not only replace the missing tissue but also encourage new bone ingrowth into the grafted area, thereby initially reinforcing the defect site, and ultimately encouraging the formation of a living bridge between the existing bone and the graft material. Moreover, with time, the graft should be replaced with healthy bone tissue via the normal bone remodeling process, exploiting bone's capacity for self-repair and regeneration, features that enable the skeleton to grow, mature, and meet different loading demands, while maintaining an optimal strength-to-weight ratio.

The importance of a bone graft's structure has been considered since the use of a porous material was first described (Smith 1963) and Hulbert et al. (1972) demonstrated that porous disks of a near-inert ceramic exhibited thinner fibrous encapsulation with faster healing in surrounding muscle and connective tissue than dense disks, as a result of a mechanical interlock which reduced motion between host tissue and implant. Subsequently, many studies have demonstrated a greater degree and faster rate of bone ingrowth or apposition with percentage porosity; however, there is still considerable dispute regarding which are the key structural features of a porous bone graft and what their optimal values might be.

Despite the fact that bone grafts are generally used in combination with fixation devices (both temporary and permanent) to ensure adequate mechanical stabilization (Zdeblick et al. 1994), early research was influenced by the fact that bone is a load-bearing tissue, and thus it was felt that a successful BGS material must have a certain degree of strength. Unfortunately, the most promising materials from the perspective of biocompatibility in the osseous environment were either glasses or ceramics, which, when fully dense, possessed reasonable levels of hardness and compressive strength but poor fracture toughness and bending strengths, especially in comparison to native bone tissue. This led to significant levels of research effort being concentrated on the development of reinforced apaite-like ceramics, the most notable being those developed by Yoshii et al. (1988), and of processing methods for porous ceramics with highly dense struts (in order to maximize their strength), such as the hydrothermal conversion of various coral structures into hydroxyapatite (Roy and Linnehan 1974; White et al. 1975). Many of these early bone grafts did not perform well and were poorly received by clinicians, particularly where especially dense structures were used, as these often contained isolated (biologically useless) pores. Fortunately, as a result of developments in the understanding of the phenomenon of stress shield in total joint replacement surgery, it became increasingly recognized that the poor performance of synthetic BGSs may be related to the fact that, like a metal implant, a dense ceramic structure is extremely stiff in comparison to native tissue, depriving the local bone tissue of mechanical stimulus.

21.4.1 Effect of Pore Structure on Graft Mechanics

The mechanical properties of a ceramic foam are highly dependent on both porosity and structural architecture (Peelen et al. 1978; Holmes et al. 1984; Le Huec et al. 1995; Hing et al. 1999a; Chu et al. 2002; Bignon et al. 2003). This is typified in a study by Hing et al. (1999a) of one type of porous hydroxyapatite BGS, converted from bovine cancellous bone, in which the ultimate compressive strength (UCS) was found to vary between 2 and 10 MPa as a function of decreasing porosity (from 80 to 50%) and the modulus was found to depend on both porosity and fabric, varying from 0.2 to 2.0 GPa for isotropic and from 0.2 to 1.0 GPa for anisotropic structures over the same porosity range. Investigation of the biological response to implantation of these BGSs at the porosity extremes of 60 and 80% demonstrated faster regeneration of bone within the higher-porosity, weaker BGS structures (Hing et al. 1999b, 2004) and, most interestingly, showed that with time, the less porous, stiffer grafts underwent a degree of mechanical adaptation, resulting at 6 months in defect properties that more closely matched the local host tissue. This adaptive remodeling of synthetic grafts has been reported in a number of studies (Holmes et al. 1984; Boyde et al. 1999; Hing et al. 2005), which show that, with time, the bulk properties of both weak and stiff grafts are modified *in situ* (both up and down) to reflect the local host tissue requirements.

It is well known that bone is functionally adaptive (i.e., that it responds to external mechanical stimuli to either reduce or increase its mass, as required; Wolff 1870) due to the mechanosensitivity of many cell types, including osteoblasts and osteocytes (Frost 1987), as well as osteoprogenitor cells (Mauney et al. 2004). Therefore, it is unsurprising that a number of studies have demonstrated that in structures in which the level of pore interconnection is sufficient to support adequate vascularization for full bony integration of internal porosity (Eggli et al. 1988), there is a degree of adaptation of bone ingrowth within the porous BGS with time (Boyde et al. 1999; Hing et al. 2004), sometimes leading to the loss of bone volume (Martin et al. 1993). This suggests that the variation of local strain in scaffold struts with macroporosity may induce or inhibit bone formation within the BGS. Additionally, it has been demonstrated that in the longer term both micro- and macroporosity influence bone adaptation (Boyde et al. 1999; Bignon et al. 2003; Hing et al. 2005). It has thus been proposed that a reduction in strut modulus associated with increasing microporosity levels is sufficient to shift the strut modulus below a threshold value, resulting in a swing in the equilibrium local bone cell

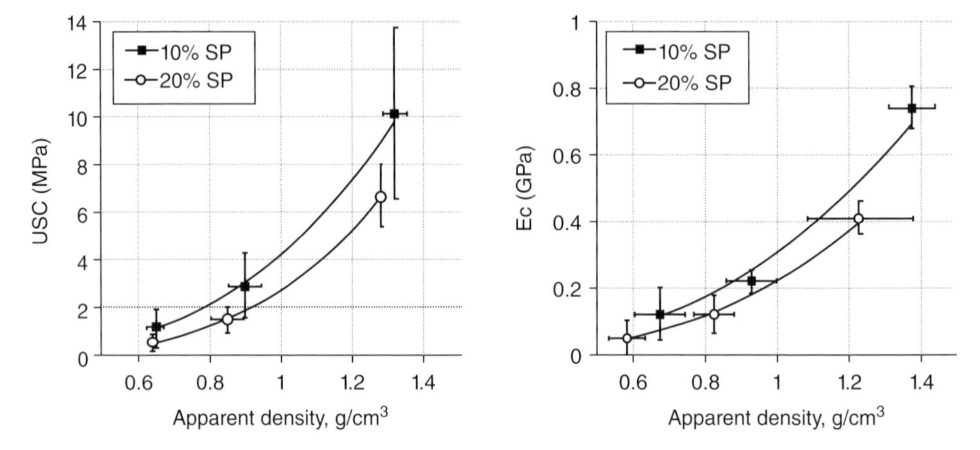

Figure 21.2 Dependence of extrinsic scaffold properties on the level of apparent density (i.e., total porosity) and strut porosity of hydroxyapatite BGS materials.

activity toward a greater degree of stable bone apposition. This is presumed to result from the sensitivity of cells associated with remodeling within normal bone to microfracturing and consequent changes in microstrain within the bone (Lanyon et al. 1982; O'Connor et al. 1982; Burr et al. 1985). Moreover, in an experiment comparing integration within a series of four BGSs with total porosities of 70 and 80% and low and high levels of microporosity, in which all materials had statistically different UCSs before implantation, all scaffolds retrieved after 24 weeks *in vivo* had UCSs that were statistically similar to one another, as well as to control bone from the same site retrieved and tested under identical conditions, suggesting that the equilibrium level of bone ingrowth within a BGS may be dependant on scaffold mechanics as a function of both macro- and microstructure (Hing et al. 2004, 2005). Inclusion of unpublished data on the mechanical properties of grafts with 60% total porosity further accentuates the relationship (Figure 21.2): after 24 weeks *in vivo*, retrieved implants with strut porosities of 20% and total porosities of 60% had a UCS of 7.3 MPa and an absolute bone volume of 21%, whereas those with a total porosity of 80% and a matched strut porosity had a UCS of 7.9 MPa and an absolute bone volume of 39%.

These results suggest that the equilibrium level of bone ingrowth attainable by a BGS may be highly sensitive to a scaffold's capacity to stress shield integrated bone, pointing to an optimal position where the scaffold mechanics either closely mimic or underperform natural bone tissue. Similar findings have been reported by researchers investigating biomechanical modulation of metaphyseal fracture healing, where they have demonstrated strain dependence in a controlled metaphyseal fracture model. In areas with interfragmentay strains below 5%, significantly less bone formation occurred compared to areas with higher strains (6–20%). For strains larger than 20%, fibrocartilage layers were observed. Moreover, low interfragmentay strain (<5%) led to intramembranous bone formation, whereas higher strains additionally provoked endochondral ossification or fibrocartilage formation (Claes et al. 2011). This has implications for observations of both spontaneous endochondral and intramembranous ossification found in close proximity – often within the same macropore – of synthetic BGSs implanted in ectopic muscle sites (Chan et al. 2012). In these studies, while the chemistry of the graft played a role in facilitating osteoinductive behavior, the inclusion of increasing levels of strut porosity was found to be the dominating factor, as even with an optimized chemistry, grafts with strut porosities of <20% were unable to support bone formation in an ectopic site (Coathup et al. 2011, 2012; Chan et al. 2012).

A number of studies *in vitro* and *in vivo* have similarly demonstrated biological sensitivity to the level of microporosity within the ceramic struts (Boyde et al. 1999; Yuan et al. 1999; Bignon et al. 2003; Annaz et al. 2004; Habibovic et al. 2005, 2006; Hing et al. 2005; Campion et al. 2011; Coathup et al. 2011, 2012; Chan et al. 2012). There is some evidence that this enhancement in bioactivity may be a direct result of a variation in the surface texture that makes it a geometrically more suitable substrate for cell attachment (Lampin et al. 1997; Dalby et al. 2000; Chong et al. 2015), resulting in enhanced cell anchorage, regulation, and/or differentiation. Alternatively, it is also postulated that the inclusion of strut porosity indirectly affects bioactivity as a function of increasing surface area, modulating ion exchange and selective sequestering and binding of adhesion proteins and growth factors and thereby increasing the quantity of these adhesion proteins and growth factors above a critical level for cell recruitment and activation (Ripamonti 1996; Lampin et al. 1997; Yuan et al. 1999, 2002; Dalby et al. 2002; Bignon et al. 2003; Annaz et al. 2004; Habibovic et al. 2005, 2006; Guth et al. 2010a,b, 2011; Campion et al. 2011; Chan et al. 2012). Publications by Barrere et al. (2003) and Barradas et al. (2011) summarize some of the proposed mechanisms for biomaterial osteoinductivity, suggesting that osteoinduction by porous calcium phosphate (CaP)-based ceramics can be attributed to: (i) the incorporation and concentration of BMPs by CaP crystals; (ii) a low oxygen tension in the central region of the implant, which triggers the pericytes of microvessels to differentiate into osteoblasts; (iii) a rough surface produced by the 3D microstructure, which causes the asymmetrical division of mesenchymal cells that produce osteoblasts; (iv) the surface charge of the substrate, which triggers cell differentiation; (v) the bonelike apatite layer formed *in vivo*, which recognizes mesenchymal cells; and/or (vi) the local high level of free Ca^{2+} provided by the CaP material, which triggers cell differentiation and bone formation. The effect on the scaffold mechanics of altering total or strut porosity is rarely considered; this may be an oversight, given the variable nature of the bone-formation pathways observed in our studies and the fact that a certain level of strut porosity is seemingly critical to the osteoinductivity of all CaP ceramic bone grafts.

Additionally, when there is a change in graft structure, not only does this affect the continuity and feature scale of the solid phase (varying the extrinsic mechanical properties of the scaffold), but it also simultaneously alters the continuity of the pore phase, changing the structure's permittivity (i.e., the resistance to fluid flow through the structure). This will have a significant effect on fluid pressure and strain fields at the internal pore surfaces of the structure.

21.4.2 Porous BGS Permittivity: Impact on Fluid Flow

It is now generally accepted that a greater volume and faster rate of bone ingrowth may be obtained by increasing BGS macroporosity (i.e., pores >50 μm in size) (Klawitter and Hulbert 1971; Roy and Linnehan 1974; Klawitter et al. 1976; Dard et al. 1994; Liu 1997); however, there is some confusion as to whether this is a reflection of a dependence between volume or rate of integration and pore size (Klawitter et al. 1976; Holmes 1979;

Holmes et al. 1984; Uchida et al. 1984; Daculsi and Passuti 1990; Martin et al. 1993; Gauthier et al. 1998) or of other structural parameters, such as pore morphology, porosity volume, and pore connectivity (Eggli et al. 1988; Kuhne et al. 1994; Hing et al. 1999b; Lu et al. 1999).

A pore diameter of 100 μm is often cited as a minimum requirement for healthy ingrowth, following the work of Klawitter et al. (1976), who actually observed mineralized bone ingrowth in pores as small as 40 μm, but reported a greater penetration of bone ingrowth into near-inert polyethylene implants with increasing *pore interconnection size* (often misquoted as "pore diameter") of "up to 100–135 mm in diameter" (Klawitter et al. 1976); that is, greater penetration was observed in polyethylene implants with pore interconnection diameters of up to 135 mm. Lu et al. (1999) demonstrated that when using either hydroxyapatite or βTCP, the critical pore interconnection size for bone ingrowth is only 50 μm, corroborating the earlier work of Holmes (1979), who found that when implanted in cortical bone, coral structures with interconnections of osteonic diameter were required for sustainable bone ingrowth.

This would suggest that pore size is not the controlling factor, but rather pore interconnection size, which is often related to both pore size and the extent of porosity (Li et al. 2003; Hing et al. 2004). This is elegantly demonstrated by the improved integration in structures with well interconnected 50–100 μm pores, as compared with less connected but larger pores of 200–400 μm with similar levels of porosity (Eggli et al. 1988). Moreover, this dependence is unsurprising when you consider that bone is a mineralized tissue that relies heavily on the presence of an internal blood supply for supply of nutrients and oxygen, which do not readily diffuse through it. Any new bone formation or repair must always be preceded by the formation of a vascular network, the rapidity and extent of which is strongly influenced by the degree of structural interconnectivity between pores (Rubin et al. 1994). Is it surprising, therefore, that a minimum interconnection size exists, in line with that of osteonal diameter?

Interestingly, pore connectivity and porosity volume cease to be such critical factors for resorbable bioceramics such as βTCP and bioglasses, as the resorption exhibited by these materials acts to open up the structure, meaning that the optimal connectivity and porosity of resorbable scaffolds may be lower than those established for nonresorbable materials. This may explain the relative insensitivity to pore interconnection size and porosity reported in the literature for scaffolds containing these materials (Gauthier et al. 1998; Lu et al. 1999).

From a mechanobiology perspective, this sensitivity to structural connectivity may reflect the fact that both osteoblasts and osteocytes are sensitive to local fluctuations in interstitial fluid flow. A number of researchers have hypothesized that flow of interstitial fluid is most likely the method by which bone cells are informed about mechanical loading (Pavalko et al. 1998; Ajubi et al. 1999; Joldersma et al. 2000; Westbroek et al. 2000; Bakker et al. 2001; Wang et al. 2003; Sikavitsas et al. 2005). Under normal physiological conditions, flow of interstitial fluid occurs because the application of mechanical strain causes the volume of some pores to decrease slightly, creating differences in bone fluid pressure, which result in fluid flow (Sikavitsas et al. 2001). While this is important in lacunar–canalicular porosity, which is on the same scale as strut porosity, it is believed to be negligible in the Haversian lumens and Volkmann canals, which are more similar in scale to BGS macroporosity, because these are 1000 times larger and their pressure is more uniform, since it must be equal to the blood pressure

(Cowin and Weinbaum 1998). However, local conditions within a recently treated defect site packed with fresh, potentially loose SBG granules are unlikely to be uniform, resulting in significant local variations in fluid flow and hence interstitial pressure, especially as the graft is incorporated in the initial inflammatory and wound-healing responses. Since osteocytes are the only cells which inhabit the lacunar–canalicular porosity, studies have postulated them to be extensively involved in mechanotransduction, and they have been found to be responsive to both pulsating fluid flow (PFF) (Klein-Nulend et al. 1995a) and intermittent hydrostatic compression (IHC) (Klein-Nulend et al. 1995b). This may have implications for the importance of strut porosity in the medium- and long-term health of regenerated bone within a BGS-treated defect.

In a healing defect site, however, where short-term conditions prevail, osteoblast sensitivity to mechanical stimuli will be more relevant. Numerous studies have demonstrated osteoblasts to be mechanoresponsive to a number of mechanical stimuli, including IHC (Klein-Nulend et al. 1995b, 1997; Roelofsen et al. 1995; Nagatomi et al. 2003), continuous fluid flow (Pavalko et al. 1998; Sikavitsas et al. 2003, 2005; Wang et al. 2003), PFF (Klein-Nulend et al. 1995a, 1998; Sterck et al. 1998; Ajubi et al. 1999; Joldersma et al. 2000; Westbroek et al. 2000; Bakker et al. 2001, 2003; Nagatomi et al. 2003; Bacabac et al. 2004; McGarry et al. 2005), piezoelectric-induced strain (Di Palma et al. 2003, 2005; Tanaka et al. 2005), four-point bending-induced strain (Mauney et al. 2004; McGarry and Prendergast 2004), and tensile-induced strain (Ignatius et al. 2005). Unsurprisingly, when mechanosensivity to fluid flow in osteoblasts was compared to mechanosensivity to direct mechanical strain, cell responses were found to vary, with both fluid flow and mechanical strain increasing nitric oxide (NO) production, but only fluid flow increasing prostaglandin production (Mullender et al. 2004). Furthermore, substrate strains enhanced the bone matrix protein collagen I twofold, whereas fluid shear caused a 50% reduction in collagen type I. Finite-element analysis (FEA) modeling has revealed that fluid flow affects all cell model components, whereas strain only affects the cell attachments (McGarry and Prendergast 2004). These observations suggest that osteoblasts have the capability to be sensitive to both changes in BGS extrinsic mechanics and permittivity.

21.5 Conclusion

In vitro experiments have come a long way toward explaining the mechanobiology of material substrates and their impact on bone cell upregulation in response to either fluid shear or direct substrate strain. However, a lot is still unknown with respect to the precise mechanisms behind the biomaterial-derived osteogenic potential of synthetic bone grafts. In addition to a permissive chemistry that supports osteogenic cell attachment, proliferation, and differentiation, studies into osteoinductivity have highlighted the importance of physical characteristics such as total and strut porosity, pore connectivity, and graft permittivity in controlling the rate, the volume, and even the formation pathway of bone regeneration. However, the precise relationship between these factors is not clear. The many different physiochemical characteristics of a biomaterial can affect multiple components of the bone-formation process, meaning that in reality, the mechanism underlying biomaterial-derived osteoinductivity is far more complex than that proposed by Barradas et al. (2011).

The structural dependence of bone regeneration in a porous BGS argues that the ideal structure of a scaffold is dictated by its ability to physically permit nutrient, cell, and vascular penetration, as well as its simultaneous possession of mechanical properties that are matched to the demands of the local environment and which will not negatively impact on local strain fields and fluid transport, suggesting that precise scaffold requirements may well vary with site of application. This is particularly apparent when the rate and the extent of bone adaptation to and remodeling of SBG materials implanted in different osteogenic sites are compared (Shors 1999).

Mechanistic studies into the pathways behind cell and tissue response to specific implant surfaces (Olivares-Navarrete et al. 2012) and mechanobiologic control of osteogenesis (Claes et al. 2011) are informing new bone-regeneration therapies in a manner once considered more appropriate to pharmacological studies of bone metabolism (Little et al. 2005) and bone regulation (Kingsmill et al. 2013), demonstrating the genetic basis for nanoscale sensitivity at the chemical, structural, and mechanical levels. Continued research in this area will be key to the development of third-generation SBGs, ensuring that these treatments are tailored to work in harmony with the host's native capacity for adaptive bone regeneration and so guide the restoration of functional bone tissue suitable to the diverse local environmental demands within the skeleton. It has been postulated that the greatest technological challenge to human space exploration is the long-term regulation of bone mass in the weightlessness of space. Understanding bone-regeneration therapies on earth may be a small step in the right direction.

References

Ajubi, N. E., J. Klein-Nulend, M. J. Alblas, E. H. Burgeret, and P. J. Nijweide. 1999. "Signal transduction pathways involved in fluid flow-induced PGE2 production by cultured osteocytes." *American Journal of Physiology* **276**(1 Pt. 1): E171–8. PMID:9886964.

Annaz, B., K. A. Hing, M. Kayser, T. Buckland, and L. Di Silvio. 2004. "Porosity variation in hydroxyapatite and osteoblast morphology: a scanning electron microscopy study." *Journal of Microscopy* **215**(Pt. 1): 100–10. doi:10.1111/j.0022-2720.2004.01354.x.

Anselme, K. 2000. "Osteoblast adhesion on biomaterials." *Biomaterials* **21**: 668–80. PMID:10711964.

Aoki, H. 1994. *Medical Applications of Hydroxyapatite.* Tokyo: Takayama Press System Center.

Athanasiou, K. A., C. Zhu, D. R. Lanctot, C. M. Agrawal, and X. Wang. 2000. "Fundamentals of biomechanics in tissue engineering of bone." *Tissue Engineering* **6**(4): 361–81. doi:10.1089/107632700418083.

Bacabac, R. G., T. H. Smit, M. G. Mullender, S. J. Dijcks, J. J. Van Loon, and J. Klein-Nulend. 2004. "Nitric oxide production by bone cells is fluid shear stress rate dependent." *Biochemical and Biophysical Research Communications* **315**(4): 823–9. doi:10.1016/j.bbrc.2004.01.138.

Bae, H., L. Zhao, D. Zhu, L. E. Kanim, J. C. Wang, and R. B. Delamarter. 2010. "Variability across ten production lots of a single demineralized bone matrix product." *Journal of Bone and Joint Surgery. American Volume* **92**(2): 427–35. doi:10.2106/JBJS.H.01400.

Bakker, A. D., K. Soejima, J. Klein-Nulend, and E. H. Burger. 2001. "The production of nitric oxide and prostaglandin E(2) by primary bone cells is shear stress dependent." *Journal of Biomechanics* **34**(5): 671–7. PMID:11311708.

Bakker, A. D., J. Klein-Nulend, and E. H. Burger. 2003. "Mechanotransduction in bone cells proceeds via activation of COX-2, but not COX-1." *Biochemical and Biophysical Research Communications* **305**(3): 677–83. PMID:12763047.

Barradas, A. M., H. Yuan, C. A. van Blitterswijk, and P. Habibovic. 2011. "Osteoinductive biomaterials: current knowledge of properties, experimental models and biological mechanisms." *European Cells & Materials* **21**: 407–29; disc. 429. PMID:21604242.

Barrere, F., C. M. van der Valk, R. A. Dalmeijer, G. Meijer, C. A. van Blitterswijk, K. de Groot, and P. Layrolle. 2003. "Osteogenecity of octacalcium phosphate coatings applied on porous metal implants." *Journal of Biomedical Materials Research. Part A* **66**(4): 779–88. doi:10.1002/jbm.a.10454.

Barriga, A., P. Diaz-de-Rada, J. L. Barroso, M. Alfonso, M. Lamata, S. Hernáez, et al. 2004. "Frozen cancellous bone allografts: positive cultures of implanted grafts in posterior fusions of the spine." *European Spine Journal* **13**(2): 152–6. doi:10.1007/s00586-003-0633-9.

Bignon, A., J. Chouteau, J. Chevalier, G. Fantozzi, J. P. Carret, P. Chavassieux, et al. 2003. "Effect of micro- and macroporosity of bone substitutes on their mechanical properties and cellular response." *Journal of Materials Science. Materials in Medicine* **14**(12): 1089–97. PMID:15348502.

Boyan, B. D., T. W. Hummert, D. D. Dean, and Z. Schwartz. 1996. "Role of material surfaces in regulating bone and cartilage cell response." *Biomaterials* **17**(2): 137–46. PMID:8624390.

Boyde, A., A. Corsi, R. Quarto, R. Cancedda, and P. Bianco. 1999. "Osteoconduction in large macroporous hydroxyapatite ceramic implants: evidence for a complementary integration and disintegration mechanism." *Bone* **24**(6): 579–89. PMID:10375200.

Burr, D. B., R. B. Martin, M. B. Schaffler, and E. L. Radin. 1985. "Bone remodeling in response to in vivo fatigue microdamage." *Journal of Biomechanics* **18**(3): 189–200. PMID:3997903.

Burstein, A. H., D. T. Reilly, and M. Martens. 1976. "Aging of bone tissue: mechanical properties." *Journal of Bone and Joint Surgery. American Volume* **58**(1): 82–6. PMID:1249116.

Cameron, D. A. 1972. *The Ultrastructure of Bone: The Biochemistry and Physiology of Bone B.* Cambridge, MA: Academic Press. pp. 191–236.

Campion, C. R., C. Chander, T. Buckland, and K. Hing. 2011. "Increasing strut porosity in silicate-substituted calcium-phosphate bone graft substitutes enhances osteogenesis." *Journal of Biomedical Materials Research. Part B, Applied Biomaterials* **97**(2): 245–54. doi:10.1002/jbm.b.31807.

Cao, W. and L. L. Hench. 1996. "Bioactive materials." *Ceramics International* **22**: 493–507. doi:10.1016/0272-8842(95)00126-3.

Carlisle, E. M. 1970. "Silicon: a possible factor in bone calcification." *Science* **167**(916): 279–80. PMID:5410261.

Carragee, E. J., E. L. Hurwitz, et al. 2011. "A critical review of recombinant human bone morphogenetic protein-2 trials in spinal surgery: emerging safety concerns and lessons learned." *Spine Journal* **11**(6): 471–91. doi:10.1016/j.spinee.2011.04.023.

Carter, D. R. and W. C. Hayes. 1976. "Bone compressive strength: the influence of density and strain rate." *Science* **194**(4270): 1174–6. PMID:996549.

Chan, O., M. J. Coathup, A. Nesbitt, C. Y. Ho, K. A. Hing, T. Buckland, et al. 2012. "The effects of microporosity on osteoinduction of calcium phosphate bone graft substitute biomaterials." *Acta Biomaterialia* **8**(7): 2788–94. doi:10.1016/j.actbio.2012.03.038.

Chau, A. M. and R. J. Mobbs. 2009. “Bone graft substitutes in anterior cervical discectomy and fusion.” *European Spine Journal* **18**(4): 449–64. doi:10.1007/s00586-008-0878-4.
Chong, D. S., L. A. Turner, N. Gadegaard, A. M. Seifalian, M. J. Dalby, and G. Hamilton. 2015. “Nanotopography and plasma treatment: redesigning the surface for vascular graft endothelialisation.” *European Journal of Vascular and Endovascular Surgery* **49**(3): 335–43. doi:10.1016/j.ejvs.2014.12.008.
Chu, T. M., D. G. Orton, S. J. Hollister, S. E. Feinberg, and J. W. Halloran. 2002. “Mechanical and in vivo performance of hydroxyapatite implants with controlled architectures.” *Biomaterials* **23**(5): 1283–93. PMID:11808536.
Claes, L., M. Reusch, M. Göckelmann, M. Ohnmacht, T. Wehner, M. Amling, et al. 2011. “Metaphyseal fracture healing follows similar biomechanical rules as diaphyseal healing.” *Journal of Orthopaedic Research* **29**(3): 425–32. doi:10.1002/jor.21227.
Coathup, M. J., S. Samizadeh, Y. S. Fang, T. Buckland, K. A. Hing, and G. W. Blunn. 2011. “The osteoinductivity of silicate-substituted calcium phosphate.” *Journal of Bone and Joint Surgery. American Volume* **93**(23): 2219–26. doi:10.2106/JBJS.I.01623.
Coathup, M. J., K. A. Hing, S. Samizadeh, O. Chan, Y. S. Fang, C. Campion, et al. 2012. “Effect of increased strut porosity of calcium phosphate bone graft substitute biomaterials on osteoinduction.” *Journal of Biomedical Materials Research. Part A* **100**(6): 1550–5. doi:10.1002/jbm.a.34094.
Cowin, S. C. and S. Weinbaum. 1998. “Strain amplification in the bone mechanosensory system.” *American Journal of Medical Science* **316**(3): 184–8. PMID:9749560.
Czitrom, A. and A. Gross. 1992. *Allografts in Orthopaedic Practice*. Baltimore, MD: Lippincott Williams & Wilkins.
Daculsi, G. and N. Passuti. 1990. “Effect of the macroporosity for osseous substitution of calcium phosphate ceramics.” *Biomaterials* **11**: 86–7. PMID:2397267.
Dalby, M. J., L. Di Silvio, G. W. Davies, and W. Bonfield. 2000. “Surface topography and HA filler volume effect on primary human osteoblasts in vitro.” *Journal of Materials Science. Materials in Medicine* **11**(12): 805–10. PMID:15348064.
Dalby, M. J., L. Di Silvio, E. J. Harper, and W. Bonfield. 2002. “Increasing hydroxyapatite incorporation into poly(methylmethacrylate) cement increases osteoblast adhesion and response.” *Biomaterials* **23**(2): 569–76. PMID:11761177.
Dard, M., A. Bauer, A. Liebendörfer, H. Wahlig, and E. Dingeldein. 1994. “Preparation physiochemical and biological evaluation of a hydroxyapatite ceramic from bovine spongiosa.” *Act Odonto Stom* **185**: 61–6.
de Jong, W. F. 1926. “Le substance minerale dans le os.” *Recueil des Travaux Chimiques des Pays-Bas* **45**: 445.
Di Palma, F., M. Douet, C. Boachon, A. Guignandon, S. Peyroche, B. Forest, et al. 2003. “Physiological strains induce differentiation in human osteoblasts cultured on orthopaedic biomaterial.” *Biomaterials* **24**(18): 3139–51. PMID:12895587.
Di Palma, F., A. Guignandon, A. Chamson, M. H. Lafage-Proust, N. Laroche, S. Peyroche, et al. 2005. “Modulation of the responses of human osteoblast-like cells to physiologic mechanical strains by biomaterial surfaces.” *Biomaterials* **26**(20): 4249–57. doi:10.1016/j.biomaterials.2004.10.041.
Dinopoulos, H., R. Dimitriou, and P. V. Giannoudis. 2012. “Bone graft substitutes: what are the options?” *Surgeon* **10**(4): 230–9. doi:10.1016/j.surge.2012.04.001.
Eggli, P. S., W. Muller, and R. K. Schenk. 1988. “Porous hydroxyapatite and tricalcium phosphate cylinders with two different pore size ranges implanted in the cancellous

bone of rabbits. A comparative histomorphometric and histologic study of bony ingrowth and implant substitution." *Clinical Orthopaedics and Related Research* **232**: 127–38. PMID:2838207.

Elliot, J. C. 1994. *Structure and Chemistry of the Apatites and other Calcium Orthophosphates*. Amsterdam: Elsevier.

Epker, B. N., R. Hattner, and H. M. Frost. 1964. "Radial rate of osteon closure: its application in the study of bone formation in metabolic bone disease." *Journal of Laboratory and Clinical Medicine* **64**: 643–53. PMID:14233153.

Frost, H. M. 1987. "Bone 'mass' and the 'mechanostat': a proposal." *Anatomical Record* **219**(1): 1–9.

Gauthier, O., J. M. Bouler, E. Aguado, P. Pilet, and G. Daculsi. 1998. "Macroporous biphasic calcium phosphate ceramics: influence of macropore diameter and macroporosity percentage on bone ingrowth." *Biomaterials* **19**(1–3): 133–9. PMID:9678860.

Giannoudis, P. V., H. Dinopoulos, and E. Tsiridis. 2005. "Bone substitutes: an update." *Injury* **36**(Suppl. 3): S20–7. doi:10.1016/j.injury.2005.07.029.

Gibson, L. J. 1985. "The mechanical behaviour of cancellous bone." *Journal of Biomechanics* **18**(5): 317–28. PMID:4008502.

Goldstein, S. A. 1987. "The mechanical properties of trabecular bone: dependence on anatomic location and function." *Journal of Biomechanics* **20**(11–12): 1055–61. PMID:3323197.

Greenspan, D. 1999. "Bioactive ceramic implant materials." *Current Opinion in Solid State & Materials Science* **4**(4): 389–93.

Guth, K., C. Campion, T. Buckland and K. A. Hing. 2010a. "Surface physiochemistry affects protein adsorption to stoichiometric and silicate-substituted microporous hydroxyapatites." *Advanced Engineering Materials* **12**(4): B113–21. doi:10.1002/adem.200980026.

Guth, K., C. Campion, T. Buckland, and K. A. Hing. 2010b. "Effect of silicate-substitution on attachment and early development of human osteoblast-like cells seeded on microporous hydroxyapatite discs." *Advanced Engineering Materials* **12**(1–2): B26–36. doi:10.1002/adem.200980003.

Guth, K., C. Campion, T. Buckland, and K. A. Hing. 2011. "Effects of serum protein on ionic exchange between culture medium and microporous hydroxyapatite and silicate-substituted hydroxyapatite." *Journal of Materials Science. Materials in Medicine* **22**(10): 2155–64. doi:10.1007/s10856-011-4409-1.

Habibovic, P., H. Yuan, C. M. van der Valk, G. Meijer, C. A. van Blitterswijk, and K. de Groot K. 2005. "3D microenvironment as essential element for osteoinduction by biomaterials." *Biomaterials* **26**(17): 3565–75. doi:10.1016/j.biomaterials.2004.09.056.

Habibovic, P., T. M. Sees, M. A. van den Doel, C. A. van Blitterswijk, and K. de Groot. 2006. "Osteoinduction by biomaterials – physicochemical and structural influences." *Journal of Biomedical Materials Research. Part A* **77**(4): 747–62. doi:10.1002/jbm.a.30712.

Healy, K. E., C. H. Thomas, A. Rezania, J. E. Kim, P. J. McKeown, B. Lom, and P. E. Hockberger. 1996. "Kinetics of bone cell organization and mineralization on materials with patterned surface chemistry." *Biomaterials* **17**(2): 195–208. PMID:8624396.

Hing, K. A. 2005. "Bioceramic bone graft substitutes: influence of porosity and chemistry." *International Journal of Applied Ceramic Technology* **2**(3): 184–99. doi:10.1111/j.1744-7402.2005.02020.x.

Hing, K. A., S. M. Best, and W. Bonfield. 1999a. "Characterization of porous hydroxyapatite." *Journal of Materials Science. Materials in Medicine* **10**(3): 135–45. PMID:15348161.

Hing, K. A., S. M. Best, K. E. Tanner, W. Bonfield, and P. A. Revell. 1999b. "Quantification of bone ingrowth within bone-derived porous hydroxyapatite implants of varying density." *Journal of Materials Science. Materials in Medicine* **10**(10/11): 663–70. PMID:15347983.

Hing, K. A., S. M. Best, K. E. Tanner, W. Bonfield, and P. A. Revell. 2004. "Mediation of bone ingrowth in porous hydroxyapatite bone graft substitutes." *Journal of Biomedical Materials Research. Part A* **68**(1): 187–200. doi:10.1002/jbm.a.10050.

Hing, K., B. Annaz, S. Saeed, P. A. Revell, and T. Buckland. 2005. "Microporosity enhances bioactivity of synthetic bone graft substitutes." *Journal of Materials Science. Materials in Medicine* **16**(5): 467–75. doi:10.1007/s10856-005-6988-1.

Hing, K. A., L. F. Wilson, and T. Buckland. 2007. "Comparative performancve of three ceramic bone graft substitutes." *Spine Journal* **7**(4): 475–90. doi:10.1016/j.spinee.2006.07.017.

Hodgskinson, R. and J. D. Currey. 1990. "The effect of variation in structure on the Young's modulus of cancellous bone: a comparison of human and non-human material." *Proceedings of the Institution of Mechanical Engineers. Part H, Journal of Engineering in Medicine* **204**(2): 115–21. PMID:2095142.

Holmes, R. E. 1979. "Bone regeneration within a coralline hydroxyapatite implant." *Plastic and Reconstructive Surgery* **63**(5): 626–33. PMID:432330.

Holmes, R., V. Mooney, R. Bucholz, and A. Tencer. 1984. "A coralline hydroxyapatite bone graft substitute. Preliminary report." *Clinical Orthopaedics and Related Research* **188**: 252–62. PMID:6147218.

Hulbert, S. F., S. J. Morrison, and J. J. Klawitter. 1972. "Tissue reaction to three ceramics of porous and non-porous structures." *Journal of Biomedical Materials Research* **6**(5): 347–74. doi:10.1002/jbm.820060505.

Hunter, A., C. W. Archer, P. S. Walker, and G. W. Blunn. 1995. "Attachment and proliferation of osteoblasts and fibroblasts on biomaterials for orthopaedic use." *Biomaterials* **16**(4): 287–95. PMID:7772668.

Ignatius, A., H. Blessing, A. Liedert, C. Schmidt, C. Neidlinger-Wilke, D. Kaspar, et al. 2005. "Tissue engineering of bone: effects of mechanical strain on osteoblastic cells in type I collagen matrices." *Biomaterials* **26**(3): 311–18. doi:10.1016/j.biomaterials.2004.02.045.

Jarcho, M. 1981. "Calcium phoshpate ceramics as hard tissue prosthetics." *Clinical Orthopaedic Related Research* **157**: 259. PMID:7018783.

Joldersma, M., E. H. Burger, C. M. Semeins, and J. Klein-Nulend. 2000. "Mechanical stress induces COX-2 mRNA expression in bone cells from elderly women." *Journal of Biomechanics* **33**(1): 53–61. PMID:10609518.

Keller, T. S. 1994. "Predicting the compressive mechanical behavior of bone." *Journal of Biomechanics* **27**(9): 1159–68. PMID:7929465.

Kim, D. H., R. Rhim, L. Li, J. Martha, B. H. Swaim, R. J. Banco, et al. 2009. "Prospective study of iliac crest bone graft harvest site pain and morbidity." *Spine Journal* **9**(11): 886–92. doi:10.1016/j.spinee.2009.05.006.

Kingsmill, V. J., I. J. McKay, P. Ryan, M. R. Ogden, and S. C. Rawlinson. 2013. "Gene expression profiles of mandible reveal features of both calvarial and ulnar bones in the adult rat." *Journal of Dentistry* **41**(3): 258–64. doi:10.1016/j.jdent.2012.11.010.

Klawitter, J. J. and S. F. Hulbert. 1971. "Application of porous ceramics for the attachment of load bearing internal orthopaedic applications." *Journal of Biomedical Materials Research* **2**(1): 161–229. doi:10.1002/jbm.820050613

Klawitter, J. J., J. G. Bagwell, A. M. Weinstein, and B. W. Sauer. 1976. "An evaluation of bone growth into porous high density polyethylene." *Journal of Biomedical Materials Research* **10**(2): 311–23. doi:10.1002/jbm.820100212.

Klein-Nulend, J., C. M. Semeins, N. E. Ajubi, P. J. Nijweide, and E. H. Burger. 1995a. "Pulsating fluid flow increases nitric oxide (NO) synthesis by osteocytes but not periosteal fibroblasts – correlation with prostaglandin upregulation." *Biochemical and Biophysical Research Communications* **217**(2): 640–8. PMID:7503746.

Klein-Nulend, J., A. van der Plas, C. M. Semeins, N. E. Ajubi, J. A. Frangos, P. J. Nijweide, and E. H. Burger. 1995b. "Sensitivity of osteocytes to biomechanical stress in vitro." *FASEB Journal* **9**(5): 441–5. PMID:7896017.

Klein-Nulend, J., J. Roelofsen, C. M. Semeins, A. L. Bronckers, and E. H. Burger. 1997. "Mechanical stimulation of osteopontin mRNA expression and synthesis in bone cell cultures." *Journal of Cell Physiology* **170**(2): 174–81. doi:10.1002/(SICI)1097-4652(199702)170:2<174::AID-JCP9>3.0.CO;2-L

Klein-Nulend, J., M. H. Helfrich, J. G. Sterck, H. MacPherson, M. Joldersma, S. H. Ralston, et al. 1998. "Nitric oxide response to shear stress by human bone cell cultures is endothelial nitric oxide synthase dependent." *Biochemical and Biophysical Research Communications* **250**(1): 108–14. doi:10.1006/bbrc.1998.9270.

Kokubo, T., S. Ito, Z. T. Huang, T. Hayashi, S. Sakka, T. Kitsugi, and T. Yamamuro. 1990a. "Ca,P-rich layer formed on high-strength bioactive glass-ceramic A-W." *Journal of Biomedical Materials Research* **24**(3): 331–43. doi:10.1002/jbm.820240306.

Kokubo, T., H. Kushitani, S. Sakka, T. Kitsugi, and T. Yamamuro. 1990b. "Solutions able to reproduce in vivo surface-structure changes in bioactive glass-ceramic A-W." *Journal of Biomedical Materials Research* **24**(6): 721–34. doi:10.1002/jbm.820240607.

Kuhne, J. H., R. Bartl, B. Frisch, C. Hammer, V. Jansson, and M. Zimmer. 1994. "Bone formation in coralline hydroxyapatite. Effects of pore size studied in rabbits." *Acta Orthopaedica Scandinavica* **65**(3): 246–52. PMID:8042473.

Lampin, M., C. Warocquier, C. Legris, M. Degrange, and M. F. Sigot-Luizard. 1997. "Correlation between substratum roughness and wettability, cell adhesion, and cell migration." *Journal of Biomedical Materials Research* **36**(1): 99–108. PMID:9212394.

Lanyon, L. E., A. E. Goodship, C. J. Pye, and J. H. MacFie. 1982. "Mechanically adaptive bone remodelling." *Journal of Biomechanics* **15**(3): 141–54. PMID:7096367.

Laurencin, C. T. 2003. *Bone Graft Substitutes.* Bridgeport, PA: ASTM international.

Le Geros, R. Z. and J. P. Le Geros. 1993. *Dense Hydroxyapatite. An Introduction to Biocramics.* Singapore: World Scientific. pp. 139–80.

Le Huec, J. C., T. Schaeverbeke, D. Clement, J. Faber, and A. Le Rebeller. 1995. "Influence of porosity on the mechanical resistance of hydroxyapatite ceramics under compressive stress." *Biomaterials* **16**(2): 113–18. PMID:7734643.

Lee, K. J., J. G. Roper, and J. C. Wang. 2005. "Demineralized bone matrix and spinal arthrodesis." *Spine Journal* **5**(6 Suppl.): 217S–23S. doi:10.1016/j.spinee.2005.02.006.

LeGeros, R. Z. 2002. "Properties of osteoconductive biomaterials: calcium phosphates." *Clinical Orthopaedics and Related Research* **395**: 81–98. PMID:11937868.

LeVier, R. R. 1975. "Distribution of silicon in the adult rat and rhesus monkey." *Bioinorganic Chemistry* **4**(2): 109–15. PMID:1125329.

Li, S., J. R. De Wijn, J. Li, P. Layrolle, and K. De Groot. 2003. "Macroporous biphasic calcium phosphate scaffold with high permeability/porosity ratio." *Tissue Engineering* **9**(3): 535–48. doi:10.1089/107632703322066714.

Linde, F., P. Norgaard, I. Hvid, A. Odgaard, and K. Soballe. 1991. "Mechanical properties of trabecular bone. Dependency on strain rate." *Journal of Biomechanics* **24**(9): 803–9. PMID:1752864.

Little, D. G., M. McDonald, R. Bransford, C. B. Godfrey, and N. Amanat. 2005. "Manipulation of the anabolic and catabolic responses with OP-1 and zoledronic acid in a rat critical defect model." *Journal of Bone and Mineral Research* **20**(11): 2044–52. doi:10.1359/JBMR.050712.

Liu, D. M. 1997. "Fabrication of hydroxyapatite ceramic with controlled porosity." *Journal of Materials Science. Materials in Medicine* **8**(4): 227–32. PMID:15348763.

Lu, J. X., B. Flautre, K. Anselme, P. Hardouin, A. Gallur, M. Descamps, and B. Thierry. 1999. "Role of interconnections in porous bioceramics on bone recolonization in vitro and in vivo." *Journal of Materials Science. Materials in Medicine* **10**(2): 111–20. PMID:15347932.

Martin, R. B., M. W. Chapman, N. A. Sharkey, S. L. Zissimos, B. Bay, and E. C. Shors. 1993. "Bone ingrowth and mechanical properties of coralline hydroxyapatite 1 yr after implantation." *Biomaterials* **14**(5): 341–8. PMID:8389612.

Martini, F. 1998. *Fundamentals of Anatomy and Phsyriology*. Upper Saddle River, NJ: Prentice Hall.

Mauney, J. R., S. Sjostorm, J. Blumberg, R. Horan, J. P. O'Leary, G. Vunjak-Novakovic, et al. 2004. "Mechanical stimulation promotes osteogenic differentiation of human bone marrow stromal cells on 3-D partially demineralized bone scaffolds in vitro." *Calcified Tissue International* **74**(5): 458–68.

McConnell, D. 1973. *Apatite*. Berlin: Springer-Verlag. pp. 68–80.

McCalden, R. W., J. A. McGeough, and C. M. Court-Brown. 1997. "Age-related changes in the compressive strength of cancellous bone. The relative importance of changes in density and trabecular architecture." *Journal of Bone and Joint Surgery. American Volume* **79**(3): 421–7. PMID:9070533.

McCann, S., J. L. Byrne, M. Rovira, P. Shaw, P. Ribaud, S. Sica, et al. 2004. "Outbreaks of infectious diseases in stem cell transplant units: a silent cause of death for patients and transplant programmes." *Bone Marrow Transplantation* **33**(5): 519–29. doi:10.1038/sj.bmt.1704380.

McGarry, J. G. and P. J. Prendergast. 2004. "A three-dimensional finite element model of an adherent eukaryotic cell." *European Cells & Materials* **7**: 27–33; disc. 33-24.

McGarry, J. G., J. Klein-Nulend, and P. J. Prendergast. 2005. "The effect of cytoskeletal disruption on pulsatile fluid flow-induced nitric oxide and prostaglandin E2 release in osteocytes and osteoblasts." *Biochemical and Biophysical Research Communications* **330**(1): 341–8. doi:10.1016/j.bbrc.2005.02.175.

Meeder, P. J. and C. Eggers. 1994. "The history of autogenous bone grafting." *Injury* **25**(Suppl. 1): A2–3. PMID:7927653.

Miyazaki, M., H. Tsumura, J. C. Wang, and A. Alanay. 2009. "An update on bone substitutes for spinal fusion." *European Spine Journal* **18**(6): 783–99. doi:10.1007/s00586-009-0924-x.

Mullender, M., A. J. El Haj, Y. Yang, M. A. van Duin, E. H. Burger, and J. Klein-Nulend. 2004. "Mechanotransduction of bone cells in vitro: mechanobiology of bone tissue." *Medical & Biological Engineering & Computing* **42**(1): 14–21. PMID:14977218.

Nagatomi, J., B. P. Arulanandam, D. W. Metzger, A. Meunier, and R. Bizios. 2003. "Cyclic pressure affects osteoblast functions pertinent to osteogenesis." *Annals of Biomedical Engineering* **31**(8): 917–23. PMID:12918906.

O'Connor, J. A., L. E. Lanyon, and H. MacFie. 1982. "The influence of strain rate on adaptive bone remodelling." *Journal of Biomechanics* **15**(10): 767–81. PMID:7153230.

Olivares-Navarrete, R., R. A. Gittens, J. M. Schneider, S. L. Hyzy, D. A. Haithcock, P. F. Ullrich, et al. 2012. "Osteoblasts exhibit a more differentiated phenotype and increased bone morphogenetic protein production on titanium alloy substrates than on poly-ether-ether-ketone." *Spine Journal* **12**(3): 265–72. doi:10.1016/j.spinee.2012.02.002.

Ouyang, J., G. T. Yang, W. Z. Wu, Q. A. Zhu, and S. Z. Zhong. 1997. "Biomechanical characteristics of human trabecular bone." *Clinical Biomechanics* **12**(7–8): 522–4. PMID:11415763.

Ozawa, S. and S. Kasugai. 1996. "Evaluation of implant materials (hydroxyapatite, glass-ceramics, titanium) in rat bone marrow stromal cell culture." *Biomaterials* **17**(1): 23–9. PMID:8962943.

Park, J. B. and J. D. Bronzino. 2003. *Biomaterials: Principles and Applications.* Boca Raton, FL: CRC Press.

Pavalko, F. M., N. X. Chen, C. H. Turner, D. B. Burr, S. Atkinson, Y. F. Hsieh, et al. 1998. "Fluid shear-induced mechanical signaling in MC3T3-E1 osteoblasts requires cytoskeleton-integrin interactions." *American Journal of Physiology* **275**(6 Pt. 1): C1591–601. PMID:9843721.

Peelen, J. G. J., B. V. Rejda, and K. de Groot. 1978. "Preparation and properties of sintered hydroxyapatite." *Ceramurgia International* **4**(2): 71–4.

Posner, A. S. 1969. "Crystal chemistry of bone mineral." *Physiological Reviews* **49**(4): 760–92. PMID:4898602.

Ravaglioli, A. 1992. *Bioceramics.* London: Chapman and Hall.

Ripamonti, U. 1996. "Osteoinduction in porous hydroxyapatite implanted in heterotopic sites of different animal models." *Biomaterials* **17**(1): 31–5. PMID:8962945.

Robinson, R. A. and M. L. Watson 1952. "Collagen-crystal relationships in bone as seen in the electron microscope." *Anatomical Record* **114**(3): 383–409. PMID:12996880.

Roelofsen, J., J. Klein-Nulend, and E. H. Burger. 1995. "Mechanical stimulation by intermittent hydrostatic compression promotes bone-specific gene expression in vitro." *Journal of Biomechanics* **28**(12): 1493–503. PMID:8666589.

Rohl, L., E. Larsen, F. Linde, A. Odgaard, and J. Jorgensen. 1991. "Tensile and compressive properties of cancellous bone." *Journal of Biomechanics* **24**(12): 1143–9. PMID:1769979.

Roy, D. M. and S. K. Linnehan. 1974. "Hydroxyapatite formed from coral skeletal carbonate by hydrothermal exchange." *Nature* **247**(438): 220–2. PMID:4149289.

Rubin, P. A., J. K. Popham, J. R. Bilyk, and J. W. Shore. 1994. "Comparison of fibrovascular ingrowth into hydroxyapatite and porous polyethylene orbital implants." *Ophthalmic Plastic and Reconstructive Surgery* **10**(2): 96–103.

Shors, E. C. 1999. "Coralline bone graft substitutes." *Orthopedic Clinics of North America* **30**(4): 599–613. PMID:10471765.

Sikavitsas, V. I., J. S. Temenoff, and A. G. Mikos. 2001. "Biomaterials and bone mechanotransduction." Biomaterials **22**(19): 2581–93. PMID:11519777.

Sikavitsas, V. I., G. N. Bancroft, H. L. Holtorf, J. A. Jansen, and A. G. Mikos. 2003. "Mineralized matrix deposition by marrow stromal osteoblasts in 3D perfusion culture increases with increasing fluid shear forces." *Proceedings of the National Academy of Sciences of the United States of America* **100**(25): 14 683–8. doi:10.1073/pnas.2434367100.

Sikavitsas, V. I., G. N. Bancroft, J. J. Lemoine, M. A. Liebschner, M. Dauner, and A. G. Mikos. 2005. "Flow perfusion enhances the calcified matrix deposition of marrow stromal cells in biodegradable nonwoven fiber mesh scaffolds." *Annals of Biomedical Engineering* **33**(1): 63–70. PMID:15709706.

Smith, L. 1963. "Ceramic-plastic material as a bone substitute." *Archives of Surgery* **87**: 653–61. PMID:14056248.

Sterck, J. G., J. Klein-Nulend, P. Lips, and E. H. Burger. 1998. "Response of normal and osteoporotic human bone cells to mechanical stress in vitro." *American Journal of Physiology* **274**(6 Pt. 1): E1113–20. PMID:9611164.

Suchanek, W., M. Yashima, M. Kakihana, and M. Yoshimura. 1996. "Processing and mechanical properties of hydroxyapatite reinforced with hydroxyapatite whiskers." *Biomaterials* **17**(17): 1715–23. PMID:8866034.

Tanaka, S. M., H. B. Sun, R. K. Roeder, D. B. Burr, C. H. Turner, and H. Yokota. 2005. "Osteoblast responses one hour after load-induced fluid flow in a three-dimensional porous matrix." *Calcified Tissue International* **76**(4): 261–71. doi:10.1007/s00223-004-0238-2.

Togawa, D., T. W. Bauer, I. H. Lieberman, and H. Sakai. 2004. "Lumbar intervertebral body fusion cages: histological evaluation of clinically failed cages retrieved from humans." *Journal of Bone and Joint Surgery. American Volume* **86-A**(1): 70–9.

Uchida, A., S. M. Nade, et al. 1984. "The use of ceramics for bone replacement. A comparative study of three different porous ceramics." *Journal of Bone and Joint Surgery. British Volume* **66**(2): 269–75. PMID:6323483.

Urist, M. R. 1965. "Bone: formation by autoinduction." *Science* **150**(698): 893–9. PMID:5319761.

Vaidya, R., R. Weir, A. Sethi, S. Meisterling, W. Hakeos, and C. D. Wybo. 2007. "Interbody fusion with allograft and rhBMP-2 leads to consistent fusion but early subsidence." *Journal of Bone and Joint Surgery. British Volume* **89**(3): 342–5. doi:10.1302/0301-620X.89B3.18270.

van Blitterswijk, C. A., D. Bakker, S. C. Hesseling, and H. K. Koerten. 1991. "Reactions of cells at implant surfaces." *Biomaterials* **12**(2): 187–93. PMID:1878452.

Wang, Y., T. Uemura, J. Dong, H. Kojima, J. Tanaka, and T. Tateishi. 2003. "Application of perfusion culture system improves in vitro and in vivo osteogenesis of bone marrow-derived osteoblastic cells in porous ceramic materials." *Tissue Engineering* **9**(6): 1205–14. doi:10.1089/10763270360728116.

Westbroek, I., N. E. Ajubi, M. J. Alblas, C. M. Semeins, J. Klein-Nulend, E. H. Burger, and P. J. Nijweide. 2000. "Differential stimulation of prostaglandin G/H synthase-2 in osteocytes and other osteogenic cells by pulsating fluid flow." *Biochemical and Biophysical Research Communications* **268**(2): 414–19. doi:10.1006/bbrc.2000.2154.

White, E. W., J. N. Weber, D. M. Roy, E. L. Owen, R. T. Chiroff, and R. A. White. 1975. "Replamineform porous biomaterials for hard tissue implant applications." *Journal of Biomedical Materials Research* **9**(4): 23–7. doi:10.1002/jbm.820090406.

Winter, M. and P. Griss. 1981. "Comparative histocompatability testing of seven calcium phoshpate ceramics." *Biomaterials* **2**: 159–61. PMID:6268208.

Wolff, J. 1870. "Uber die innrer Architektur der knochen und ihre Bedeutung fur die Fragen vom Knochenwachsthum." *Virchows Archiv fur pathologische Anatomie und Physiologie und fur klinische Medizin* **50**: 389–450.

Wong, D. A., A. Kumar, S. Jatana, G. Ghiselli, and K. Wong. 2008. "Neurologic impairment from ectopic bone in the lumbar canal: a potential complication of off-label PLIF/TLIF use of bone morphogenetic protein-2 (BMP-2)." *Spine Journal* **8**(6): 1011–18. doi:10.1016/j.spinee.2007.06.014.

Yoshii, S., Y. Kakutani, T. Yamamuro, T. Nakamura, T. Kitsugi, M. Oka, et al. 1988. "Strength of bonding between A-W glass-ceramic and the surface of bone cortex." *Journal of Biomedical Materials Research* **22**(3 Suppl.): 327–38. PMID:3235467.

Yuan, H., K. Kurashina, J. D. de Bruijn, Y. Li, K. de Groot, and X. Zhang. 1999. "A preliminary study on osteoinduction of two kinds of calcium phosphate ceramics." *Biomaterials* **20**(19): 1799–806. PMID:10509190.

Yuan, H., M. Van Den Doel, S. Li, C. A. Van Blitterswijk, K. De Groot, and J. D. De Bruijn. 2002. "A comparison of the osteoinductive potential of two calcium phosphate ceramics implanted intramuscularly in goats." *Journal of Materials Science. Materials in Medicine* **13**(12): 1271–5.

Zdeblick, T. A., M. E. Cooke, D. N. Kunz, D. Wilson, and R. P. McCabe. 1994. "Anterior cervical discectomy and fusion using a porous hydroxyapatite bone graft substitute." *Spine* **19**(20): 2348–57.

22

Exploitation of Mechanobiology for Cardiovascular Therapy

Winston Elliott[1], Amir Keshmiri[2,3], and Wei Tan[1]

[1] Department of Mechanical Engineering, University of Colorado at Boulder, Boulder, CO, USA
[2] Engineering and Materials Research Centre, Manchester Metropolitan University, Manchester, UK
[3] School of Mechanical, Aerospace and Civil Engineering, the University of Manchester, Manchester, UK

22.1 Introduction

Cardiovascular disease (CVD) is the most common cause of death in the United States, accounting for 27% of all deaths each year (Anderson and Smith 2003). Globally, it is projected to rise above 20% by 2030 (Mathers and Loncar 2006). Traditional CVD biomedical research studies have largely focused on biochemical signaling cascades, leading to improved understanding of CVD pathogenesis and thus innovations in CVD treatment. In the last 2 decades, increasing attention has been paid to cardiovascular mechanobiology, and researchers have started the journey of exploiting it for medical benefit, including modeling vascular diseases and vascular treatments to unravel insightful understandings and to achieve optimal therapeutic results, respectively. This chapter reviews recent trends in research related to cardiovascular mechanobiology. It contains two main topics: (i) arterial wall mechanics and mechanobiology, and their exploitation in understanding arterial disease progression and therapeutics; and (ii) the role of computational hemodynamics in cardiovascular mechanobiology, and its exploitation in vascular graft application. We will discuss each topic only in its essentials, with no intention of exhausting it.

The cardiovascular system is a dynamic, cyclically loaded, pressure-driven flow system that involves ongoing mechanical feedback loops directing cell response and/or extracellular matrix (ECM) remodeling, which sustain homeostasis of the system, guide adaptive remodeling of the vessel wall, and perpetuate disease progression. Mechanotransduction thus represents an important mechanism in determining vascular health and disease response. Mechanical signals within the vasculature have been shown to be atheroprotective in healthy blood vessels, and to initiate or exacerbate vascular lesions in diseased vessels. Current knowledge about mechanical signaling in the cardiovascular system has been extended to improve treatment options; for example, the performance of many artificial grafting and stenting devices used for restenosis may be enhanced when the local impact of their mechanics on the biomechanical environments in the vasculature of neighboring blood vessels, as well as the global

Mechanobiology: Exploitation for Medical Benefit, First Edition. Edited by Simon C. F. Rawlinson.

impact on the upstream and downstream vascular system, is considered for design optimization. To address how the interactions of vascular flow and vessel wall have an impact on vascular health and disease, we begin with an overview of the importance of mechanotransduction to vascular health and the development of various mechanical signals that lead to disease states. We then give examples showing that the pulsatile artery flow interacts with artery structural mechanics to alter mechanical signals, making the healthy vasculature progress to a diseased one. We also review the use of computational fluid dynamics (CFD) to explore vascular mechanobiology and guide the design of vascular implants, and suggest potential avenues for future work using fluid–structure interaction, combining CFD and finite element models (FEMs) with the bench mechanobiology models to provide a better understanding of clinical measures and improve the design of treatment interventions.

22.2 Arterial Wall Mechanics and Mechanobiology

In general, mechanical signaling in the vasculature originates from the intermittent (or pulsatile) blood flow as a result of rhythmic ventricular contractions. The flow travels through the vasculature, interacting with each vascular section and altering its waveform along the way. In particular, the structure and mechanical properties of arteries, including conduit elastic arteries and muscular arteries, play important roles in modulating the blood flow, thereby influencing arterial cell biology, ventricular loading, and downstream vascular function (Figure 22.1).

Increased resistance in conduit arteries may reduce transport downstream and increase ventricular work, which is tied to many vascular diseases. Muscular arteries, such as the femoral and coronary arteries, are also important in maintaining physiologic hemodynamics and preventing CVD. Such influences result from the impact of vessel material mechanics or vessel geometry on blood flow in the vessel, which is translated into dynamic shear signaling on cell response, promoting vessel homeostasis or vessel adaptive remodeling toward dysfunctional conditions. Arterial stiffening and narrowing (or stenosis), for example, are both important indicators for CVD conditions, including myocardial infarction (MI), heart failure, and overall mortality

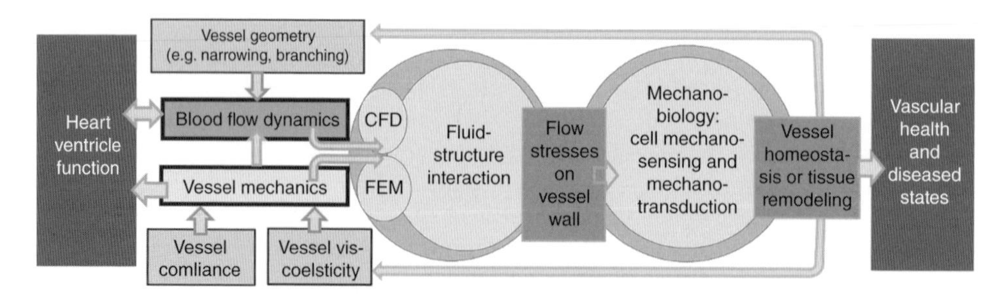

Figure 22.1 Impact of vessel mechanics and geometry on blood flow dynamics through the interaction of flow with vessel wall. This promotes mechanosensing and mechanotransduction signaling cascades within vascular cells, which determines tissue homeostasis or remodeling and thus vascular healthy or diseased states.

(Zieman et al. 2005). Therefore, understanding how fluid–structure interactions influence flow stresses on vessel walls and the resulting cell mechanosensing and mechanotransduction mechanism of the flow stresses is crucial to modeling and predicting pathological progression of CVD, as well as to determining new avenues for treatment optimization.

22.3 Mechanical Signal and Mechanotransduction on the Arterial Wall

Historically, chemokine release and reception was the first paradigm studied in order to gain a physiologic and pathologic understanding of vascular tissue response (Gimbrone et al. 1997). Further work in biochemical signaling examined the influence of ECM ligands and integrin on cellular response (Clark and Brugge 1995; Alenghat and Ingber 2002; Davis and Senger 2005). Within the vasculature, this has progressed to focus on mechanical signaling, another important initiator of cell and tissue adaptation within the artery (Chien 2006; Birukov 2009; Chiu and Chien 2011; Lan et al. 2013). Like biochemical signals, mechanical signals applied on the blood vessel wall alter cellular activities, through signaling cascades that lead to morphological, functional, and behavioral changes in cells, and thereby to remodeling of vascular tissue. Mechanical signals applied on the blood vessel wall result from three different stresses: compressive stress from hydrostatic pressure, cyclical circumferential stress from wall stretching, and fluid shear stress. Mechanotransduction occurs in response to application of stresses on the vascular cell, as well as traction forces applied by the cell on the ECM (Davis and Senger 2005; Huynh et al. 2011; Mason et al. 2013). Additionally, applied stress and resulting strain may be shared among cells through cell–cell connections, such as vascular endothelial cadherin (Chien 2006; Birukov 2009; Chiu and Chien 2011; Shav et al. 2014). This results not only in localized disease response, but likely also in systemic responses within the vasculature.

From the viewpoint of clinical measures, the determination of the mean and/or dynamic pressure and cyclic stretch occurs indirectly through a pressure cuff; alternatively, cyclic stretch and fluid velocity values may be measured directly through Doppler ultrasound (Mitchell 2008; Mitchell et al. 2010, 2011; McArthur et al. 2011). Often, clinical measured hemodynamic results focus on the mean arterial pressure (MAP), peak-to-peak pressure values, and the pulsatility index (PI):

$$PI = \left(\frac{V_{Max} - V_{Min}}{V_{Mean}} \right) \tag{22.1}$$

each of which may be differentiated between physiological and pathological values in determining disease states (Panaritis et al. 2005; Armentano et al. 2006; Arribas et al. 2006; Mitchell 2008; Lemarié et al. 2010; Mitchell et al. 2011).

At the cell and tissue level, studies on mechanical forces applied to the blood vessel have mainly focused on wall stretch stress and fluid shear stress. While hydrostatic pressure exerts compressive stress to affect vascular endothelial cells (Shin et al. 2002), its most important role lies in regulating fluid shear and cyclic stretch through radial expansion of the arterial wall (Chien 2006; Ando and Yamamoto 2009; Birukov 2009; Chiu and Chien 2011). Endothelial cells are especially susceptible to shear forces, through numerous mechanosignaling transduction methods: ion channels, surface glycocalyx, primary cilia, tyrosine kinase receptors, G-proteins, caveolae, adhesive proteins, and, simply, the cytoskeleton (Chien 2006; Ando and Yamamoto 2009; Chiu and Chien 2011). Cyclic stretch of the vascular wall is also important in determining cell response, for both endothelial cells (Birukov et al. 1995; Birukov 2009) and smooth-muscle cells (SMCs) (Leung et al. 1976; Birukov et al. 1995; Kim et al. 1999; Mata-Greenwood et al. 2005; Haga et al. 2007; Birukov 2009). These may induce the release of molecules that facilitate paracrine signaling, such as nitric oxide (NO), reactive oxygen species (ROS), and protein or steroidal paracrine molecules (Leung et al. 1976; Chien 2006; Birukov 2009; Chiu and Chien 2011; Egorova et al. 2011b; Scott et al. 2013; Tan et al. 2014; Elliott et al. 2015). These biomolecular signals may in turn cause changes in vascular tone, ECM production, or degradation, which effectively alters matrix composition and stiffness, and changes the hemodynamic environment (Noris et al. 1995; Davis and Senger 2005; Chien 2006; Birukov 2009; Chiu and Chien 2011). Some signal cascades may only occur when mechanical and chemical signals are applied together (Resnick and Gimbrone 1995; Mata-Greenwood et al. 2005; Bergh et al. 2009; Birukov 2009; Shav et al. 2014), indicating a complicated signaling network and interplay between chemical and mechanical signaling.

At the subcellular level, vascular cell cytoskeletal structure, cellular protrusion (i.e., primary cilia), and cell membrane proteins or macromolecules (e.g., PECAM, VE-cadherin, surface glycocalyx, and ion channels) play key roles in sensing or measuring mechanical signals applied on the vessel wall. Cell cytoskeletal structure aids in controlling the cell cycle and regulating proliferation or apoptosis (Chien 2006; Chiu and Chien 2011). In particular, cell shape is an indicator of focal adhesion (FA) location and density (De Caterina 2000), cytoskeletal restructuring in response to stress (Chien 2006), and cytoskeletal actin tension. Flow stresses, as well as substrate or ECM stiffness and fiber alignment, have been shown to play an important role in determining cell spreading and orientation (Chien 2006; Uttayarat et al. 2010; Huynh et al. 2011; Mason et al. 2013; Chaudhuri et al. 2015). Additionally, the chemical structure of the ECM aids in the migration of epithelial cells to denuded areas (Herbst et al. 1988; Uttayarat et al. 2010) and regulates epithelial cell activity through focal adhesion kinase (FAK) (Wu 2005; Chien 2006; Chiu and Chien 2011; Lu and Rounds 2012; Zebda et al. 2012). Also, specific matrix ligands and their corresponding integrins may further alter cell mechano-sensing and the mechanotransduction of flow stress conditions (Wu 2005). Regarding intracellular signaling molecules involved in flow mechanotransduction, several transcriptional factors and epigenetic mechanisms have been identified in recent studies as critical players in endothelial cells that respond to mechanical signals applied on the vessel wall. The regulatory hierarchy of endothelial function includes control at the epigenetic level; very recent reports have highlighted three epigenetic mechanisms in cell mechanotransduction of flow: microRNAs, histone modifications, and DNA methylation to endothelial gene expression. The discovery of a connection between endothelial cell structures (e.g., cilia) and signaling mechanisms at the genetic and epigenetic levels would open a new chapter in our understanding of the molecular mechanisms regulating vascular responses to changes in flow.

22.4 Physiological and Pathological Responses to Mechanical Signals

Mechanical signaling is an important mediator of healthy vascular activities, affecting both vascular epithelial cells and SMCs in concert with paracrine signaling. Fluid shear and cyclic wall stretch signaling to cells within the vasculature may preserve a healthy, physiological state (Zarins et al. 1983; Ku et al. 1985; Asakura and Karino 1990; Takada et al. 1994; Birukov et al. 1995; Traub and Berk 1998; Berk 2008; Bergh et al. 2009; Birukov 2009; Uttayarat et al. 2010) or may induce a pathological response (Sakao 2006; Birukov 2009; Egorova et al. 2011b; Scott et al. 2013; Tan et al. 2014; Elliott et al. 2015). Importantly, these mechanical signals can exhibit a dose-dependent response (Takada et al. 1994; Noris et al. 1995; Ando and Yamamoto 2009; Bergh et al. 2009; Birukov 2009; Uttayarat et al. 2010; Shav et al. 2014), or they may be optimized toward physiological ranges (Shin et al. 2002; Birukova et al. 2008; Birukov 2009; Shi et al. 2009; Scott et al. 2013). Additionally, vascular cells have been found to be phenotypically specific between different vascular branches (Liu et al. 2008; Chiu and Chien 2011), in that beneficial shear and cyclic stretch values in arteries can initiate inflammation and remodeling response in the venous system (Liu et al. 2008; Owens 2010). Wall shear and stretch stresses may work in concert with each other, or with matrix signals, paracrine signals, or other biochemical signals (Bergh et al. 2009; Egorova et al. 2011b).

For normal vascular physiology, mechanical signaling plays a critical role in maintaining homeostasis of arteries, with both cyclic stretch and fluid shear sustaining the healthy state of the blood vessel. The vascular lumen is lined with a confluent, interconnected, epithelial cell monolayer, protecting it from occlusion. Laminar flow occurs throughout the vasculature, and induces a quiescent state in arteries (Zarins et al. 1983; Ku et al. 1985; Asakura and Karino 1990; Traub and Berk 1998; Chien 2006; Berk 2008; Chiu and Chien 2011). Expression of antithrombogenic factors (Takada et al. 1994; Chien et al. 1998; Traub and Berk 1998; Ando and Yamamoto 2009), cell alignment (Resnick and Gimbrone 1995; García-Cardeña et al. 2001; Chien 2006; Chiu and Chien 2011), cell migration (Hsu et al. 2001; Urbich 2002; Chien 2006), epithelial cell–cell connections (Levesque et al. 1986; Traub and Berk 1998; Kladakis and Nerem 2004; Chien 2006), and NO (Noris et al. 1995; Uematsu et al. 1995; Chiu and Chien 2011) are all upregulated by exposure to laminar shear stresses, while monocyte chemotactic protein and lipid metabolism are downregulated (Chien 2006; Chiu and Chien 2011). Luminal laminar flow is additionally effective in preventing SMC and adventitial fibroblast migration (Garanich 2005; Garanich et al. 2007), and counters inflammatory paracrine signals (Bergh et al. 2009). Additionally, cyclic stretch within the physiological range results in aligned actin and tubulin fibers within the epithelial cell (Chien 2006; Birukov 2009; Chiu and Chien 2011), as well as SMC contractility (Leung et al. 1976; Birukov et al. 1995; Kim et al. 1999; Birukov 2009). Alignment of the epithelial cell cytoskeleton, by shear and cyclic stretch, has been shown to prevent proliferation and apoptosis (Chien 2006). Cell–cell connections and crosstalk between epithelial cells and SMCs are important to consider in maintaining homeostasis (Liu and Goldman 2001; Sho et al. 2002b; Wang et al. 2006; Scott et al. 2013; Shav et al. 2014; Elliott et al. 2015).

Pathological conditions may also be progressed by mechanical signaling, through reduction in atheroprotective expression (Ku et al. 1985; Asakura and Karino 1990; Hsu et al. 2001; Chiu et al. 2003), paracrine signaling by endothelial cells (Resnick and

Gimbrone 1995; Chien et al. 1998; Chien 2006; Scott et al. 2013; Shav et al. 2014), or apoptosis of endothelial cells (Shin et al. 2002; Sakao 2006; Elliott et al. 2015). While high laminar shear stress (e.g., a mean value of 10–20 dyn/cm^2, versus a low shear stress of 1–2 dyn/cm^2) has been shown to be atheroprotective, shear above threshold values for the cell type (e.g., 60–100 dyn/cm^2) may induce apoptosis, endothelial–mesenchymal transition (EMT), and remodeling (Sho et al. 2002a; Chien 2006; Egorova et al. 2011a, 2012). Alternatively, interstitial flow or low shear stress may induce SMC and adventitial fibroblast migration (Liu and Goldman 2001; Sho et al. 2002b; Sakamoto et al. 2006; Shi et al. 2009), which, in conjunction with increased monolayer permeability, can lead to intimal hyperplasia (IH). In addition to the mean flow shear conditions, dynamic flow profiles have great effects on disease progression, with increased PI associated with systemic fibrosis (Panaritis et al. 2005; Mitchell 2008; Mitchell et al. 2011; Scott et al. 2013), EMT (Elliott et al. 2015), increased lipid metabolism (Chiu and Chien 2011), plaque deposition (Ku et al. 1985; Asakura and Karino 1990; Hsu et al. 2001), and inflammation (Chiu et al. 2003; Chiu and Chien 2011; Scott et al. 2013; Tan et al. 2014; Elliott et al. 2015). Most notably, clinically measured high PI indicates systemic fibrosis (Panaritis et al. 2005; Mitchell 2008; Mitchell et al. 2011), particularly at branching points within the vascular tree, where reciprocal or disturbed flow occurs (Ku et al. 1985; Asakura and Karino 1990; Chien 2006; Chiu and Chien 2011). This may account for shear-induced apoptosis and epithelial cell permeability (Chien 2006; Chiu and Chien 2011; Elliott et al. 2015), resulting in pathogenic cell migration or material deposition on the subendothelial ECM. Cyclic stretch outside physiologic ranges may also increase epithelial cell monolayer permeability – exacerbated by ROS and vascular endothelial growth factor (VEGF) release (Mata-Greenwood et al. 2005; Birukov 2009) – further perpetuating pathological conditions.

22.5 The Role of Vascular Mechanics in Modulating Mechanical Signals

Mechanical signaling within the vasculature is initiated and driven by the volumetric contraction of the heart, which produces a complex flow wave that transmits throughout the hierarchical vascular system. The heart contraction causes a pulsatile pressure wave, which precedes the volumetric flow wave in the vascular system. Localized pressure-induced stretch stress and flow-induced shear stress on the blood vessel are both reliant on local mechanical properties of the vessel. Radial expansion from the resulting pressures increases the cross-sectional area, decreasing the centerline velocity of blood flow and resulting in fluid shear at the wall. Therefore, localized remodeling of vessel ECM has an impact on local hemodynamics via changes in vessel mechanics, and matrix remodeling may thus exacerbate the disease process.

Arterial compliance has often been related to the maintenance of healthy hemodynamics (Nichols et al. 2011). Arterial compliance likely results from changes in elastin and/or collagen content or in vascular tone, and its importance is evident in the steady shear values maintained throughout the arterial tree (Humphrey 2007). The ECM content changes along the arterial tree, with higher elastin content within the aorta, and greater SMC and proteoglycan content in branching muscular arteries and arterioles (Kawasaki et al. 2005; Valdez-Jasso et al. 2011). This results in different mechanical properties among the arteries and varied arterial response to flow, with an elastic response from elastin content, creating a "windkessel" damping effect (Wagenseil and Mecham 2009; Nichols et al. 2011), and a viscoelastic response from proteoglycan content (Armentano et al. 1995). Compliance, alongside the "windkessel" effect associated with aortic damping of pressure and fluid shear, is the most commonly examined effect of arterial mechanics on hemodynamics, and has been shown *in vivo* and *in vitro* to decrease PI levels (Humphrey 2007; Scott et al. 2013; Tan et al. 2014; Elliott et al. 2015). While cell mechanotransduction with steady shear has been studied thoroughly (Chien 2006; Chiu and Chien 2011), dynamic flow shear conditions have mostly been examined in clinical settings or computational studies (Mitchell et al. 2010, 2011), and have only recently been studied using benchtop models (Scott et al. 2013; Tan et al. 2014; Elliott et al. 2015). Pathological dynamic flow profiles may be related by PI values, with normal physiologic pulsatile flow marked by $PI < 1$, high pulsatility flow marked by an increased PI, and oscillating or reciprocating flows marked by even higher PIs. The occurrence of high-pulsatility flow is related to systemic fibrosis within the arterial tree (Mitchell 2008; McArthur et al. 2011; Mitchell et al. 2011), while recursive flow is often indicated in atherosclerotic lesions (Ku et al. 1985; Asakura and Karino 1990). To further understand the dynamic interaction and relationship of hemodynamics and arterial wall mechanics, studies should be performed which relate wall mechanics to hemodynamics at different vascular disease states. This will also advise treatment options. To keep the amount of information provided manageable, we will limit ourselves to localized responses within the aorta and arteries, leaving systemic effects to other reviews.

Within the aorta, the highly elastic and relatively soft elastin layering within the media acts as an energy reservoir. During systole, the aorta undergoes significant strain, which decreases during aging (Redheuil et al. 2010). By contrast, the muscular arteries, such as the femoral and coronary arteries, exhibit stiffer wall properties and greater viscous material response (Bergel, 1961; Armentano et al. 1995). Such difference in arterial wall mechanics may be explained by the histological comparison of ECM and cellular content, in that muscular arteries exhibit much greater SMC and proteoglycan content. These mechanical properties may directly be related to flow and pressure wave attenuation in healthy vascular physiology, decreasing pulsatility and improving cell response downstream. The effect of wall mechanics becomes more apparent in diseased conditions. In chronic hypertension, for example, increased systemic pressures result in arterial remodeling, with initial ECM breakdown and weakening, allowing for luminal expansion and thereby decreasing mean fluid shear; this is followed by wall thickening and ECM stiffening, which counters the increased pressure (Arribas et al. 2006; Lemarié et al. 2010). Eventually, ECM reconstruction results in decreased elastin content and increased collagen type I content, with greater crosslinking of the ECM protein, or even calcification of the tissue (Lemarié et al. 2010). Arterial wall stiffening in hypertension has been associated with downstream vascular dysfunction, enhancing the risk of stroke and of kidney failure. Additionally, increased collagen-I deposition and calcification may be seen in the lumen under recursive flow (Ku et al. 1985; Asakura and Karino 1990; Chien 2006; Chiu and Chien 2011; Peloquin et al. 2011). Atherosclerotic plaque may also build at and distal to branching points (Ku et al. 1985; Asakura and Karino 1990), identified by altered flow pulsatility. The stiffness of the plaque depositions is much greater (100–300-fold) than that of healthy ECM (Ebenstein et al. 2009; Peloquin et al. 2011), affecting epithelial cell health (Huynh et al. 2011; Mason et al. 2013) and creating surface discontinuities which exacerbate local PI values (Ojha 1994).

Generally speaking, arterial tissue is moderately viscoelastic (Bergel 1961; Fung 1967; Tanaka and Fung 1974; Bia et al. 2006), with significant hysteresis curves present during *in situ* dynamic measurement (Armentano et al. 1995, 2006; Giannattasio et al. 2008). Hysteresis marks energy loss and wave attenuation in flow, and is predominantly located within the muscular arteries, which further reduce pulsatility before flow enters the arterial tree. While the impact of viscous material response within the vasculature is not fully understood, some previous modeling efforts using *in vivo* pressure/diameter relationships have exhibited significant pressure and flow wave attenuation, especially with the increased frequencies associated with exercise (Holenstein et al. 1980; Raghu et al. 2011). Remodeling in mild-to-moderate hypertensive patients also suggests the importance of viscosity, when the ratio of viscous to elastic damping is still maintained even though wall stiffening through media thickening occurs (Armentano et al. 1998, 2006). This provides an interesting new avenue for understanding the fluid–structure relationship and future treatment design.

22.6 Therapeutic Strategies Exploiting Mechanobiology

With the rapid increase in research efforts devoted to vascular mechanobiology, the next decade may see a big surge in the exploitation of new findings in the identification of novel therapeutic targets, development of new therapeutic drugs targeting mechanosensing or mechanotransduction pathways, and enhancement of disease prognosis and management of CVD. Thus, the future application of vascular mechanobiology is likely to expand beyond the realm of its current uses, toward the design of vascular implant devices. Current solutions in small-vessel occlusion often result in compliance mismatch, a common cause of restenosis (Abbott et al. 1987; Okuhn et al. 1989; Stewart and Lyman 1992). The gold standard in vascular grafting, for example, is the autologous graft – specifically, the saphenous vein; however, saphenous veins may not be available for transplant, and they have a 30–50% failure rate, specifically due to arterial hemodynamics (Morinaga et al. 1987; Owens 2010). Anastomosis has long been a source of disturbed flow or high-pulsatility flow (Kim et al. 1993; Chiu and Chien 2011), and new directions aim to cuff the graft anastomosis in order to reduce disturbed flow and compliance mismatch (Chiu and Chien 2011). The only US Food and Drug Administration (FDA)-approved prosthetic graft materials are polyethylene terephthalate (PET) and expanded polytetrafluoroethylene (ePTFE), which only show satisfactory outcomes for arteries >6 mm in diameter (Deutsch et al. 2009) and can exhibit a 10–40-fold increase in construct stiffness from native arteries (Bia et al. 2006). Alternatively, the design of tissue-engineered vascular substitutes has often centered on improving compliance through incorporation of active biomolecules (Soletti et al. 2010; Wise et al. 2011; McKenna et al. 2012; Neufurth et al. 2015). While some may achieve similar values to native arteries (Wise et al. 2011), these studies often do not examine the hemodynamic response of the proposed graft. Also, testing of compliance focuses on steady-state material responses, assuming little or no frequency dependence. Similar compliance issues exist in vascular stents, giving enhanced wall stiffness and thus flow pulsatility and tissue remodeling (Vernhet et al. 2001; Greil et al. 2003). All of this indicates the necessity and difficulty of matching artery mechanical properties to proper hemodynamics and its beneficial effects for vascular cells. Additionally, current device designs do not address the dynamic viscoelastic response of arterial tissue. The next section provides a review and examples of current work examining arterial flow environments, using computational methods to aid in parametric design optimization of vascular implant devices.

22.7 The Role of Hemodynamics in Mechanobiology

22.7.1 Computational Fluid Dynamics

The equations governing fluid flows are a set of coupled, nonlinear partial differential equations, including:

$$\text{Continuity}: \frac{\partial \rho}{\partial t} + \frac{\partial \rho U_i}{\partial x_i} = 0 \tag{22.2}$$

$$\text{Momentum}: \frac{\partial \rho U_i}{\partial t} + \frac{\partial \rho U_i U_j}{\partial x_j} = \frac{\partial P}{\partial x_j} + \frac{\partial}{\partial x_j}\left(\mu \frac{\partial U_i}{\partial x_j}\right) \tag{22.3}$$

These are known as Navier–Stokes equations. Many real problems include additional terms and/or equations, governing heat transfer, chemical species, and so on. Analytical solutions are known only for a few very simple flow cases. An alternative is to solve the governing equations numerically, on a computer. The process of obtaining numerical approximations to the solution of the governing fluid flow equations is known as CFD. Figure 22.2 provides an overview of the steps involved in a typical arterial bypass CFD flow problem.

One could view CFD as a *numerical experiment*. In a typical fluids experiment, an experimental model has to be built and the flow interacting with that model needs to be measured using various measurement devices; the results can then be analyzed. In CFD, the building of the model is replaced with the formulation of the governing equations and the development of the numerical algorithm. The process of obtaining measurements is replaced with running an algorithm on the computer to simulate the flow interaction. The analysis of the results is the same for both techniques, however.

Despite its many advantages, CFD does not remove the need for experiments: numerical models must be validated to ensure they produce reliable and accurate results. In clinical applications in particular, there is a pressing need for rigorous model validation against detailed laboratory data. With the growth of available computing power and the advent of powerful user-friendly graphical user interfaces and automated options/features in commercial CFD codes, it has become possible for a wide range of users to apply CFD to even very complex flowfields, giving detailed information about the velocity field, pressure, temperature, and so on. This sometimes results in solutions that are hemodynamically irrelevant and which fail to capture even the most basic flow features.

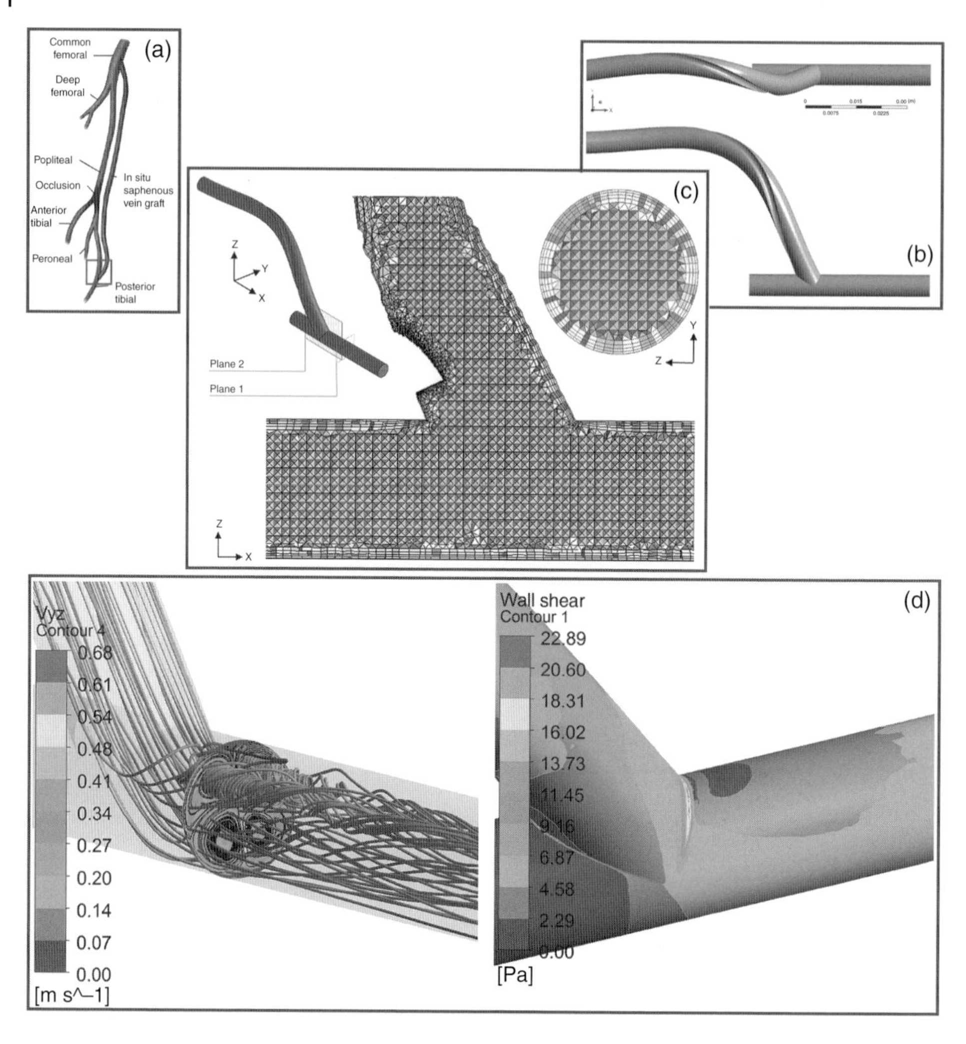

Figure 22.2 Typical CFD process in a biomedical application. (a) Typical peripheral bypass grafting. (b) Solid computer-aided model (CAD) for the computational domain. (c) Hybrid mesh consisting of quadrilateral cells near the wall boundaries and tetrahedral cells further from the wall. (d) Post-processing of the CFD simulation results using streamlines and contours of the hemodynamic parameters. (*See insert for color representation of the figure.*)

22.7.2 Biomedical Applications of CFD

Recent advances in vascular biology, biomechanics, medical imaging, and computational techniques, including CFD, have provided the research community with a unique opportunity to analyze the progression of vascular diseases from a new angle and to improve the design of medical devices and develop new strategies for intervention. The increasing power/cost ratio of computers and the advent of methods for subject-specific modeling of cardiovascular mechanics have made CFD-based modeling sometimes more reliable than methods based solely on *in vivo* measurement. In fact, numerical simulations have played an important role in expanding our understanding of the hemodynamics in several clinical cases involving bypass grafting, cardiovascular treatment planning, cerebrovascular flow, the effects of exercise on aortic flow conditions, congenital heart disease (CHD), and coronary stents.

In addition, recent developments in patient-specific computer simulations have provided a means of assessing new surgeries and interventions that poses no risk to the patient. As in other engineering fields, such as aerospace and automotive, design optimization can now be applied to predictive tools and methods in order to optimize surgeries for individual patients. Therefore, in the new paradigm of predictive medicine, surgeons (for example) may use advanced imaging tools alongside computational techniques such as CFD to create a patient-specific model and predict the outcome of a particular treatment for an individual patient. However, to be effective and attractive to the medical community, these simulation-based medical planning systems must be quick and efficient and should require minimum user intervention.

22.7.3 Vascular Grafting

In general, there are two broad applications of vascular grafting:

1) *Arterial bypass grafts (ABGs):* Examples include peripheral vascular disease (PVD) and coronary artery disease (CAD). Each year, over 1 million vascular grafts (excluding valves) are used in current medical practice. Problems that require the use of a graft include occluded vessels due to stenosis, damaged vessels due to trauma or aneurysm, and the formation of a new tissue structure through regenerative therapies. In ABGs, the gold standard is currently to use naturally occurring ("autologous") vessels; however, there a number of problems inherent to this technique, including the need for additional surgery and the frequent unsuitability or limited availability of patient veins, due to systemic disease. There is also a lack of viable treatment options when the blood vessel is <6 mm in diameter (Sarkar et al. 2007). Hence, prosthetic grafts, following either biomaterial or tissue-engineered approaches, are utilized. Current prosthetic surgical options commonly include Dacron (PET) and Teflon (ePTFE). Unfortunately, prosthetic grafts are known to exhibit unsatisfactory long-term performance (Haruguchi and Teraoka 2003); much research is thus underway aimed at reducing failure rates and improving patency rates, particularly for vessels <6 mm in diameter.
2) *Arteriovenous access grafts (AVGs):* The world is currently facing a huge increase in the number of people with diabetes. For example, in the United Kingdom, the number of people diagnosed with diabetes has nearly doubled since 1996, and diabetes is the single most common cause of end-stage renal disease (Roberts 2007). AVGs are mainly used to create an "access point" for hemodialysis treatment in patients with renal disease. The ideal vascular access for patients undergoing hemodialysis is an arteriovenous fistula in the forearm. However, the recent increase in the incidence of diabetes mellitus (DM) and the need for hemodialysis in patients with this disorder has expanded the number of patients who require implantation ePTFE graft for vascular access.

Generally, for prosthetic grafts, there are two separate strands of research: the first tends to focus on tissue-engineering and biomaterials science, while the second is concerned with biomechanics, flow field augmentations, and hemodynamic forces.

The attention of the present chapter is restricted to the latter strand, and in particular the role of CFD in investigating and designing novel grafts with improved patency rates.

22.7.4 Vascular Graft Failure

Graft failure is currently a major concern for medical practitioners engaged in treating PVD and CAD. Almost 35 000 coronary artery bypass graft (CABG) procedures are performed each year in the United Kingdom, according to the British Heart Foundation (BHF), but over 50% fail within 10 years. In 1999, an estimated 688 000 bypass surgeries were performed in the United States, but up to 10% failed within 30 days (Ku et al. 2005). Stenosis at the graft–vein junction caused by IH is the major cause of failure of AVGs used for hemodialysis. For example, in the United States alone, 175 000 ePTFE grafts are currently used for permanent vascular access, with 1- and 2-year primary patency rates of 50 and 25%, respectively. These low patency rates have limited the use of AVGs and have resulted in arteriovenous fistula in the forearm the remaining the preferred route of vascular access for patients undergoing hemodialysis.

Early graft failure (within 30 days) is attributable to surgical technical errors (such as choosing a poor location for the distal anastomosis) and resulting thrombosis, while late graft failures are mainly caused by progression of atherosclerosis and IH (Whittemore et al. 1981; Bryan and Angelini 1994).

It is now widely accepted that hemodynamic factors play an important role in the formation and development of IH (Archie et al. 2001; Ghista and Kabinejadian 2013). The intimal thickening (IT) and restenosis due to IH is normally characterized by unsteady shear stress, recirculation regions, pulsatile stress, and graft deformations (Sottiurai et al. 1983). These hemodynamic factors include low and oscillating wall shear stress (Ethier et al. 1998), large spatial wall shear stress gradients (WSSGs) (DePaola et al. 1992), and long residence times among blood cells (Perktold et al. 1997). Several studies have shown a correlation between these hemodynamic factors and localized sites of IT, which in a conventional end-to-side (ETS) configuration occur predominantly at the heel and toe of the anastomosis, on the artery floor opposite the anastomosis, and on the suture line (Figure 22.3). Consequently, much research in the past few decades has been aimed at designing grafts that will remain patent for far longer – ideally, for longer than the life-span of the patient. For example, for CABGs, in order to achieve higher patency rates, numerous studies have attempted to improve the

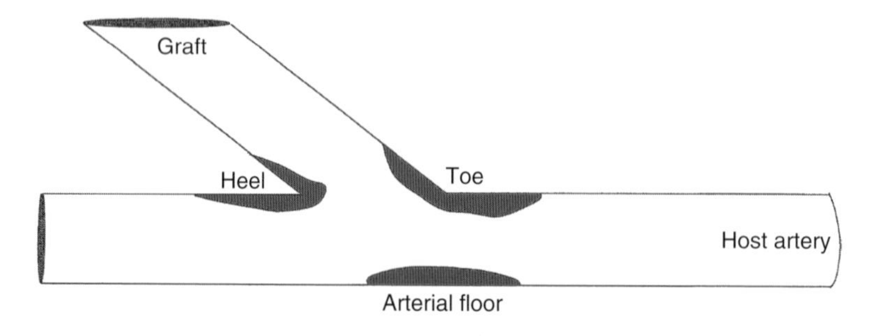

Figure 22.3 Spatial distribution and localized sites of IT/IH in a typical ABG.

hemodynamics at the anastomosis by examining the effects of anastomotic angle, distal anastomosis shape, out-of-plane graft, graft-to-host artery diameter ratio, competitive flow, and grafting distance (Ghista and Kabinejadian 2013).

22.7.5 Novel Graft Designs

As alluded to earlier, it is now widely accepted that both IH and acute thrombosis, which are the main causes of ABG and arteriovenous graft failures, have close correlations with hemodynamic forces, including low and oscillating wall shear stress. Therefore, designs for new prosthetic grafts have started to take move toward improving patency via novel flow field augmentations.

One of the most significant contributions to this area was based on a study showing that the "spiral flow" is a natural phenomenon throughout the arterial system (Stonebridge and Brophy 1991; Stonebridge et al. 1996). The spiral flow in arteries is caused by the rotational compressive pumping of the heart (Jung et al. 2006) and is supported by the tapered, curved, and nonplanar geometry of the arterial system (Caro et al. 1996; Stonebridge 2011). In other words, the rotational motion of the blood flow is induced by the twisting of the left ventricle during contraction and is accentuated upon entering the aortic arch (Murphy and Boyle 2012).

The spiral flow has also been used in many industrial applications, mainly to enhance mixing and turbulence for heat-transfer augmentation. For example, as shown in Figure 22.4, the fuel pins in the UK's fleet of advanced gas-cooled reactors (AGRs) have been designed to induce helical flow by creating a series of protrusions in the form of spiral ridges on their outer surfaces.

These configurations have significantly enhanced heat transfer compared to that in conventional pins. Biomedical prosthetics could potentially use this concept to modify the hemodynamic forces in the cardiovascular system in order to mimic the

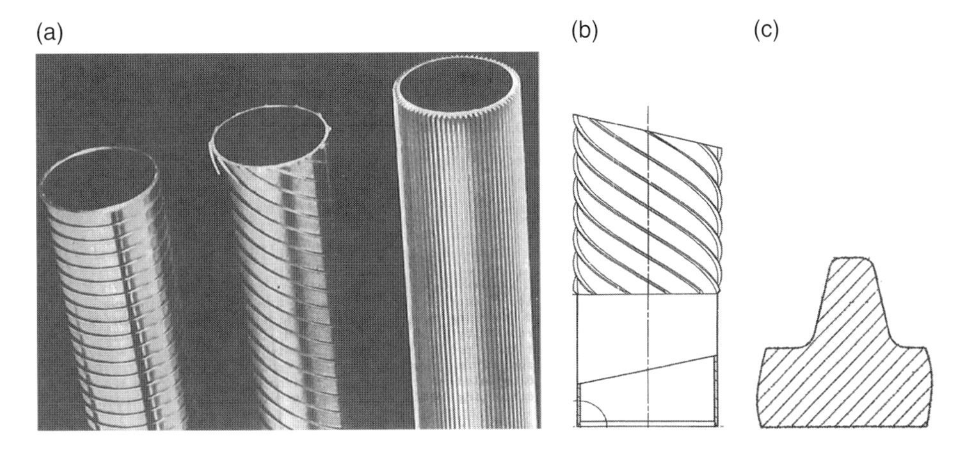

Figure 22.4 Industrial examples of the induction of helical flow. (a) Schematics of "transverse-ribbed," "multi-start ribbed," and "longitudinally finned" nuclear fuel pins used in the UK's advanced gas-cooled reactors (AGRs). (b) Schematic of helical ridges on the outer surface of "multi-start" fuel pins. (c) Profile of the rib/ridge in the "multi-start" design.

physiological spiral blood flow in arteries and grafts. In fact, two of the most innovative and successful graft designs have used this principal in order to induce spiral flow in arteries:

1) *SwirlGraft:* Developed by Caro et al. (2005) at Veryan Medical Ltd., SwirlGraft is a new arterioventricular shunt graft with a helical out-of-plane geometric feature incorporating "small amplitude helical technology" (SMAHT). Compared to a conventional ePTFE graft, the animal experiments reported by Caro et al. (2005) demonstrated less thrombosis in the SwirlGraft. The difference became even more remarkable after 8 weeks of implementation, when occlusion occurred in the conventional grafts, and less thrombosis and IH occurred in the SwirlGraft. Recently, a few attempts have been made to numerically simulate the blood flow in out-of-plane graft geometries that induce 3D swirling flows (Cookson et al. 2009; Zheng et al. 2009; Sun et al. 2010; Lee et al. 2011) with the aim of understanding the flow physics and improving the current graft designs. In addition, Veryan Medical Ltd. has recently designed a peripheral stent known as the "BioMimics 3D Helical Stent," which appears to have evolved from SwirlGraft and SMAHT and is currently under clinical trials. However, there are some concerns associated with both SwirlGraft and 3D helical stents, including an increase in the pressure drop across the helical segment. The helical stent can also impose a severe geometric change on an artery, causing damage to the arterial wall. The details of this design and its results are beyond the scope of this chapter.
2) *Spiral Flow Peripheral Vascular Graft:* Introduced by Vascular Flow Technologies (VFT) Ltd., this is a prosthetic ePTFE graft design that is engineered to induce spiral flow through an internal ridge within its distal end. This design is primarily based on research carried out by Stonebridge et al. (2012) at Ninewells Hospital in Dundee, Scotland into the naturally occurring helical flow found in studies of the cardiovascular system using Doppler ultrasound. The results of an early clinical nonrandomized study for the VFT peripheral bypass graft are promising, showing primary patency rates of 81% for above-the-knee bypasses and 57.3% for below-the-knee bypasses at 30 months of follow-up, and secondary patency rates of 81 and 64%, respectively. In an unpublished study, similar improvements were also found when using the spiral flow graft for arterioventricular access for hemodialysis ("Spiral Flow AV Access Graft"). While these initial results highlight the potential for the use of spiral-induced grafts in bypass surgeries, they have a number of limitations. For example, this design cannot currently be used for CABG because satisfactory prosthetic grafts have been shown to be unsuccessful for small-caliber applications due to their poor long-term patency rates (Yellin et al. 1991) – autologous internal thoracic arteries, radial arteries, and saphenous veins remain the most widely used conduits for CABG (Angelini and Newby 1989; Cameron et al. 1996; Desai et al. 2011). Another disadvantage of this type of prosthetic graft is its limitation to ePTFE, which has been shown to have poor patency rates due to its poor mechanical characteristics (i.e., low compliance) and the lack of endothelial cells lining its lumen (Salacinski et al. 2001).

While the physiological importance of secondary motion in circulation has clearly been highlighted in the literature (Stonebridge and Brophy 1991; Stonebridge et al. 1996), the benefits of helical/spiral prostheses in vascular conduits have yet to be firmly established. Moreover, the range of possible design configurations is large, and the effects of different design/geometrical parameters on hemodynamic forces are unclear. Therefore, there is scope to investigate existing and innovative flow field augmentations using computational techniques in order to inform the design of novel prostheses and surgical vascular reconstructions.

22.7.6 Hemodynamic Metrics

One of the main attractions of using CFD in biomedical problems is its ability to calculate several different hemodynamic parameters/metrics, which could potentially have important clinical implications. For example, previous studies have shown that: localized distribution of low wall shear stress and high oscillatory shear index (OSI) strongly correlates with the focal locations of atheroma (He and Ku 1996); large spatial WSSG contributes to elevated wall permeability and atherosclerotic lesions (DePaola et al. 1992); a combination of high shear stress and long exposure times may induce platelet activation (Ramstack et al. 1979; Wurzinger et al. 1985, 1986; Hellums 1994); and stagnant and recirculation flow regions can cause platelet aggregation and thrombogenesis.

These hemodynamic parameters can be directly derived from the flow fields obtained by CFD-based simulation tools. It is worth noting that the reason we have to define different metrics for the variation in blood flow characteristics is mainly due to anatomic and physiologic variations and the systemic complexity of the fluid flow, as well as its interaction with the vessel wall and tissue. In addition, single-feature hemodynamic metrics are generally unable to capture the multidirectionality of the flow field (Peiffer et al. 2013).

The distributions of hemodynamic parameters, including time-averaged wall shear stress (TAWSS), TAWSS gradient (Buchanan et al. 2003), OSI (He and Ku 1996), and relative residence time (RRT) (Lee et al. 2009), can be calculated according to the following equations:

$$TAWSS = \frac{1}{T}\int_0^T \left|\vec{\tau}_W\right| dt \tag{22.4}$$

$$TAWSSG = \frac{1}{T}\int_0^T \sqrt{\left(\frac{\partial \tau_x}{\partial x}\right)^2 + \left(\frac{\partial \tau_y}{\partial y}\right)^2 + \left(\frac{\partial \tau_z}{\partial z}\right)^2}\, dt \tag{22.5}$$

$$OSI = \frac{1}{2}\left(1 - \frac{\left|\int_0^T \vec{\tau}_W dt\right|}{\int_0^T \left|\vec{\tau}_W\right| dt}\right) \tag{22.6}$$

$$RRT = \frac{1}{(1 - 2\times OSI)\times TAWSS} = \frac{1}{\frac{1}{T}\left|\int_0^T \vec{\tau}_W dt\right|} \tag{22.7}$$

where $\vec{\tau}_W$ is the wall shear stress vector and T is the time period of the flow cycle.

22.7.7 Hemodynamics of Spiral and Helical Grafts

The main configuration studied in this section represents a typical ETS distal graft anastomosis, which can be found in the following three graft configurations: (i) peripheral artery bypass graft; (ii) CABG; and (iii) AVG. The former has been selected for our purposes here because both of the novel flow field augmentation designs discussed earlier have only been tested for the peripheral artery bypass configuration.

It is important to note that while some resemblances can be found among these three ETS graft configurations, the hemodynamic patterns and consequent locations of IH formation are different. For instance, the blood flow rate in arteriovenous grafts is about 5–10 times higher than that in peripheral artery bypass grafts or CABGs; consequently, the spatial distribution of IH is different in each configuration (Haruguchi and Teraoka 2003).

The dimensions and the schematics of the models tested here are given in Figure 22.5.

- *Model 1 (Control Graft):* This represents a baseline and a conventional ETS distal graft anastomosis for a peripheral artery bypass configuration.
- *Model 2 (Spiral Graft):* This has an internal spiral inducer within the distal end of the graft in the form of an internal ridge. The ridge, which is one pitch long, provides a ridged cross-sectional geometry to engender and deliver a single-spiral flow pattern.
- *Model 3 (Helical Graft):* The distal end of the bypass graft consists of a small-amplitude out-of-plane (nonplanar) helical configuration. The helical section of this model involves a one-turn helix with pitch and amplitude approximately 14D and 0.5D, respectively (where D is the internal diameter of the graft).

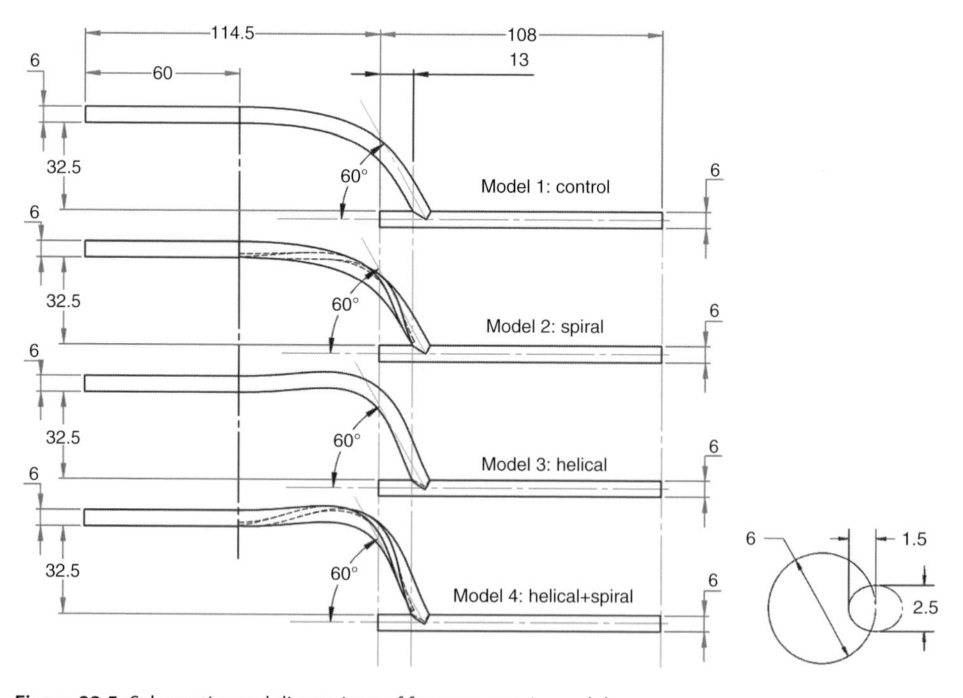

Figure 22.5 Schematics and dimensions of four geometric models.

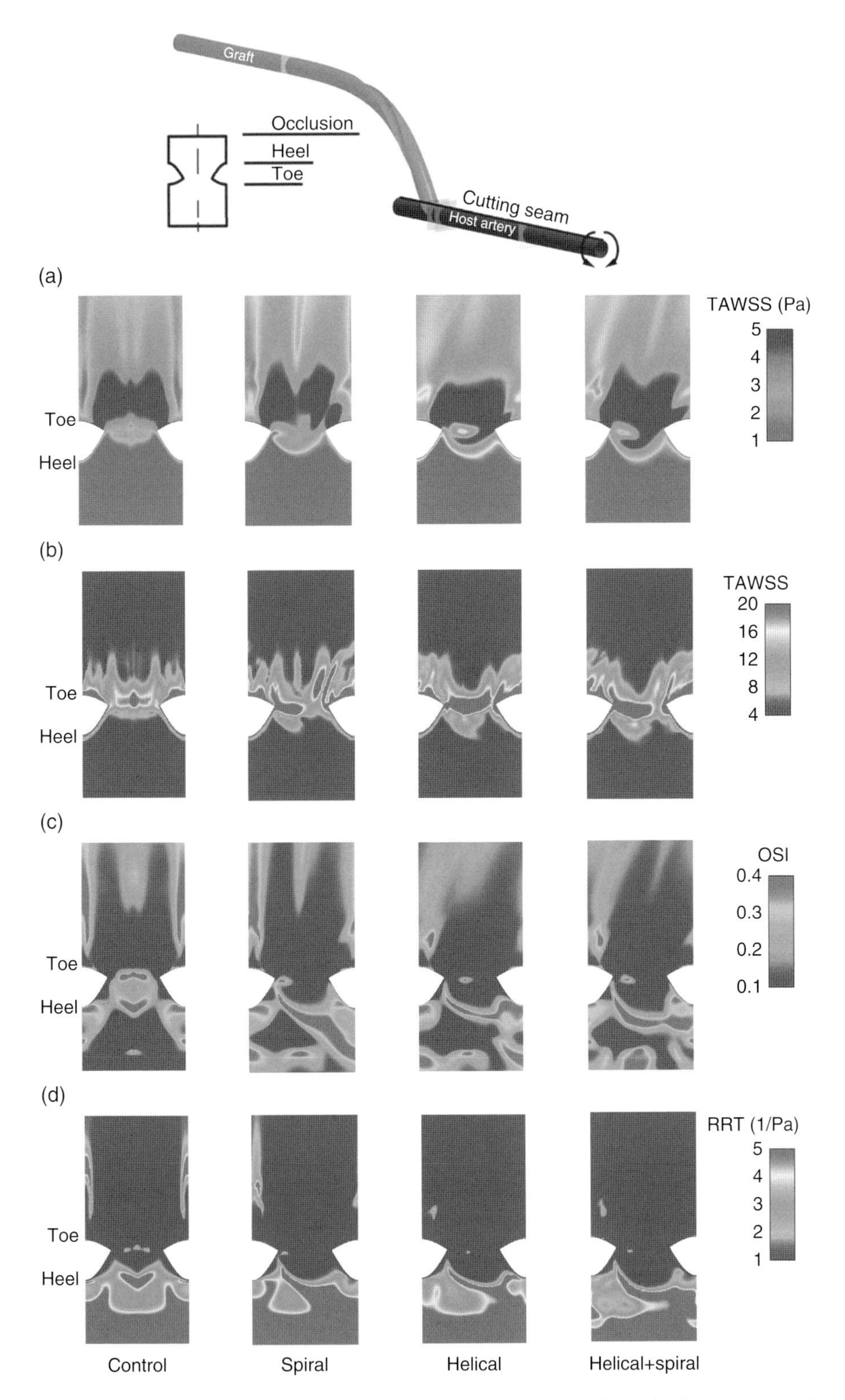

Figure 22.6 Distributions of different hemodynamic parameters viewed from the host arterial wall, opened ventrally and shown *en face*: (a) time-averaged wall shear stress (TAWSS); (b) TAWSS gradient; (c) oscillatory shear index (OSI); and (d) relative residence time (RRT). Note that the color scale of the TAWSS map is inverted for ease of comparison.

- *Model 4 (Helical + Spiral Graft):* The distal end of the bypass graft combines the geometrical features present in Models 2 and 3. Note that in this model, the out-of-plane helical and the internal ridge have the same pitch and turn in the same direction.

Figure 22.6 shows the distribution of these metrics for all four geometrical models on an unfolded model of the host artery, which has been opened ventrally. The maps shown in Figure 22.6 should be interpreted such that the vertical axis represents distance along the host artery (with blood flow from bottom to top) and the horizontal axis represents the circumferential distance along the wall of the host vessel (the bed region appears at the centre).

Investigation of the hemodynamic parameters distribution in Figure 22.6 shows that the TAWSS increases on the arterial bed and around the anastomosis upon the addition of the nonplanar helicity (while the effect of including the spiral ridge is minimal). There are indications that the blood monocytes are more likely to adhere to the endothelial layer at regions with low TAWSS (Pritchard et al. 1995); hence, introducing the nonplanar curvature to the graft may reduce the spatial extent of the early wall lesion. Also, significant correlations have been reported between the preferred sites of IT and the regions of low wall shear stress (White et al. 1993; Sunamura et al. 2007); as such, the observed altered flow patterns may be perceived as a beneficial feature of the helical grafts that positively impacts the graft patency rate.

The elevated OSI region on the host arterial wall (distal to the toe) reduces in size considerably in helical graft models (i.e., Models 3 and 4) and moderately in the spiral graft model (i.e., Model 2). Hence, a low-TAWSS/high-OSI region in the control model is replaced by a high-TAWSS/low-OSI area in the helical graft models. This is another positive feature of the introduction of nonplanar helicity to the bypass grafts (and, to a lesser extent, the inclusion of the spiral ridge), since the combination of high TAWSS and low OSI is believed to contribute to IT and atherosclerosis development, as well as increasing the risk of the aggregation of red blood cells (RBCs) (Ku et al. 1985; He and Ku 1996; Malek et al. 1999).

In addition, the swirling flow induced by the helical graft in Models 3 and 4 reduces the RRT distal to the toe of the anastomosis, by eliminating the flow separation in this region (which is present in Model 1, and to a lesser extent in Model 2). This could reduce the chance of platelet aggregation and thrombus formation (Hellums 1994), and consequently it enhances the patency of the bypass graft.

22.8 Conclusion

Cardiovascular biomechanics and mechanobiology is a pivotal area of research for the improvement of existing therapeutic strategies and the development of new therapeutics for CVD. For example, inducing swirling flow into bypass grafts can bring about positive flow features and a more favorable distribution of hemodynamic parameters in the graft, the anastomotic region, and the host artery, which may enhance the patency and longevity of the bypass graft. For this purpose, graft helicity has been found to be significantly more effective than a spiral ridge. It has also been found that a combination of graft helicity and spiral ridge can further enhance this swirling effect. However, in order to optimize these features for use in different types of bypass graft, further studies are warranted.

Equally important to future parametric improvement is the creation of more realistic models of the vessel wall. Instead of treating the wall as a nonlinear orthotropic elastic cylindrical membrane for use in CFD analysis, the experimentally observed material damping (elastic) or viscoelastic properties of the wall tissues should be incorporated in the analysis in order to reflect more realistic flow environments *in vivo* and thus provide a more accurate prediction of cell and tissue responses based on our existing knowledge of the cell mechanobiology of flow. By using models that take into account fluid–structure interaction, analytical expressions for the displacement and shear stresses developed in the wall, and velocity distribution, we can derive the fluid acceleration and volume flow rate of blood computationally.

Future vascular mechanobiology studies will greatly benefit from highly integrated approaches that use computational analysis together with *in vitro* and *in vivo* measures and benchtop flow models containing vascular cells for the exploration of mechanosensing and mechanotransduction mechanisms.

References

Abbott, W. M., J. Megerman, J. E. Hasson, G. L'Italien, and D. F. Warnock. 1987. "Effect of compliance mismatch on vascular graft patency." *Journal of Vascular Surgery* **5**(2): 376–82. PMID:3102762.

Alenghat, F. J. and D. E. Ingber. 2002. "Mechanotransduction: all signals point to cytoskeleton, matrix, and integrins." *Science Signaling* **2002**(119): pe6. doi:10.1126/stke.2002.119.pe6.

Anderson, R. N. and B. L. Smith. 2003. "Deaths: leading causes for 2001." Centers for Disease Control and Prevention. Available from: https://www.cdc.gov/nchs/data/nvsr/nvsr52/nvsr52_09.pdf (last accessed June 30, 2016).

Ando, J. and K. Yamamoto. 2009. "Vascular mechanobiology endothelial cell responses to fluid shear stress." *Circulation Journal* **73**(11): 1983–92. PMID:19801852.

Angelini, G. D. and A. C. Newby. 1989. "The future of saphenous vein as a coronary artery bypass conduit." *European Heart Journal* **10**(3): 273–80. PMID:2565234.

Archie, J. P., S. Hyun, C. Kleinstreuer, P. W. Longest, G. A. Truskey, and J. R. Buchanan. 2001. "Hemodynamic parameters and early intimal thickening in branching blood vessels." *Critical Reviews in Biomedical Engineering* **29**(1): 1–64. PMID:11321642.

Armentano, R. L., J. G. Barra, J. Levenson, A. Simon, and R. H. Pichel. 1995. "Arterial wall mechanics in conscious dogs. Assessment of viscous, inertial, and elastic moduli to characterize aortic wall behavior." *Circulation Research* **76**(3): 468–78. PMID:7859392.

Armentano, R. L., S. Graf, J. G. Barra, G. Velikovsky, H. Baglivo, R. Sánchez, et al. 1998. "Carotid wall viscosity increase is related to intima-media thickening in hypertensive patients." *Hypertension* **31**(1): 534–39. doi:10.1161/01.HYP.31.1.534.

Armentano, R. L., J. G. Barra, D. B. Santana, F. M. Pessana, S. Graf, D. Craiem, et al. 2006. "Smart damping modulation of carotid wall energetics in human hypertension: effects of angiotensin-converting enzyme inhibition." *Hypertension* **47**(3): 384–90. doi:10.1161/01.HYP.0000205915.15940.15.

Arribas, S. M., A. Hinek, and M. C. González. 2006. "Elastic fibres and vascular structure in hypertension." *Pharmacology & Therapeutics* **111**(3): 771–91. doi:10.1016/j.pharmthera.2005.12.003.

Asakura, T. and T. Karino. 1990. "Flow patterns and spatial distribution of atherosclerotic lesions in human coronary arteries." *Circulation Research* **66**(4): 1045–66. doi:10.1161/01.RES.66.4.1045.

Bergel, D. H. 1961. "The dynamic elastic properties of the arterial wall." *Journal of Physiology* **156**(3): 458–69. doi:10.1113/jphysiol.1961.sp006687.

Bergh, N., E. Ulfhammer, K. Glise, S. Jern, and L. Karlsson. 2009. "Influence of TNF-α and biomechanical stress on endothelial anti- and prothrombotic genes." *Biochemical and Biophysical Research Communications* **385**(3): 314–18. doi:10.1016/j.bbrc.2009.05.046.

Berk, B. C. 2008. "Atheroprotective signaling mechanisms activated by steady laminar flow in endothelial cells." *Circulation* **117**(8): 1082–9. doi:10.1161/CIRCULATIONAHA.107.720730.

Bia, D., Y. Zócalo, F. Pessana, R. Armentano, H. Pérez, E. Cabrera, et al. 2006. "[Viscoelastic and functional similarities between native femoral arteries and fresh or cryopreserved arterial and venous homografts.]" *Revista Española de Cardiología* **59**(7): 679–87. doi:10.1016/S1885-5857(07)60027-9.

Birukov, K. G. 2009. "Cyclic stretch, reactive oxygen species, and vascular remodeling." *Antioxidants & Redox Signaling* **11**(7): 1651–67. doi:10.1089/ars.2008.2390.

Birukov, K. G., V. P. Shirinsky, O. V. Stepanova, V. A. Tkachuk, A. W. A. Hahn, T. J. Resink, and V. N. Smirnov. 1995. "Stretch affects phenotype and proliferation of vascular smooth muscle cells." *Molecular and Cellular Biochemistry* **144**(2): 131–9. doi:10.1007/BF00944392.

Birukova, A. A., N. Moldobaeva, J. Xing, and K. G. Birukov. 2008. "Magnitude-dependent effects of cyclic stretch on HGF- and VEGF-induced pulmonary endothelial remodeling and barrier regulation." *American Journal of Physiology. Lung Cellular and Molecular Physiology* **295**(4): L612–23. doi:10.1152/ajplung.90236.2008.

Bryan, A. J. and G. D. Angelini. 1994. "The biology of saphenous vein graft occlusion: etiology and strategies for prevention." *Current Opinion in Cardiology* **9**(6): 641–9. PMID:7819622.

Buchanan, J. R., C. Kleinstreuer, S. Hyun, and G. A. Truskey. 2003. "Hemodynamics simulation and identification of susceptible sites of atherosclerotic lesion formation in a model abdominal aorta." *Journal of Biomechanics* **36**(8): 1185–96. doi:10.1016/S0021-9290(03)00088-5.

Cameron, A., K. B. Davis, G. Green, and H. V. Schaff. 1996. "Coronary bypass surgery with internal-thoracic-artery grafts – effects on survival over a 15-year period." *New England Journal of Medicine* **334**(4): 216–19. doi:10.1056/NEJM199601253340402.

Caro, C. G., O. J. Doorly, M. Tarnawski, K. T. Scott, Q. Long, C. L. Dumoulin, and D. J. Doorly. 1996. "Non-planar curvature and branching of arteries and non-planar-type flow." *Proceedings of the Royal Society A: Mathematical, Physical and Engineering Sciences* **452**(1944): 185–97. doi:10.1098/rspa.1996.0011.

Caro, C. G., N. J. Cheshire, and N. Watkins. 2005. "Preliminary comparative study of small amplitude helical and conventional ePTFE arteriovenous shunts in pigs." *Journal of the Royal Society* **2**(3): 261–6. doi:10.1098/rsif.2005.0044.

Chaudhuri, O., L. Gu, M. Darnell, D. Klumpers, S. A. Bencherif, J. C. Weaver, et al. 2015. "Substrate stress relaxation regulates cell spreading." *Nature Communications* **6**: 6364. doi:10.1038/ncomms7365.

Chien, S. 2006. "Mechanotransduction and endothelial cell homeostasis: the wisdom of the cell." *American Journal of Physiology. Heart and Circulatory Physiology* **292**(3): H1209–24. doi:10.1152/ajpheart.01047.2006.

Chien, S., S. Li, and J. Y. J. Shyy. 1998. "Effects of mechanical forces on signal transduction and gene expression in endothelial cells." *Hypertension* **31**(1 Pt. 2): 162–9. PMID:9453297.

Chiu, J.-J. and S. Chien. 2011. "Effects of disturbed flow on vascular endothelium: pathophysiological basis and clinical perspectives." *Physiological Reviews* **91**: 327–87. doi:10.1152/physrev.00047.2009.

Chiu, J.-J., C.-N. Chen, P.-L. Lee, C. T. Yang, H. S. Chuang, S. Chien, and S. Usami. 2003. "Analysis of the effect of disturbed flow on monocytic adhesion to endothelial cells." *Journal of Biomechanics* **36**(12): 1883–95. doi:10.1016/S0021-9290(03)00210-0.

Clark, E. A. and J. S. Brugge. 1995. "Integrins and signal transduction pathways: the road taken." *Science* **268**(5208): 233–9. doi:10.1126/science.7716514.

Cookson, A. N., D. J. Doorly, and S. J. Sherwin. 2009. "Mixing through stirring of steady flow in small amplitude helical tubes." *Annals of Biomedical Engineering* **37**(4): 710–21. doi:10.1007/s10439-009-9636-y.

Davis, G. E. and D. R. Senger. 2005. "Endothelial extracellular matrix: biosynthesis, remodeling, and functions during vascular morphogenesis and neovessel stabilization." *Circulation Research* **97**(11): 1093–107. doi:10.1161/01.RES.0000191547.64391.e3.

De Caterina, R. 2000. "Endothelial dysfunctions: common denominators in vascular disease." *Current Opinion in Lipidology* **11**(1): 9–23. PMID:10750689.

DePaola, N., M. A. Gimbrone, P. F. Davies, and C. F. Dewey. 1992. "Vascular endothelium responds to fluid shear stress gradients." *Arteriosclerosis and Thrombosis* **12**(11): 1254–7. PMID:1420084.

Desai, M., A. M. Seifalian, and G. Hamilton. 2011. "Role of prosthetic conduits in coronary artery bypass grafting." *European Journal of Cardio-thoracic Surgery* **40**(2): 394–8. doi:10.1016/j.ejcts.2010.11.050.

Deutsch, M., J. Meinhart, P. Zilla, N. Howanietz, M. Gorlitzer, A. Froeschl, et al. 2009. "Long-term experience in autologous in vitro endothelialization of infrainguinal ePTFE grafts." *Journal of Vascular Surgery* **49**(2): 352–62. doi:10.1016/j.jvs.2008.08.101.

Ebenstein, D. M., D. Coughlin, J. Chapman, C. Li, and L. A. Pruitt. 2009. "Nanomechanical properties of calcification, fibrous tissue, and hematoma from atherosclerotic plaques." *Journal of Biomedical Materials Research Part A* **91**(4): 1028–37. doi:10.1002/jbm.a.32321.

Egorova, A. D., P. P. S. J. Khedoe, M.-J. T. H. Goumans, B. K. Yoder, S. M. Nauli, P. ten Dijke, et al. 2011a. "Lack of primary cilia primes shear-induced endothelial-to-mesenchymal transition." *Circulation Research* **108**(9): 1093–101. doi:10.1161/CIRCRESAHA.110.231860.

Egorova, A. D., K. Van der Heiden, S. Van de Pas, P. Vennemann, C. Poelma, M. C. DeRuiter, et al. 2011b. "Tgfβ/Alk5 signaling is required for shear stress induced klf2 expression in embryonic endothelial cells." *Developmental Dynamics* **240**(7): 1670–80. doi:10.1002/dvdy.22660.

Egorova, A. D., K. van der Heiden, R. E. Poelmann, and B. P. Hierck. 2012. "Primary cilia as biomechanical sensors in regulating endothelial function." *Differentiation* **83**(2): S56–61. doi:10.1016/j.diff.2011.11.007.

Elliott, W. H., Y. Tan, M. Li, and W. Tan. 2015. "High pulsatility flow promotes vascular fibrosis by triggering endothelial EndMT and fibroblast activation." *Cellular and Molecular Bioengineering* **8**(2): 285–95. doi:10.1007/s12195-015-0386-7.

Ethier C. R., D. A. Steinman, X. Zhang, S. R. Karpik, and M. Ojha. 1998. "Flow waveform effects on end-to-side anastomotic flow patterns." *Journal of Biomechanics* **31**(7): 609–17. doi:10.1016/S0021-9290(98)00059-1.

Fung, Y. C. 1967. "Elasticity of soft tissues in simple elongation." *American Journal of Physiology* **213**(6): 1532–44. PMID:6075755.

Garanich, J. S. 2005. "Shear stress inhibits smooth muscle cell migration via nitric oxide-mediated downregulation of matrix metalloproteinase-2 activity." *American Journal of Physiology. Heart and Circulatory Physiology* **288**(5): H2244–52. doi:10.1152/ajpheart.00428.2003.

Garanich, J. S., R. A. Mathura, Z.-D. Shi, and J. M. Tarbell. 2007. "Effects of fluid shear stress on adventitial fibroblast migration: implications for flow-mediated mechanisms of arterialization and intimal hyperplasia." *American Journal of Physiology. Heart and Circulatory Physiology* **292**(6): H3128–35. doi:10.1152/ajpheart.00578.2006.

García-Cardeña, G., J. Comander, K. R. Anderson, B. R. Blackman, and M. A. Gimbrone. 2001. "Biomechanical activation of vascular endothelium as a determinant of its functional phenotype." *Proceedings of the National Academy of Sciences* **98**(8): 4478–85. doi:10.1073/pnas.071052598.

Ghista, D. N. and F. Kabinejadian. 2013. "Coronary artery bypass grafting hemodynamics and anastomosis design: a biomedical engineering review." *Biomedical Engineering Online* **12**: 129. doi:10.1186/1475-925X-12-129.

Giannattasio, C., P. Salvi, F. Valbusa, A. Kearney-Schwartz, A. Capra, M. Amigoni, et al. 2008. "Simultaneous measurement of beat-to-beat carotid diameter and pressure changes to assess arterial mechanical properties." *Hypertension* **52**(5): 896–902. doi:10.1161/HYPERTENSIONAHA.108.116509.

Gimbrone, M. A., T. Nagel, and J. N. Topper. 1997. "Biomechanical activation: an emerging paradigm in endothelial adhesion biology." *Journal of Clinical Investigation* **99**(8): 1809–13. doi:10.1172/JCI119346.

Greil, O., G. Pflugbeil, K. Weigand, W. Weiss, D. Liepsch, P. C. Maurer, and H. Berger. 2003. "Changes in carotid artery flow velocities after stent implantation: a fluid dynamics study with laser doppler anemometry." *Journal of Endovascular Therapy* **10**(2): 275–84. doi:10.1583/1545-1550(2003)010<0275:CICAFV>2.0.CO;2.

Haga, J. H., Y.-S. J. Li, and S. Chien. 2007. "Molecular basis of the effects of mechanical stretch on vascular smooth muscle cells." *Journal of Biomechanics* **40**(5): 947–60. doi:10.1016/j.jbiomech.2006.04.011.

Haruguchi, H. and S. Teraoka. 2003. "Intimal hyperplasia and hemodynamic factors in arterial bypass and arteriovenous grafts: a review." *Journal of Artificial Organs* **6**(4): 227–35. doi:10.1007/s10047-003-0232-x.

He, X. and D. N. Ku. 1996. "Pulsatile flow in the human left coronary artery bifurcation: average conditions." *Journal of Biomechanical Engineering* **118**(1): 74–82. PMID:8833077.

Hellums, J. D. 1994. "1993 Whitaker Lecture: biorheology in thrombosis research." *Annals of Biomedical Engineering* **22**(5): 445–55. PMID:7825747.

Herbst, T. J., J. B. McCarthy, E. C. Tsilibary, and L. T. Furcht. 1988. "Differential effects of laminin, intact type IV collagen, and specific domains of type IV collagen on endothelial cell adhesion and migration." *The Journal of Cell Biology* **106**(4): 1365–73. PMID:3360855.

Holenstein, R., P. Niederer, and M. Anliker. 1980. "A viscoelastic model for use in predicting arterial pulse waves." *Journal of Biomechanical Engineering* **102**(4): 318–25. PMID:6965195.

Hsu, P.-P., S. Li, Y.-S. Li, S. Usami, A. Ratcliffe, X. Wang, and S. Chien. 2001. "Effects of flow patterns on endothelial cell migration into a zone of mechanical denudation." *Biochemical and Biophysical Research Communications* **285**(3): 751–9. doi:10.1006/bbrc.2001.5221.

Humphrey, J. D. 2007. "Vascular adaptation and mechanical homeostasis at tissue, cellular, and sub-cellular levels." *Cell Biochemistry and Biophysics* **50**(2): 53–78. doi:10.1007/s12013-007-9002-3.

Huynh, J., N. Nishimura, K. Rana, J. M. Peloquin, J. P. Califano, C. R. Montague, et al. 2011. "Age-related intimal stiffening enhances endothelial permeability and leukocyte transmigration." *Science Translational Medicine* **3**(112): 112ra122. doi:10.1126/scitranslmed.3002761.

Jung, B., M. Markl, D. Föll, and J. Hennig. 2006. "Investigating myocardial motion by MRI using tissue phase mapping." *European Journal of Cardio-thoracic Surgery* **29**(Suppl. 1): S150–7. doi:10.1016/j.ejcts.2006.02.066.

Kawasaki, M., Y. Ito, H. Yokoyama, M. Arai, G. Takemura, A. Hara, et al. 2005. "Assessment of arterial medial characteristics in human carotid arteries using integrated backscatter ultrasound and its histological implications." *Atherosclerosis* **180**(1): 145–54. doi:10.1016/j.atherosclerosis.2004.11.018.

Kim, Y. H., K. B. Chandran, T. J. Bower, and J. D. Corson. 1993. "Flow dynamics across end-to-end vascular bypass graft anastomoses." *Annals of Biomedical Engineering* **21**(4): 311–20. PMID:8214816.

Kim, B.-S., J. Nikolovski, J. Bonadio, and D. J. Mooney. 1999. "Cyclic mechanical strain regulates the development of engineered smooth muscle tissue." *Nature Biotechnology* **17**(10): 979–83. doi:10.1038/13671.

Kladakis, S. M. and R. M. Nerem. 2004. "Endothelial cell monolayer formation: effect of substrate and fluid shear stress." *Endothelium* **11**(1): 29–44. doi:10.1080/10623320490432461.

Ku, D. N., D. P. Giddens, C. K. Zarins, and S. Glagov. 1985. "Pulsatile flow and atherosclerosis in the human carotid bifurcation. positive correlation between plaque location and low and oscillating shear stress." *Arteriosclerosis* **5**(3): 293–302. PMID:3994585.

Ku, J. P., C. J. Elkins, and C. A. Taylor. 2005. "Comparison of CFD and MRI flow and velocities in an in vitro large artery bypass graft model." *Annals of Biomedical Engineering* **33**(3): 257–69. doi:10.1007/s10439-005-1729-7.

Lan, T.-H., X.-Q. Huang, and H.-M. Tan. 2013. "Vascular fibrosis in atherosclerosis." *Cardiovascular Pathology* **22**(5): 401–7. doi:10.1016/j.carpath.2013.01.003.

Lee, S.-W., L. Antiga, and D. A. Steinman. 2009. "Correlations among indicators of disturbed flow at the normal carotid bifurcation." *Journal of Biomechanical Engineering* **131**(6): 061013. doi:10.1115/1.3127252.

Lee, K. E., J. S. Lee, and J. Y. Yoo. 2011. "A numerical study on steady flow in helically sinuous vascular prostheses." *Medical Engineering & Physics* **33**(1): 38–46. doi:10.1016/j.medengphy.2010.09.005.

Lemarié, C. A., P.-L. Tharaux, and S. Lehoux. 2010. "Extracellular matrix alterations in hypertensive vascular remodeling." *Journal of Molecular and Cellular Cardiology* **48**(3): 433–9. doi:10.1016/j.yjmcc.2009.09.018.

Leung, D. Y., S. Glagov, and M. B. Mathews. 1976. "Cyclic stretching stimulates synthesis of matrix components by arterial smooth muscle cells in vitro." *Science* **191**(4226): 475–7. doi:10.1126/science.128820.

Levesque, M. J., D. Liepsch, S. Moravec, and R. M. Nerem. 1986. "Correlation of endothelial cell shape and wall shear stress in a stenosed dog aorta." *Arteriosclerosis, Thrombosis, and Vascular Biology* **6**(2): 220–9. doi:10.1161/01.ATV.6.2.220.

Liu, S. Q. and J. Goldman. 2001. "Role of blood shear stress in the regulation of vascular smooth muscle cell migration." *IEEE Transactions on Biomedical Engineering* **48**(4): 474–83. doi:10.1109/10.915714.

Liu, M., M. S. Kluger, A. D'Alessio, G. García-Cardeña, and J. S. Pober. 2008. "Regulation of arterial-venous differences in tumor necrosis factor responsiveness of endothelial cells by anatomic context." *American Journal of Pathology* **172**(4): 1088–99. doi:10.2353/ajpath.2008.070603.

Lu, Q. and S. Rounds. 2012. "Focal adhesion kinase and endothelial cell apoptosis." *Microvascular Research* **83**(1): 56–63. doi:10.1016/j.mvr.2011.05.003.

Malek, Adel M., S. L. Alper, and S. Izumo. 1999. "Hemodynamic shear stress and its role in atherosclerosis." *Journal of the American Medical Association* **282**(21): 2035–42. doi:10.1001/jama.282.21.2035.

Mason, B. N., A. Starchenko, R. M. Williams, L. J. Bonassar, and C. A. Reinhart-King. 2013. "Tuning three-dimensional collagen matrix stiffness independently of collagen concentration modulates endothelial cell behavior." *Acta Biomaterialia* **9**(1): 4635–44. doi:10.1016/j.actbio.2012.08.007.

Mata-Greenwood, E., A. Grobe, S. Kumar, Y. Noskina, and S. M. Black. 2005. "Cyclic stretch increases VEGF expression in pulmonary arterial smooth muscle cells via TGF-β1 and reactive oxygen species: a requirement for NAD(P)H oxidase." *American Journal of Physiology. Lung Cellular and Molecular Physiology* **289**(2): L288–9. doi:10.1152/ajplung.00417.2004.

Mathers, C. D. and D. Loncar. 2006. "Projections of global mortality and burden of disease from 2002 to 2030." *PLoS Medicine* **3**(11): e442. doi:10.1371/journal.pmed.0030442.

McArthur, C., C. C. Geddes, and G. M. Baxter. 2011. "Early measurement of pulsatility and resistive indexes: correlation with long-term renal transplant function." *Radiology* **259**(1): 278–85. doi:10.1148/radiol.10101329.

McKenna, K. A., M. T. Hinds, R. C. Sarao, P.-C. Wu, C. L. Maslen, R. W. Glanville, et al. 2012. "Mechanical property characterization of electrospun recombinant human tropoelastin for vascular graft biomaterials." *Acta Biomaterialia* **8**(1): 225–33. doi:10.1016/j.actbio.2011.08.001.

Mitchell, G. F. 2008. "Effects of central arterial aging on the structure and function of the peripheral vasculature: implications for end-organ damage." *Journal of Applied Physiology* **105**(5): 1652–60. doi:10.1152/japplphysiol.90549.2008.

Mitchell, G. F., S.-J. Hwang, R. S. Vasan, M. G. Larson, M. J. Pencina, N. M. Hamburg, et al. 2010. "Arterial stiffness and cardiovascular events the Framingham Heart Study." *Circulation* **121**(4): 505–11. doi:10.1161/CIRCULATIONAHA.109.886655.

Mitchell, G. F., M. A. van Buchem, S. Sigurdsson, J. D. Gotal, M. K. Jonsdottir, O. Kjartansson, et al. 2011. "Arterial stiffness, pressure and flow pulsatility and brain structure and function: the Age, Gene/Environment Susceptibility – Reykjavik Study." *Brain* **134**(Pt. 11): 3398–407. doi:10.1093/brain/awr253.

Morinaga, K., H. Eguchi, T. Miyazaki, K. Okadome, and K. Sugimachi. 1987. "Development and regression of intimal thickening of arterially transplanted autologous vein grafts in dogs." *Journal of Vascular Surgery* **5**(5): 719–30. doi:10.1016/0741-5214(87)90160-1.

Murphy, E. A. and F. J. Boyle. 2012. "Reducing in-stent restenosis through novel stent flow field augmentation." *Cardiovascular Engineering and Technology* **3**(4): 353–73. doi:10.1007/s13239-012-0109-3.

Neufurth, M., X. Wang, E. Tolba, B. Dorweiler, H. C. Schröder, T. Link, et al. 2015. "Modular small diameter vascular grafts with bioactive functionalities." *PLoS One* **10**(7): e0133632. doi:10.1371/journal.pone.0133632.

Nichols, W., O'Rourke, M., and Vlachopoulos, C. 2011. *McDonald's Blood Flow in Arteries, Sixth Edition: Theoretical, Experimental and Clinical Principles*. Boca Raton, FL: CRC Press.

Noris, M., M. Morigi, R. Donadelli, S. Aiello, M. Foppolo, M. Todeschini, et al. 1995. "Nitric oxide synthesis by cultured endothelial cells is modulated by flow conditions." *Circulation Research* **76**(4): 536–43. doi:10.1161/01.RES.76.4.536.

Ojha, M. 1994. "Wall shear stress temporal gradient and anastomotic intimal hyperplasia." *Circulation Research* **74**(6): 1227–31. doi:10.1161/01.RES.74.6.1227.

Okuhn, S. P., D. P. Connelly, N. Calakos, L. Ferrell, P. Man-Xiang, and J. Goldstone. 1989. "Does compliance mismatch alone cause neointimal hyperplasia?" *Journal of Vascular Surgery* **9**(1): 35–45. PMID:2911141.

Owens, C. D. 2010. "Adaptive changes in autogenous vein grafts for arterial reconstruction: clinical implications." *Journal of Vascular Surgery* **51**(3): 736–46. doi:10.1016/j.jvs.2009.07.102.

Panaritis, V., A. V. Kyriakidis, M. Pyrgioti, L. Raffo, E. Anagnostopoulou, G. Gourniezaki, and E. Koukou. 2005. "Pulsatility index of temporal and renal arteries as an early finding of arteriopathy in diabetic patients." *Annals of Vascular Surgery* **19**(1): 80–3. doi:10.1007/s10016-004-0134-2.

Peiffer, V., S. J. Sherwin, and P. D. Weinberg. 2013. "Computation in the rabbit aorta of a new metric – the transverse wall shear stress – to quantify the multidirectional character of disturbed blood flow." *Journal of Biomechanics* **46**(15): 2651–8. doi:10.1016/j.jbiomech.2013.08.003.

Peloquin, J., J. Huynh, R. M. Williams, and C. A. Reinhart-King. 2011. "Indentation measurements of the subendothelial matrix in bovine carotid arteries." *Journal of Biomechanics* **44**(5): 815–21. doi:10.1016/j.jbiomech.2010.12.018.

Perktold, K, M. Hofer, G. Rappitsch, M. Loew, B. D. Kuban, and M. H. Friedman. 1997. "Validated computation of physiologic flow in a realistic coronary artery branch." *Journal of Biomechanics* **31**(3): 217–28. doi:10.1016/S0021-9290(97)00118-8.

Pritchard, W. F., P. F. Davies, Z. Derafshq, D. C. Polacek, R. Tsao, R. Dull, et al. 1995. "Effects of wall shear stress and fluid recirculation on the localization of circulating monocytes in a three-dimensional flow model." *Journal of Biomechanics* **28**(12): 1459–69. PMID:8666586.

Raghu, R., I. E. Vignon-Clementel, C. A. Figueroa, and C. A. Taylor. 2016. "Comparative study of viscoelastic arterial wall models in nonlinear one-dimensional finite element simulations of blood flow." *Journal of Biomechanical Engineering* **133**(8): 081003. doi: 10.1115/1.4004532.

Ramstack, J. M., L. Zuckerman, and L. F. Mockros. 1979. "Shear-induced activation of platelets." *Journal of Biomechanics* **12**: 113–25. PMID:422576.

Redheuil, A., W.-C. Yu, C. O. Wu, E. Mousseaux, A. de Cesare, R. Yan, et al. 2010. "Reduced ascending aortic strain and distensibility earliest manifestations of vascular aging in humans." *Hypertension* **55**(2): 319–26. doi:10.1161/HYPERTENSIONAHA.109.141275.

Resnick, N. and M. A. Gimbrone. 1995. "Hemodynamic forces are complex regulators of endothelial gene expression." *FASEB Journal* **9**(10): 874–82. PMID:7615157.

Roberts, S. 2007. "The way ahead: the local challenge. Improving diabetes services: the NSF four years on." Department of Health. Available from: https://www.bipsolutions.com/docstore/pdf/16198.pdf (last accessed June 30, 2016).

Sakamoto, N., T. Ohashi, and M. Sato. 2006. "Effect of fluid shear stress on migration of vascular smooth muscle cells in cocultured model." *Annals of Biomedical Engineering* **34**(3): 408–15. doi:10.1007/s10439-005-9043-y.

Sakao, S. 2006. "Apoptosis of pulmonary microvascular endothelial cells stimulates vascular smooth muscle cell growth." *American Journal of Physiology. Lung Cellular and Molecular Physiology* **291**(3): L362–8. doi:10.1152/ajplung.00111.2005.

Salacinski, H. J., S. Goldner, A. Giudiceandrea, G. Hamilton, A. M. Seifalian, A. Edwards, and R. J. Carson. 2001. "The mechanical behavior of vascular grafts: a review." *Journal of Biomaterials Applications* **15**(3): 241–78. doi:10.1106/NA5T-J57A-JTDD-FD04.

Sarkar, S., T. Schmitz-Rixen, G. Hamilton, and A. M Seifalian. 2007. "Achieving the ideal properties for vascular bypass grafts using a tissue engineered approach: a review." *Medical & Biological Engineering & Computing* **45**(4): 327–36. doi:10.1007/s11517-007-0176-z.

Scott, D., Y. Tan, R. Shandas, K. R. Stenmark, and W. Tan. 2013. "High pulsatility flow stimulates smooth muscle cell hypertrophy and contractile protein expression." *American Journal of Physiology. Lung Cellular and Molecular Physiology* **304**(1): L70–81. doi:10.1152/ajplung.00342.2012.

Shav, D., R. Gotlieb, U. Zaretsky, D. Elad, and S. Einav. 2014. "Wall shear stress effects on endothelial-endothelial and endothelial-smooth muscle cell interactions in tissue engineered models of the vascular wall." *PLoS One* **9**(2): e88304. doi:10.1371/journal.pone.0088304.

Shin, H. Y., M. E. Gerritsen, and R. Bizios. 2002. "Regulation of endothelial cell proliferation and apoptosis by cyclic pressure." *Annals of Biomedical Engineering* **30**(3): 297–304. doi:10.1114/1.1458595.

Shi, Z.-D., X.-Y. Ji, H. Qazi, and J. M. Tarbell. 2009. "Interstitial flow promotes vascular fibroblast, myofibroblast, and smooth muscle cell motility in 3-D collagen I via upregulation of MMP-1." *American Journal of Physiology. Heart and Circulatory Physiology* **297**(4): H1225–34. doi:10.1152/ajpheart.00369.2009.

Sho, E., M. Sho, T. M. Singh, H. Nanjo, M. Komatsu, C. Xu, et al. 2002a. "Arterial enlargement in response to high flow requires early expression of matrix metalloproteinases to degrade extracellular matrix." *Experimental and Molecular Pathology* **73**(2): 142–53. doi:10.1006/exmp.2002.2457.

Sho M., E. Sho, T. M. Singh, M. Komatsu, A. Sugita, C. Xu, et al. 2002b. "Subnormal shear stress-induced intimal thickening requires medial smooth muscle cell proliferation and migration." *Experimental and Molecular Pathology* **72**(2): 150–60. doi:10.1006/exmp.2002.2426.

Soletti, L., Y. Hong, J. Guan, J. J. Stankus, M. S. El-Kurdi, W. R. Wagner, and D. A. Vorp. 2010. "A bilayered elastomeric scaffold for tissue engineering of small diameter vascular grafts." *Acta Biomaterialia* **6**(1): 110–22. doi:10.1016/j.actbio.2009.06.026.

Sottiurai, V. S., J. S. T. Yao, W. R. Flinn, and R. C. Batson. 1983. "Intimal hyperplasia and neointima: an ultrastructural analysis of thrombosed grafts in humans." *Surgery* **93**(6): 809–17. PMID:6222499.

Stewart, S. F. C. and D. J. Lyman. 1992. "Effects of a vascular graft/natural artery compliance mismatch on pulsatile flow." *Journal of Biomechanics* **25**(3): 297–310. doi:10.1016/0021-9290(92)90027-X.

Stonebridge, P. A. 2011. "Three-dimensional blood flow dynamics: spiral/helical laminar flow." *Methodist DeBakey Cardiovascular Journal* 7(1): 21–6. PMID:21490550.

Stonebridge, P. A. and C. M. Brophy. 1991. "Spiral laminar flow in arteries?" *Lancet* **338**(8779): 1360–1. doi:10.1016/0140-6736(91)92238-W.

Stonebridge, P. A., P. R. Hoskins, P. L. Allan, and J. F F Belch. 1996. "Spiral laminar flow in vivo." *Clinical Science* **91**(1): 17–21. PMID:8774255.

Stonebridge, P. A., F. Vermassen, J. Dick, J. J. F. Belch, and G. Houston. 2012. "Spiral laminar flow prosthetic bypass graft: medium-term results from a first-in-man structured registry study." *Annals of Vascular Surgery* **26**(8): 1093–9. doi:10.1016/j.avsg.2012.02.001.

Sunamura, M., H. Ishibashi, and T. Karino. 2007. "Flow patterns and preferred sites of intimal thickening in diameter-mismatched vein graft interpositions." *Surgery* **141**(6): 764–76. doi:10.1016/j.surg.2006.12.019.

Sun, A., Y. Fan, and X. Deng. 2010. "Numerical comparative study on the hemodynamic performance of a new helical graft with noncircular cross section and SwirlGraft." *Artificial Organs* **34**(1): 22–7. doi:10.1111/j.1525-1594.2009.00797.x.

Takada, Y., F. Shinkai, S. Kondo, S. Yamamoto, H. Tsuboi, R. Korenaga, and J. Ando. 1994. "Fluid shear stress increases the expression of thrombomodulin by cultured human endothelial cells." *Biochemical and Biophysical Research Communications* **205**(2): 1345–52. doi:10.1006/bbrc.1994.2813.

Tanaka, T. T. and Y.-C. Fung. 1974. "Elastic and inelastic properties of the canine aorta and their variation along the aortic tree." *Journal of Biomechanics* 7(4): 357–70. PMID:4413195.

Tan, Y., P.-O. Tseng, D. Wang, H. Zhang, K. Hunter, J. Hertzberg, et al. 2014. "Stiffening-induced high pulsatility flow activates endothelial inflammation via a TLR2/NF-κB pathway." *PLoS One* **9**(7): e102195. doi:10.1371/journal.pone.0102195.

Traub, O. and B. C. Berk. 1998. "Laminar shear stress: mechanisms by which endothelial cells transduce an atheroprotective force." *Arteriosclerosis, Thrombosis, and Vascular Biology* **18**(5): 677–85. doi:10.1161/01.ATV.18.5.677.

Uematsu, M., Y. Ohara, J. P. Navas, K. Nishida, T. J. Murphy, R. W. Alexander, et al. 1995. "Regulation of endothelial cell nitric oxide synthase mRNA expression by shear stress." *American Journal of Physiology. Cell Physiology* **269**(6 Pt. 1): C1371–8. PMID:8572165.

Urbich, C. 2002. "Shear stress-induced endothelial cell migration involves integrin signaling via the fibronectin receptor subunits alpha5 and beta1." *Arteriosclerosis, Thrombosis, and Vascular Biology* **22**(1): 69–75. doi:10.1161/hq0102.101518.

Uttayarat, P., A. Perets, M. Li, P. Pimton, S. J. Stachelek, I. Alferiev, et al. 2010. "Micropatterning of three-dimensional electrospun polyurethane vascular grafts." *Acta Biomaterialia* **6**(11): 4229–37. doi:10.1016/j.actbio.2010.06.008.

Valdez-Jasso, D., D. Bia, Y. Zócalo, R. L. Armentano, M. A. Haider, and M. S. Olufsen. 2011. "Linear and nonlinear viscoelastic modeling of aorta and carotid pressure-area dynamics under in vivo and ex vivo conditions." *Annals of Biomedical Engineering* **39**(5): 1438–56. doi:10.1007/s10439-010-0236-7.

Vernhet, H., R. Demaria, J. M. Juan, M. C. Oliva-Lauraire, J. P. Sénac, and M. Dauzat. 2001. "Changes in wall mechanics after endovascular stenting in the rabbit aorta: comparison of three stent designs." *American Journal of Roentgenology* **176**(3): 803–7. doi:10.2214/ajr.176.3.1760803.

Wagenseil, J. E. and R. P. Mecham. 2009. "Vascular extracellular matrix and arterial mechanics." *Physiological Reviews* **89**(3): 957–89. doi:10.1152/physrev.00041.2008.

Wang, H. Q., L. X. Huang, M. J. Qu, Z. Q. Yan, B. Liu, B. R. Shen, and Z. L. Jiang. 2006. "Shear stress protects against endothelial regulation of vascular smooth muscle cell migration in a coculture system." *Endothelium* **13**(3): 171–80. doi:10.1080/10623320600760282.

White, S. S., C. K. Zarins, D. P. Giddens, H. Bassiouny, F. Loth, S. A. Jones, and S. Glagov. 1993. "Hemodynamic patterns in two models of end-to-side vascular graft anastomoses: effects of pulsatility, flow division, reynolds number, and hood length." *Journal of Biomechanical Engineering* **115**(1): 104–11. PMID:8445887.

Whittemore, A. D., A. W. Clowes, N. P. Couch, and J. A. Mannick. 1981. "Secondary femoropopliteal reconstruction." *Annals of Surgery* **193**(1): 35–42. PMID:7458449.

Wise, S. G., M. J. Byrom, A. Waterhouse, P. G. Bannon, M. K. C. Ng, and A. S. Weiss. 2011. "A multilayered synthetic human elastin/polycaprolactone hybrid vascular graft with tailored mechanical properties." *Acta Biomaterialia* **7**(1): 295–303. doi:10.1016/j.actbio.2010.07.022.

Wu, M. H. 2005. "Endothelial focal adhesions and barrier function." *Journal of Physiology* **569**(2): 359–66. doi:10.1113/jphysiol.2005.096537.

Wurzinger, L. J., R. Optiz, M. Wolf, and H. Schmid-Schonbein. 1985. "'Shear induced platelet activation' – a critical reappraisal." *Biorheology* **22**(5): 399–413. PMID:2937464.

Wurzinger, L. J., R. Opitz, and H. Eckstein. 1986. "Mechanical blood trauma – an overview." *Angeiologie* **38**: 81–97.

Yellin, A. E., D. N. Jones, R. B. Rutherford, T. B. Ikezawa, N. B. Nishikimi, and H. B. Ishibashi. 1991. "Factors affecting the patency of small-caliber prostheses: observations in a suitable canine model." *Journal of Vascular Surgery* **14**(4): 441–51. doi:10.1067/mva.1991.31015.

Zarins, C. K., D. P. Giddens, B. K. Bharadvaj, V. S. Sottiurai, R. F. Mabon, and S. Glagov. 1983. "Carotid bifurcation atherosclerosis. Quantitative correlation of plaque localization with flow velocity profiles and wall shear stress." *Circulation Research* **53**(4): 502–14. doi:10.1161/01.RES.53.4.502.

Zebda, N., O. Dubrovskyi, and K. G. Birukov. 2012. "Focal adhesion kinase regulation of mechanotransduction and its impact on endothelial cell functions." *Microvascular Research* **83**(1): 71–81. doi:10.1016/j.mvr.2011.06.007.

Zheng, T., Y. Fan, Y. Xiong, W. Jiang, and X. Deng. 2009. "Hemodynamic performance study on small diameter helical grafts." *ASAIO Journal* **55**(3): 192–9. doi:10.1097/MAT.0b013e31819b34f2.

Zieman, S. J., V. Melenovsky, and D. A. Kass. 2005. "Mechanisms, pathophysiology, and therapy of arterial stiffness." *Arteriosclerosis, Thrombosis, and Vascular Biology* **25**(5): 932–43. doi:10.1161/01.ATV.0000160548.78317.29.

Index

Mechanobiology: Exploitation for Medical Benefit, First Edition. Edited by Simon C. F. Rawlinson.